Electrodynamics of Particles and Plasmas

Electrodynamics of Particles and Plasmas

P.C. CLEMMOW
University of Cambridge

J.P. DOUGHERTY
University of Cambridge

CRC Press
Taylor & Francis Group
Boca Raton London New York

CRC Press is an imprint of the
Taylor & Francis Group, an **informa** business

Originally published in 1969 by Addison-Wesley Publishing Company, Inc.

Published 1990 by Westview Press

Published 2018 by CRC Press
Taylor & Francis Group
6000 Broken Sound Parkway NW, Suite 300
Boca Raton, FL 33487-2742

CRC Press is an imprint of the Taylor & Francis Group, an informa business

Visit the Taylor & Francis Web site at
http://www.taylorandfrancis.com

and the CRC Press Web site at
http://www.crcpress.com

Library of Congress Cataloging-in-Publication Data

Clemmow, P.C.
Electrodynamics of particles and plasmas / P.C. Clemmow and J.P. Dougherty.
p. cm. -- (Advanced book classics series)
Includes bibliographical references.
1. Plasma (Ionized gases) 2. Electrodynamics. I. Dougherty, J. P. (John P.) II. Title. III. Series.
QC718.5.E4C54 1990 537.6--dc20 89-37705

ISBN 13: 978-0-201-47986-7 (pbk)
ISBN 13: 978-0-201-51500-8 (hbk)

This unique image was created with special-effects photography. Photographs of a broken road, an office building, and a rusted object were superimposed to achieve the effect of a faceted pyramid on a futuristic plain. It originally appeared in a slide show called "Fossils of the Cyborg: From the Ancient to the Future," produced by Synapse Productions, San Francisco. Because this image evokes a fusion of classicism and dynamism, the future and the past, it was chosen as the logo for the Advanced Book Classics series.

Publisher's Foreword

"Advanced Book Classics" is a reprint series which has come into being as a direct result of public demand for the individual volumes in this program. That was our initial criterion for launching the series. Additional criteria for selection of a book's inclusion in the series include:

- Its intrinsic value for the current scholarly buyer. It is not enough for the book to have some historic significance, but rather it must have a timeless quality attached to its content, as well. In a word, "uniqueness."
- The book's global appeal. A survey of our international markets revealed that readers of these volumes comprise a boundaryless, worldwide audience.
- The copyright date and imprint status of the book. Titles in the program are frequently fifteen to twenty years old. Many have gone out of print, some are about to go out of print. Our aim is to sustain the lifespan of these very special volumes.

We have devised an attractive design and trim-size for the "ABC" titles, giving the series a striking appearance, while lending the individual titles unifying identity as part of the "Advanced Book Classics" program. Since "classic" books demand a long-lasting binding, we have made them available in hardcover at an affordable price. We envision them being purchased by individuals for reference and research use, and for personal and public libraries. We also foresee their use as primary and recommended course materials for university level courses in the appropriate subject area.

The "Advanced Book Classics" program is not static. Titles will continue to be added to the series in ensuing years as works meet the criteria for inclusion which we've imposed. As the series grows, we naturally anticipate our book buying audience to grow with it. We welcome your support and your suggestions concerning future volumes in the program and invite you to communicate directly with us.

Advanced Book Classics

V.I. Arnold and A. Avez, *Ergodic Problems of Classical Mechanics*

E. Artin and J. Tate, *Class Field Theory*

Michael F. Atiyah, *K-Theory*

David Bohm, *The Special Theory of Relativity*

P.C. Clemmow and J.P. Dougherty, *Electrodynamics of Particles and Plasmas*

Ronald C. Davidson, *Theory of Nonneutral Plasmas*

P.G. deGennes, *Superconductivity of Metals and Alloys*

Bernard d'Espagnat, *Conceptual Foundations of Quantum Mechanics, 2nd Edition*

Richard Feynman, *Photon-Hadron Interactions*

Dieter Forster, *Hydrodynamic Fluctuations, Broken Symmetry, and Correlation Functions*

William Fulton, *Algebraic Curves: An Introduction to Algebraic Geometry*

Kurt Gottfried, *Quantum Mechanics*

Leo Kadanoff and Gordon Baym, *Quantum Statistical Mechanics*

I.M. Khalatnikov, *An Introduction to the Theory of Superfluidity*

George W. Mackey, *Unitary Group Representations in Physics, Probability and Number Theory*

A.B. Migdal, *Qualitative Methods in Quantum Theory*

Phillipe Nozières and David Pines, *The Theory of Quantum Liquids, Volume II* - new material, 1989 copyright

David Pines and Phillipe Nozières, *The Theory of Quantum Liquids, Volume I: Normal Fermi Liquids*

F. Rohrlich, *Classical Charged Particles - Foundations of Their Theory*

David Ruelle, *Statistical Mechanics: Rigorous Results*

Julian Schwinger, *Particles, Sources and Fields, Volume I*

Julian Schwinger, *Particles, Sources and Fields, Volume II*

Julian Schwinger, *Particles, Sources and Fields, Volume III* - new material, 1989 copyright

Jean-Pierre Serre, *Abelian ℓ-Adic Representations and Elliptic Curves*

R.F. Streater and A.S. Wightman, *PCT Spin and Statistics and All That*

René Thom, *Structural Stability and Morphogenesis*

Vita

Phillip C. Clemmow is a retired Lecturer, Cambridge University. Dr. Clemmow shared his time between the Department of Physics (Cavendish Laboratory) and the Department of Applied Mathematics and Theoretical Physics. He received both his undergraduate degree and Ph.D. from Cambridge. His professional interests have included diffraction theory (starting with work in the development of radar during World War II), ionospheric physics and plasma physics. At Cambridge, he has also been a Fellow of Sidney Sussex College for most of his career, and served as Vice-Master of the college for two years.

John P. Dougherty is Stokes Lecturer in Applied Mathematics and Theoretical Physics at Cambridge University. He received his Ph.D. from Cambridge, and is a Fellow of Pembroke College there. Dr. Dougherty is also Editor of the Journal of Plasma Physics, which he helped establish in 1967. His current research interests are in general formulations of the foundations of non-equilibrium statistical mechanics, a result of the considerations of plasma kinetic theory.

Special Preface

This book is reissued, twenty years after publication, the only changes being the correction of typographical and minor errors.

As indicated in the original preface, the book was written at the level of the beginning graduate student. It did not claim then, and still less would it claim now, to be a research monograph. Neither of us is able to undertake extensive rewriting, so we are rather pleased that the publishers feel it appropriate to reprint the book virtually unaltered

The subject has naturally advanced enormously since 1969. One need only scan the pages of the Physics of Fluids, Journal of Plasma Physics and Controlled Fusion, and others, to discover the extend of that advance. But what, in 1989, should be included in a book for the beginning graduate student in plasma physics? The first point to make in answer to that question is that work on the subject is now more specialized than in 1969, and such a student would need more advanced knowledge of a particular area than could be included in a general text. But he would be well advised to be acquainted to some extent with the broader subject which we endeavoured to present, and we hope that substantial parts of our book would fill his needs. If we were starting again, we might well discard some of the details given here, in favor of other subjects, some of which are now mentioned briefly.

The study of nonlinear effects in plasmas was already well launched before 1969. It has several aspects. One is the study of exact solutions, for example, waves of finite amplitude and individual pulses (solitons). This is an activity which plasma physics has shared with other disciplines such as fluid dynamics and particle physics. Then there is the study of the interaction of three or more waves, and their reaction onto the medium itself (ponderomotive effect). Finally there is the difficult field of plasma turbulence. More recent work has benefited greatly from the expansion of computer science and in computer hardware.

A very useful early exposition of nonlinear plasma physics was that of Davidson (1972). For plasma turbulence, see Tsytovich (1977).

The availability of microcomputers has also stimulated new ideas on the teaching of theoretical physics. Referring to Chapter 5 of the present book, today's graduate student will have no difficulty in programming for himself the means to display on his screen the dispersion curves and polar diagrams for refractive index surfaces, using the formulas we present. We suggest this as a valuable exercise from which one can learn so much more easily the results which had to be hard won by the founding fathers of magnetoionic theory.

Although we touch on plasma stability and the classification of instabilities in Chapters 6 and 10, we can do no more than give the reader a sample of that vast subject. A recent survey (Cap, 1982) runs to three volumes, and these should be consulted for more details.

The recent study of chaotic dynamics and stochasticity, and new developments in Hamiltonian dynamics have infused new ideas and methodology into plasma physics. The reconsideration of Hamiltonian dynamics has led to more powerful methods of dealing with the guiding center approximation to particle motion treated in Chapter 4 (see for an example Littlejohn, 1983).

It has continued to be the case that fusion research has been the main impetus (and source of funds) for plasma physics research. A useful introduction to the area of magnetic confinement is the book by Wesson (1987); besides this, Mercier (1974) may be consulted. For laser plasmas, and laser fusion, see Hora (1981).

The subject of transport processes in plasmas, the classical theory of which is treated in Chapter 11 of our book, is crucial to fusion research. This topic is extensively covered in a recent book by Balescu (1988).

For the reader interested in applications to geophysical and astrophysical plasmas, the three volumes edited by Kennel, Lanzerotti and Parker (1979) may be recommended.

We hope our book, having returned to print, will be of service to another generation of apprentice plasma physicists.

P.C.C.
J.P.D.

References

R.Balescu, *Transport Processes in Plasmas* (2 vols), North-Holland (1988)
F.F.Cap, *Handbook on Plasma Instabilities* (3 vols), Academic (1982)
R.C.Davidson, *Methods in Nonlinear Plasma Theory*, Academic (1972)
H.Hora, *Physics of Laser Driven Plasmas*, Wiley (1981)
C.F.Kennel, L.J.Lanzarotti and E.N.Parker (eds), *Solar System Plasma Physics* (3 vols), North-Holland (1979)

R.G.Littlejohn, *Variational Problems on Guiding Centre Motion*, J. Plasma physics 29, 111 (1983)
C.Mercier, *The Magnetic Approach to the Problem of Plasma Confinement in Closed Magnetic Configurations*, Commission of the European Communities (1974)
V.N.Tsytovich, *Theory of Turbulent Plasma*, Academic (1972)
J.Wesson, *Tokamaks*, Oxford University Press (1987)

Preface

The classical electrodynamics of point charges has a long history. Only comparatively recently, though, has an appreciation of its importance in a broad range of physical phenomena given theorists encouragement to explore all its consequences by removing the criticism that such exercises are purely academic. In this sense there has arisen a new discipline, now usually called plasma physics. Its literature is extensive; and since in the last decade it has given rise to numerous books and also invaded the traditional territory of texts on general electromagnetic theory, it may be helpful to indicate briefly the aim, scope, and level of the present book.

Its aim is to present an account that would give the reader a firm theoretical grounding in the subject. The contents are, in summary, the classical theory of radiation applied to point charges, the dynamics of individual point charges, and the description of plasmas. The latter are treated by proceeding first from the simplifying approximations of magneto-ionic theory, and subsequently from the Boltzmann-Vlasov equations, leaving to the final chapter an introduction to fundamental kinetic theory.

The description throughout relies essentially on mathematical deduction from basic equations, but it is hoped that the indications of orders of magnitude and applications are sufficient to remind the reader that the theories are relevant to the real world and are indeed being widely applied. Moreover, the mathematics itself does not go beyond standard ideas in Fourier analysis, tensors and complex integration; and the steps in each deduction are for the most part given explicitly.

The level at which we have aimed can be most nearly described by saying that we hope the book will be suitable for beginning graduate students. There is, however, some unevenness in the difficulty of the work, inherent in the nature of the subject. It is hoped that the book may also be found to be a useful reference source for other categories of readers; and although there is for the most part no serious claim to originality, it is hoped that more senior research workers may find some of our material of interest.

In keeping with the aim of the book, no attempt has been made to compile comprehensive lists of references. Those given at the end of each chapter refer on the one hand to accounts in other books, and on the other to comparatively recent papers that are either particularly relevant to the matters discussed or else illustrative of further work; references occur in the body of the text only where it seems natural in the interests of clarity. Problems are given at the end of each chapter; they include both exercises on the specific material of the chapter, and example on other aspects of the theory.

The delivery in recent years of several courses of lectures by each of us, and the reception accorded to those courses by the classes of students concerned, have considerably influenced our selection of material and the choice of presentation, and also led us to construct some of the problems. Parts of the manuscript have been read and commented on by several people to whom our thanks are tendered; in particular Mr. S.R. Watson has read the whole of the manuscript at least once and some of it several times, and we are grateful to him for his helpful suggestions.

P.C.C.
J.P.D.

Contents

Chapter 12 Kinetic Equations for Plasmas

Electrodynamics of Particles and Plasmas

CHAPTER 1

INTRODUCTION

1.1 ORIGIN AND SCOPE OF THE SUBJECT

Before the introduction of quantum theory one of the main areas of interest in theoretical physics was the development of the Maxwell–Lorentz equations. Following the discovery of the electron it seemed possible that these equations might account for all known physical phenomena, at least in principle. But difficulties soon appeared, and when the need for quantum hypotheses was recognized there was naturally some abatement of interest in the application of the classical theory of charged particles. More recently, however, there has been a revival of interest, due to the growth of fields of study to which such application is both legitimate and useful. In ionospheric physics, radio astronomy and various other aspects of geophysics, astrophysics and "space" physics, as well as laboratory plasma physics and accelerating machines, the classical theory of charged particles is allowed full scope. The purpose of this book is to give a general mathematical account of the theory and of certain definitive aspects of its development.

Just as the classical treatment of neutral particles can in the main be divided into a consideration, on the one hand, of the dynamics of individual particles, and on the other, of the behavior of a group of particles sufficient in number to be treated in some manner collectively, so for charged particles both the individual and collective characteristics have been extensively examined. The latter study constitutes what is now usually called plasma physics, the word *plasma*, introduced in 1928, being taken broadly to mean any assemblage of particles, some or all of which are charged, that can from the macroscopic view be regarded as a gas which in its equilibrium state is everywhere locally neutral. The fact that individual particles of the gas are charged, and therefore generate and interact with electromagnetic fields, gives rise to many distinctive phenomena; the state in which there is an appreciable degree of ionization has indeed been called the fourth state of matter.

Apart from the difficulty of allowing for interaction with its own electromagnetic field, an interaction which can often safely be neglected, the dynamics of a single charged particle offers problems basically similar to those of a neutral particle, but wider in scope because of the great importance of electromagnetic forces. The effects of a magnetic field are particularly noteworthy.

Peculiar to the charged particle alone, of course, is the generation of an electromagnetic field. In the pre-quantum era the field radiated by a charged particle in uniform circular motion was considered in relation to an electron in orbit about a nucleus. Whilst the calculation is now of no interest in that context, it has acquired great significance in certain situations where particles, otherwise free, gyrate under the influence of a magnetic field. The sources of many of the observations in radio astronomy are currently thought to have their origin in the "synchrotron" radiation produced in this way by particles of such high energy that their speeds are comparable with the vacuum speed of light.

Another particular situation giving rise to a distinctive type of radiation, which is now recognized to play a part in many phenomena other than that through which it originally attracted attention in 1934, is that where a charged particle in uniform rectilinear motion travels in a medium with a speed exceeding the phase speed, for some frequencies, of an electromagnetic wave in the medium. The particle does then radiate, even though it has no acceleration. Radiation by such a process is called Čerenkov radiation. Since a necessary condition for its appearance is that the refractive index of the medium is greater than the ratio of the vacuum speed of light to the speed of the particle, it is relevant that the refractive index of a plasma can be greater than unity.

Plasmas, in considerable variety, are the rule rather than the exception in "outer space." Interstellar space itself, most stars, and in particular the sun, afford examples of the fourth state. Nearer home, the earth's ionosphere exhibits a significant degree of ionization from about 60 km above the earth's surface out to as far as has been explored, that is to distances many times the radius of the earth. Another illustration of a natural plasma is a lightning flash, and in the laboratory too the earliest studies were made of plasmas in the form of electrical discharges in gases. Examples of man-made plasmas are now commonplace. Only comparatively recently, however, has there been a major effort in laboratory plasma physics; the prime objective is controlled extraction of energy from a thermonuclear fusion reaction, and the properties of the fourth state are also being exploited in other technical projects.

It is natural that the theoretical treatment of plasmas should have affinities with that of neutral particle gases. In what respects are the treatments similar or dissimilar?

For the plasma perhaps the first consideration is whether there are circumstances in which the complication of the processes of ionization and deionization can be ignored. This is indeed often permissible. The plasma may simply be in thermal equilibrium at a high enough temperature for the degree of ionization to be appreciable. Or the ionization produced initially by some external agency may alter little in a given period if the recombination rate is low enough. Or again, there may be a steady external source of ionization which is in balance with the recombination processes; this is the case, for example, in the ionosphere, which is largely maintained through photo detachment by ultraviolet light from the sun. In these circumstances the appropriate model is one in which the number of each type of particle is conserved. This model only is treated in the present book.

Next, it may be recalled that neutral particle gases can be discussed in two ways, either as a branch of hydrodynamics, or in terms of kinetic theory. The same is true of plasmas, but the relative importance of the long-range nature of the electromagnetic interaction between charged particles, in particular the Coulomb inverse square force, does mean that both the justification for the use of a quasi-hydrodynamic treatment and also the development of the kinetic approach are different in major respects from their counterparts for neutral particle gases.

It is because the behavior of neutral particle gases is dominated by collisions that such gases can be readily represented by a continuum model. In a plasma, however, the more distinctive situation is that in which *close* collisions have negligible influence, and the validity of any continuum model must then be based on other considerations. By a close collision is meant either a collision in which at least one of the partners is uncharged, or an interaction between charged particles that results in an appreciable deflection of their tracks. The more a plasma is collision dominated the closer is the parallel with conventional hydrodynamics; the main trend of applications is then to phenomena that vary comparatively slowly in time, and the new effects are mostly those associated with the presence of a magnetic field—this is the realm of *magnetohydrodynamics*. When, on the other hand, a plasma is approximately collision free, as is often the case of interest, novel high frequency phenomena are involved, and the associated problems fall into the domain of *radiophysics*. The treatment of plasmas in this book is mainly concerned with the collisionless model, the chief exceptions occurring in the final two chapters.

1.2 ORDERS OF MAGNITUDE

For the sake of easy reference Table 1 gives the notation for and magnitudes of various relevant physical constants, and Table 2 gives the conversion to mks units of certain other units in frequent use. The following statements and calculations all refer implicitly to electrons, since in the present context they are much the most important charged particles.

Table 1

Vacuum speed of light c	$= 2.998 \times 10^8$ m sec^{-1}
Vacuum permeability μ_0	$= 4\pi \times 10^{-7}$ henry m^{-1}

Vacuum permittivity $\varepsilon_0 = \dfrac{1}{36\pi} \times 10^{-9}$ farad m^{-1}

Electron charge $e = 1.60 \times 10^{-19}$ coulomb (4.80×10^{-10} e.s.u.)

Electron (rest) mass $m = 0.911 \times 10^{-30}$ kg

Electron charge/mass $e/m = 1.76 \times 10^{11}$ coulomb kg^{-1}

Electron rest energy $mc^2 = 0.82 \times 10^{-13}$ joule (0.51×10^6 eV)

Proton mass $1837m = 1.67 \times 10^{-27}$ kg

Planck's constant $2\pi\hbar = 6.63 \times 10^{-34}$ joule sec (4.14×10^{-15} eV sec)

Boltzmann's constant $K = 1.38 \times 10^{-23}$ joule °K^{-1} (0.86×10^{-4} eV °K^{-1})

Classical electron radius $r_e = \dfrac{e^2}{4\pi\varepsilon_0 mc^2} = 2.82 \times 10^{-15}$ m

Fine structure constant $\eta = \dfrac{e^2}{4\pi\varepsilon_0 c\hbar} = \dfrac{1}{137}$

Bohr atomic radius $r_a = \dfrac{4\pi\varepsilon_0\hbar^2}{me^2} = \dfrac{r_e}{\eta^2} = 0.53 \times 10^{-10}$ m

Table 2

1 e.s.u. = $\frac{1}{3} \times 10^{-9}$ coulomb	1 Å = 10^{-10} m
1 erg = 10^{-7} joule	1 gauss = 10^{-4} weber m^{-2}
1 eV = 1.60×10^{-19} joule	1 $\gamma = 10^{-5}$ gauss

Consider first the sort of particle energies that might be of interest. Some phenomena involve highly relativistic particles for which the speeds are very close indeed to the vacuum speed of light. The rest energy of an electron is

$$mc^2 = 0.51 \times 10^6 \text{ eV},$$

so that the speed v corresponding to an energy of 10 mev is given by

$$\frac{mc^2}{\sqrt{(1 - v^2/c^2)}} = \frac{10}{0.51} mc^2,$$

from which

$$v/c = 0.9987.$$

This sort of energy is unlikely to represent an average thermal energy; for

$$KT = 0.86 \times 10^{-4} T \text{ eV},$$

where T is measured in °K, so that 10 meV would correspond to a temperature above 10^{11} °K. For comparison it may be remarked that the temperature required to sustain the fusion of heavy isotopes of hydrogen is of the order of 5×10^7 °K.

The importance of the part played by magnetic fields in influencing the motion of charged particles can hardly be over-emphasized, and it is worth giving some indication of the magnitudes of the fields that may be encountered. Perhaps the most familiar magnetic field is that of the earth; from the earth's surface out to a distance of some ten earth radii this is represented, to a first approximation, by the field of a dipole located at the center of the earth. At the earth's surface the magnitude of the field is of the order of 0.5 gauss, and it falls off with distance proportionally to the inverse cube, so that at ten earth radii it is about 50 γ.

Strengths of the order of 1 gauss are intermediate between the very much weaker and very much stronger fields that may need to be considered. The general solar magnetic field is thought to be of this order or somewhat larger, whereas the fields associated with sunspots may well exceed 10^3 gauss. Yet stronger fields of the order of 10^4 gauss are available through laboratory electromagnets, and are also present in magnetic variable stars. At the other extreme, the field in the vicinity of the earth but beyond ten earth radii is less than 50 γ, and the interstellar magnetic field is thought to be about 1 γ or even as low as 0.1 γ. The effect of such minute fields is by no means necessarily negligible, because of the vast distances across which they are operative.

A non-relativistic electron moving in a magnetic field gyrates with angular velocity eB/m; that is

$$1.76 \times 10^7 B \text{ radian sec}^{-1},$$

where B is measured in gauss. The associated radius of gyration for an electron with speed v is

$$\frac{v}{1.76 \times 10^7 B} \text{ m}.$$

Thus for an electron in the ionosphere with $B = 0.5$ gauss the angular gyro frequency is 0.88×10^7 radian sec^{-1}, in the central region of the radio spectrum; and if the electron has thermal speed 2×10^5 m sec^{-1}, corresponding to a temperature of about 1000 °K, the radius of gyration is 2.3 cm.

For a high energy electron the corresponding relativistic formula for the gyro frequency is $(eB/m)\sqrt{(1 - v^2/c^2)}$, and this can be expressed as

$$0.90 \times 10^{13} \frac{B}{\mathscr{E}} \text{ radian sec}^{-1},$$

where B is measured in gauss, and $\mathscr{E}$ is the relativistic energy $mc^2/\sqrt{(1 - v^2/c^2)}$ measured in eV. Thus for a 10 meV electron in a 10^3 gauss field the gyro frequency is 0.90×10^9 radian sec^{-1}.

Turn now to consider some of the parameters typifying a plasma, supposing that the charged particles, say electrons of charge $-e$ and positive ions of charge e, can be treated classically. The point has been made that the more distinctive plasma characteristics are revealed when cooperative phenomena arising from the long-range nature of the electromagnetic interaction between charged particles dominate over the effects of close collisions; when, in fact, in the limit the plasma can be regarded as collision free. In the following discussion of this point collisions involving neutral particles are not at issue; the plasma may be imagined to be fully ionized.

In considering collisions between charged particles it is convenient to introduce the distance l, sometimes called the Landau length, at which the mutual potential energy $e^2/(4\pi\varepsilon_0 l)$ of two charged particles is equal to the characteristic kinetic energy KT of the thermal motion. That is

$$l = \frac{e^2}{4\pi\varepsilon_0 KT} = \frac{1.67 \times 10^{-5}}{T} \text{ m.}$$

A binary collision between charged particles can then be considered close if the impact parameter (that is, the distance of minimum separation that there would be in the absence of mutual interaction) is less than l, for this results in an appreciable deflection of the particle tracks. Now the average distance between charged particles is $N^{-1/3}$, where N is the number of charged particles per unit volume (cubic metre). Thus close collisions can be comparatively unimportant only if $\alpha \ll 1$, where

$$\alpha = \frac{l}{N^{-1/3}} = \frac{e^2}{4\pi\varepsilon_0 N^{-1/3}} \bigg/ KT = 1.67 \times 10^{-5} \frac{N^{1/3}}{T}.$$

The cross-section for close collisions between charged particles is of the order πl^2, so the corresponding mean free path is of the order

$$L = \frac{1}{\pi N l^2} = \frac{N^{-1/3}}{\pi\alpha^2} = 1.1 \times 10^9 \frac{T^2}{N} \text{ m,}$$

which exceeds the interparticle distance by the factor $1/(\pi\alpha^2)$.

Another important length parameter is the Debye length

$$h = \sqrt{\left(\frac{\varepsilon_0 KT}{Ne^2}\right)} = \frac{N^{-1/3}}{\sqrt{(4\pi\alpha)}} = 0.69 \times 10^2 \sqrt{\left(\frac{T}{N}\right)} \text{ m,}$$

which makes an explicit appearance in a number of contexts. For example, it typifies the thickness of the electron sheath that can form where a plasma is in juxtaposition with a solid boundary. Again, the potential of a point charge e' immersed in the plasma is, at distance r,

$$\frac{e'}{4\pi\varepsilon_0 r} \exp\left(-\sqrt{2}r/h\right);$$

the exponential term exhibits the strong shielding effect of the plasma at distances beyond h. These results are established in §7.6. They are closely connected with the proclivity of a plasma to maintain local electrical neutrality by virtue of the strong electrostatic restoring forces brought into play by even a small degree of charge separation. A rough argument shows that regions with volume exceeding h^3 will be quasi-neutral; for Poisson's equation indicates that appreciable charge separation over a length scale ξ gives rise to a potential of order $Ne\xi^2/\varepsilon_0$, and the associated potential energy of the charges can be counteracted by their kinetic energy of thermal agitation only if $KT > Ne^2\xi^2/\varepsilon_0$, that is $\xi^2 < h^2$.

To summarize, in terms of the parameter α, the ratios of the four characteristic lengths are given by

$$l : N^{-1/3} : \quad h \quad : L$$
$$= \alpha : \quad 1 \quad : \frac{1}{2\sqrt{(\pi\alpha)}} : \frac{1}{\pi\alpha^2}.$$

In Table 3 rough orders of magnitude are listed for the number of electrons per cubic metre and the temperature of various plasmas (of which only the ionosphere is weakly ionized), and the corresponding values of α are noted. It is seen that the latter are all indeed small. In contrast, α for the free electrons in a metal with $N = 3 \times 10^{28}$ and $T = 300$ °K is 170.

Table 3

	N (m^{-3})	T (°K)	$\alpha = 1.7 \times 10^{-5} N^{1/3}/T$
Interstellar gas	10^6	10^2	1.7×10^{-5}
Ionosphere	10^8–10^{12}	10^3	0.8×10^{-6}–1.7×10^{-4}
Solar corona	10^{13}	10^6	3.5×10^{-7}
Solar atmosphere	10^{18}	10^4	1.7×10^{-3}
Laboratory plasma	10^{18}–10^{24}	10^6	1.7×10^{-5}–1.7×10^{-3}

If characteristic time intervals, or more conveniently, frequencies, are considered, the most prominent is the *plasma frequency*. This is the frequency of the natural oscillation associated with the electrostatic restoring force arising from a small displacement of charge. Expressed as an angular frequency ω_p it is given by

$$\omega_p^2 = \frac{Ne^2}{\varepsilon_0 m},$$

so that

$$\omega_p = 56\sqrt{N} \text{ radian sec}^{-1},$$

and

$$f_p = \frac{\omega_p}{2\pi} = 9.0\sqrt{N} \text{ c/s}.$$

It is thus evident from Table 3 that for the ionosphere the plasma frequency covers the broadcasting spectrum of radio waves, whereas for laboratory plasmas it is in the microwave and infrared region. For metals it is in the ultraviolet region. We observe that

$$\omega_p = \sqrt{(KT/m)}/h,$$

the ratio of a typical thermal speed to the Debye length.

If the plasma is in a magnetic field another significant frequency is, of course, that of gyration of an electron. From what has already been said it appears that there will be cases of interest in which the plasma and gyro frequencies are quite comparable, as well as those in which one or other is much the greater.

When close collisions are considered their importance can be assessed by relating their frequency to other relevant frequencies. For close collisions between charged particles a mean collision frequency can be defined by

$$\nu = \sqrt{\left(\frac{3KT}{m}\right)}\Big/L = \frac{\sqrt{3}e^4N}{16\pi\varepsilon_0^2 m^{1/2}(KT)^{3/2}} = 0.6 \times 10^5 \frac{N}{T^{3/2}} \text{ sec}^{-1}.$$

We note that

$$\nu/\omega_p = \sqrt{3}h/L = \tfrac{1}{2}\sqrt{(3\pi)}\alpha^{3/2}.$$

In a weakly ionized gas close collisions may well be mostly those between electrons and neutral particles; in the ionosphere these are sufficiently numerous commonly to give rise to significant damping of radio waves.

Finally, it is instructive to indicate briefly the limits within which a classical (that is, non-quantum) treatment of the problems is valid. The familiar broad distinction arises, that the generation of electromagnetic waves in the radio spectrum can be treated classically, whilst those in the optical spectrum require quantum methods. In the interaction of an electromagnetic field of frequency $f(=\omega/2\pi)$ with an otherwise quite free electron the classical treatment is valid provided the photon energy $\hbar\omega$, where $2\pi\hbar$ is Planck's constant, is much less than the rest energy of the electron; that is

$$\hbar\omega \ll mc^2 \simeq 0.5 \times 10^6 \text{ eV},$$

giving

$$f \ll 1.2 \times 10^{20} \text{ c/s},$$

a hard X-ray frequency.

If the processes of absorption and radiation associated with "collisions" between charged particles are considered, it may first be noted that the exchange of field energy can only involve the possibility of a transition between free and bound states at again comparatively high frequencies. For since ionization energies are of the order of 10 eV, the inequality

$$\hbar\omega \ll \text{ionization energy}$$

gives roughly

$$f \ll 3 \times 10^{15} \text{ c/s},$$

a frequency in the ultraviolet region of the spectrum. The calculation of the field radiated by a charged particle in an orbit determined by the nature of the collision may be carried through classically provided the photon energy is much less than the kinetic energy of the particle. Taking KT to typify the latter quantity the condition is

$$\hbar\omega \ll KT,$$

which gives

$$f \ll 2.6 \times 10^{10} T;$$

for $T = 400$ °K,

$$f \ll 10^{13} \text{ c/s},$$

an infrared frequency.

A rough criterion as to when it is permissible to treat particle orbits classically is obtained by expressing the fact that $1/(2\pi)$ times the de Broglie wavelength of the particles is much less than the interparticle distance. Since the de Broglie wavelength is $2\pi\hbar$ divided by the momentum, it may be typified by

$$\frac{2\pi\hbar}{\sqrt{(mKT)}}.$$

Then with the interparticle distance interpreted as $N^{-1/3}$ the criterion is

$$N^{2/3} \ll mKT/\hbar^2.$$

Apart from a factor $1/(2\pi)$ on the right-hand side this is just the well known condition for a gas not to be degenerate, and is convincingly satisfied by all the plasmas listed in Table 3. For $T = 300$ °K it gives roughly

$$N \ll 10^{24},$$

and so the conduction electrons in metals, which number about 3×10^{28} m^3, are highly degenerate.

If, on the other hand, the interparticle distance characteristic of a close collision is used, namely the Landau length l, the criterion reads

$$\frac{\hbar}{\sqrt{(mKT)}} \ll \frac{e^2}{4\pi\varepsilon_0 KT}$$

or

$$KT \ll \frac{e^4 m}{16\pi^2\varepsilon_0^2\hbar^2}.$$

This is conveniently interpreted by introducing the Bohr radius of the hydrogen atom

$$r_a = \frac{4\pi\varepsilon_0\hbar^2}{me^2},$$

and writing

$$KT \ll \frac{e^2}{4\pi\varepsilon_0 r_a}.$$

The quantity on the right-hand side of the inequality is the potential energy of two charges e at distance r_a apart; it would be expected to be of the same order as an ionization energy, and is in fact about 27 eV. Thus the treatment of close collisions entirely by classical methods is suspect when the thermal energy becomes comparable with or greater than the ionization energy; that is, roughly when $T > 10^5$ °K.

1.3 FUNDAMENTAL EQUATIONS

For our purposes the basic equations of the theory of classical electrodynamics are conveniently taken to be Maxwell's equations

$$\begin{aligned}
\operatorname{curl} \mathbf{E} &= -\dot{\mathbf{B}}, \\
\operatorname{curl} \mathbf{H} &= \dot{\mathbf{D}} + \mathbf{j}, \\
\operatorname{div} \mathbf{D} &= \rho, \\
\operatorname{div} \mathbf{B} &= 0,
\end{aligned}$$

with the associated charge conservation relation

$$\operatorname{div} \mathbf{j} + \dot{\rho} = 0.$$

To these must be added the Lorentz expression for the force density, namely

$$\rho\mathbf{E} + \mathbf{j} \times \mathbf{B}.$$

In all but one application in this book the only media that are considered are the plasmas themselves, in which case their electromagnetic behavior is completely described by the explicit charge and current densities ρ and $\mathbf{j}$. In effect, then, we are considering charge and current densities in a vacuum, and

$$\mathbf{D} = \varepsilon_0\mathbf{E}, \qquad \mathbf{B} = \mu_0\mathbf{H}.$$

Conventionally **E** (rather than **D**) is retained in Maxwell's equations, and one of **H**, **B**.

The way in which these equations are applied in the following pages may be summarized in the broadest terms thus. First, the situation envisaged is essentially that in which **j** is given, and the task is to find the field; the special case in which **j** is due to the motion of a point charge (representing, for example, an electron) is examined in detail. Next, it is supposed that the field is given, and the task is to find the way in which a point charge moves under the influence of the field; if the contribution to the Lorentz force made by the field generated by the charge itself is neglected the formulation of the problem is in principle a matter of standard particle dynamics, but things are far from straightforward when we attempt to take account of the "self-force." Finally, an assemblage of charged particles forming a plasma is treated. Here, of course, there is in general no question of any of the current or field vectors being given (though contributions to them may well be specified, as, for example, a magnetostatic field). The essence of the problem is, therefore, to find the relation between **E**, **B** and **j**; this relation is then fed into Maxwell's equations and self-consistent solutions are sought.

Before considering the different levels of sophistication at which the **j**, **E**, **B** relation can be investigated it is important to point out that, since we are now discussing a vast number of particles, we must deal in terms of average quantities. The symbols ρ and **j** now stand for charge and current densities averaged over a volume small on the environmental scale but nevertheless containing a large number of particles. Likewise, the **E** and **B** now to appear in our equations are the average fields associated with the "smeared out" charge and current densities, and not the actual fields at each point of space, which latter can fluctuate appreciably in distances of the order of the interparticle distance $N^{-1/3}$. This is perhaps an obvious point, and is naturally built into any treatment of the plasma problem; but there is one aspect which, though long recognized, is of some subtlety, and is conveniently introduced here. The relation between the average vectors **E**, **B** and **j** must come from the equations governing the motion of the charged particles in the electromagnetic field. The question is whether the average, over particles in a small volume, of the actual field acting on the individual particles (called for brevity the *effective* field) is approximately the same as the average field. Doubts are raised by parallel results from the apparently closely related theory of dielectrics; there, in expressing the permittivity of the dielectric in terms of the polarizability of the individual atoms, it is commonly accepted that, for certain types of atomic lattice, the effective field acting on and thereby polarizing the atoms is not **E**, the average field, but rather $\mathbf{E} + \mathbf{P}/(3\varepsilon_0)$, where **P** is the average polarization vector. The extra term $\mathbf{P}/(3\varepsilon_0)$ was originally derived by Lorentz, and in the subsequent development of magneto-ionic theory (the theory of radio waves in the ionosphere under the influence of the earth's magnetic field) there was some uncertainty as to whether or not the so-called "Lorentz term" should be included in the treatment of plasmas. The question was of more than academic interest, since the issue could make an appreciable difference to predicted results. There were, indeed, attempts to settle the matter by experiment, and the outcome of various findings tended to favor the exclusion of the Lorentz term. This conclusion has theoretical support. In the present book its justification is delayed until Chapter 12, where it follows automatically from the derivation of the kinetic equations. However, since it is tacitly accepted earlier on, we offer here the following comment.

In a dielectric the electrons are bound to the atoms, which, apart from thermal agitation, form a fixed lattice. The atoms thus occupy privileged positions in relation to the actual field associated with any displacement of the electrons, and so there is no reason to expect the effective field to be approximately the same as the average field. In a plasma, however, the electrons have comparative freedom of movement, and so the effective and average fields may well be nearly the same. This is certainly likely to be the case if, during a characteristic time interval, the thermal speed of a particle carries it through a distance well in excess of the interparticle spacing. If the characteristic time interval be taken as $2\pi/\omega_p$, where ω_p is the plasma (angular) frequency, the criterion is

$$2\pi\sqrt{\left(\frac{\varepsilon_0 m}{Ne^2}\right)}\sqrt{\left(\frac{KT}{m}\right)} \gg N^{-1/3},$$

that is

$$2\pi h \gg N^{-1/3},$$

where h is the Debye length; and this is once again the ubiquitous condition $\alpha \ll 1$, shown in §1.2 to be rather generally satisfied. If the disturbance in the plasma has a time scale much shorter than $2\pi/\omega_p$ the Lorentz term is unlikely to have any marked effect on the calculations.

We now return to consideration of the way in which the **j**, **E**, **B** relation is obtained, and for simplicity of description confine the discussion to electron motions only. In the least sophisticated approach no account is taken of thermal velocities, and the particles are accorded a local mean velocity **v** which is determined by the appropriate hydrodynamic equation, most simply

$$\frac{D\mathbf{v}}{Dt} = \mathbf{F},$$

where **F** is the force per unit mass. If electromagnetic forces only are relevant **F** is $-(e/m)(\mathbf{E} + \mathbf{v} \times \mathbf{B})$, where, for the reasons just given, **E** and **B** are the average fields appearing in Maxwell's equations. The corresponding current density is just

$$\mathbf{j} = -Ne\mathbf{v},$$

where N is the local mean number of electrons per unit volume. This is the approach of magneto-ionic theory; it is sometimes said to describe a "cold" plasma, because it gives results which would be obtained from a treatment of thermal velocities if the temperature were ultimately set zero. It can, however, be generalized to make some allowance for temperature by adding a pressure term to the hydrodynamic equation of motion, though this procedure should be viewed with some caution. Another generalization is to include a "drag" term proportional to $\mathbf{v}$ to represent the effect of collisions.

A much more nearly rigorous approach is to use a kinetic theory based on the concept of a distribution function $f(\mathbf{x}, \mathbf{v}, t)$ of phase space, where $\mathbf{x}$ and $\mathbf{v}$ refer to the position and velocity of a particle, respectively, and t is time. The function is defined by the statement that

$$f(\mathbf{x}, \mathbf{v}, t)\, d^3x\, d^3v$$

is the number of particles, at time t, in the volume of phase space $d^3x\, d^3v$ embraced by the range $\mathbf{x}$ to $\mathbf{x} + d\mathbf{x}$ and $\mathbf{v}$ to $\mathbf{v} + d\mathbf{v}$; and in a collisionless plasma it satisfies the Boltzmann equation

$$\frac{\partial f}{\partial t} + \mathbf{v}\cdot\frac{\partial f}{\partial \mathbf{x}} + \mathbf{F}\cdot\frac{\partial f}{\partial \mathbf{v}} = 0,$$

where again $\mathbf{F} = -(e/m)(\mathbf{E} + \mathbf{v} \times \mathbf{B})$ if only electromagnetic forces are operative. The number of particles per unit volume is

$$N = \int f(\mathbf{x}, \mathbf{v}, t)\, d^3v,$$

and the charge and current densities are

$$\rho = -e\int f(\mathbf{x}, \mathbf{v}, t)\, d^3v,$$

$$\mathbf{j} = -e\int \mathbf{v}f(\mathbf{x}, \mathbf{v}, t)\, d^3v.$$

The word "collisionless" is not intended to imply that interaction between particles is neglected. On the contrary, the most important aspects of the interaction are included by virtue of the terms involving $\mathbf{F}$, provided the electromagnetic field includes the part due to ρ and $\mathbf{j}$ representing the electrons (and the other plasma particles). We shall devote much attention to the mathematical treatment of the collisionless Boltzmann equation combined with Maxwell's equations, especially in connection with waves and instabilities in plasmas.

In using the smoothed-out densities and electromagnetic fields some error is committed, and in some refined problems it is necessary to begin with equations which express the dynamics of discrete charged particles more accurately than is possible with the collision-free Boltzmann equation. In the closing chapter we therefore undertake a fundamental investigation of what should be the appropriate governing equations.

PROBLEMS

1. An electron has relativistic energy $\mathscr{E}$ meV and speed v m sec^{-1}. Plot $\mathscr{E}$ as a function of v/c.
2. Assuming that the electrons in the ionosphere have a temperature of 1000°K, give a rough estimate of the mean radius of gyration in the earth's magnetic field at distances of three, five and ten earth radii (a) at the equator, (b) at the poles.
3. Find the electric field needed to keep an electron moving with constant velocity 10^4 m sec^{-1} at an angle of 30° to a uniform magnetic field of 10 gauss.
4. Estimate roughly the Landau length, mean distance between electrons, Debye length and mean free path for collisions between charged particles for the plasmas listed in Table 3.
5. The neutral molecules of a gas have polarizability p; that is, when acted on by an electric field **E** they become electric dipoles of moment $p\mathbf{E}$. Write down the mutual potential energy of an electron and a molecule, and also the distance between the particles at which this energy is equal to the mean thermal energy.
6. An electron, of charge $-e$ and mass m, executes a hyperbolic trajectory in the field of a fixed positive ion of charge e. The electron's speed at infinity is V, and the impact parameter is b. Show that the angle χ through which the electron's motion is deflected by the collision is given by

$$\tan(\tfrac{1}{2}\chi) = e^2/(4\pi\varepsilon_0 bmV^2).$$

7. From Maxwell's equations establish Poynting's theorem

$$-\int \mathbf{E}\cdot\mathbf{j}\,d\tau = \int (\mathbf{E}\cdot\dot{\mathbf{D}} + \mathbf{H}\cdot\dot{\mathbf{B}})\,d\tau + \int (\mathbf{E}\times\mathbf{H})\cdot d\mathbf{S},$$

and give its physical interpretation.

REFERENCES

A classic book still worth reading is

H. A. Lorentz, *Theory of Electrons*, Leipzig (1909).

An early discussion of gyro radiation is given by

O. Heaviside, *Nature*, 342 (1904).

which is reproduced in

O. Heaviside, *Electromagnetic Theory*, **III**, The Electrician Printing and Publishing Co. (1912).

Its role in setting an upper limit to the performance of the betatron was pointed out by

D. Iwanenko and I. Ya Pomeranchuk, *Phys. Rev.*, **65**, 343 (1944).

For its importance in radio astronomy see

I. S. Shklovsky, *Cosmic Radio Waves* (translated), Harvard University Press (1960).

Čerenkov described the experimental investigations of the radiation now named after him in a series of papers of which the first was

P. A. Čerenkov, *Dokl. Akad. Nauk, SSSR*, **2**, 451 (1934).

Earlier experiments on the same phenomenon were reported by

L. Mallett, *C.R. Acad. Sci.* (Paris), **183**, 274 (1926).

The term "plasma" was first introduced in the present context by

I. Langmuir, *Proc. Nat. Acad. Sci.*, **14**, 627 (1928).

Short books giving purely descriptive accounts of plasmas are

L. A. Arzimovich, *Elementary Plasma Physics*, Blaisdell (1965).
E. J. Hellund, *The Plasma State*, Reinhold (1961).

CHAPTER 2

ELECTRODYNAMICS

2.1 FIELDS AND SOURCES

2.1.1 Harmonic time variation; field in terms of current

This chapter is concerned with the study of the problem of the field generated by the motion of charges in a vacuum, and in particular with that due to the motion of a single point charge.

The investigation is based on Maxwell's vacuum equations

$$\text{curl}\,\mathbf{E} = -\mu_0\dot{\mathbf{H}}, \tag{2.1}$$

$$\text{curl}\,\mathbf{H} = \varepsilon_0\dot{\mathbf{E}} + \mathbf{j}, \tag{2.2}$$

and the first objective is to express the field vectors **E** and **H** in terms of the current density **j**. In the discussion of the general theory it is assumed that **j** vanishes identically outside some sufficiently large sphere; and also, when necessary, that **j** and its partial derivatives are continuous. The subsequent application of the analysis to current distributions with singularities takes for granted the validity of implicit limiting processes; in particular, free use is made of the delta function representation of a point charge.

It is helpful to think of **j**, no matter what the origin of the current, as the *source* of the electromagnetic field, and in this sense the aim is to find the appropriate solution of equations (2.1), (2.2). If **j** is prescribed the result is obviously useful in that, in principle at least, it enables the field vectors to be obtained by direct integration over the region occupied by the current. If, on the other hand, as would strictly be the case in most practical situations, **j** as well as **E** and **H** was initially an unknown, then the result can be regarded as an alternative equivalent statement of Maxwell's equations; it leads to the formulation of problems in terms of integral equations rather than differential equations.

To begin with we assume that the field varies harmonically in time with angular frequency ω. Corresponding results for arbitrary time variation are subsequently obtained by a frequency synthesis. The mathematical device of the introduction of complex representations **E** and **H**, with time factor $\exp(i\omega t)$ understood, gives equations

$$\text{curl}\,\mathbf{E} = -i\omega\mu_0\,\mathbf{H}, \tag{2.3}$$

$$\text{curl}\,\mathbf{H} = i\omega\varepsilon_0\,\mathbf{E} + \mathbf{j}. \tag{2.4}$$

Evidently **E** can be eliminated by taking the curl of (2.4) and then substituting for curl **E** from (2.3). The result is

$$\text{curl curl } \mathbf{H} = \omega^2 \varepsilon_0 \mu_0 \mathbf{H} + \text{curl } \mathbf{j}. \tag{2.5}$$

The mathematical identity

$$\text{curl curl} \equiv \text{grad div} - \nabla^2, \tag{2.6}$$

together with

$$\text{div } \mathbf{H} = 0, \tag{2.7}$$

enables (2.5) to be written as

$$\nabla^2 \mathbf{H} + k_0^2 \mathbf{H} = -\text{ curl } \mathbf{j} \tag{2.8}$$

where

$$k_0 = \omega \sqrt{(\varepsilon_0 \mu_0)} = \omega / c. \tag{2.9}$$

Each cartesian component of **H** therefore satisfies a time-harmonic inhomogeneous wave equation; that is,

$$\nabla^2 f + k^2{}_0 f = -s, \tag{2.10}$$

where f is one of the Cartesian components of **H** and s is the corresponding component of the source term curl **j**. It is well known that a solution of (2.10) is

$$F(x, y, z) = \frac{1}{4\pi} \int s \frac{e^{-ik_0 R}}{R} \, d\tau, \tag{2.11}$$

where the volume integral is taken over any region which includes all the points at which s differs from zero, and R is the distance from the volume element $d\tau$ to the field point P at (x, y, z).

That (2.11) does satisfy (2.10) can be seen from the analogous result in potential theory. If P is outside the region where s differs from zero it is permissible to carry differentiation of (2.11) under the integral sign, so that $(\nabla^2 + k_0^2)F = 0$ is an immediate consequence of the standard and easily verified result that $(\nabla^2 + k_0^2)[\exp(-ik_0 R)/R] = 0$. If P is within the region where s differs from zero it may first be observed that although (2.11) is an improper integral by virtue of the singularity of the integrand at $R = 0$, nevertheless, the integral exists and is continuous; this follows from precisely the same considerations as in potential theory. Furthermore, $(\nabla^2 + k_0^2)F = (\nabla^2 + k_0^2)F_1$, where F_1 is given by the right-hand side of (2.11) when the volume of integration is restricted to the interior of a sphere centered on P of arbitrarily small radius ε; and as $\varepsilon \to 0$ it can readily be shown both that $F_1 \to 0$ and that $\nabla^2 F_1 \to -s$, the latter by an analysis parallel to that in potential theory.

The solution of (2.8) corresponding to (2.11) is

$$\frac{1}{4\pi}\int \psi \operatorname{curl}' \mathbf{j}\, d\tau, \tag{2.12}$$

where

$$\psi = \frac{e^{-ik_0 R}}{R}, \tag{2.13}$$

and the dash is used to emphasize that the partial differentiation of the curl operator is with respect to the coordinates (x', y', z') of the volume element $d\tau$. If, then, (2.12) has zero divergence, it can be identified with **H** to give a solution of Maxwell's equations, with **E** determined by (2.4). However, because the integrand of (2.12) involves derivatives of **j**, the integral is awkward to evaluate in such common, if idealized, cases as those in which **j** has a surface of discontinuity. An alternative form, obtainable from (2.12) by a simple transformation, is not only free from this disadvantage but also puts in evidence the fact that the divergence of (2.12) is indeed zero.

To derive the alternative form it is noted that

$$\int \psi \operatorname{curl}' \mathbf{j}\, d\tau = \int \operatorname{curl}'(\psi\, \mathbf{j}) d\tau + \int \mathbf{j} \times \operatorname{grad}' \psi\, d\tau.$$

The first integral on the right-hand side vanishes, since it can be transformed into the surface integral

$$-\int \psi \mathbf{j} \times d\mathbf{S}$$

with the entire surface of integration lying outside the current distribution. Thus (2.12) can be replaced by

$$\frac{1}{4\pi}\int \mathbf{j} \times \operatorname{grad}' \psi\, d\tau. \tag{2.14}$$

No derivatives of **j** appear in (2.14). Furthermore, since

$$\operatorname{curl}(\psi \mathbf{j}) = -\,\mathbf{j} \times \operatorname{grad} \psi = \mathbf{j} \times \operatorname{grad}' \psi,$$

(2.14) can in turn be written

$$\frac{1}{4\pi} \operatorname{curl} \int \mathbf{j} \psi\, d\tau. \tag{2.15}$$

This proves that it has zero divergence, and can therefore be identified with **H** to give a solution of Maxwell's equations.

Explicitly,

$$\operatorname{grad}' \psi = \left(ik_0 + \frac{1}{R}\right) \psi \hat{\boldsymbol{R}}, \tag{2.16}$$

where $\hat{\mathbf{R}}$ is the unit vector specified by

$$R\hat{\mathbf{R}} = \mathbf{R} = (x - x', y - y', z - z'). \tag{2.17}$$

In summary, then,

$$\mathbf{H} = \frac{1}{4\pi} \int \left(ik_0 + \frac{1}{R}\right) \frac{e^{-ik_0 R}}{R} \mathbf{j} \times \hat{\mathbf{R}}\, d\tau, \tag{2.18}$$

$$\mathbf{E} = \frac{Z_0}{ik_0} (\text{curl}\, \mathbf{H} - \mathbf{j}), \tag{2.19}$$

where $Z_0 = \sqrt{(\mu_0/\varepsilon_0)}$ is the vacuum impedance, gives a solution of Maxwell's equations. It is now *postulated* that this is indeed the solution that corresponds to physical reality. The postulate is made on the basis of experience, the decisive feature of the solution being its 'outgoing' character, which is exhibited in the next section.

An important example of an idealized current distribution is that of a dipole. A dipole of strength $\mathbf{p}$ amp m, situated at the origin, is specified by

$$\mathbf{j} = \mathbf{p}\, \delta(x)\, \delta(y)\, \delta(z).$$

In this case (2.18) gives at once

$$\mathbf{H} = \frac{ik_0^2}{4\pi} \mathbf{p} \times \hat{\mathbf{r}} \left(1 - \frac{i}{k_0 r}\right) \frac{e^{-ik_0 r}}{k_0 r}, \tag{2.20}$$

where $\mathbf{r} = r\hat{\mathbf{r}}$ is the radius vector from the origin to the field point. In conventional spherical polar coördinates r, θ, ϕ, where θ is the angle between $\mathbf{r}$ and $\mathbf{p}$, this reads

$$H_r = H_\theta = 0,$$

$$H_\phi = \frac{pk_0^2}{4\pi} i \sin\theta \left(1 - \frac{i}{k_0 r}\right) \frac{e^{-ik_0 r}}{k_0 r}, \tag{2.21}$$

and it follows from (2.19) that

$$E_\phi = 0,$$

$$E_\theta = Z_0 \frac{pk_0^2}{4\pi} i \sin\theta \left[1 - \frac{i}{k_0 r}\left(1 - \frac{i}{k_0 r}\right)\right] \frac{e^{-ik_0 r}}{k_0 r}, \tag{2.22}$$

$$E_r = Z_0 \frac{pk_0^2}{4\pi} 2\cos\theta \frac{1}{k_0 r} \left(1 - \frac{i}{k_0 r}\right) \frac{e^{-ik_0 r}}{k_0 r}. \tag{2.23}$$

By integrating $\frac{1}{2}\,\text{Re}\,(E_\theta H_\phi^*)$ over a sphere center the origin it is straightforward to show, most easily by letting the radius of the sphere tend to infinity, that the time-averaged radiated power is

$$Z_0 \frac{p^2 k_0^2}{12\pi}. \tag{2.24}$$

That **H** can be written in the form (2.15) is often expressed by introducing the so-called electric Hertz vector

$$\mathbf{\Pi} = \frac{1}{4\pi i\omega\varepsilon_0}\int \mathbf{j}\,\psi\,d\tau, \tag{2.25}$$

so that

$$\mathbf{H} = i\omega\varepsilon_0 \operatorname{curl} \mathbf{\Pi}, \tag{2.26}$$

$$\mathbf{E} = \operatorname{curl}\operatorname{curl} \mathbf{\Pi} - \frac{1}{i\omega\varepsilon_0}\,\mathbf{j}. \tag{2.27}$$

Evidently $\mathbf{\Pi}$ satisfies

$$\nabla^2\mathbf{\Pi} + k_0^2\mathbf{\Pi} = -\frac{1}{i\omega\varepsilon_0}\,\mathbf{j}, \tag{2.28}$$

and this gives as an alternative to (2.27)

$$\mathbf{E} = k_0^2\mathbf{\Pi} + \operatorname{grad}\operatorname{div}\mathbf{\Pi}. \tag{2.29}$$

Another standard representation is in terms of the scalar potential ϕ and the vector potential **A**, which are related to $\mathbf{\Pi}$ through the equations

$$\mathbf{A} = i\omega\varepsilon_0\mu_0\mathbf{\Pi}, \tag{2.30}$$

$$\phi = -\operatorname{div}\mathbf{\Pi}, \tag{2.31}$$

and are thus seen to satisfy the Lorentz relation

$$\operatorname{div}\mathbf{A} + i\omega\varepsilon_0\mu_0\phi = 0. \tag{2.32}$$

Substitution from (2.30) and (2.31) into (2.26) and (2.29) gives

$$\mathbf{H} = \frac{1}{\mu_0}\operatorname{curl}\mathbf{A}, \tag{2.33}$$

$$\mathbf{E} = -\operatorname{grad}\phi - i\omega\mathbf{A}. \tag{2.34}$$

Also, from (2.25) and (2.30),

$$\mathbf{A} = \frac{\mu_0}{4\pi}\int \mathbf{j}\psi\,d\tau \tag{2.35}$$

and from (2.25) and (2.31), by a simple transformation and use of the charge conservation relation

$$\operatorname{div}\mathbf{j} + i\omega\rho = 0, \tag{2.36}$$

where ρ is the charge density,

$$\phi = \frac{1}{4\pi\varepsilon_0}\int \rho\psi\,d\tau. \tag{2.37}$$

Of course ϕ and $\mathbf{A}$ satisfy inhomogeneous time-harmonic wave equations with respective source terms $-\rho/\varepsilon_0$, $-\mu_0\,\mathbf{j}$.

Finally it may be remarked that the substitution of (2.37) and (2.35) into (2.34) gives $\mathbf{E}$ as an integral over the charge and current distribution, namely

$$\mathbf{E} = \frac{Z_0}{4\pi}\int\left[\left(ik_0 + \frac{1}{R}\right)\hat{\mathbf{R}}c\rho - ik_0\mathbf{j}\right]\frac{e^{-ik_0R}}{R}\,d\tau. \tag{2.38}$$

This is analogous to the form (2.18) for $\mathbf{H}$, and could have been obtained in a similar way by eliminating $\mathbf{H}$ from (2.3), (2.4) and transforming the solution (2.11) of the resulting inhomogeneous wave equation for $\mathbf{E}$.

2.1.2 The radiation field

In many circumstances a knowledge of the field is required only at great distances from the source. If r denotes distance from some origin O located in the vicinity of the current distribution a quite cursory inspection of (2.18), (2.19) reveals that as $r \to \infty$ in any given direction the magnitudes of $\mathbf{E}$ and $\mathbf{H}$ are in general of order $1/r$. The terms of this order constitute what is known as the radiation field. In this section it is proved that the radiation field takes the form

$$\mathbf{H} \sim -(\mathbf{G} \times \hat{\mathbf{r}})\frac{e^{-ik_0r}}{r}, \tag{2.39}$$

$$\mathbf{E} \sim Z_0\hat{\mathbf{r}} \times (\mathbf{G} \times \hat{\mathbf{r}})\frac{e^{-ik_0r}}{r}, \tag{2.40}$$

where

$$\mathbf{G} = -\frac{ik_0}{4\pi}\int \mathbf{j}\,e^{ik_0\hat{\mathbf{r}}\cdot\mathbf{r}'}\,d\tau, \tag{2.41}$$

$\mathbf{r}'$ being the radius vector from O to the volume element $d\tau$ (see Fig. 2.1).

For a specified current distribution the vector $\mathbf{G}$ depends on the direction $\hat{\mathbf{r}}$ of the field point, but not on its distance r; the dependence of the radiation field on r is exhibited explicitly in (2.39) and (2.40). Evidently the actual vectors $\mathbf{E}$ and $\mathbf{H}$ of the radiation field are mutually orthogonal

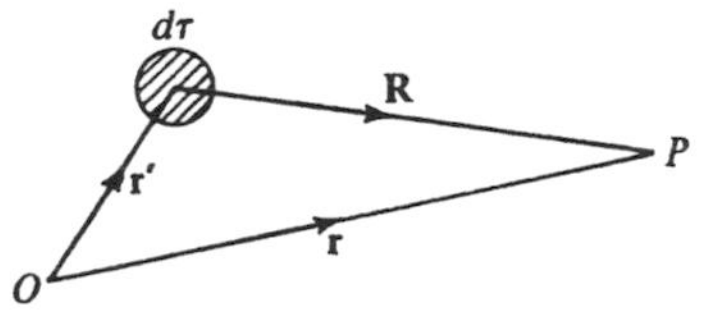

Fig. 2.1. Relative configuration of origin O, field point P, and volume element $d\tau$.

and also orthogonal to $\hat{\mathbf{r}}$, with the Poynting vector $\mathbf{E} \times \mathbf{H}$ directed in the same sense as $\hat{\mathbf{r}}$. The ratio of the magnitudes of $\mathbf{E}$ and $\mathbf{H}$ is the constant Z_0, and the polarization is in general elliptical.

To derive the required results from (2.18) the straightforward procedure can be adopted of expanding the integrand as a power series in inverse powers of r, and then reversing the orders of integration and summation. In fact

$$\mathbf{R} = \mathbf{r} - \mathbf{r}', \tag{2.42}$$

so that

$$R^2 = r^2(1 - 2\hat{\mathbf{r}}\cdot\mathbf{r}'/r + r'^2/r^2),$$

and the binomial theorem gives

$$R = r - \mathbf{r}\cdot\mathbf{r}' + \sum_{n=1}^{\infty} \frac{p_{n+1}}{r^n}, \tag{2.43}$$

where the p_n, expressible in terms of $\hat{\mathbf{r}}\cdot\mathbf{r}'$ and r'^2, are certain homogeneous polynomials of degree n in x', y', z', whose precise specification is not required. It follows that

$$e^{-ik_0R} = e^{-ik_0r}\, e^{ik_0\hat{\mathbf{r}}\cdot\mathbf{r}'}\left(1 - \frac{ik_0p_2}{r} + \cdots\right),$$

$$\frac{1}{R} = \frac{1}{r}\left(1 + \frac{\hat{\mathbf{r}}\cdot\mathbf{r}'}{r} + \cdots\right),$$

$$\hat{\mathbf{R}} = (\mathbf{r} - \mathbf{r}')\frac{1}{R} = \left(\hat{\mathbf{r}} - \frac{\mathbf{r}'}{r}\right)\left(1 + \frac{\hat{\mathbf{r}}\cdot\mathbf{r}'}{r} + \cdots\right),$$

and so

$$\left(ik_0 + \frac{1}{R}\right)\frac{e^{-ik_0R}}{R}\,\hat{\mathbf{R}} = \frac{e^{-ik_0r}}{r}\, e^{ik_0\hat{\mathbf{r}}\cdot\mathbf{r}'}\left(\hat{\mathbf{r}} - \frac{\mathbf{r}'}{r}\right)\left(ik_0 + \sum_{n=1}^{\infty}\frac{q_{n+1}}{r^n}\right), \tag{2.44}$$

where the q_n are certain polynomials in x', y', z', not now homogeneous, of degree n.

The expansion indicated in (2.44) is absolutely and uniformly convergent throughout the region exterior to any sphere center O which encloses the entire current distribution. When substituted into (2.18), therefore, the order of integration and summation may legitimately be reversed, with the result

$$\mathbf{H} = \frac{e^{-ik_0r}}{r}\sum_{n=0}^{\infty}\frac{\mathbf{A}_n}{r^n}, \tag{2.45}$$

where

$$\mathbf{A}_0 = -\,\mathbf{G} \times \hat{\mathbf{r}}, \tag{2.46}$$

in the notation of (2.41), and

$$\mathbf{A}_n = \frac{1}{4\pi}\left\{\int q_{n+1}\,\mathbf{j}\,e^{ik_0\hat{\mathbf{r}}\cdot\mathbf{r}'}\,d\tau\right\}\times\hat{\mathbf{r}} - \frac{1}{4\pi}\int q_n\,\mathbf{j}\times\mathbf{r}'\,e^{ik_0\hat{\mathbf{r}}\cdot\mathbf{r}'}\,d\tau \tag{2.47}$$

for $n = 1, 2, 3 \ldots$, in which q_1 is written for ik_0. For a specified current distribution each $\mathbf{A}_n$ depends on direction $\hat{\mathbf{r}}$ but not on distance r.

The series for **E** corresponding to (2.45) is obtained, according to (2.19), on taking the curl and multiplying by $-iZ_0/k_0$. As regards the radiation field it turns out that in this operation $\mathbf{A}_0$ can be treated as a constant, and the expressions (2.39), (2.40) are thereby established.

Three further observations are made in connection with these results. First, if in conventional spherical polar coordinates r, θ, ϕ

$$\mathbf{G} = (G_r, G_\theta, G_\phi),$$

then

$$\mathbf{A}_0 = (0, -G_\phi, G_\theta),$$

and the radiation field appears in the form

$$\mathbf{H} \sim (0, -G_\phi, G_\theta)\,\frac{e^{-ik_0 r}}{r}, \tag{2.48}$$

$$\mathbf{E} \sim Z_0(0, G_\theta, G_\phi)\,\frac{e^{-ik_0 r}}{r}. \tag{2.49}$$

The flux of $\mathrm{Re}\,\frac{1}{2}\,\mathbf{E}\times\mathbf{H}^*$ out of a large sphere gives the time-averaged radiated power, which is therefore

$$\tfrac{1}{2}Z_0\int (G_\theta G_\theta^* + G_\phi G_\phi^*)\,d\Omega, \tag{2.50}$$

where $d\Omega$ is the element of solid angle.

Secondly, it is observed that the radiation field can in general be represented as that of a dipole only if the linear dimensions of the current distribution are very much less than a vacuum wavelength at the angular frequency ω. If this is so, $k_0\hat{\mathbf{r}}\cdot\mathbf{r}'$ in the integrand in (2.41) is small, and

$$\mathbf{G} = -\frac{ik_0}{4\pi}\int \mathbf{j}\,d\tau \tag{2.51}$$

approximately. The radiation field (2.39) is then approximately that of a dipole of strength

$$\mathbf{p} = \int \mathbf{j}\,d\tau, \tag{2.52}$$

as can be verified from (2.20). That the dipole approximation does not appear solely as a consequence of retreating far enough from the source is by virtue of the finite wavelength, and contrasts with the electrostatic analog, which can be recovered by setting $k_0 = 0$ (infinite wavelength). Another contrast with electrostatics resides in the fact that it is not possible for $\mathbf{A}_0$ in (2.45) to be identically zero without all the other $\mathbf{A}_n$ also being identically zero; the field of a current distribution which on a time average radiates no power must therefore vanish at all points outside the current distribution. The proof is not simple and is offered as a problem at the end of the chapter.

The third observation concerns the distance to which it is necessary to go before the radiation field (2.39), (2.40) becomes a good approximation to the total field. An inspection of the expansions leading up to (2.44) reveal the two obvious conditions $k_0 r \gg 1$, $r'/r \ll 1$, that is, the distance must be much greater than both the wavelength and the linear dimensions of the current distribution; and also a third condition, perhaps less obvious, $k_0 r'^2 \ll r$. This last states that the distance must be much greater than the square of the linear extent of the source divided by the wavelength, and its importance is recognized by calling the quantity so defined the Rayleigh distance. When the extent of the source exceeds a wavelength it is the Rayleigh distance which determines the range beyond which the radiation field is a good approximation to the total field.

2.1.3 General time variation; retarded potentials

The discussion of monochromatic fields, that is, fields with a harmonic time variation, in addition to being important in itself, also enables corresponding results to be written down for fields with a general time variation. For by frequency Fourier (spectral) analysis any field can be represented as a superposition of monochromatic fields of different frequencies, so that a linear relation derived for a monochromatic field of angular frequency ω if restated in a form in which ω does not appear explicitly is then applicable to a general field.

With this objective, the relations (2.33), (2.34) are written

$$\mathbf{H} = \frac{1}{\mu_0} \operatorname{curl} \mathbf{A}, \tag{2.53}$$

$$\mathbf{E} = -\operatorname{grad} \phi - \dot{\mathbf{A}}; \tag{2.54}$$

and (2.53), (2.54) give the representation of the actual field vectors of any electromagnetic field in terms of a scalar potential ϕ and a vector potential $\mathbf{A}$. Furthermore, in a similar way, expressions for ϕ and $\mathbf{A}$ in terms of the charge and current distributions can be obtained from (2.37) and (2.35). If it is remembered that $k_0 = \omega/c$ and that the time t appears solely through the factor exp ($i\omega t$), the corresponding general results are evidently

$$\phi(x, y, z, t) = \frac{1}{4\pi\varepsilon_0}\int \frac{\rho(x', y', z', t - R/c)}{R} d\tau, \tag{2.55}$$

$$\mathbf{A}(x, y, z, t) = \frac{\mu_0}{4\pi}\int \frac{\mathbf{j}(x', y', z', t - R/c)}{R} d\tau. \tag{2.56}$$

In words, the contribution to ϕ and $\mathbf{A}$ at the field point at the time t from the volume element $d\tau$ comes from the charge and current densities, respectively, which exist at $d\tau$ at time $t - R/c$.

Conventionally, $t - R/c$ is called the retarded time, and (2.55), (2.56) are called the retarded potentials. Their interpretation in terms of a disturbance which travels from $d\tau$ to the field point with speed c is self-evident, and reinforces the view that these solutions of Maxwell's equations are the correct expression of physical experience, in contrast to those represented by means of the advanced potentials in which $t + R/c$ replaces $t - R/c$. It is perhaps worth remarking that the retarded potentials, thought of as generalizations of the potentials of electrostatics and steady current flow to situations where effects are propagated with a finite speed, were indeed much in the minds of theorists some time before Maxwell formulated the equations which bear his name.

The Lorentz relation (2.32) now reads

$$\operatorname{div} \mathbf{A} + \dot{\phi}/c^2 = 0, \tag{2.57}$$

and this could be established directly from (2.55), (2.56) with the help of the charge conservation relation

$$\operatorname{div} \mathbf{j} + \dot{\rho} = 0. \tag{2.58}$$

The partial differential equations satisfied by ϕ and $\mathbf{A}$ are, of course,

$$\nabla^2\phi - \ddot{\phi}/c^2 = -\rho/\varepsilon_0, \tag{2.59}$$

$$\nabla^2\mathbf{A} - \ddot{\mathbf{A}}/c^2 = -\mu_0 \mathbf{j}. \tag{2.60}$$

It is sometimes convenient to use heavy square brackets [ ] to indicate that the quantity that they embrace is to be evaluated at the retarded time. Thus (2.55), (2.56) appear more briefly as

$$\phi = \frac{1}{4\pi\varepsilon_0}\int \frac{[\rho]}{R} d\tau, \tag{2.61}$$

$$\mathbf{A} = \frac{\mu_0}{4\pi}\int \frac{[\mathbf{j}]}{R} d\tau, \tag{2.62}$$

where the various arguments are understood from the context.

Explicit expressions for **H** and **E** corresponding to (2.18) and (2.38) can likewise be written down, namely

$$\mathbf{H} = \frac{1}{4\pi}\int\left(\frac{[\dot{\mathbf{j}}]}{cR} + \frac{[\mathbf{j}]}{R^2}\right) \times \hat{\mathbf{R}}\, d\tau, \tag{2.63}$$

$$\mathbf{E} = \frac{Z_0}{4\pi}\int\left\{\left(\frac{[\dot{\rho}]}{R} + \frac{c[\rho]}{R^2}\right)\hat{\mathbf{R}} - \frac{[\dot{\mathbf{j}}]}{cR}\right\} d\tau. \tag{2.64}$$

It is of interest to examine the radiation field, and in particular the dipole approximation. From (2.63) the radiation field is evidently specified by

$$\mathbf{H} \sim -\frac{1}{4\pi c}\frac{\hat{\mathbf{r}}}{r} \times \int [\dot{\mathbf{j}}]\, d\tau, \tag{2.65}$$

where $\mathbf{r} = r\hat{\mathbf{r}}$ is the radius vector to the field point from some origin in the vicinity of the current distribution. Now for an idealized dipole, of arbitrary time-varying strength $\mathbf{p}$, located at the origin,

$$\mathbf{j} = \mathbf{p}\, \delta(x)\, \delta(y)\, \delta(z),$$

and the complete magnetic vector from (2.63) is

$$\mathbf{H} = \frac{1}{4\pi}\left(\frac{\mathbf{p}}{r^2} + \frac{\dot{\mathbf{p}}}{cr}\right) \times \hat{\mathbf{r}}, \tag{2.66}$$

with radiation field

$$\mathbf{H} \sim \frac{1}{4\pi c}\frac{\dot{\mathbf{p}} \times \hat{\mathbf{r}}}{r}, \tag{2.67}$$

in which $\mathbf{p}$ and $\dot{\mathbf{p}}$ are evaluated at $t - r/c$. The radiation field (2.65) can therefore be regarded as that of a dipole of strength $\int \mathbf{j}\, d\tau$ if the current distribution is such that the various retardations R/c associated with the volume elements $d\tau$ can all validly be approximated by the common retardation r/c.

Now for any function $f(t)$ a Taylor expansion gives, as $r \to \infty$,

$$f(t - R/c) = f - \frac{\partial f}{\partial t}\frac{\hat{\mathbf{r}}\cdot\mathbf{r}'}{c} + \cdots, \tag{2.68}$$

where f and its derivatives on the right-hand side are evaluated at $t - r/c$. It appears, therefore, that an order of magnitude criterion for the validity of the dipole approximation is

$$\frac{a}{cT} \ll 1, \tag{2.69}$$

where a is the maximum linear dimension of the current distribution and T represents the time scale of the variation of current density. For monochromatic fields with angular frequency $\omega = ck_0$ the inequality corresponding to (2.69) is $k_0a \ll 1$, which has already been noted as the condition under which (2.51) is a good approximation to (2.41).

2.1.4 The power flux spectrum

The Poynting vector $\mathbf{E} \times \mathbf{H}$, where $\mathbf{E}$ and $\mathbf{H}$ are the actual vectors of a field with general time variation, gives the power flux density at any point, in the sense that $(\mathbf{E} \times \mathbf{H})\cdot d\mathbf{S}$ is the rate at which energy crosses the infinitesimal vector area $d\mathbf{S}$. The spectral resolution of the field into its frequency components gives a corresponding resolution of the power flux density, and there are several important associated formulas which are now derived.

Since the vector character of the field has no relevance to the spectral analysis the essential mathematics is stated in terms of arbitrary (possibly complex) functions $F(t)$, $G(t)$ of the real variable t with Fourier representations

$$[F(t), G(t)] = \int_{-\infty}^{\infty} [f(\omega), g(\omega)]\, e^{i\omega t}\, d\omega. \tag{2.70}$$

The product representation

$$F(t)G^*(t) = \int_{-\infty}^{\infty}\int_{-\infty}^{\infty} f(\omega)g^*(\omega')\, e^{i(\omega-\omega')t}\, d\omega\, d\omega', \tag{2.71}$$

where the star denotes the complex conjugate, yields two results. One is obtained by replacing the integration variable ω by $\omega + \omega'$; this gives

$$F(t)G^*(t) = \int_{-\infty}^{\infty}\int_{-\infty}^{\infty} f(\omega+\omega')g^*(\omega')\, e^{i\omega t}\, d\omega\, d\omega', \tag{2.72}$$

so that the complex spectrum of $F(t)G^*(t)$ is the *convolution*

$$\int_{-\infty}^{\infty} f(\omega+\omega')g^*(\omega')\, d\omega'$$

of $f(\omega)$ and $g^*(\omega)$. The other result is obtained by integrating (2.71) over all t; since $\delta(\omega - \omega')$ then appears on the right-hand side, this gives

$$\int_{-\infty}^{\infty} F(t)G^*(t)\, dt = 2\pi \int_{-\infty}^{\infty} f(\omega)g^*(\omega)\, d\omega. \tag{2.73}$$

When $F(t)$ and $G(t)$ are real, the representations (2.70) imply that $f^*(\omega) = f(-\omega)$, $g^*(\omega) = g(-\omega)$, and can be written

$$(F, G) = \mathrm{Re}\, 2\int_{0}^{\infty} (f, g)\, e^{i\omega t}\, d\omega, \tag{2.74}$$

in which only positive frequencies are involved. Correspondingly, the formula (2.73) can be restated

$$\int_{-\infty}^{\infty} F(t)G(t)\, dt = \mathrm{Re}\, 4\pi \int_{0}^{\infty} f(\omega)g^*(\omega)\, d\omega. \tag{2.75}$$

The left-hand side of (2.75), which is the total energy if $F(t)G(t)$ is power, does not converge in those applications where it is otherwise convenient to assume that $F(t)$ and $G(t)$ are not sensibly zero for all t outside some finite interval. Periodic functions furnish a familiar example. Suppose that

$$(F, G) = \sum_{n=-\infty}^{\infty} (f_n, g_n)\, e^{in\Omega t}, \tag{2.76}$$

where

$$(f_n, g_n) = \frac{\Omega}{2\pi} \int_{-\pi/\Omega}^{\pi/\Omega} (F, G)\, e^{-in\Omega t}\, dt. \tag{2.77}$$

Then

$$FG^* = \sum_{m,\, n=-\infty}^{\infty} f_m\, g_n^*\, e^{i(m-n)t}, \tag{2.78}$$

and the quantity of interest is now the average value of FG^* over a period; this is

$$\frac{\Omega}{2\pi}\int_{-\pi/\Omega}^{\pi/\Omega} FG^*\, dt = \sum_{n=-\infty}^{\infty} f_n g_n^*. \tag{2.79}$$

Alternatively, when F and G are real, (2.76) can be written

$$(F, G) = (f_0, g_0) + \mathrm{Re}\, 2 \sum_{n=1}^{\infty} (f_n, g_n)\, e^{in\Omega t}, \tag{2.80}$$

and (2.79) is

$$\frac{\Omega}{2\pi} \int_{-\pi/\Omega}^{\pi/\Omega} FG\, dt = f_0 g_0 + \mathrm{Re}\, 2 \sum_{n=1}^{\infty} f_n g_n^*, \tag{2.81}$$

which shows that the average value over a period of the product of two real periodic functions is the sum of the average values of the products of their corresponding harmonic components.

Another example where it is the average power which is of interest occurs when $F(t)$ and $G(t)$ represent real stationary stochastic processes. The result can be formally derived from (2.73) by supposing in the first place that F and G vanish for t outside the interval $-\frac{1}{2}T$ to $\frac{1}{2}T$, and subsequently letting T tend to infinity. This gives the average power

$$\frac{1}{T}\int_{-\frac{1}{2}T}^{\frac{1}{2}T} FG\,dt = 2\pi\int_{-\infty}^{\infty} f'(\omega)g'^*(\omega)\,d\omega, \tag{2.82}$$

where

$$(f', g') = \lim_{T\to\infty} \frac{1}{2\pi\sqrt{T}}\int_{-\frac{1}{2}T}^{\frac{1}{2}T} (F, G)\,e^{-i\omega t}\,dt. \tag{2.83}$$

An alternative formal derivation is provided by (2.79) when Ω is allowed to tend to zero; the sum is replaced by an integral in much the same way as in the familiar demonstration of the relation between the series representation (2.76) and the integral representation (2.70).

To put the foregoing results into the context of electromagnetic theory it is only necessary to introduce the complex spectra $\mathbf{E}^{\omega}(\omega)$, $\mathbf{H}^{\omega}(\omega)$ of $\mathbf{E}(t)$ and $\mathbf{H}(t)$, and to replace FG by $\mathbf{E}\times\mathbf{H}$. From (2.75), for example, it then follows that the total energy that crosses infinitesimal area $d\mathbf{S}$ is the scalar product of $d\mathbf{S}$ with

$$\mathrm{Re}\, 4\pi\int_{0}^{\infty} \mathbf{E}^{\omega}\times\mathbf{H}^{\omega *}\,d\omega. \tag{2.84}$$

Again, from (2.79), it follows that, for a time-periodic field

$$(\mathbf{E}, \mathbf{H}) = \sum_{n=-\infty}^{\infty} (\mathbf{E}_n, \mathbf{H}_n)\,e^{in\Omega t}, \tag{2.85}$$

the power flux density averaged over a period is

$$\sum_{n=-\infty}^{\infty} \mathbf{E}_n\times\mathbf{H}_n^*. \tag{2.86}$$

And finally, from (2.82), (2.83), it follows that, for a statistically stationary fluctuating field, the time-averaged power flux density is

$$2\pi\int_{-\infty}^{\infty} \mathbf{E}^{\omega\prime}\times\mathbf{H}^{\omega\prime *}\,d\omega \tag{2.87}$$

where

$$(\mathbf{E}^{\omega\prime}, \mathbf{H}^{\omega\prime}) = \lim_{T\to\infty} \frac{1}{2\pi\sqrt{T}}\int_{-\frac{1}{2}T}^{\frac{1}{2}T} (\mathbf{E}, \mathbf{H})\,e^{-i\omega t}\,dt. \tag{2.88}$$

It should be noted that the statement leading to (2.84) has a particularly useful corollary when $d\mathbf{S}$ is in the radiation field. For $\mathbf{E}^{\omega}$, $\mathbf{H}^{\omega}$ can then be written in the form (2.48), (2.49), where $\mathbf{G}$ is given by (2.41) with $\mathbf{j}$ replaced by the complex spectrum $\mathbf{j}^{\omega}$ of the current density, and the corollary reads: the total energy that crosses a distant surface element that subtends an infinitesimal solid angle $d\Omega$ at the origin is

$$4\pi Z_0\, d\Omega \int_0^{\infty} (G_\theta G_\theta^* + G_\phi G_\phi^*)\, d\omega. \tag{2.89}$$

2.1.5 The general Fourier method

It has been shown that the general physical solution for the field in terms of the charge and current densities can be expressed by (2.63), (2.64), or by equivalent forms involving potential functions. A particular case is the idealized dipole field (2.66); and conversely, apart from a purely electrostatic field, it is legitimate to regard the general formulas as representing the superposition of the fields of dipoles of strengths $\mathbf{j}\, d\tau$ located at the positions of the various volume elements $d\tau$. There are, however, other useful ways of describing the field. The value of Fourier analysis in the time variable has already been manifested, and Fourier analysis in the space variables also can be a profitable mathematical technique. The source is then conceived as a superposition of space and time harmonic current distributions, rather than as a collection of dipoles; and the relative simplicity of the field of each harmonic distribution is an advantage which may offset the extra complication of the preliminary Fourier resolution.

Suppose, then, that the current density is represented in the form

$$\mathbf{j} = \iiiint \mathbf{j}'\, e^{i(\omega t - \mathbf{k}\cdot\mathbf{r})}\, d^3k\, d\omega \tag{2.90}$$

where if $\mathbf{k} = (l, m, n)$, d^3k stands for $dl\, dm\, dn$, and the integration limits are from $-\infty$ to $+\infty$ for each of the four variables. The transform $\mathbf{j}'$ is of course a function of $\mathbf{k}$ and ω, and if $\mathbf{E}'$, $\mathbf{H}'$ are the corresponding transforms of $\mathbf{E}$, $\mathbf{H}$, Maxwell's equations (2.1), (2.2) give

$$\mathbf{k} \times \mathbf{E}' = \omega\mu_0\mathbf{H}', \tag{2.91}$$

$$\mathbf{k} \times \mathbf{H}' = -\omega\varepsilon_0\mathbf{E}' + i\,\mathbf{j}'. \tag{2.92}$$

Clearly, in fact, all the previously obtained linear differential relations between various field quantities have their counterpart in algebraic relations between the transforms of the field quantities. For example, the relation corresponding to (2.60) is

$$(k^2 - k_0^2)\mathbf{A}' = \mu_0\,\mathbf{j}', \tag{2.93}$$

where k stands for $|\,\mathbf{k}\,| = \sqrt{(l^2 + m^2 + n^2)}$, and it is recalled that $k_0 = \omega/c$. The solution of (2.93) for $\mathbf{A}'$ in terms of $\mathbf{j}'$ leads to the representation

$$\mathbf{A} = \mu_0 \iiiint \frac{\mathbf{j}'}{k^2 - k_0{}^2}\, e^{i(\omega t - \mathbf{k}\cdot\mathbf{r})}\, d^3k\, d\omega, \tag{2.94}$$

where the way in which the paths of integration avoid the poles is explained a little later on. Of course $\mathbf{j}'$ itself is specified by the inverse of (2.90), namely

$$\mathbf{j}' = \frac{1}{16\pi^4}\iiiint \mathbf{j}\, e^{-i(\omega t - \mathbf{k}\cdot\mathbf{r})}\, d\tau\, dt. \tag{2.95}$$

Similar representations are readily obtained for **E**, **H**, ϕ and **Π**.

It is instructive to make a direct comparison of (2.94), (2.95) with (2.62). Evidently (2.95) can be rewritten as the volume integral

$$\mathbf{j}' = \frac{1}{8\pi^3}\int \mathbf{j}^{\omega}\, e^{i\mathbf{k}\cdot\mathbf{r}'}\, d\tau, \tag{2.96}$$

where

$$\mathbf{j}^{\omega} = \frac{1}{2\pi}\int \mathbf{j}\, e^{-i\omega t}\, dt \tag{2.97}$$

is the complex frequency spectrum of the current density, and $\mathbf{r}'$ has been reintroduced to denote the position vector of the volume element $d\tau$ as in Fig. 2.1. The substitution of (2.96) into (2.94) gives

$$\mathbf{A} = \frac{\mu_0}{8\pi^3}\iiiint\!\!\int \frac{\mathbf{j}^{\omega}\, e^{i\omega t}\, e^{-i\mathbf{k}\cdot\mathbf{R}}}{k^2 - k_0^{\,2}}\, d^3k\, d\omega\, d\tau, \tag{2.98}$$

where **R** is the vector from the volume element to the field point as in Fig. 2.1 But since

$$\mathbf{j} = \int \mathbf{j}^{\omega}\, e^{i\omega t}\, d\omega \tag{2.99}$$

it follows that in the integrand of (2.62)

$$[\mathbf{j}] = \int \mathbf{j}^{\omega}\, e^{i\omega(t - R/c)}\, d\omega, \tag{2.100}$$

and (2.62) can therefore be written

$$\mathbf{A} = \frac{\mu_0}{4\pi}\iint \frac{\mathbf{j}^{\omega}\, e^{i\omega t}\, e^{-ik_0 R}}{R}\, d\omega\, d\tau, \tag{2.101}$$

which is, of course, nothing other than an explicit statement of the step by which (2.62) was deduced from (2.35). The identity of (2.98) and (2.101) is thus seen to be synonymous with the result

$$\frac{1}{2\pi^2}\iiint \frac{e^{-i\mathbf{k}\cdot\mathbf{R}}}{k^2 - k_0^{\,2}}\, d^3k = \frac{e^{-ik_0 R}}{R}. \tag{2.102}$$

To complete the discussion we must return to (2.94) and explain how the poles of the integrand, given by $k^2 = k_0^2\,(= \omega^2/c^2)$, are situated in relation to the paths of integration. The matter is perhaps most simply decided by allowing ω to have a small imaginary part, and insisting that this imaginary part be negative on the physical grounds that the associated quasi-harmonic disturbance must tend to zero (rather than infinity) as $t \to -\infty$. In other words, the ω path of integration in (2.94) can be conceived as running parallel to, and just below, the real axis. With this prescription (2.94), and related integrals, are unambiguously defined. We note, in particular, that a direct proof of (2.102) by contour integration (which is helped by the observation that the integral depends on **R** only through the magnitude R) shows the prescription Im $\omega < 0$ to be necessary and sufficient to give an outgoing wave.

The grounds for taking Im $\omega < 0$ can be restated in terms of causality. Consider the situation in which there is no disturbance until a current source is switched on at $t = 0$. Then (2.94) must be zero for $t < 0$. Mathematically this implies that the integrand, regarded as a function of ω, is free of poles and branch points in the region of the complex ω-plane below the ω path of integration. The latter must therefore lie below the poles at $\omega = \pm ck$, that is, since k takes all real values, below the real axis.

The general Fourier method can also be applied to the case when media other than vacuum (such as plasmas) are considered. In fact it then seems the only comprehensive technique available, and the ideas outlined in the present subsection are relevant to much of the analysis in the later chapters.

2.2 THE FIELD OF A MOVING POINT CHARGE

2.2.1 The non-relativistic approximation

In §2.1 the field of a general charge and current distribution is discussed. It is now proposed to consider in some detail the field of a single point charge in arbitrary motion. For this particular case the Eulerian picture of §2.1, in which variations of charge and current density at each point in space are envisaged, is appropriately replaced by a Lagrangian picture in which the path of the particle is followed. The corresponding reformulation of the mathematics results in expressions for the field in terms of the position, velocity and acceleration of the point charge.

The general formulas are quite complicated, and it is helpful to inquire first into the dipole approximation. The notation to be used is e for the charge, $\mathbf{v}$ and $\dot{\mathbf{v}}$ for its velocity and acceleration (where in this context the dot, of course, denotes time differentiation following the particle), and $\mathbf{r} = r\hat{\mathbf{r}}$ for the vector from the charge to the field point (in contrast to its use in §2.1 as the vector from a fixed origin to the field point).

Then

$$\int \mathbf{j}\, d\tau = e\mathbf{v},$$

and the dipole approximation to (2.63) gives for the magnetic field at time t

$$\mathbf{H} = \frac{e}{4\pi}\left(\frac{\mathbf{v}}{\mathbf{r}^2} + \frac{\dot{\mathbf{v}}}{cr}\right) \times \hat{\mathbf{r}}, \tag{2.103}$$

where r and the quantities on the right-hand side are evaluated at time $t - [r]/c$. This retarded time is only specified implicitly, since the interval r/c by which it lags on t is itself a function of time and has to be evaluated at the retarded time, as indicated by the bold square brackets.

The formula (2.103) is valid only if the differences between the time retardations associated with different parts of the charge are unimportant. The condition for this to be the case must be the appropriate restatement of (2.69), and in the present context the requirement that suggests itself is

$$v/c \ll 1.$$

Thus even for a charge distribution conceived as located at a point it is to be expected for a relativistic particle that time retardation effects need detailed examination, and this is confirmed explicitly by the rigorous development given subsequently.

In the approximation (2.103) for a non-relativistic particle the $1/r^2$ term is recognized as the Biot–Savart expression, and the other term is the contribution of the radiation field. The corresponding approximation to **E** is not so easy to derive directly from (2.64), because for the $[\dot{\rho}]$ term in the integrand it is necessary to retain two terms in the expansion (2.68). However, in the radiation field, **E** can of course be written down from a knowledge of **H**, and (2.64) shows that the only other term is simply the Coulomb field. Thus for a non-relativistic particle

$$\mathbf{E} = \frac{e}{4\pi\varepsilon_0}\left\{\frac{\hat{\mathbf{r}}}{r^2} - \frac{\hat{\mathbf{r}} \times (\dot{\mathbf{v}} \times \hat{\mathbf{r}})}{c^2 r}\right\} \tag{2.104}$$

evaluated at time $t - [r]/c$.

The rate at which energy crosses a distant surface element that subtends an infinitesimal solid angle $d\Omega$ at the retarded position of the charge is given by the value in the radiation field of

$$|\mathbf{E} \times \mathbf{H}|\, r^2\, d\Omega.$$

To the present approximation this is, per unit solid angle,

$$\frac{e^2}{16\pi^2\varepsilon_0 c^3}\,|\dot{\mathbf{v}} \times \hat{\mathbf{r}}|^2 = \frac{e^2 \dot{\mathbf{v}}^2 \sin^2\theta}{16\pi^2\varepsilon_0 c^3} \tag{2.105}$$

evaluated at the retarded time, where θ is the angle between the acceleration and the direction of the surface element. The formula is, of course, that for a dipole of strength $e\mathbf{v}$. The radiation pattern described by $\sin^2\theta$ has a zero in the direction of the acceleration and a maximum at right angles to it. The rate at which energy crosses a closed spherical surface is

$$\frac{e^2\dot{\mathbf{v}}^2}{8\pi\varepsilon_0 c^3}\int_0^\pi \sin^3\theta\, d\theta = \frac{e^2\dot{\mathbf{v}}^2}{6\pi\varepsilon_0 c^3}\,. \tag{2.106}$$

These statements about the power that at time t crosses a surface distant from the charge, which involve a knowledge of the behavior of the charge at time $t - [r]/c$, can be rephrased to give an interpretation of the *contemporary* value of expressions (2.105) and (2.106). For with $v/c \ll 1$ (and not otherwise, as is shown in detail later) the energy that crosses a surface element in a time interval δt at time t can be identified with that emitted by the charge in a time interval δt at time $t - [r]/c$. The value of (2.105) at any instant must therefore be the rate at which, at that instant, the charge radiates energy, per unit solid angle, in a direction making an angle θ with its acceleration. Likewise, (2.106) gives the total radiated power.

2.2.2 The Liénard–Wiechert potentials

The task is now to derive exact expressions for the field of a moving point charge. As has been indicated in §2.2.1, careful consideration must be given to time retardation effects, and the results for **E** and **H** are expected to conform to (2.103) and (2.104) only when v/c is negligible. In fact, it is found that important modifications do arise when v/c approaches unity, and the field generated by a particle moving at relativistic speeds is quite different in character from that generated by a slow particle.

There are various ways of doing the analysis. The oldest method, due to Liénard and Wiechert, is first to derive expressions for ϕ and **A** for a moving point charge by a direct evaluation of the integrals in (2.61) and (2.62); and then to obtain **E** and **H** from (2.53) and (2.54). This method is now given.

The time retardation effect in an integral such as (2.61) is perhaps most easily interpreted by picturing a spherical surface S, whose radius diminishes with speed c, collapsing towards its center at the field point P, which it reaches at time t. Then the value of the numerator of the integrand in (2.61) corresponding to each point in space is the charge density at that point at the instant the sphere S passes through it.

Now apply this interpretation to a small rigid element of charge moving with velocity **v**. Only when the element is intersected by the surface S does it make any contribution to (2.61); and it may be noted that if v remains always less than c', where c' is some fixed speed less than c by an arbitrarily small amount, then each surface S associated with a given time t crosses the charge element once and only once. Moreover, the component of **v** tangential to S during the crossing period has no effect on the value of (2.61), since it does not alter the distance r of the charge from P; and for the purposes of calculation the charge may therefore be assumed to be moving radially with speed $\dot{r}$.

For ease of visualization it is convenient to consider the charge element subdivided, by planes perpendicular to the line from P to the charge, into thin slices which move with the charge. Let the cross-section in any plane through P of any particular slice be represented in Fig. 2.2 by $ABCD$ when S contains BC, and by $A'B'C'D'$ when S contains $A'D'$; the figure is drawn for the case in which the charge is moving away from P, and no error is introduced by treating the cross-section as rectangular and the portion of S that passes through the charge as plane. It is then evident that the contribution of the slice to the charge density at any point in space at the instant that S passes through that point differs from zero only in the region specified by the cross-section $A'BCD'$; and the volume of this region is $c/(c+\dot{r})$ times that of the charge slice, since the radius of S reduces by BA' while the slice moves a distance AA'.

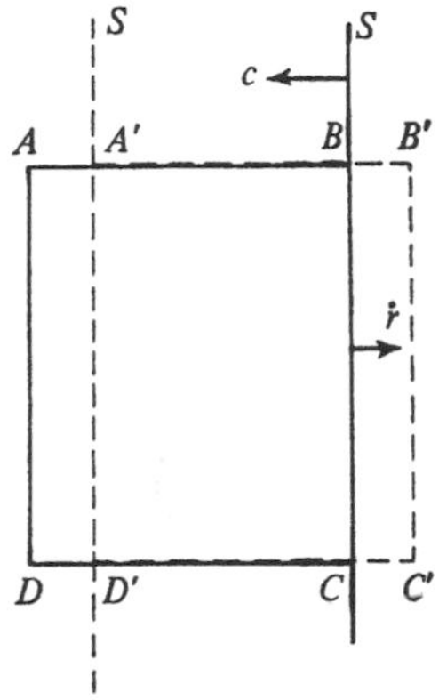

Fig. 2.2. Illustrating the spherical surface S crossing the charge slice.

It follows at once that, if e is the total charge in the element, now thought of as located at a point, (2.61) gives ϕ at P at time t as

$$\frac{1}{4\pi\varepsilon_0}\frac{ec}{c+\dot{r}}\frac{1}{r}$$

evaluated at the retarded time $t-[r]/c$. As remarked after (2.103), the retarded time is only specified implicitly, since the interval r/c by which it lags on t is itself a function of time and has to be evaluated at the retarded time.

Since current density is charge density times velocity, exactly the same argument can be used to obtain $\mathbf{A}$ from (2.62). In summary

$$\phi = \frac{e}{4\pi\varepsilon_0}\left[\frac{1}{\kappa r}\right], \tag{2.107}$$

$$\mathbf{A} = \frac{\mu_0 e}{4\pi}\left[\frac{\mathbf{v}}{\kappa r}\right], \tag{2.108}$$

where

$$\kappa = 1 + \dot{r}/c = 1 - \hat{\mathbf{r}}\cdot\mathbf{v}/c, \tag{2.109}$$

$\mathbf{r} = r\hat{\mathbf{r}}$ being the radius vector *from* the charge *to* the field point P, as in Fig. 2.3. Expressions (2.107) and (2.108) are known as the Liénard–Wiechert potentials.

2.2.3 The field vectors

The apparent simplicity of the forms (2.107), (2.108) is deceptive. Their direct evaluation for any but a few quite special prescribed motions of the point charge is by no means straightforward because of the complication of the equation which determines the retarded time. And even in the determination of the corresponding formulas for **E** and **H** through (2.53) and (2.54), which now follows, there is need for care in the partial differentiation of retarded quantities.

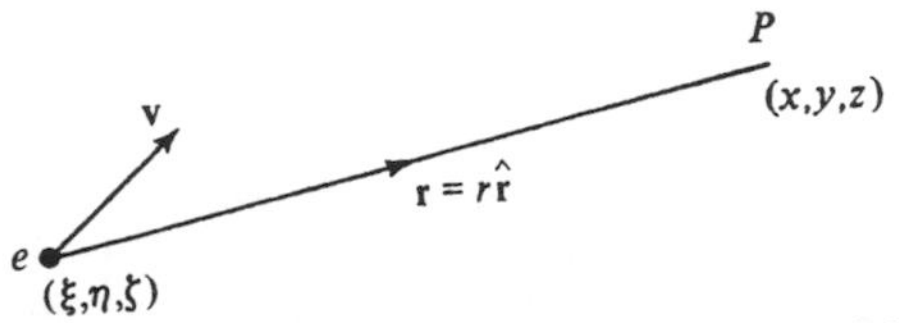

Fig. 2.3. Relative configuration of point charge e and field point P.

The coordinates of the field point are (x, y, z), and those of the point charge are (ξ, η, ζ); the latter are functions of t, with $(\dot{\xi}, \dot{\eta}, \dot{\zeta}) = \mathbf{v}$ (see Fig. 2.3), and the quantities $1/(\kappa r)$ and $\mathbf{v}/(\kappa r)$ in (2.107) and (2.108) are therefore functions of x, y, z and t.

Now for any such function, $\psi(x, y, z, t)$ say,

$$[\psi(x, y, z, t)] = \psi(x, y, z, t - [r]/c), \tag{2.110}$$

so that

$$\operatorname{grad}[\psi] = [\operatorname{grad}\psi] - \frac{1}{c}\left[\frac{\partial\psi}{\partial t}\right]\operatorname{grad}[r]. \tag{2.111}$$

But if ψ itself is taken as r (2.111) gives, with the notation of (2.109),

$$\operatorname{grad}[r] = [\hat{\mathbf{r}}/\kappa]; \tag{2.112}$$

and therefore

$$\operatorname{grad}[\psi] = \left[\operatorname{grad}\psi - \frac{\hat{\mathbf{r}}}{c\kappa}\frac{\partial\psi}{\partial t}\right]. \tag{2.113}$$

Again, from (2.110),

$$\frac{\partial[\psi]}{\partial t} = \left[\frac{\partial\psi}{\partial t}\right]\left(1 - \frac{1}{c}\frac{\partial[r]}{\partial t}\right). \tag{2.114}$$

Here the identification of ψ with r gives

$$\frac{\partial[r]}{\partial t} = [\dot{r}/\kappa], \tag{2.115}$$

and therefore

$$\frac{\partial[\psi]}{\partial t} = \left[\frac{1}{\kappa}\frac{\partial\psi}{\partial t}\right]. \tag{2.116}$$

With the help of (2.113) and (2.116) the substitution of (2.107) and (2.108) into (2.54) is seen to give

$$\mathbf{E} = -\frac{e}{4\pi\varepsilon_0}\left[\operatorname{grad}\left(\frac{1}{\kappa r}\right) - \frac{\hat{\mathbf{r}}}{c\kappa}\frac{\partial}{\partial t}\left(\frac{1}{\kappa r}\right) + \frac{1}{c^2\kappa}\frac{\partial}{\partial t}\left(\frac{\mathbf{v}}{\kappa r}\right)\right], \tag{2.117}$$

and it only remains to express the retarded quantity in a form which shows explicitly the separation into the radiation field and the near field. Evidently

$$\kappa r = r - \mathbf{r}\cdot\mathbf{v}/c, \tag{2.118}$$

$$\operatorname{grad}(\kappa r) = \hat{\mathbf{r}} - \mathbf{v}/c, \tag{2.119}$$

$$\frac{\partial(\kappa r)}{\partial t} = \dot{r} + v^2/c - \mathbf{r}\cdot\dot{\mathbf{v}}/c; \tag{2.120}$$

and after some rearrangement of the terms in (2.117) it appears as

$$\mathbf{E} = \frac{e}{4\pi\varepsilon_0}\left[\frac{1}{\kappa^3 r^2}\left(\hat{\mathbf{r}} - \frac{\mathbf{v}}{c}\right)\left(1 - \frac{v^2}{c^2}\right) - \frac{1}{c^2\kappa^3 r}\left\{\kappa\dot{\mathbf{v}} - \left(\hat{\mathbf{r}} - \frac{\mathbf{v}}{c}\right)(\hat{\mathbf{r}}\cdot\dot{\mathbf{v}})\right\}\right]. \tag{2.121}$$

In the alternative version

$$\mathbf{E} = \frac{e}{4\pi\varepsilon_0}\left[\frac{1}{\kappa^3 r^2}\left(\hat{\mathbf{r}} - \frac{\mathbf{v}}{c}\right)\left(1 - \frac{v^2}{c^2}\right) - \frac{1}{c^2\kappa^3 r}\hat{\mathbf{r}} \times \left\{\dot{\mathbf{v}} \times \left(\hat{\mathbf{r}} - \frac{\mathbf{v}}{c}\right)\right\}\right], \tag{2.122}$$

it is immediately apparent that the radiation field is perpendicular to $[\hat{\mathbf{r}}]$, and that when v/c is negligible compared with unity the complete expression reduces to (2.104).

The corresponding expression for **H** can be found in a similar way by substituting (2.108) into (2.53). Thus

$$\mathbf{H} = \frac{e}{4\pi}\operatorname{curl}\left[\frac{\mathbf{v}}{\kappa r}\right]$$

$$= \frac{e}{4\pi}\left[\operatorname{curl}\left(\frac{\mathbf{v}}{\kappa r}\right) - \frac{\hat{\mathbf{r}}}{c\kappa} \times \frac{\partial}{\partial t}\left(\frac{\mathbf{v}}{\kappa r}\right)\right],$$

using the result for curl analogous to (2.113); and

$$\operatorname{curl}\left(\frac{\mathbf{v}}{\kappa r}\right) = -\mathbf{v} \times \operatorname{grad}\left(\frac{1}{\kappa r}\right)$$

$$= -c\hat{\mathbf{r}} \times \operatorname{grad}\left(\frac{1}{\kappa r}\right)$$

from (2.119), so that

$$\mathbf{H} = -\frac{ec}{4\pi}\left[\hat{\mathbf{r}} \times \operatorname{grad}\left(\frac{1}{\kappa r}\right) + \frac{\hat{\mathbf{r}}}{c^2\kappa} \times \frac{\partial}{\partial t}\left(\frac{\mathbf{v}}{\kappa r}\right)\right]. \tag{2.123}$$

But (2.123) is precisely the vector product of $Y_0[\hat{\mathbf{r}}]$ with (2.117). Thus

$$\mathbf{H} = Y_0[\hat{\mathbf{r}}] \times \mathbf{E} \tag{2.124}$$

exactly.

If the form (2.121) is used for **E**, (2.124) gives

$$\mathbf{H} = \frac{e}{4\pi}\left[\frac{1}{\kappa^3 r^2}\left(1 - \frac{v^2}{c^2}\right)\mathbf{v} + \frac{1}{c\kappa^3 r}\left\{\kappa\dot{\mathbf{v}} + (\hat{\mathbf{r}}\cdot\dot{\mathbf{v}})\frac{\mathbf{v}}{c}\right\}\right] \times [\hat{\mathbf{r}}]. \tag{2.125}$$

When v/c is neglected compared with unity this expression reduces to (2.103).

Before examining further the implications of these rigorous formulas for the field of a moving point charge an alternative method of representation is discussed.

2.2.4 Alternative formulation

An alternative method of analysis to that just given makes use of the frequency spectrum concept. For example, the expression (2.101) is used for **A**, with $\mathbf{j}^\omega$ given by (2.97). Likewise the scalar potential is written

$$\phi(x, y, z, t) = \frac{1}{4\pi\varepsilon_0}\iint \frac{\rho^\omega(x', y', z', \omega)}{R} e^{i\omega(t-R/c)}\, d\omega\, d\tau, \tag{2.126}$$

where

$$\rho^\omega(x', y', z', \omega) = \frac{1}{2\pi}\int \rho(x', y', z', s)\, e^{-i\omega s}\, ds, \tag{2.127}$$

in the notation of Fig. 2.1. In (2.127) s has been introduced as integration variable in preference to t so that the results can be combined into the single statement

$$\phi = \frac{1}{8\pi^2\varepsilon_0}\iiint \frac{\rho(x', y', z', s)}{R} e^{i\omega(t-R/c-s)}\, d\tau\, ds\, d\omega. \tag{2.128}$$

A representation for the moving point charge is then obtained by expressing the charge density in terms of delta functions as

$$\rho(x', y', z', s) = e\delta(x' - \xi)\delta(y' - \eta)\delta(z' - \zeta), \tag{2.129}$$

where (ξ, η, ζ) are the coordinates of the charge at time s. If the now trivial space integration in (2.128) is carried out the result is

$$\phi = \frac{e}{8\pi^2\varepsilon_0}\iint \frac{e^{i\omega(t-r/c-s)}}{r}\, ds\, d\omega, \tag{2.130}$$

in which r denotes the distance between the field point and the position of the charge at time s.

The corresponding representation of the vector potential is evidently

$$\mathbf{A} = \frac{e\mu_0}{8\pi^2}\iint \frac{\mathbf{v}}{r}\, e^{i\omega(t-r/c-s)}\, ds\, d\omega, \tag{2.131}$$

where $\mathbf{v}$ is the velocity of the charge at time s; and those for the field vectors $\mathbf{E}$ and $\mathbf{H}$ can be obtained directly from (2.63) and (2.64) in just the same way. For example, (2.63) gives

$$\mathbf{H} = \frac{e}{8\pi^2}\iint \left(\frac{i\omega}{cr} + \frac{1}{r^2}\right)\mathbf{v} \times \hat{\mathbf{r}}\, e^{i\omega(t-r/c-s)}\, ds\, d\omega. \tag{2.132}$$

The present formulas, in contrast to those of the previous section, are particularly useful if it is the frequency spectrum of the field that is required, rather than the field itself. However, they do also furnish an alternative derivation of the previous results. The first step in this derivation is to introduce the delta function representation of the ω integral. For example (2.130) becomes

$$\phi = \frac{e}{4\pi\varepsilon_0}\int \frac{\delta(s + r/c - t)}{r}\, ds, \tag{2.133}$$

an expression which is seen, in retrospect, to come directly from (2.55) on writing

$$\rho(x', y', z', t - R/c) = \int \rho(x', y', z', s)\delta(s + R/c - t)\, ds,$$

then using (2.129) and doing the space integration. Next, the variable of integration is changed to

$$s' = s + r/c,$$

so that

$$ds' = ds\left(1 + \frac{1}{c}\frac{dr}{ds}\right).$$

Then (2.133) is

$$\phi = \frac{e}{4\pi\varepsilon_0}\int \frac{\delta(s' - t)}{\left(1 + \frac{1}{c}\frac{dr}{ds}\right)r}\, ds',$$

$$= \frac{e}{4\pi\varepsilon_0}\,\frac{1}{\left(1 + \frac{1}{c}\frac{dr}{ds}\right)r} \tag{2.134}$$

evaluated at $s' = t$. But at $s' = t$ the value of s is given by $s = t - r(s)/c$, and is therefore the retarded time. Thus (2.134) is identical with (2.107).

The same technique applied to (2.132) gives

$$\mathbf{H} = \frac{e}{4\pi}\left\{\left[\frac{\mathbf{v}\times\hat{\mathbf{r}}}{\kappa r^2}\right] + \frac{\partial}{\partial t}\left[\frac{\mathbf{v}\times\hat{\mathbf{r}}}{c\kappa r}\right]\right\}$$

$$= \frac{e}{4\pi}\left[\frac{\mathbf{v}\times\hat{\mathbf{r}}}{\kappa r^2} + \frac{1}{c\kappa}\frac{\partial}{\partial t}\left(\frac{\mathbf{v}\times\hat{\mathbf{r}}}{\kappa r}\right)\right] \tag{2.135}$$

from (2.116). By noting that

$$\frac{\partial\hat{\mathbf{r}}}{\partial t} = \frac{\partial}{\partial t}\left(\frac{\mathbf{r}}{r}\right) = -\frac{\mathbf{v}}{r} - \frac{\dot{r}}{r}\hat{\mathbf{r}}$$

and using (2.120) it is not difficult to check that (2.135) is in agreement with (2.125).

2.2.5 Special cases

In this section attention is drawn to some of the particular features of the expressions (2.122) for **E** and (2.125) for **H**. The first thing to notice is that separation into the radiation field, of order $1/[r]$, and the near field, of order $1/[r^2]$, effects equally a separation into those terms which do and those which do not contain the acceleration as a factor; the near field is unaffected by the acceleration, whereas there must be acceleration (at the retarded time) to produce a radiation field.

If the acceleration is permanently zero the charge moves uniformly in a straight line and

$$\mathbf{E} = \frac{e}{4\pi\varepsilon_0}\left(1 - \frac{v^2}{c^2}\right)\left[\frac{1}{\kappa^3 r^2}\left(\hat{\mathbf{r}} - \frac{\mathbf{v}}{c}\right)\right], \tag{2.136}$$

$$\mathbf{H} = \frac{ec}{4\pi}\left(1 - \frac{v^2}{c^2}\right)\left[\frac{1}{\kappa^3 r^2}\frac{\mathbf{v}}{c}\times\hat{\mathbf{r}}\right]. \tag{2.137}$$

Evidently **E** lies in the plane containing the field point and the line of motion, and **H** is perpendicular to this plane. This case is examined in detail later on, and it appears that **E** at any instant does in fact point directly away from the position the charge occupies at that same instant.

If there is acceleration, but it remains parallel to the velocity, for example the point charge oscillating rectilinearly, the radiation field is comparatively simple. From (2.122)

$$\mathbf{E} \sim \frac{e}{4\pi\varepsilon_0 c^2}\left[\frac{(\dot{\mathbf{v}}\times\hat{\mathbf{r}})\times\hat{\mathbf{r}}}{\kappa^3 r}\right], \tag{2.138}$$

and then from (2.124)

$$\mathbf{H} \sim \frac{e}{4\pi c}\left[\frac{\dot{\mathbf{v}}\times\hat{\mathbf{r}}}{\kappa^3 r}\right]. \tag{2.139}$$

These expressions differ from those of a dipole of strength $e\mathbf{v}$ by virtue of the factor κ^3 in the denominator. Since κ is $1 - \frac{v}{c}\cos\theta$, where θ is the angle between the radius vector and the common direction of the velocity and acceleration, the factor is only important for a relativistic particle. When v/c is near unity the radiation pattern differs markedly from that of a dipole, because of the prominence given to directions close to $\theta = 0$ by the comparative smallness of the denominator. This effect is, indeed, present in the general case, and is examined quantitatively to some extent in the next section in a discussion of the radiated power.

2.2.6 Radiated power

In the radiation field **E**, **H** and $[\hat{\mathbf{r}}]$ form an orthogonal triad, so the Poynting vector is

$$\mathbf{E} \times \mathbf{H} = Y_0 \mathbf{E}^2 [\hat{\mathbf{r}}].$$

If the expression for **E** in the radiation field is taken from (2.121) the magnitude of this, with removal of the factor $[r^{-2}]$, is easily seen to be

$$\frac{e^2}{16\pi^2\varepsilon_0 c^3}\left[\frac{1}{\kappa^6}\left\{\kappa^2\dot{\mathbf{v}}^2 + 2\kappa(\hat{\mathbf{r}}\cdot\dot{\mathbf{v}})\frac{\mathbf{v}\cdot\dot{\mathbf{v}}}{c} - \left(1 - \frac{v^2}{c^2}\right)(\hat{\mathbf{r}}\cdot\dot{\mathbf{v}})^2\right\}\right] \tag{2.140}$$

Therefore (2.140) represents the rate at which energy crosses a distant surface element, per unit solid angle that the element subtends at the position of the charge at the retarded time.

To examine the variation with direction of the retarded quantity in

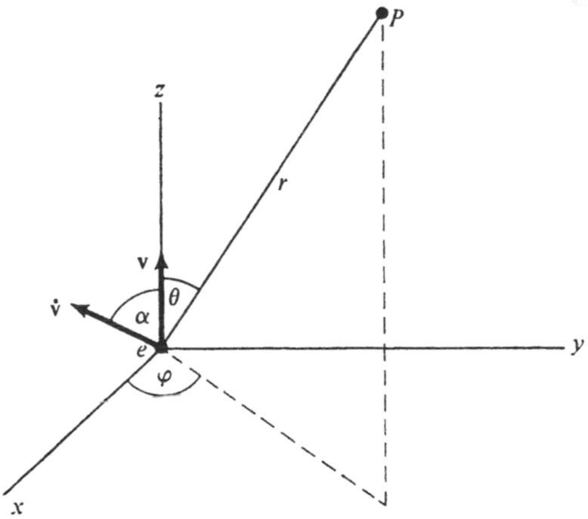

Fig. 2.4. Coordinate system for examining the directional dependence of the distant power flux.

(2.140) the coordinate system depicted in Fig. 2.4 is adopted. The origin is at the charge, Oz is along $\mathbf{v}$, Ox lies in the plane of $\mathbf{v}$ and $\dot{\mathbf{v}}$, and r, θ, ϕ are spherical polar coordinates. The angle between $\mathbf{v}$ and $\dot{\mathbf{v}}$ is called α. Then, apart from the factor $\dot{\mathbf{v}}^2$, the retarded quantity in (2.140) is

$$\frac{1}{\kappa^4} + 2\cos\alpha(\sin\alpha\sin\theta\cos\phi + \cos\alpha\cos\theta)\frac{v/c}{\kappa^5}$$
$$- (\sin\alpha\sin\theta\cos\phi + \cos\alpha\cos\theta)^2\frac{1 - v^2/c^2}{\kappa^6}, \tag{2.141}$$

where

$$\kappa = 1 - \frac{v}{c}\cos\theta.$$

To avoid lengthy algebra the explicit discussion is now confined to two special cases; first, $\mathbf{v}$ parallel to $\dot{\mathbf{v}}$, secondly $\mathbf{v}$ perpendicular to $\dot{\mathbf{v}}$.

When $\alpha = 0$, (2.141) is just

$$\frac{\sin^2\theta}{\left(1 - \dfrac{v}{c}\cos\theta\right)^6}. \tag{2.142}$$

Since the chief interest is in the pronounced maximum which this expression assumes for a small value of θ when v/c is near unity, the notation

$$v/c = 1 - \eta, \quad \theta^2 = 2\chi\eta, \quad \eta \ll 1,$$

is introduced, and (2.142) is approximated for small values of η by

$$\frac{2\chi}{(1+\chi)^6}\frac{1}{\eta^5}. \tag{2.143}$$

As χ increases from zero (2.143) rises from zero to a maximum when $\chi = \frac{1}{5}$, and then falls off rapidly, dropping below 10^{-1} times its maximum value before $\chi = 2$.

When $\alpha = \frac{1}{2}\pi$, (2.141) is

$$\frac{1}{\left(1 - \dfrac{v}{c}\cos\theta\right)^4} - \frac{\left(1 - \dfrac{v^2}{c^2}\right)\sin^2\theta\cos^2\phi}{\left(1 - \dfrac{v}{c}\cos\theta\right)^6}, \tag{2.144}$$

to which the corresponding approximation is

$$\frac{1 + 2\chi(1 - 2\cos^2\phi) + \chi^2}{(1+\chi)^6}\frac{1}{\eta^4}. \tag{2.145}$$

In this case the expression has its greatest value when χ is zero and becomes comparatively very small when χ exceeds unity.

It appears, therefore, that the radiation from a highly relativistic particle is thrown forward into a narrow beam in the direction in which the particle is traveling. The precise pattern of this beam depends on the angle between the acceleration and velocity of the particle, but its angular width is of the order $\sqrt{(2\eta)}$, that is, of order $\sqrt{(1 - v^2/c^2)}$. This beam width is conveniently expressed as the ratio of the rest energy of the particle to its total relativistic energy. As an illustration, radiation from an electron with energy 10^9 eV is concentrated into a beam of width about 1.8 minutes.

Finally, the question of the *total* power radiated by the moving charge is considered. Following the point made at the end of §2.2.1, the analysis of the present section is now to be interpreted in terms of the *contemporary* value of an expression like (2.140). But in the exact treatment, unlike the non-relativistic approximation, a distinction must be drawn between the rate at which the charge loses energy to radiation and the rate at which the radiated energy subsequently crosses a distant closed spherical surface. The latter quantity is of no particular significance or interest, but the former is of fundamental importance in the determination of the effect of radiation loss on the dynamics of the motion of the charge.

The reason for the distinction is just this. During an infinitesimal time interval δt the charge increases its distance r from a field point P by the amount $\dot{r}\,\delta t$, with the result that radiation emitted at the end of the time interval reaches P at a time $\delta t + \frac{\dot{r}}{c}\delta t$ later than that emitted at the beginning of the time interval. In other words, the energy emitted throughout a time interval δt crosses a distant surface element throughout a time interval $\kappa\,\delta t$. It is therefore the contemporary value of (2.140) when multiplied by κ, namely

$$\frac{e^2}{16\pi^2\varepsilon_0 c^3}\left\{\frac{\dot{\mathbf{v}}^2}{\kappa^3} + 2(\hat{\mathbf{r}}\cdot\dot{\mathbf{v}})\frac{\mathbf{v}\cdot\dot{\mathbf{v}}/c}{\kappa^4} - (\hat{\mathbf{r}}\cdot\dot{\mathbf{v}})^2\,\frac{1 - v^2/c^2}{\kappa^5}\right\}, \tag{2.146}$$

which gives the power radiated by the charge, per unit solid angle, in the direction $\hat{\mathbf{r}}$.

To get the total power radiated it is only necessary to integrate (2.146) over all solid angles. This can be done by introducing the notation of Fig. 2.4 as before. Using (2.141) multiplied by κ, the total radiated power appears in the form

$$\frac{e^2\dot{v}^2}{16\pi^2\varepsilon_0 c^3}\int_{\phi=0}^{2\pi}\int_{\theta=0}^{\pi}\left\{\frac{1}{\kappa^3} + 2\frac{v}{c}\cos\alpha(\sin\alpha\sin\theta\cos\phi + \cos\alpha\cos\theta)/\kappa^4\right.$$
$$\left. - \left(1 - \frac{v^2}{c^2}\right)(\sin\alpha\sin\theta\cos\phi + \cos\alpha\cos\theta)^2/\kappa^5\right\}\sin\theta\,d\theta\,d\phi. \tag{2.147}$$

The ϕ integration is trivial, and when it is completed (2.147) is

$$\frac{e^2\dot{\mathbf{v}}^2}{8\pi\varepsilon_0 c^3}\int_0^\pi\left\{\frac{1}{\kappa^3}+2\frac{v}{c}\cos^2\alpha\,\frac{\cos\theta}{\kappa^4} - \frac{1}{2}\left(1-\frac{v^2}{c^2}\right)\frac{\sin^2\alpha+(3\cos^2\alpha-1)\cos^2\theta}{\kappa^5}\right\}\sin\theta\,d\theta. \tag{2.148}$$

If the variable of integration is now changed from θ to κ, where

$$\cos\theta = \frac{c}{v}(1-\kappa), \qquad \sin\theta\,d\theta = \frac{c}{v}\,d\kappa,$$

the integrand is simply the sum of three terms proportional to κ^{-3}, κ^{-4} and κ^{-5}, respectively, and the limits are $1 - v/c$, $1 + v/c$. The integration is therefore entirely straightforward, and when the terms are collected together the result can be written as

$$\frac{e^2\dot{\mathbf{v}}^2}{6\pi\varepsilon_0 c^3}\,\frac{1}{(1-v^2/c^2)^2}\left\{1+\frac{v^2/c^2}{1-v^2/c^2}\cos^2\alpha\right\}, \tag{2.149}$$

or in vector form

$$\frac{e^2}{6\pi\varepsilon_0 c^3}\,\frac{1}{(1-v^2/c^2)^2}\left\{\dot{\mathbf{v}}^2+\frac{(\mathbf{v}\cdot\dot{\mathbf{v}}/c)^2}{1-v^2/c^2}\right\}. \tag{2.150}$$

If v/c is negligible compared with unity these expressions revert to (2.106).

2.3 THE TECHNIQUES OF SPECIAL RELATIVITY

2.3.1 The Lorentz transformation

In practice it often happens that particles move with sufficiently high speeds for relativistic effects to be important. With regard to the purely electromagnetic equations, such effects are automatically included in a rigorous treatment, since the principles of special relativity do not require the equations to be modified in any way. With the dynamical equations, on the other hand, the standard Newtonian form must be discarded if relativistic effects are to be evaluated properly.

This chapter concludes with a brief resumé of those technical results of special relativity which are particularly relevant to various aspects of particle electrodynamics as developed in the present book. The purpose is twofold: to have available the basis for consolidating or looking afresh at purely electromagnetic results and also a statement of the elementary parts of special relativity dynamics that are required subsequently.

It is a tenet of special relativity that if the same event is located at (x, y, z) at time t in a frame of reference S and at (x', y', z') at time t' in a frame of reference S' that is moving relative to S with uniform velocity u along the x-axis, where the space axes of S and S' coincide at the common time zeros of both frames, then

$$x' = \frac{x - ut}{\sqrt{(1 - u^2/c^2)}}, \quad y' = y, \quad z' = z, \quad t' = \frac{t - ux/c^2}{\sqrt{(1 - u^2/c^2)}}. \tag{2.151}$$

The relations (2.151) specify what is known as the Lorentz transformation. If the uniform velocity of S' relative to S is $\mathbf{u} = u\hat{\mathbf{u}}$, not necessarily along Ox, the equivalent transformation can be written

$$\mathbf{r}'_{\parallel} = \frac{\mathbf{r}_{\parallel} - \mathbf{u}t}{\sqrt{(1 - u^2/c^2)}}, \quad \mathbf{r}'_{\perp} = \mathbf{r}_{\perp}, \quad t' = \frac{t - \mathbf{u}\cdot\mathbf{r}/c^2}{\sqrt{(1 - u^2/c^2)}}, \tag{2.152}$$

where $\mathbf{r} = (x, y, z)$ and the suffices $\parallel$ and $\perp$ denote components parallel and perpendicular to $\mathbf{u}$; or, alternatively,

$$\mathbf{r}' = \mathbf{r} - (\hat{\mathbf{u}}\cdot\mathbf{r})\hat{\mathbf{u}} + \frac{(\hat{\mathbf{u}}\cdot\mathbf{r})\hat{\mathbf{u}} - \mathbf{u}t}{\sqrt{(1 - u^2/c^2)}}, \quad t' = \frac{t - \mathbf{u}\cdot\mathbf{r}/c^2}{\sqrt{(1 - u^2/c^2)}}. \tag{2.153}$$

The Lorentz transformation leaves invariant the interval ds defined by

$$ds^2 = c^2\, dt^2 - dx^2 - dy^2 - dz^2, \tag{2.154}$$

and conforms with the physical law that the measured vacuum speed of light is independent of the frame of reference.

The most convenient expression of the theory of relativity is in terms of four-dimensional vectors and tensors. In suffix notation, x^i, where

$$(x^0, x^1, x^2, x^3) \equiv (ct, x, y, z) \equiv (ct, \mathbf{r}),$$

is the prototype of a contravariant 4-vector. Four quantities a^i form a contravariant vector if on a change of reference frame from S to S' they transform according to the prescription (2.153).

With the introduction of the metric 4-tensor

$$g_{ij} = \begin{pmatrix} 1 & 0 & 0 & 0 \\ 0 & -1 & 0 & 0 \\ 0 & 0 & -1 & 0 \\ 0 & 0 & 0 & -1 \end{pmatrix}, \tag{2.155}$$

the invariant (2.154) can be written

$$ds^2 = g_{ij}\, dx^i\, dx^j,$$

where the summation convention is adopted. Corresponding to any contravariant vector a^i there is a covariant vector

$$a_i = g_{ij}a^j = (a^0, -a^1, -a^2, -a^3).$$

Then (2.154) can be written even more compactly as

$$ds^2 = dx^i \, dx_i.$$

Note also that

$$a^i = g^{ij} a_j,$$

where g^{ij} has the same components as (2.155).

In the case where (x, y, z) specify the position at time t of a particle with velocity $\mathbf{v}$, (2.154) gives

$$\frac{1}{c^2}\left(\frac{ds}{dt}\right)^2 = 1 - v^2/c^2. \tag{2.156}$$

In connection with the motion of the particle an important 4-vector is

$$\frac{dx^i}{ds} = \left\{\frac{1}{\sqrt{(1 - v^2/c^2)}}, \frac{\mathbf{v}/c}{\sqrt{(1 - v^2/c^2)}}\right\}. \tag{2.157}$$

2.3.2 Four-dimensional representation of the electromagnetic field

The equations of electromagnetic theory can be expressed in forms whose invariance under a Lorentz transformation is self-evident, provided only that certain combinations of the field quantities constitute 4-vectors or 4-tensors. By this means the equations are seen to conform with the principle of the theory of relativity, and the way in which the field quantities transform with a change of reference frame is revealed. The transformation formulas can be put to good use in the analysis of various problems.

It is convenient to start from the hypothesis that quantity of charge is invariant. Then for a specified quantity of charge, of density ρ, occupying a volume element $d\tau$, $\rho \, d\tau$ is invariant. In particular

$$\rho \, d\tau = \rho_0 \, d\tau_0,$$

where the zero suffix denotes the values of the quantities measured in a frame moving with the velocity $\mathbf{v}$ of the charge. But by virtue of the so-called Fitzgerald contraction in the linear dimension in the direction of motion, which is readily deduced from (2.153),

$$d\tau = \sqrt{(1 - v^2/c^2)} \, d\tau_0; \tag{2.158}$$

from which it follows that

$$\rho = \rho_0/\sqrt{(1 - v^2/c^2)}. \tag{2.159}$$

Multiplication of (2.157) by ρ_0 therefore gives $(c\rho, \rho\mathbf{v})$, and establishes that the charge density and current density constitute a 4-vector

$$j^i = (c\rho, \mathbf{j}). \tag{2.160}$$

The charge conservation relation (2.58) appears in the explicitly invariant form

$$\frac{\partial j^i}{\partial x^i} = 0. \tag{2.161}$$

Furthermore, the presumption that the potentials ϕ and $\mathbf{A}$ constitute a 4-vector,

$$A^i = (\phi/c, \mathbf{A}), \tag{2.162}$$

enables the wave equations that they satisfy, (2.59) and (2.60), to be combined in the invariant form

$$\frac{\partial^2 A^i}{\partial x^j \, \partial x_j} = \mu_0 j^i. \tag{2.163}$$

From (2.162) the transformation formulas for the potentials are

$$\mathbf{A}'_\parallel = \frac{\mathbf{A}_\parallel - \mathbf{u}\phi c^2}{\sqrt{(1 - u^2/c^2)}}, \qquad \mathbf{A}'_\perp = \mathbf{A}_\perp, \qquad \phi' = \frac{\phi - \mathbf{u}\cdot\mathbf{A}}{\sqrt{(1 - u^2/c^2)}} \tag{2.164}$$

in the notation of (2.152).

The procedure for obtaining invariant forms of Maxwell's equations (2.1), (2.2) is indicated by the representations (2.53) and (2.54) of $\mathbf{H}$ and $\mathbf{E}$ in terms of the potentials. These representations suggest the introduction of the tensor

$$f_{ij} = \frac{1}{\mu_0}\left(\frac{\partial A_j}{\partial x^i} - \frac{\partial A_i}{\partial x^j}\right), \tag{2.165}$$

which in full is

$$f_{ij} = \begin{pmatrix} 0 & Y_0E_x & Y_0E_y & Y_0E_z \\ -Y_0E_x & 0 & -H_z & H_y \\ -Y_0E_y & H_z & 0 & -H_x \\ -Y_0E_z & -H_y & H_x & 0 \end{pmatrix}, \tag{2.166}$$

where $Y_0 = \sqrt{(\varepsilon_0/\mu_0)}$. Then it can be verified that the equation

$$\frac{\partial f_{jk}}{\partial x^i} + \frac{\partial f_{ki}}{\partial x^j} + \frac{\partial f_{ij}}{\partial x^k} = 0 \tag{2.167}$$

has just four independent components, and is the invariant form of div $\mathbf{H} = 0$, curl $\mathbf{E} = -\mu_0 \partial\mathbf{H}/\partial t$. Again

$$f^{ij} = g^{ik} f_{kl} g^{lj} = \begin{pmatrix} 0 & -Y_0E_x & -Y_0E_y & -Y_0E_z \\ Y_0E_x & 0 & -H_z & H_y \\ Y_0E_y & H_z & 0 & -H_x \\ Y_0E_z & -H_y & H_x & 0 \end{pmatrix}; \tag{2.168}$$

and evidently, therefore,

$$\frac{\partial f^{ji}}{\partial x^j} = j^i \tag{2.169}$$

is the invariant form of div $\mathbf{E} = \rho/\varepsilon_0$, curl $\mathbf{H} = \varepsilon_0\, \partial\mathbf{E}/\partial t + \mathbf{j}$.

The transformation formulae for the components of **E** and **H** follow from (2.168) by the transformation rule for tensors. If (2.152) is expressed in the form

$$x^{i\prime} = L^i_j x^j, \tag{2.170}$$

then

$$f^{ij\prime} = L^i_k L^j_l f^{kl}, \tag{2.171}$$

which can be seen to give

$$\begin{aligned} E'_{\parallel} &= E_{\parallel}, \qquad H'_{\parallel} = H_{\parallel}, \\ \mathbf{E}'_{\perp} &= \frac{1}{\sqrt{(1-u^2/c^2)}}\left(\mathbf{E}_{\perp} + Z_0 \frac{\mathbf{u}}{c} \times \mathbf{H}_{\perp}\right), \\ \mathbf{H}'_{\perp} &= \frac{1}{\sqrt{(1-u^2/c^2)}}\left(\mathbf{H}_{\perp} - Y_0 \frac{\mathbf{u}}{c} \times \mathbf{E}_{\perp}\right). \end{aligned} \tag{2.172}$$

2.3.3 Particle dynamics; the Lorentz force

In an inertial frame of reference Newton's law of motion for a particle of mass m and velocity $\mathbf{v}$ is

$$\frac{d}{dt}(m\mathbf{v}) = \mathbf{F}, \tag{2.173}$$

where **F** is the force acting on the particle. This suggests that the relativistic equation of motion, invariant in form under a Lorentz transformation, is

$$c^2 \frac{d}{ds}\left(m_0 \frac{dx^i}{ds}\right) = \mathscr{F}^i, \tag{2.174}$$

where m_0 is an invariant. For if the 4-vector $\mathscr{F}^i$ is written $(\mathscr{F}^0, \boldsymbol{\mathscr{F}})$ the space-like components of (2.174) give

$$\frac{d}{dt}\left\{\frac{m_0\mathbf{v}}{\sqrt{(1-v^2/c^2)}}\right\} = \sqrt{(1-v^2/c^2)}\boldsymbol{\mathscr{F}}, \tag{2.175}$$

which can be matched with (2.173) when v/c is negligible.

If (2.175) is accepted, the relativistic prescription differs from the Newtonian when v/c is not negligible; but it is still possible to think in terms of the familiar Newtonian form (2.173) provided, first, that in the latter the mass is taken to be

$$m = \frac{m_0}{\sqrt{(1 - v^2/c^2)}}, \tag{2.176}$$

the invariant m_0 being called the rest mass, and secondly, that after division by $\sqrt{(1 - v^2/c^2)}$ the force **F** is the space-like part of a certain 4-vector. The second point can be made more precise. For if the quantities

$$\mathscr{E} = \frac{m_0 c^2}{\sqrt{(1 - v^2/c^2)}}, \tag{2.177}$$

$$\mathbf{p} = \frac{m_0 \mathbf{v}}{\sqrt{(1 - v^2/c^2)}} \tag{2.178}$$

are introduced, (2.175) is

$$\frac{d\mathbf{p}}{dt} = \sqrt{(1 - v^2/c^2)}\boldsymbol{\mathscr{F}}, \tag{2.179}$$

and the time-like ($i = 0$) part of (2.174) is

$$\frac{d\mathscr{E}}{dt} = \sqrt{(1 - v^2/c^2)}c\mathscr{F}^0. \tag{2.180}$$

Now the identity

$$\frac{d\mathscr{E}}{dt} = \mathbf{v}\cdot\frac{d\mathbf{p}}{dt} \tag{2.181}$$

follows trivially from (2.177) and (2.178), assuming $dm_0/dt = 0$. The relativistic equation of motion therefore implies

$$\mathscr{F}^0 = \mathbf{v}\cdot\boldsymbol{\mathscr{F}}/c. \tag{2.182}$$

Thus the force **F** used in the Newtonian form (2.173) must be such that

$$\left\{\frac{\mathbf{v}\cdot\mathbf{F}/c}{\sqrt{(1 - v^2/c^2)}}, \frac{\mathbf{F}}{\sqrt{(1 - v^2/c^2)}}\right\} \tag{2.183}$$

is a 4-vector.

In the interpretation of relativistic particle mechanics $\mathscr{E}$ is identified with the energy of the particle; it has been seen that the time-like part of (2.174) is the energy equation, $d\mathscr{E}/dt = \mathbf{v}\cdot\mathbf{F}$, and follows from the space-like part on scalar multiplication by **v**. Also

$$p^i = m_0 c\frac{dx^i}{ds} = \left(\frac{\mathscr{E}}{c}, \mathbf{p}\right) \tag{2.184}$$

is called the 4-momentum. Differentiation of the invariant $p^i p_i$, which itself is just the rest energy m_0c^2, gives $p^i dp_i/ds = 0$, the same result as (2.181).

Much the most important force on a charged particle e is the Lorentz force

$$e(\mathbf{E} + \mathbf{v} \times \mathbf{B}), \tag{2.185}$$

and it is readily shown that this fits into the relativistic scheme. For, from (2.159), (2.160) and (2.168),

$$\frac{\mu_0}{\rho_0} f^{ij} j_j = \left\{ \frac{\mathbf{E}\cdot\mathbf{v}/c}{\sqrt{(1 - v^2/c^2)}}, \frac{\mathbf{E} + \mathbf{v} \times \mathbf{B}}{\sqrt{(1 - v^2/c^2)}} \right\}; \tag{2.186}$$

and the expression (2.183) with $\mathbf{F}$ given by (2.185) is precisely the right-hand side of (2.186), and is therefore a 4-vector. The relativistic equation of motion of a point charge in an electromagnetic field is thus

$$\frac{d}{dt}\left\{ \frac{m_0\mathbf{v}}{\sqrt{(1 - v^2/c^2)}} \right\} = e(\mathbf{E} + \mathbf{v} \times \mathbf{B}). \tag{2.187}$$

Also, $dm_0/dt = 0$ is implied, since this condition is necessary and sufficient for the correctness of (2.181).

2.3.4 Invariance of radiated power

The rate of radiation of energy by a point charge in arbitrary motion, given by the contemporary value of the expression (2.150), is invariant under a Lorentz transformation. To show this it is perhaps simplest to write down the explicitly invariant form that reduces to (2.106) when $\mathbf{v} = 0$, and then to verify that it is indeed identical with (2.150). The required invariant is

$$-\frac{e^2}{6\pi\varepsilon_0 c m_0^2} \frac{dp^i}{ds} \frac{dp_i}{ds}, \tag{2.188}$$

where p^i is the 4-momentum (2.184); for in terms of the relativistic energy and momentum, (2.177) and (2.178), it is

$$\frac{e^2}{6\pi\varepsilon_0 m_0^2 c^3}\left(\frac{\mathscr{E}}{m_0c^2}\right)^2 \left\{ \left(\frac{d\mathbf{p}}{dt}\right)^2 - \frac{1}{c^2}\left(\frac{d\mathscr{E}}{dt}\right)^2 \right\}, \tag{2.189}$$

which when $\mathbf{v} = 0$ is seen at once to be the same as (2.106).

Moreover,

$$\frac{d\mathscr{E}}{dt} = \frac{m_0\mathbf{v}\cdot\dot{\mathbf{v}}}{(1 - v^2/c^2)^{3/2}},$$

$$\frac{d\mathbf{p}}{dt} = \frac{m_0\dot{\mathbf{v}}}{\sqrt{(1 - v^2/c^2)}} + \frac{m_0(\mathbf{v}\cdot\dot{\mathbf{v}}/c)\mathbf{v}/c}{(1 - v^2/c^2)^{3/2}},$$

so that

$$\left(\frac{d\mathbf{p}}{dt}\right)^2 - \frac{1}{c^2}\left(\frac{d\mathscr{E}}{dt}\right)^2 = \frac{m_0^2}{1 - v^2/c^2}\left\{\dot{\mathbf{v}}^2 + \frac{(\mathbf{v}\cdot\dot{\mathbf{v}}/c)^2}{1 - v^2/c^2}\right\},$$

which establishes the identity of (2.189) and (2.150).

In conclusion it is remarked that it is possible to deduce the invariance of the radiated power without prior knowledge of the explicit expression for it. However, the proof requires an investigation of the transformation properties of the energy and momentum of the electromagnetic field that takes the analysis somewhat beyond the outline given in §§2.3.1–2.3.3.

PROBLEMS

1. Find the equation for $\mathbf{E}$ analogous to (2.8). Write down the solution in the form (2.11), and show that it is equivalent to (2.38).
2. Calculate the actual Poynting vector $\mathbf{E} \times \mathbf{H}$ for the dipole field (2.21). Write down its time average, and obtain the mean radiated power (2.24).
3. From the representation (2.45) obtain $\mathbf{E}$, $\operatorname{div}\mathbf{H}$ and $\operatorname{Re}\frac{1}{2}\mathbf{E} \times \mathbf{H}^*$ as inverse power series in r. Write down the relations implied by $\operatorname{div}\mathbf{H} = 0$ and by the fact that the time-averaged power flux out of a sphere $r = R$ is independent of R. Use these relations to show by induction that if $\mathbf{A}_0$ is zero so also are all other $\mathbf{A}_n$.
4. Verify the Lorentz relation (2.57) directly: (a) from the retarded potentials (2.55), (2.56), (b) from the Liénard–Wiechert potentials (2.107), (2.108).
5. Establish the result (2.102) by a direct evaluation of the integral.
6. Consider the field produced by a uniform, plane, surface current impulse. Represent the current density by $\mathbf{j} = (\delta(z)\delta(t), 0, 0)$, and show from (2.94) that the vector potential is $\mathbf{A} = (A, 0, 0)$, where

$$A = \frac{\mu_0}{4\pi^2}\iint_{-\infty}^{\infty}\frac{e^{i(\omega t - nz)}}{n^2 - \omega^2/c^2}\,dn\,d\omega.$$

Evaluate A by doing first the n integration and then the ω integration. Confirm that the same result is obtained by taking the integrations in the reverse order.
7. Show that

$$-\operatorname{grad}\left[\frac{1}{\kappa r}\right] = \left[\frac{\hat{\mathbf{r}}}{\kappa r^2}\right] + \frac{1}{c}\frac{\partial}{\partial t}\left[\frac{\hat{\mathbf{r}}}{\kappa r}\right],$$

and with the help of this result obtain from the Liénard–Wiechert potentials (2.107), (2.108) the expression

$$\mathbf{E} = \frac{e}{4\pi\varepsilon_0}\left\{\left[\frac{\hat{\mathbf{r}}}{r^2}\right] + \frac{[r]}{c}\frac{\partial}{\partial t}\left[\frac{\hat{\mathbf{r}}}{r^2}\right] + \frac{1}{c^2}\frac{\partial^2[\hat{\mathbf{r}}]}{\partial t^2}\right\},$$

equivalent, of course, to (2.122).

8. A point charge moves along the x-axis in an electric field which is in the x-direction and of magnitude $E_0 \exp(-kx)$. At time $t = -\infty$ the charge is at $x = \infty$ and has velocity V towards the origin. Show that, in a non-relativistic treatment, the total energy radiated is $e^2kV^3/(9\pi\varepsilon_0c^3)$. [For the relativistic case see Chapter 4, Problem 9.]

9. Show that the power radiated by a point charge e, with speed v, can be written

$$\frac{e^2}{6\pi\varepsilon_0c^3}\frac{1}{(1-v^2/c^2)^2}\left(\omega^2v^2+\frac{\dot{v}^2}{1-v^2/c^2}\right),$$

where ω is the charge's instantaneous angular velocity about the principal center of curvature of its trajectory.

A 100 meV electron moves in a circular orbit with uniform angular velocity 10^6 rad sec^{-1}. Calculate the power radiated.

10. Deduce from (2.172) that F^{ij} constructed from (2.168) by the transformation

$$\mathbf{E}\to\mathbf{H},\quad \mathbf{H}\to-\mathbf{E},\quad Y_0\leftrightarrow Z_0,$$

is a tensor. Confirm that (2.167) can be written

$$\frac{\partial F^{ij}}{\partial x^j}=0.$$

Show that $\mathbf{E}\cdot\mathbf{H}$ is an invariant.

11. In a frame of reference S there are uniform, orthogonal electrostatic and magnetostatic fields $\mathbf{E}$ and $\mathbf{H}$. The corresponding fields in frame S' are $\mathbf{E}'$ and $\mathbf{H}'$. Show that it is possible either to find a velocity of S' relative to S for which $\mathbf{E}'=0$, or one for which $\mathbf{H}'=0$; and state the conditions under which the velocities are attainable.

REFERENCES

The vigor of classical electromagnetic theory is indicated by the continuing appearance of advanced textbooks on the subject, many of which deal with the topics covered in this chapter. Particularly relevant, among comparatively recent books, are

J. D. JACKSON, *Classical Electrodynamics*, Wiley (1962).

D. S. JONES, *Theory of Electromagnetism*, Pergamon (1964).

L. D. LANDAU and E. M. LIFSHITZ, *Classical theory of fields*, Addison–Wesley (1962).

W. K. H. PANOFSKY and M. PHILLIPS, *Classical Electricity and Magnetism*, 2nd edn, Addison–Wesley (1962).

A comprehensive mathematical treatise on the field of a moving point charge is

G. A. SCHOTT, *Electromagnetic Radiation and the Mechanical Reactions Arising From It*, Cambridge University Press (1912).

The form of the field of a moving point charge stated in Problem 9 is discussed in Vol. I, Chapter 28, and Vol. II, Chapter 21 of

R. P. FEYNMAN, *Lectures on Physics*, Addison–Wesley (1964).

It seems first to have been obtained by

O. HEAVISIDE, *Nature*, 342 (1904),

reproduced in

O. HEAVISIDE, *Electromagnetic Theory*, **III**, The Electrician Printing and Publishing Co. (1912).

CHAPTER 3

ČERENKOV AND GYRO RADIATION

3.1 UNIFORM RECTILINEAR MOTION IN A VACUUM

3.1.1 Deduction from the Liénard–Wiechert potentials

The previous chapter contained a rather general account of the theory of the field generated by a moving charged particle. In the present chapter the fields associated with two relatively simple types of motion are considered in some detail, partly because they offer clear cut illustrations of the general theory and partly because they are the key to the understanding of important physical phenomena. The respective motions are uniform rectilinear and uniform circular. In the former there is no acceleration, and, of course, the particle does not radiate if it travels in a vacuum; but it can radiate by traveling in a medium of refractive index μ with a speed in excess of c/μ, and this is the so-called Čerenkov radiation process. In the latter motion the acceleration is that pıoduced by a magnetostatic field, and the corresponding radiation may be called gyro radiation.

First the vacuum field of a point charge e moving along the x-axis with constant speed v is obtained from the Liénard–Wiechert potentials (2.107), (2.108). The field at (x, y, z) at time t is associated with the position $(vs, 0, 0)$ of the charge at the retarded time

$$s = t - [r]/c, \tag{3.1}$$

where $[r]$ is the distance from $(vs, 0, 0)$ to (x, y, z). Thus

$$[r]^2 = (vs - x)^2 + \rho^2, \tag{3.2}$$

where $\rho = \sqrt{(y^2 + z^2)}$ is the distance of the field point from the line of motion; and substitution for s from (3.1) into (3.2) gives

$$(1-v^2/c^2)[r^2] + 2\frac{v}{c}(vt - x)[r] - (vt - x)^2 - \rho^2 = 0. \tag{3.3}$$

In this particular case, therefore, the retarded distance is expressible as a simple algebraic function of the "instantaneous" time t. Explicitly,

$$(1 - v^2/c^2)[r] = -\frac{v}{c}(vt - x) + \sqrt{\{(vt - x)^2 + (1 - v^2/c^2)\rho^2\}}, \tag{3.4}$$

where, since $[r]$ is positive, only the solution with the positive square root is permitted by the condition $v < c$. It is recalled that for any motion (in vacuum) for which v is always less than some fixed value c' which itself is less than c there is one and only one retarded time s corresponding to each time t.

Actually required is

$$[\kappa r] = [r] + \frac{[r]}{c}\frac{d[r]}{ds}. \tag{3.5}$$

From (3.2), $[r]\, d[r]/ds = v(vs - x)$, and substitution for s from (3.1) gives

$$[\kappa r] = (1 - v^2/c^2)[r] + \frac{v}{c}(vt - x);$$

and hence, from (3.4),

$$[\kappa r] = \sqrt{\{(vt - x)^2 + (1 - v^2/c^2)\rho^2\}}. \tag{3.6}$$

Therefore

$$\phi = \frac{e}{4\pi\varepsilon_0}\frac{1}{\sqrt{\{(vt - x)^2 + (1 - v^2/c^2)\rho^2\}}}, \tag{3.7}$$

and $\qquad \mathbf{A} = (A, 0, 0),$

where

$$A = \frac{e\mu_0}{4\pi}\frac{v}{\sqrt{\{(vt - x)^2 + (1 - v^2/c^2)\rho^2\}}}. \tag{3.8}$$

It is now an easy matter to find $\mathbf{E}$ and $\mathbf{H}$ from (2.53) and (2.54). The results are most conveniently expressed in terms of cylindrical coordinates ρ, θ, x. Then $\mathbf{H} = (1/\mu_0)$ curl $\mathbf{A}$ has only a θ-component $-(1/\mu_0)\, \partial A/\partial \rho$, so that

$$H_x = H_\rho = 0,\ H_\theta = \frac{e}{4\pi}\frac{v(1 - v^2/c^2)\rho}{\{(vt - x)^2 + (1 - v^2/c^2)\rho^2\}^{3/2}}. \tag{3.9}$$

Also $E_\rho = -\partial\phi/\partial\rho$, $E_\theta = 0$, $E_x = -\partial\phi/\partial x - \partial A/\partial t$, which give

$$E_\rho = \frac{e}{4\pi\varepsilon_0}\frac{(1 - v^2/c^2)\rho}{\{(vt - x)^2 + (1 - v^2/c^2)\rho^2\}^{3/2}}, \tag{3.10}$$

$$E_x = -\frac{e}{4\pi\varepsilon_0}\frac{(1 - v^2/c^2)(vt - x)}{\{(vt - x)^2 + (1 - v^2/c^2)\rho^2\}^{3/2}}. \tag{3.11}$$

It may be noted that

$$E_\rho/E_x = \rho/(x - vt), \tag{3.12}$$

which shows that the direction of $\mathbf{E}$ is along the "instantaneous" radius vector from the position of the charge at time t to the field point. It is easy to verify that $\mathbf{H} = \sqrt{(\varepsilon_0/\mu_0)}[\hat{\mathbf{r}}]\times\mathbf{E}$.

That there is no radiation associated with the field is confirmed by the observation that, for any given values of v ($<c$), x and t, the magnitudes of $\mathbf{E}$ and $\mathbf{H}$ tend to zero as ρ tends to infinity in proportion to $1/\rho^2$. The nature of the field is, however, dependent on whether v/c is small or close to unity. In the former case the situation is described approximately by the Coulomb law $\mathbf{E} = e\mathbf{r}/(4\pi\varepsilon_0 r^3)$ and the Biot–Savart law $\mathbf{H} = e\mathbf{v} \times \mathbf{r}/(4\pi r^3)$. In the latter case, at any fixed point, the field strength maximum when $t = x/v$ becomes very pronounced, and the field can be pictured as localized near a plane normal to the direction of motion and carried along with the particle.

As a prelude to discussing Čerenkov radiation it is instructive to see what is predicted by the present analysis on the hypothetical assumption that v exceeds c; but this is deferred until §3.1.3 in order to interpose an alternative derivation by relativistic arguments of the results just obtained.

3.1.2 Deduction from the Lorentz transformation

From the point of view of special relativity the derivation of the field of a point charge in uniform rectilinear motion from the Coulomb field of a point charge at rest is a straightforward exercise in the transformation of 4-vectors and tensors.

Suppose that S is the laboratory frame in which the charge has constant velocity v along the x-axis and passes through the origin $x = y = z = 0$ at time $t = 0$; and that S' is the frame moving with the charge, which in S' is located at the origin $x' = y' = z' = 0$ for all t'. Then from (2.151) the position x, y, z at time t as measured in S has coordinates

$$x' = \frac{x - vt}{\sqrt{(1 - v^2/c^2)}}, \qquad y' = y, \qquad z' = z \tag{3.13}$$

at time

$$t' = \frac{t - vx/c^2}{\sqrt{(1 - v^2/c^2)}} \tag{3.14}$$

as measured in S'. Also the transformation (2.164) of the potentials ϕ, $\mathbf{A}$ reads

$$A_x = \frac{A_x' + v\phi'/c^2}{\sqrt{(1 - v^2/c^2)}}, \qquad A_y = A_y', \qquad A_z = A_z', \quad \phi = \frac{\phi' + vA_x}{\sqrt{(1 - v^2/c^2)}}. \tag{3.15}$$

But in S' the charge is at rest at the origin, so

$$\mathbf{A}' = 0, \qquad \phi' = \frac{e}{4\pi\varepsilon_0}\frac{1}{r'}, \tag{3.16}$$

where

$$r'^2 = x'^2 + y'^2 + z'^2.$$

From (3.13)

$$(1 - v^2/c^2)r'^2 = (x - vt)^2 + (1 - v^2/c^2)(y^2 + z^2), \tag{3.17}$$

and substitution from (3.16) into (3.15) therefore gives

$$\phi = \frac{e}{4\pi\varepsilon_0} \frac{1}{\sqrt{\{(x - vt)^2 + (1 - v^2/c^2)(y^2 + z^2)\}}}, \tag{3.18}$$

$$\mathbf{A} = (v\phi/c^2, 0, 0), \tag{3.19}$$

which are identical with (3.7), (3.8).

It is worth mentioning here that very much the same technique as that just employed can in fact lead to yet another derivation of the Liénard–Wiechert potentials for general motion of the point charge. In discussing general motion the above argument need only be modified to the extent that S' is now that Lorentz frame, in uniform motion with respect to S, in which the charge is *instantaneously* at rest at the origin at time $t' = 0$. Then (3.16) is valid, but, and this is the vital point, only at time $t' = r'/c$; a statement for which formal justification is provided by the retarded potential expression (2.61) evaluated for a point charge at rest at the retarded time. But

$$c^2t^2 - r^2 = c^2t'^2 - r'^2, \tag{3.20}$$

so the solutions for ϕ and $\mathbf{A}$ in the frame S in which, instantaneously, the charge has velocity v along the x-axis, are therefore (3.18) and (3.19) at time $t = r/c$. And since, when $t = r/c$,

$$(x - vt)^2 + (1 - v^2/c^2)(y^2 + z^2) = (r - vx/c)^2, \tag{3.21}$$

where

$$r^2 = x^2 + y^2 + z^2,$$

(3.18) is then

$$\phi = \frac{e}{4\pi\varepsilon_0} \frac{1}{r - vx/c} = \frac{e}{4\pi\varepsilon_0} \frac{1}{r - \mathbf{r}\cdot\mathbf{v}/c}. \tag{3.22}$$

The statement that (3.22) is the scalar potential at distance r from the origin at time $t = r/c$ for a point charge which is at the origin with velocity $\mathbf{v}$ at time $t = 0$ is indeed equivalent to (2.107), (2.109).

Returning now to the point charge in uniform motion, the derivation of the vectors $\mathbf{E}$ and $\mathbf{H}$ is given as an example of the transformations (2.172). Since

$$\mathbf{E}' = \frac{e}{4\pi\varepsilon_0} \frac{\mathbf{r}'}{r'^3}, \qquad \mathbf{H}' = 0,$$

the inverse of (2.172) leads to $H_{\|} = 0$, that is $H_x = 0$, and

$$\mathbf{H}_{\perp} = \frac{e}{4\pi} \frac{1}{\sqrt{(1 - v^2/c^2)}} \frac{\mathbf{v} \times \mathbf{r}'}{r'^3}; \tag{3.23}$$

and (3.23) is equivalent to

$$H_y = -\frac{e}{4\pi}\frac{v}{\sqrt{(1-v^2/c^2)}}\frac{z'}{r'^3}, \qquad H_z = \frac{e}{4\pi}\frac{v}{\sqrt{(1-v^2/c^2)}}\frac{y'}{r'^3},$$

which, with (3.13) and (3.17), are in turn seen to be equivalent to (3.9). Likewise (3.11) and (3.10) follow from $E_{\parallel} = E'_{\parallel}$ and $E_{\perp} = E'_{\perp}/\sqrt{(1 - v^2/c^2)}$ respectively.

3.1.3 The case $v > c$

In this section the method of Liénard–Wiechert potentials is applied, much as in §3.1.1, to the same problem in the hypothetical case $v > c$. Clearly the formal solution thus obtained is virtually equivalent to that of the more physically realistic situation in which a point charge travels in a straight line with constant speed $v > c/\mu$ through a lossless dielectric whose refractive index μ is independent of frequency. This latter problem gives some insight into the nature of Čerenkov radiation, but is still too idealized to lead to quantitative predictions because, as is shown in §3.2, an adequate theory must take account of the frequency dependence of μ.

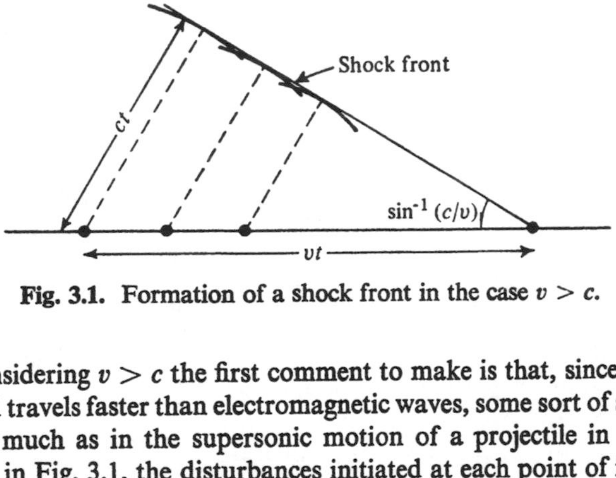

Fig. 3.1. Formation of a shock front in the case $v > c$.

In considering $v > c$ the first comment to make is that, since the source of the field travels faster than electromagnetic waves, some sort of shock front is set up, much as in the supersonic motion of a projectile in a gas. As illustrated in Fig. 3.1, the disturbances initiated at each point of its track by the moving charge arrive simultaneously at the surface of a right circular cone whose vertex is at the charge. The semi-vertical angle of the cone is evidently $\sin^{-1}(c/v)$. The field at any point is, of course, zero up to the instant that the front passes through it.

The field at any point after the front has passed through it can be obtained from the Liénard–Wiechert potentials; for the derivation of (2.107) and (2.108), by, for example, the "collapsing sphere" argument immediately preceding these expressions, is not in principle affected by the supposition that

v can exceed c. Indeed, only two observations need be made further to what has already been said. First that, depending on the specific motion, any number of retarded times $t - [r]/c$, not just one, may be associated with a given time t; this is so since, with $v > c$ allowed, the charge is capable of weaving in and out of the "collapsing sphere". Secondly, that $[\kappa]$, as defined by (2.109), may be negative, and must then clearly be replaced in (2.107), (2.108) by its modulus.

In the problem under consideration there are, at any point, two retarded times corresponding to any time subsequent to the passage of the shock front through the point, and none corresponding to any time prior to the passage of the front. In fact, the retarded time is $t - [r]/c$, where, by the analysis that led to (3.4),

$$(v^2/c^2 - 1)[r] = \frac{v}{c}(vt - x) \pm \sqrt{\{(vt - x)^2 - (v^2/c^2 - 1)\rho^2\}}; \quad (3.24)$$

this is the same formula as (3.4) except that neither sign of the radical may be discarded. For (3.24) to give a solution, $vt - x$ must be positive (since $[r]$ is positive) and must also exceed $\sqrt{(v^2/c^2 - 1)}\rho$, in which case either sign is valid. Since the semi-vertical angle of the shock cone is

$$\sin^{-1}(c/v) = \tan^{-1}(v^2/c^2 - 1)^{-1/2},$$

the instant given by

$$vt - x = \sqrt{(v^2/c^2 - 1)}\rho \quad (3.25)$$

is precisely that instant at which the shock front passes the field point.

Where there are two retarded times the same analysis that led to (3.6) shows that the two associated values of $[\kappa r]$ are

$$[\kappa r] = \pm\sqrt{\{(vt - x)^2 - (v^2/c^2 - 1)\rho^2\}}. \quad (3.26)$$

Each value (3.26) makes the same contribution to ϕ and $\mathbf{A}$. The field behind the shock front is therefore determined by

$$\phi = \frac{e}{4\pi\varepsilon_0}\frac{2}{\{(vt - x)^2 - (v^2/c^2 - 1)\rho^2\}^{3/2}}, \quad (3.27)$$

$$\mathbf{A} = (v\phi/c^2, 0, 0). \quad (3.28)$$

Although in this way the field can be exhibited in terms of simple algebraic expressions, it is not easy, because of the infinities in the field in the shock front corresponding to the vanishing of the denominator in (3.27), to deduce whether or not there is radiation. A far more complete treatment of Čerenkov radiation is presented in the next section.

3.2 ČERENKOV RADIATION

3.2.1 The field representation

The theory developed from Maxwell's equations in §§2.1.1 and 2.1.2 for determining the vacuum field of a time-harmonic current distribution applies with trivial modification to the field in a homogeneous medium characterized by values of permittivity and permeability different from ε_0 and μ_0 respectively. The case of the permeability differing appreciably from μ_0 is ignored as being of no interest in the present context, and the only modification required, therefore, is to write the appropriate value ε in place of ε_0 throughout. It must be remembered that ε_0 appears in c and k_0, and these parameters have to be replaced by c/μ and $k_0\mu$ respectively, where

$$\mu = \sqrt{(\varepsilon/\varepsilon_0)} \tag{3.29}$$

is the refractive index.

Although ε may be almost constant over some quite wide range of frequencies, it is fundamentally a function of frequency. This has, of course, no effect on the discussion of a time-harmonic field, except in so far as the value of the angular frequency determines the particular value of ε; but it does mean that an investigation of the field generated by a current distribution which varies with time other than harmonically must proceed via a frequency spectrum analysis.

In the present problem the current distribution is a point charge e traveling along the x-axis with constant speed v and passing through the origin at $t = 0$. This is a mathematically singular distribution, but the current density can be represented by delta functions. For the charge density is

$$\rho = e\,\delta(x - vt)\,\delta(y)\,\delta(z), \tag{3.30}$$

being a special case of (2.129), and with this charge density is associated the current density

$$\mathbf{j} = (j, 0, 0) \tag{3.31}$$

where

$$j = ev\,\delta(x - vt)\,\delta(y)\,\delta(z). \tag{3.32}$$

The corresponding complex frequency spectrum, defined in (2.99) or (2.97), is

$$\mathbf{j}^{\omega} = (j^{\omega}, 0, 0), \tag{3.33}$$

where

$$j^{\omega} = \frac{ev}{2\pi}\int_{-\infty}^{\infty} \delta(x - vt)\,\delta(y)\,\delta(z)\,e^{-i\omega t}\,dt\,; \tag{3.34}$$

or, on integrating,

$$j^{\omega} = \frac{e}{2\pi}\,\delta(y)\,\delta(z)\,e^{-i\omega x/v}. \tag{3.35}$$

The procedure therefore is to find the time-harmonic field generated in a medium with refractive index $\mu(\omega)$ by the current density (3.35); the actual fields, if required, can then be constructed by restoring the integration over ω in the way typified by (2.99). Since no explicit form of the function $\mu(\omega)$ is introduced in the general analysis, the final result for the energy radiated per unit length of path traversed by the particle appears as an integral over ω, the integrand of which displays the power spectrum.

The infinite extent in the x-direction, and the periodicity in x, of (3.35) strongly suggest that a convenient representation of the associated field is in terms of the spatial Fourier integrals described in §2.1.5. The complete spectrum function $\mathbf{j}'$, introduced in (2.90), is here, using (2.96) and (3.35),

$$\mathbf{j}' = (j', 0, 0) \tag{3.36}$$

with

$$j' = \frac{e}{16\pi^4}\int \delta(y)\,\delta(z)\,e^{-i\omega x/v}\,e^{i\mathbf{k}\cdot\mathbf{r}}\,d\tau,$$

where $\mathbf{r} = (x, y, z)$, $\mathbf{k} = (l, m, n)$ and the volume integral is taken over all space. Doing the trivial y and z integration leaves

$$j' = \frac{e}{16\pi^4}\int_{-\infty}^{\infty} e^{i(l-\omega/v)x}\,dx,$$

which itself represents a delta function, namely

$$j' = \frac{e}{8\pi^3}\,\delta(l - \omega/v). \tag{3.37}$$

To obtain the complex frequency spectra $\mathbf{E}^{\omega}$ and $\mathbf{H}^{\omega}$ of the field vectors $\mathbf{E}$ and $\mathbf{H}$ it may be noted first that the vector potential $\mathbf{A}$ is determined from $\mathbf{j}'$ through (2.94), where it must be remembered that $k_0^2 = \omega^2/c^2$ is to be replaced by $\omega^2\mu^2/c^2$. Since $\mathbf{H} = (1/\mu_0)$ curl $\mathbf{A}$, the corresponding representation of $\mathbf{H}^{\omega}$ is

$$\mathbf{H}^{\omega} = -i\iiint \frac{\mathbf{k}\times\mathbf{j}'}{k^2 - \omega^2\mu^2/c^2}\,e^{-i\mathbf{k}\cdot\mathbf{r}}\,d^3k, \tag{3.38}$$

in which the positions of the poles in relation to the path of integration can be determined by giving ω a small negative imaginary part, as explained in §2.1.5. Using (3.36) and (3.37) the l integration is trivial. If also, for subsequent convenience, the integration variables m and n are then replaced by $\omega m/c$, $\omega n/c$ respectively, the result is

$$\mathbf{H}^\omega = \frac{ie\omega}{8\pi^3 c} e^{-i\omega x/v} \int_{-\infty}^{\infty} \int_{-\infty}^{\infty} \frac{(0, -n, m)}{m^2 + n^2 - \mu^2 + c^2/v^2} e^{-i\omega(my+nz)/c}\, dm\, dn. \tag{3.39}$$

Either the m or the n integration can be effected by closing the path of integration with an infinite semi-circle and using the residue theorem. Choosing the latter, the integrand is seen to have poles in the complex n-plane at $n = \pm n_1$, where

$$n_1 = \sqrt{(\mu^2 - c^2/v^2 - m^2)}, \tag{3.40}$$

and it is at this stage in the analysis that the important distinction between the cases v less than or greater than c/μ makes itself felt. If $v < c/\mu$ the expression under the square root in (3.40) is negative for all real values of m. The poles $\pm n_1$ are therefore on the imaginary axis, and just one of these poles will be captured by closing the path of integration with an infinite semi-circle. If, on the other hand, $v > c/\mu$, the expression under the square root in (3.40) is positive for some real values of m, and for these values of m the poles $\pm n_1$ are on the real axis. However, by appeal to the comment following (3.38) it is clear that in effect one pole remains above and one below the path of integration, and the formal mathematics is unaffected. The results for both cases can be written in the form

$$\mathbf{H}^\omega = \frac{e\omega}{8\pi^2 c} e^{-i\omega x/v} \int_{-\infty}^{\infty} \left(0, \mp 1, \frac{m}{n_1}\right) e^{-i\omega(my \pm n_1 z)/c}\, dm, \tag{3.41}$$

with the upper or lower sign according to whether $z > 0$ or < 0 respectively, on the understanding that ω is positive and that n_1, defined in (3.40), is positive when real and otherwise negative pure imaginary. The signs are most easily determined by the requirement that the waves implicit in the integrand of (3.41) are either progressing, or decaying, away from the plane $z = 0$; and it is sufficient to state the results for $\omega > 0$.

The corresponding expressions for the components of $\mathbf{E}^\omega$ are readily obtained from curl $\mathbf{H}^\omega = i\omega\varepsilon_0\mu^2\mathbf{E}^\omega$. They are, for $\omega > 0$,

$$E_x^\omega = -Z_0 \frac{e\omega}{8\pi^2 c}\left(1 - \frac{c^2}{v^2\mu^2}\right) e^{-i\omega x/v} \int_{-\infty}^{\infty} \frac{1}{n_1} e^{-i\omega(my \pm n_1 z)/c}\, dm, \tag{3.42}$$

$$E_y^\omega = Z_0 \frac{c}{v\mu^2} H_z^\omega, \qquad E_z^\omega = -Z_0 \frac{c}{v\mu^2} H_y^\omega, \tag{3.43}$$

where $Z_0 = \sqrt{(\mu_0/\varepsilon_0)}$.

The integrals in (3.41) and (3.42) can in fact be evaluated in terms of Bessel functions. Such evaluation makes explicit the symmetry about the

x-axis, and, with the help of asymptotic approximations to the Bessel functions, can be used to calculate the radiated power by examining the Poynting vector at great distances from the x-axis. It is, however, quite straightforward to make the power calculation directly from the integral forms themselves, and this is done in the next section.

The present section ends with the reminder that the expressions for the actual vectors **E** and **H** are obtained by restoring the ω integration. Since (3.41), (3.42) hold for $\omega > 0$, the procedure, as indicated in (2.74), is first to multiply by $2 \exp(i\omega t)$, then to integrate over ω from 0 to infinity, and finally to take the real part. As an example,

$$E_x = \mathrm{Re} -\frac{Z_0 e}{4\pi^2 c} \int_{\omega=0}^{\infty} \int_{m=-\infty}^{\infty} \left(1 - \frac{c^2}{v^2 \mu^2}\right) \frac{\omega}{n_1} e^{i\omega\left(t - \frac{x}{v} - \frac{my}{c} \mp \frac{n_1 z}{c}\right)} \, dm \, d\omega, \tag{3.44}$$

where n_1 is given by (3.40).

3.2.2 The radiated power

The representation typified by (3.44) is formed by the superposition of time-harmonic plane waves. If n_1 is purely imaginary these waves are evanescent in the z-direction, and on average carry no energy away from the plane $z = 0$. But if n_1 is real the waves are propagated and do carry energy away from the plane. As already observed, the necessary and sufficient condition for some values of n_1 to be real is that

$$\mu(\omega) > c/v \tag{3.45}$$

for some values of ω; this then is the necessary and sufficient condition for radiation.

Consider now a strip of a plane $z = z_0\,(>0)$, the strip being of width d and bounded by two straight lines parallel to the central line $x = x_0$. The total energy, $\frac{1}{2}Wd$ say, that crosses the strip is obviously independent of x_0; it is also independent of z_0, since if it were not the total energy crossing the complete plane $z = z_0$ would depend on z_0. Allowing for equal radiation above and below $z = 0$, W evidently represents the energy radiated by the particle per unit length of its track, and this is the quantity commonly quoted in the theory. In terms of the Poynting vector

$$W = 2\int_{t=-\infty}^{\infty} \int_{y=-\infty}^{\infty} E_x H_y \, dy \, dt, \tag{3.46}$$

and from §2.1.4, specifically (2.84), this is

$$W = \mathrm{Re}\; 8\pi \int_{\omega=0}^{\infty} \int_{y=-\infty}^{\infty} E_x^{\omega} H_y^{\omega *} \, dy \, d\omega. \tag{3.47}$$

In evaluating (3.47) it is natural to aim for simplicity by setting x_0 and z_0 zero. Then from (3.41)

$$H_y^w = -\frac{e\omega}{4\pi c}\,\delta(\omega y/c), \tag{3.48}$$

and from (3.42)

$$E_x^\omega = -Z_0\frac{e\omega}{8\pi^2 c}\left(1-\frac{c^2}{v^2\mu^2}\right)\int_{-\infty}^{\infty}\frac{1}{n_1}e^{-i\omega my/c}\,dm, \tag{3.49}$$

which give

$$W = \mathrm{Re}\,\frac{e^2\mu_0}{4\pi^2}\int_{\omega=0}^{\infty}\left(1-\frac{c^2}{v^2\mu^2}\right)\omega\int_{-\infty}^{\infty}\frac{dm}{n_1}\,d\omega, \tag{3.50}$$

where $\mu_0 = Z_0/c$ is the vacuum permeability.

Now it may properly be assumed, on both theoretical and experimental grounds, that $\mu(\omega) \to 1$ as $\omega \to \infty$. There is, therefore, an upper range for ω, beyond ω_0 say, in which (3.45) does not hold, and possibly other ranges also. Values of ω in these ranges do not contribute to (3.50), since they make n_1 purely imaginary; and for each other value of ω the m integration may, for the same reason, be confined between $\pm\sqrt{(\mu^2 - c^2/v^2)}$. Since, for any m_0,

$$\int_{-m_0}^{m_0}\frac{dm}{\sqrt{(m_0^2 - m^2)}} = \pi,$$

the final result is

$$W = \frac{e^2\mu_0}{4\pi}\int_0^{\omega_0}\left[1-\frac{c^2}{v^2\mu^2(\omega)}\right]\omega\,d\omega, \tag{3.51}$$

where it is to be understood that the range of integration is restricted to those values of ω for which the integrand is positive.

The formula (3.51) is standard in the theory of Čerenkov radiation, and two remarks are made in connection with the present derivation. First, it is noted that use of the intermediate expression (3.50) could be criticized on the grounds that the inner integral is divergent. This feature is not obtrusive because the divergent part is purely imaginary; and in any case the difficulty is only apparent, since it enters solely through the expedient of setting z_0 zero in the analysis. If the representations (3.41), (3.42) with a finite value z_0 of z are substituted into (3.47) the evaluation leads to (3.51) without encountering any divergent integrals; it is only somewhat more elaborate to write out.

The second remark bears some relation to the first. A plausible alternative way of finding the radiated power is to equate it to the rate at which the field does work against the charge; for in general these quantities must balance on average, and here they are, of course, both constants. This suggests that

$$W = -eE_x,$$

where E_x is evaluated at $x = vt$, $y = z = 0$; and use of (3.44) then leads precisely to the statement (3.50). The fact that this method necessarily introduces a divergent integral is not surprising, since the attempt is being made to write down the force on a point charge due to its own field, a procedure that really calls for a much closer examination than has been offered here (see §4.6).

3.2.3 Finite path length

There is a number of other situations, similar to that just described, in which radiation can be generated by a charged particle in uniform rectilinear motion; one example is a particle moving in a vacuum past the plane face of a semi-infinite dielectric. The problem has also been treated when the dielectric medium is anisotropic, and again when the moving particle is replaced by a dipole, static or oscillatory. Although many of these cases are amenable to the type of analysis given in §§3.2.1, 3.2.2, their development is not pursued here; instead, this account of Čerenkov radiation is concluded by examining the effect of confining the particle to a finite length of path.

If it be supposed that the particle starts at $x = -a$ and then proceeds with uniform speed v to $x = a$, where it stops, having passed through $x = 0$ at $t = 0$, then it is evident from (3.35) that the complex frequency spectrum of the current density is $\mathbf{j}^\omega = (j^\omega, 0, 0)$, where

$$j^\omega = \begin{cases} \dfrac{e}{2\pi}\,\delta(y)\,\delta(z)\,e^{-i\omega x/v}, & \text{for } |x| < a, \\ 0, & \text{for } |x| > a. \end{cases} \tag{3.52}$$

That $\mathbf{j}^\omega$ is now confined in all directions means that the conventional treatment of the radiation field, given in §2.1.2, is applicable. In fact, from (2.39) and (2.41), with k_0 replaced by $\omega\mu/c$,

$$\mathbf{H}^\omega \sim -(G, 0, 0) \times \hat{\mathbf{r}}\,\frac{e^{-i\omega\mu r/c}}{r} \tag{3.53}$$

as $r \to \infty$, where

$$G = -\frac{i\omega\mu}{4\pi c}\int j^\omega\, e^{i\omega\mu\hat{\mathbf{r}}\cdot\mathbf{r}'/c}\,d\tau. \tag{3.54}$$

It is convenient to write

$$\hat{\mathbf{r}} = (l, m, n), \tag{3.55}$$

and then substitution from (3.52) into (3.54) gives at once

$$G = -\frac{e}{8\pi^2 c}\, i\omega\mu \int_{-a}^{a} e^{i\omega(\mu l - c/v)x'/c}\, dx', \tag{3.56}$$

that is

$$G = -\frac{e}{4\pi^2}\, i\, \frac{\sin\left[\dfrac{\omega\mu a}{c}\left(l - \dfrac{c}{v\mu}\right)\right]}{l - \dfrac{c}{v\mu}}. \tag{3.57}$$

The total energy $\mathscr{E}$ that passes out of a large sphere of radius r is readily found by appealing to the formula (2.89). For in the present context this states that the total energy that crosses a distant surface element that subtends a solid angle $d\Omega$ at the origin is

$$d\Omega\, \frac{4\pi Z_0}{\mu} \int_0^\infty (1 - l^2)\, |G|^2\, d\omega. \tag{3.58}$$

Because of the symmetry about the x-axis, $d\Omega$ can be replaced by $2\pi\, dl$; and integration of (3.58) over a sphere, with G given by (3.57), then yields

$$\mathscr{E} = \int_0^\infty \mathscr{E}'(\omega)\, d\omega, \tag{3.59}$$

where

$$\mathscr{E}'(\omega) = \frac{e^2 Z_0}{2\pi^2 \mu} \int_{-1}^{1} \left\{ \frac{\sin\left[\dfrac{\omega\mu a}{c}\left(l - \dfrac{c}{v\mu}\right)\right]}{l - \dfrac{c}{v\mu}} \right\}^2 (1 - l^2)\, dl. \tag{3.60}$$

It only remains to evaluate and interpret (3.60). With the integration variable changed to

$$\lambda = \frac{a\omega\mu}{c}\left(l - \frac{c}{v\mu}\right) \tag{3.61}$$

it is seen to be

$$\mathscr{E}' = 2a\, \frac{e^2 \mu_0 \omega}{4\pi^2} \int_{-\frac{a\omega\mu}{c}\left(1 + \frac{c}{v\mu}\right)}^{\frac{a\omega\mu}{c}\left(1 - \frac{c}{v\mu}\right)} \left(1 - \frac{c^2}{v^2\mu^2} - \frac{2c^2}{av\omega\mu^2}\lambda - \frac{c^2}{a^2\omega^2\mu^2}\lambda^2\right) \frac{\sin^2\lambda}{\lambda^2}\, d\lambda. \tag{3.62}$$

If the path length $2a$ is treated as a large parameter, it is fairly obvious, and not difficult to establish rigorously, that, when $v > c/\mu$,

$$\mathscr{E}' = 2a\frac{e^2\mu_0\omega}{4\pi^2}\left(1 - \frac{c^2}{v^2\mu^2}\right)\int_{-\infty}^{\infty}\frac{\sin^2\lambda}{\lambda^2}\,d\lambda + O(1), \tag{3.63}$$

which gives

$$\frac{\mathscr{E}'}{2a} = \frac{e^2\mu_0}{4\pi}\omega\left(1 - \frac{c^2}{v^2\mu^2}\right) + O\left(\frac{1}{a}\right). \tag{3.64}$$

This formula, in combination with (3.59), is in evident agreement with (3.51). On the other hand, when $v < c/\mu$, both limits in (3.62) are large and negative, with the result that $\mathscr{E}'/(2a)$ is $O(1/a)$.

It might perhaps be wondered why (3.64) is not exact without the $O(1/a)$ term, and why there is any radiation for $v < c/\mu$. The answer is that there is delta-function acceleration at the beginning and end of the path of the particle, associated with the instantaneous changes first from rest to motion and then from motion to rest. These accelerations are separated by the distance $2a$, and therefore give rise to radiation fields whose combined contribution to $\mathscr{E}'$ depends on a, though it does not, of course, contain a part proportional to a.

3.3 GYRO RADIATION

3.3.1 Uniform circular motion; the far field

Radiation from a charged particle in uniform circular motion is of considerable importance in various fields of study. It is thought to be the basis of the explanation of many radio astronomical and allied observations, and can also be significant in the laboratory as a factor affecting the performance of high energy particle accelerators.

Since the field is clearly periodic, the fundamental angular frequency Ω being that of the particle in its orbit, it is natural to express each field vector as the sum of a series of time-harmonic terms. The procedure differs from that described in §2.2.4 only in that these sums replace the integrals over the angular frequency ω. Thus if, as in (2.85),

$$\mathbf{H} = \sum_{n=-\infty}^{\infty}\mathbf{H}_n e^{in\Omega t}, \tag{3.65}$$

it is easy to appreciate that the representation corresponding to (2.132) is

$$\mathbf{H}_n = -\frac{e\Omega}{8\pi^2}\int_0^{2\pi/\Omega}\left(\frac{in\Omega}{cr} + \frac{1}{r^2}\right)\hat{\mathbf{r}} \times \mathbf{v}\, e^{-in\Omega(s+r/c)}\,ds, \tag{3.66}$$

where it is recalled that in this formula $\mathbf{r} = r\hat{\mathbf{r}}$ is the radius vector from the particle to the field point at time s, and $\mathbf{v} = -d\mathbf{r}/ds$ is the associated velocity.

In most contexts, and certainly the present, only the radiation field is of interest. This can be written down from (3.66) with the help of the sort of approximations introduced in §2.1.2. First, however, it is desirable to make a change of notation from that in (3.66) by reverting as of now to the use of $\mathbf{r}$ for the radius vector from a fixed origin O to the field point. If, in addition, the radius vector from O to the particle is called $\boldsymbol{\rho}$, the radiation field is specified by

$$\mathbf{H}_n \sim -\frac{e\Omega}{4\pi c}\, in\, \mathbf{I}_n \frac{e^{-in\Omega r/c}}{r}, \tag{3.67}$$

where

$$\mathbf{I}_n = \frac{\Omega}{2\pi}\int_0^{2\pi/\Omega} \hat{\mathbf{r}} \times \mathbf{v}\, e^{-in\Omega(s-\hat{\mathbf{r}}\cdot\boldsymbol{\rho}/c)}\, ds. \tag{3.68}$$

The integrals $\mathbf{I}_n$ are now evaluated for the specific case of motion with constant speed v in a circle of radius $\rho = v/\Omega$. The origin O is naturally taken at the center of the circle, and the position of the field point is specified by spherical polar coordinates r, θ, ϕ, with axis perpendicular to the plane of the circle. The corresponding coordinates of the particle are ρ, $\frac{1}{2}\pi$, $\phi + \chi$, say; and with appropriate choice of time zero $\chi = \Omega s$ at time s (see Fig. 3.2).

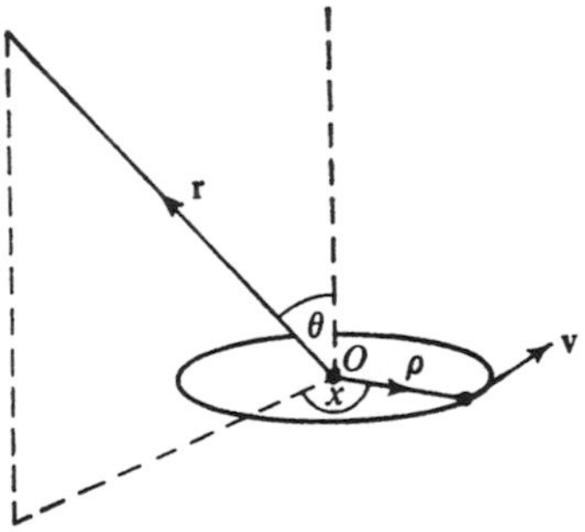

Fig. 3.2. The configuration for a point charge in uniform circular motion.

It is easily seen that

$$\hat{\mathbf{r}}\cdot\boldsymbol{\rho} = \rho \sin\theta \cos\chi,$$

and also that the θ- and ϕ-components of $\hat{\mathbf{r}} \times \mathbf{v}$ are

$$(\hat{\mathbf{r}} \times \mathbf{v})_\theta = -v\cos\chi, \qquad (\hat{\mathbf{r}} \times \mathbf{v})_\phi = -v\cos\theta\sin\chi.$$

The θ-component of (3.68) is therefore

$$I_{n\theta} = -\frac{\Omega v}{2\pi}\int_0^{2\pi/\Omega} \cos\chi\, e^{-in\Omega[s-(\rho/c)\sin\theta\cos\chi]}\, ds$$

$$= -\frac{v}{2\pi}\int_0^{2\pi} \cos\chi\, e^{-in[\chi-(v/c)\sin\theta\cos\chi]}\, d\chi. \tag{3.69}$$

Likewise the ϕ-component of (3.68) is

$$I_{n\phi} = -\frac{v\cos\theta}{2\pi}\int_0^{2\pi} \sin\chi\, e^{-in[\chi-(v/c)\sin\theta\cos\chi]}\, d\chi$$

$$= c\,\frac{\cot\theta}{2\pi i n}\int_0^{2\pi} e^{-in\chi}\frac{d}{d\chi}\, e^{in(v/c)\sin\theta\cos\chi}\, d\chi$$

$$= c\cot\theta\,\frac{1}{2\pi}\int_0^{2\pi} e^{-in[\chi-(v/c)\sin\theta\cos\chi]}\, d\chi. \tag{3.70}$$

Now a standard representation of the Bessel function of the first kind, of integral order n and argument z, is

$$J_n(z) = \frac{e^{-in\pi/2}}{2\pi}\int_0^{2\pi} e^{-i(n\chi - z\cos\chi)}\, d\chi, \tag{3.71}$$

and it follows that

$$J_n'(z) = \frac{ie^{-in\pi/2}}{2\pi}\int_0^{2\pi} \cos\chi\, e^{-i(n\chi-z\cos\chi)}\, d\chi, \tag{3.72}$$

where the dash denotes differentiation with respect to the argument. Hence (3.69) and (3.70) give

$$I_{n\theta} = i\, e^{in\pi/2} v\, J_n'\!\left(n\frac{v}{c}\sin\theta\right), \tag{3.73}$$

$$I_{n\phi} = e^{in\pi/2}\, c\cot\theta\, J_n\!\left(n\frac{v}{c}\sin\theta\right). \tag{3.74}$$

The substitution of these expressions into (3.67) gives, with (3.65), the magnetic field components

$$(H_\theta, H_\varphi) = \sum_{-\infty}^{\infty} (H_{n\theta}, H_{n\phi})e^{in\Omega t}, \tag{3.75}$$

where in the radiation field

$$(H_{n\theta}, H_{n\varphi}) = \frac{e\Omega}{4\pi}\left\{\frac{v}{c}J_n'\!\left(n\frac{v}{c}\sin\theta\right), -i\cot\theta J_n\!\left(n\frac{v}{c}\sin\theta\right)\right\} n e^{in\pi/2}\,\frac{e^{-in\Omega r/c}}{r}. \tag{3.76}$$

The way in which the expressions (3.76) vary with n is considered in the next section, where it is shown that the relative importance of the respective harmonic terms depends markedly on the value of v/c. A general comment can, however, be made about the polarization of each harmonic component, which may be specified by

$$\frac{H_{n\phi}}{H_{n\theta}} = -i\frac{c}{v}\cot\theta\,\frac{J_n\left(n\frac{v}{c}\sin\theta\right)}{J_n'\left(n\frac{v}{c}\sin\theta\right)}. \tag{3.77}$$

For $\theta = \frac{1}{2}\pi$, that is at points in the plane of the particle's orbit, $H_{n\phi}$ is of course zero, and the polarization is therefore linear. At points away from the plane of the orbit, the polarization is elliptical. Moreover, since

$$J_n(z) = \frac{z^n}{2^n n!} + O(z^{n+2}) \tag{3.78}$$

for $|z| \ll 1$, when θ is nearly zero

$$\frac{H_{n\phi}}{H_{n\theta}} \simeq -i\cos\theta \tag{3.79}$$

and the polarization is nearly circular, being strictly circular at points on the line perpendicular to and through the center of the orbit. In fact, at $\theta = 0$ only the fundamental frequency is present, with

$$(H_{1\theta}, H_{1\phi}) = \frac{e\Omega v}{8\pi c}(i, 1)\frac{e^{-i\Omega r/c}}{r}. \tag{3.80}$$

3.3.2 The spectrum of the power radiated in a given direction

In the radiation field $\mathbf{E}_n = -Z_0\hat{\mathbf{r}} \times \mathbf{H}_n$, where $Z_0 = \sqrt{(\mu_0/\varepsilon_0)}$ is the vacuum impedance, and from (2.86) the power flux density averaged over a period is then

$$Z_0\hat{\mathbf{r}} \sum_{n=-\infty}^{\infty} \mathbf{H}_n \cdot \mathbf{H}_n^*.$$

The time-averaged power per unit solid angle radiated in a direction making θ with the normal to the plane of motion is therefore seen from (3.76) to be

$$\frac{e^2\Omega^2}{8\pi^2\varepsilon_0 c}\sum_{n=1}^{\infty} n^2\left[\frac{v^2}{c^2}J_n'^2\left(n\frac{v}{c}\sin\theta\right) + \cot^2\theta\, J_n^2\left(n\frac{v}{c}\sin\theta\right)\right]. \tag{3.81}$$

Bessel functions are copiously tabulated, and for numerical computation the formula (3.81) may well suffice as it stands. However, considerable insight into the nature of the formula can be gained with the help of some standard results from Bessel function theory.

The fundamental expansion, used in (3.78), is

$$J_n(z) = \sum_{m=0}^{\infty} \frac{(-)^m}{m!(m+n)!} \left(\frac{z}{2}\right)^{n+2m}, \tag{3.82}$$

giving also (for $n \neq 0$)

$$J_n'(z) = \sum_{m=0}^{\infty} \frac{(-)^m(n+2m)}{2m!(n+m)!} \left(\frac{z}{2}\right)^{n+2m-1}. \tag{3.83}$$

Much more useful here, however, is an asymptotic approximation for "large" n which can be written

$$J_n\left(\frac{n}{\cosh \alpha}\right) \sim \left(\frac{2}{n}\right)^{1/3} \mathrm{Ai}(n^{2/3}\zeta), \tag{3.84}$$

where

$$\zeta^{3/2} = \tfrac{3}{2}(\alpha - \tanh \alpha), \tag{3.85}$$

and Ai is the Airy function, defined, for example, by

$$\mathrm{Ai}(z) = \frac{1}{\pi} \int_0^{\infty} \cos\left(\tfrac{1}{3}t^3 + zt\right) dt. \tag{3.86}$$

Differentiation of (3.84) with respect to α gives

$$J_n'\left(\frac{n}{\cosh \alpha}\right) \sim -\left(\frac{2}{n}\right)^{1/3} \frac{\sinh \alpha}{n^{1/3}\,\zeta^{1/2}} \mathrm{Ai}'(n^{2/3}\zeta). \tag{3.87}$$

In applying (3.84) and (3.87) to (3.81) the parameter α enters, of course, through the identification of the respective Bessel function arguments; that is

$$\frac{1}{\cosh \alpha} = \frac{v}{c} \sin \theta. \tag{3.88}$$

The Airy function and its derivatives are tabulated, and are shown graphically in Fig. 3.3. The asymptotic approximation (3.84) improves in accuracy as n increases, but it is quite adequate for present purposes even down to $n = 1$; for example, to three figures

$$\mathrm{Ai}(0) = 3^{-2/3}(-\tfrac{1}{3})! = 0.355,$$

so that the values given by the right-hand side of (3.84) for $\alpha = 0$ and $n = 1$, 2 are 0.447 and 0.355 respectively, whereas $J_1(1) = 0.440$, $J_2(2) = 0.353$.

It is observed in (3.84) and (3.87) that for $n \geqslant 1$, alone relevant here, the

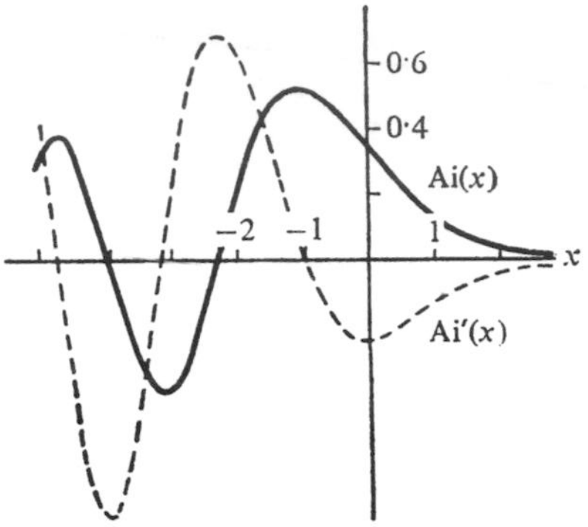

Fig. 3.3. The function Ai and its derivative Ai′.

argument of the Airy function can only be small by virtue of α being small. If the argument is "large" the asymptotic approximations

$$\mathrm{Ai}(z) \sim \frac{1}{2\sqrt{\pi}z^{1/4}}\,e^{-\frac{2}{3}z^{3/2}}, \qquad \mathrm{Ai}'(z) \sim -\frac{z^{1/4}}{2\sqrt{\pi}}\,e^{-\frac{2}{3}z^{3/2}}, \tag{3.89}$$

are available; they improve in accuracy as z increases, but remain useful down to values of z of the order of unity. Insertion of (3.89) into (3.84) and (3.87) yields the corresponding approximations

$$nJ_n\left(\frac{n}{\cosh\alpha}\right) \sim \frac{n^{1/2}}{2^{2/3}\sqrt{\pi}\zeta^{1/4}}\,e^{-\frac{2}{3}n\zeta^{3/2}} \sim \frac{n}{\sinh\alpha}\,J_n'\left(\frac{n}{\cosh\alpha}\right). \tag{3.90}$$

Against this background consider, first, the case $(v/c)\sin\theta \ll 1$. The ratio of argument to order of the Bessel functions in (3.81) is then small, and it is readily shown that the terms decrease rapidly with increasing n. For when α is large, (3.85) reads

$$\zeta^{3/2} \simeq \tfrac{3}{2}(\alpha - 1),$$

and the asymptotic forms (3.90) are valid and give

$$nJ_n\left(\frac{n}{\cosh\alpha}\right) \simeq \frac{n^{1/2}}{\sqrt{(2\pi)}3^{1/6}(\alpha-1)^{1/6}}\,e^{-n(\alpha-1)}. \tag{3.91}$$

The dependence on n exhibited by (3.91) is confirmed by noting that, with $z = n/\cosh\alpha$ and α large, the use of Stirling's formula

$$n! = \sqrt{(2\pi)}n^{n+\frac{1}{2}}\,e^{-n}[1 + O(1/n)] \tag{3.92}$$

puts n times the $m = 0$ term of the series (3.82) in the form

$$\frac{n^{1/2}}{\sqrt{(2\pi)}}\left(\frac{e}{2\cosh\alpha}\right)^n.$$

The conclusion is that the radiation is mainly at the fundamental frequency either for a slow particle ($v/c \ll 1$), or for a fast particle in directions close to the normal to the plane of the orbit ($\sin\theta \ll 1$). From the $n = 1$ term in the series (3.81), and using (3.78), the time-averaged power per unit solid angle radiated in the fundamental is

$$\frac{e^2\Omega^2 v^2}{32\pi^2\varepsilon_0 c^3}(1 + \cos^2\theta). \tag{3.93}$$

If this is integrated over a complete sphere the result is

$$\frac{e^2\Omega^2 v^2}{6\pi\varepsilon_0 c^3}, \tag{3.94}$$

which agrees with the general non-relativistic formula $e^2\dot{\mathbf{v}}^2/(6\pi\varepsilon_0 c^3)$ for the radiated power, since for uniform circular motion $|\dot{\mathbf{v}}| = \Omega v$. The motion is, of course, such that the "instantaneous" radiated power is the same at all times, and it must therefore be equal to the time-averaged power flux across a distant sphere.

Consider next the somewhat more taxing case $(v/c)\sin\theta \simeq 1$, which implies both a high-speed particle and points of observation close to the plane of motion. It is known from the discussion of §2.2.6 that the radiation is now strongly beamed in the direction of the velocity of the particle, so the effect is similar to that of a lighthouse; the observer sufficiently near the plane of motion sees a regular succession of identical pulses, the duration of each pulse being much shorter than the repetition period $2\pi/\Omega$. The frequency spectrum of the field is characterized by that of the individual pulse, and the dominant angular frequencies are much higher than Ω. This picture is confirmed and put on a quantitative basis by the Bessel function analysis, as is shown in what follows.

The parameter α is now small, and series expansions of (3.85) and (3.88) give the approximations

$$2^{-\frac{1}{3}}\zeta \simeq \tfrac{1}{2}\alpha^2 \simeq 1 - \frac{v}{c}\sin\theta. \tag{3.95}$$

The argument of the Airy function in (3.84) and (3.87) is approximately $(\frac{1}{2}n)^{2/3}\alpha^2$, and therefore small, of order unity, or large according to whether n is appreciably less than, comparable with, or appreciably greater than $1/\alpha^3$, respectively. It is clear from the graph of Ai(z) in Fig. 3.3 that n times the right-hand side of (3.84), regarded as a function of n, rises to a single maximum as n increases from unity, and then decays towards zero. It is convenient to write n_m for the value of n at which the maximum of $nJ_n(n/\cosh\alpha)$ is achieved. If $n\alpha^3$ is moderate or large the approximation (3.90) is valid, and takes the form

$$nJ_n\left(\frac{n}{\cosh\alpha}\right) \simeq \sqrt{\left(\frac{n}{2\pi\alpha}\right)}\, e^{-\frac{1}{3}\alpha^3 n}. \tag{3.96}$$

The right-hand side of (3.96) has a maximum at $n = 3/(2\alpha^3)$, and since for this value of n (3.96) does represent an adequate approximation to the Bessel function it follows that

$$n_m \simeq \frac{3}{2\alpha^3}. \tag{3.97}$$

From (3.90) the same result holds for $nJ_n'(n/\cosh \alpha)$. By substituting for α from (3.95) into (3.97) it is therefore concluded that the order of the harmonic that radiates most power in the direction θ is

$$n_m(\theta) \simeq \frac{3}{4\sqrt{2}}\left(1 - \frac{v}{c}\sin\theta\right)^{-3/2} \simeq \frac{3}{2}\left(1 - \frac{v^2}{c^2}\sin^2\theta\right)^{-3/2}. \tag{3.98}$$

The value (3.98) is, apart from numerical factors, much what would be expected from the investigation of §2.2.6 into the width of the beam thrown forward from a high-speed particle, for the angular width of beam was estimated as $\sqrt{[2(1 - v/c)]}$, which implies that each pulse of radiation seen by an observer at $\theta = \frac{1}{2}\pi$ is emitted by the particle during a time interval $\sqrt{[2(1 - v/c)]}/\Omega$ and is therefore detectable at the point of observation during a time interval $\sqrt{2}(1 - v/c)^{3/2}/\Omega$. The reciprocal of this latter interval is of the order of magnitude of the dominant frequency.

The relative dominance of harmonics in the vicinity of n_m is conveniently exhibited by taking $\theta = \frac{1}{2}\pi$. Then the power per unit solid angle, (3.81), is

$$\frac{e^2\Omega^2 v^2}{2^{5/3}\pi^2\varepsilon_0 c^3}\sum_{n=1}^{\infty} n^{2/3}\,\mathrm{Ai}'^2(2^{-2/3}\gamma), \tag{3.99}$$

where (3.87) and (3.95) have been used, and

$$\gamma = n^{2/3}(1 - v^2/c^2). \tag{3.100}$$

Since, from (3.98),

$$n_m(\tfrac{1}{2}\pi) \simeq \frac{3}{2(1 - v^2/c^2)^{3/2}} = n_M, \text{ say,} \tag{3.101}$$

the largest terms in the series (3.99) are those for which $2^{-2/3}\gamma$ is about $(\frac{3}{4})^{2/3}$. But $-\mathrm{Ai}'(z)$ varies, monotonically, only between 0.26 and 0.18 as z varies from 0 to $(\frac{3}{4})^{2/3}$, as seen in Fig. 3.3. For a high-energy particle, therefore, the relative contributions at $\theta = \frac{1}{2}\pi$ from the harmonics up to the vicinity of $n = n_M$ behave predominantly as $n^{2/3}$. Beyond $n = n_M$ the decline in the relative contributions is specified by $n^{2/3}\gamma^{1/2}\exp(-\frac{4}{3}\gamma^{3/2})$, as derived from (3.89).

3.3.3 The spectrum of the total power radiated

The total power radiated is displayed in spectral form by multiplying (3.81) by $2\pi \sin\theta$ and integrating over θ from 0 to π. Rather than attempt to perform this operation on the approximations to the terms of (3.81) derived in the previous section it is preferable to proceed some way with the exact analysis before introducing approximations. To this end it is convenient first to transform (3.81), by noting that

$$J_n^2(z) = \frac{1}{\pi}\int_0^\pi J_0(2z\sin\tau)\, e^{-2in\tau}\,d\tau, \tag{3.102}$$

a result easily derived from the representation (3.71), and that

$$J_n'^2(z) = \tfrac{1}{2}[J_{n-1}^2(z) + J_{n+1}^2(z)] - \frac{n^2}{z^2}J_n^2(z), \tag{3.103}$$

which follows at once from the familiar recurrence relations

$$J_n'(z) = \tfrac{1}{2}[J_{n-1}(z) - J_{n+1}(z)],$$

$$\frac{n}{z}J_n(z) = \tfrac{1}{2}[J_{n-1}(z) + J_{n+1}(z)].$$

Hence (3.81) can be written

$$\frac{e^2\Omega^2}{8\pi^3\varepsilon_0 c}\sum_{n=1}^{\infty} n^2\int_0^\pi\left[\frac{v^2}{c^2}\cos(2\tau) - 1\right]e^{-2in\tau}J_0\left(2n\frac{v}{c}\sin\theta\sin\tau\right)d\tau. \tag{3.104}$$

If (3.104) is multiplied by $2\pi\sin\theta$, and integrated over θ from 0 to π, the result

$$\int_0^\pi J_0(z\sin\theta)\sin\theta\,d\theta = 2\frac{\sin z}{z}$$

can be used, and the total power appears in the form

$$\frac{e^2\Omega^2}{4\pi^2\varepsilon_0 v}\sum_{n=1}^{\infty} n\int_0^\pi\left[\frac{v^2}{c^2}\cos(2\tau) - 1\right]\frac{\sin\left(2n\frac{v}{c}\sin\tau\right)}{\sin\tau}e^{-2in\tau}\,d\tau. \tag{3.105}$$

The expression (3.105) can be simplified by replacing $\cos(2\tau)$ in the integrand by $1 - 2\sin^2\tau$, and noting that (cf. (3.72))

$$\frac{1}{\pi}\int_0^\pi \sin\tau\sin\left(2n\frac{v}{c}\sin\tau\right)e^{-2in\tau}\,d\tau = -J_{2n}'\left(2n\frac{v}{c}\right),$$

and that

$$\frac{1}{\pi}\int_0^\pi\frac{\sin\left(2n\frac{v}{c}\sin\tau\right)}{\sin\tau}e^{-2in\tau}\,d\tau = \frac{2n}{\pi}\int_0^\pi\int_0^{v/c}\cos(2nz\sin\tau)\,dz\,e^{-2in\tau}\,d\tau$$

$$= 2n\int_0^{v/c}J_{2n}(2nz)\,dz.$$

The total power radiated is therefore seen to be

$$\frac{e^2\Omega^2}{4\pi\varepsilon_0 v}\sum_{n=1}^{\infty} n\left\{2\frac{v^2}{c^2}J'_{2n}\left(2n\frac{v}{c}\right)-2n\left(1-\frac{v^2}{c^2}\right)\int_0^{v/c}J_{2n}(2nz)\,dz\right\}. \quad (3.106)$$

At this stage the two cases $v/c \ll 1$ and $v/c \simeq 1$ are treated separately, and appropriate approximations to the exact expressions (3.106) are obtained.

For $v/c \ll 1$ the result is particularly simple, since the $n = 1$ term is dominant. If in this term $J_2(2v/c)$ and $J_2(2v/c)$ are replaced by the first terms in the respective expansions (3.82), (3.83) it is easily confirmed that the power radiated in the fundamental is indeed (3.94), as previously determined.

For $v/c \simeq 1$ the approximations (3.84) and (3.87) are invoked. By changing the variable of integration from z to $\tau = (2n)^{2/3}\zeta$, the former gives

$$2n\int_0^{v/c}J_{2n}(2nz)\,dz \simeq \int_\gamma^\infty \frac{2^{1/3}\,\zeta^{1/2}}{\sinh\alpha}\,\mathrm{Ai}(\tau)\,d\tau\,, \quad (3.107)$$

where ζ and α are related by (3.85), and the lower limit can be taken as

$$\gamma = n^{2/3}(1 - v^2/c^2) = \left[\frac{3n}{2n_M}\right]^{2/3} \quad (3.108)$$

because $v/c \simeq 1$. Furthermore, for a reason to be explained in a moment, the factor $\zeta^{1/2}/\sinh\alpha$ in the integrand in (3.107) can be replaced by its value when ζ is small, namely unity. The approximation (3.106) is thus

$$\frac{e^2\Omega^2}{4\pi\varepsilon_0 v}\sum_{n=1}^{\infty} n^{1/3}\left\{-\frac{2v^2}{c^2}\mathrm{Ai}'(\gamma)-\gamma\int_\gamma^\infty \mathrm{Ai}(\tau)\,d\tau\right\}. \quad (3.109)$$

This indicates that until n reaches the vicinity of n_M the relative power in the harmonics behaves in the main like $n^{1/3}$. For greater values of n the approximation

$$\gamma\int_\gamma^\infty \mathrm{Ai}(\tau)\,d\tau \simeq \int_\gamma^\infty \tau\mathrm{Ai}(\tau)\,d\tau$$

is valid, since the rapid decrease of $\mathrm{Ai}(\tau)$ when τ progresses much beyond unity ensures that the dominant contribution to the integrals comes from values of τ close to γ; and since

$$\mathrm{Ai}''(\tau) = \tau\mathrm{Ai}(\tau),$$

the result is simply

$$\gamma\int_\gamma^\infty \mathrm{Ai}(\tau)\,d\tau \simeq -\mathrm{Ai}'(\gamma),$$

giving spectral components in (3.109) proportional to

$$-n^{1/3}\,\mathrm{Ai}'(\gamma),$$

which in turn, from (3.89), are proportional to

$$n^{1/3}\,\gamma^{1/4}\,e^{-\frac{2}{3}\gamma^{3/2}}.$$

It should be noted that the two terms inside the curly bracket in (3.109) are of the same order of magnitude except when γ is small, that is $n \ll n_M$, in which case the second term is unimportant compared with the first. This explains why $\zeta^{1/2}/\sinh\alpha$ in (3.107) can be evaluated at small ζ. For the dominant contribution to the integral comes from values of τ either less than unity, or close to γ if γ exceeds unity; thus if γ is not small,

$$\zeta = (1 - v^2/c^2)\tau/(2^{2/3}\gamma)$$

does indeed remain small over the significant range of integration; and if γ is small, (3.107) is of no consequence anyway.

A very rough check on the predominantly $n^{1/3}$ behavior of the terms up to n_M is afforded by assessing the order of magnitude of the sum in (3.109) as $n_M \times n_M^{1/3}$; that is, essentially $(1 - v^2/c^2)^{-2}$, giving the order of magnitude of (3.109) as

$$\frac{e^2\Omega^2}{4\pi\varepsilon_0 v}\,\frac{1}{(1 - v^2/c^2)^2}.$$

This is to be compared, for $v/c \simeq 1$, with the exact result, known from formula (2.150) to be

$$\frac{e^2\Omega^2}{6\pi\varepsilon_0 c}\,\frac{v^2/c^2}{(1 - v^2/c^2)^2}. \tag{3.110}$$

An exact summation of (3.106) must, of course, give (3.110). The verification requires the use of the series summations

$$\sum_{n=1}^{\infty} nJ'_{2n}(2nx) = \frac{x}{2(1 - x^2)^2},$$

$$\sum_{n=1}^{\infty} n^2 J_{2n}(2nx) = \frac{x^2(1 + x^2)}{2(1 - x^2)^4} = \frac{1}{6}\frac{d}{dx}\left[\frac{x^3}{(1 - x^2)^3}\right],$$

both of which are readily derived, with the help of Bessel's equation, by successive differentiation with respect to x of the Kapteyn series summation

$$\sum_{n=1}^{\infty} \frac{1}{n^2} J_{2n}(2nx) = \tfrac{1}{2}x^2.$$

PROBLEMS

1. Show that the field (3.41) can be expressed in terms of the vector potential $\mathbf{A}^\omega = (A^\omega, 0, 0)$, where

$$A^\omega = -\frac{ie\mu_0}{8\pi}e^{-i\omega x/v}H_0^{(2)}\left[\frac{\omega}{c}\sqrt{(\mu^2 - c^2/v^2)}\rho\right],$$

with $\rho = \sqrt{(y^2 + z^2)}$.

Write down the asymptotic form of A^ω for $\rho \to \infty$. Derive corresponding asymptotic expressions for $\mathbf{E}^\omega$ and $\mathbf{H}^\omega$, and hence obtain (3.51).

2. Calculate the energy W radiated per unit length of path by a point charge e moving with constant speed v along a straight line in a medium whose refractive index μ is given by

$$\mu^2 = 1 + \omega_p^2/(\omega_0^2 - \omega^2).$$

Show that, in the ultra relativistic limit $v \to c$,

$$W = \frac{e^2\mu_0}{8\pi}\omega_p^2 \log(1 + \omega_0^2/\omega_p^2).$$

3. The refractive index of an idealized medium is unity for $\omega > \omega_0$, and has a value μ, greater than unity and independent of ω, for $\omega < \omega_0$. A highly relativistic electron (energy much greater than m_0c^2), free from external forces, moves in the medium. Estimate the time taken to lose through Čerenkov radiation energy of amount m_0c^2.

Confirm that, with $\omega_0 \sim 10^{16}$, the effect of radiation associated specifically with the deceleration is quite negligible.

4. A point charge executes the uniform circular motion described in §3.3.1. Use the formula (2.121) to find the spherical polar components of the far field in terms of the retarded time. Confirm that a frequency analysis then gives the results described by (3.67), (3.69) and (3.70).

5. A point charge e executes the uniform helical motion

$$x = \frac{v}{\Omega}\cos(\Omega t), \qquad y = \frac{v}{\Omega}\sin(\Omega t), \qquad z = ut.$$

By applying a Lorentz transformation to the far field (3.75), (3.76), show that the components of $\mathbf{H}$ at a great distance from the particle are given by

$$(H_\theta, H_\phi) = \sum_{-\infty}^{\infty}(H_{n\theta}, H_{n\phi})e^{-in\Omega_1 t},$$

with

$$(H_{n\theta}, H_{n\phi}) = \frac{e\Omega_1}{4\pi}\frac{ne^{\frac{1}{2}in\pi}}{1-(u/c)\cos\theta}\left\{\frac{v}{c}J_n'\left(\frac{nv\sin\theta}{c-u\cos\theta}\right),\right.$$
$$\left. -i\frac{\cos\theta - u/c}{\sin\theta}J_n\left(\frac{nv\sin\theta}{c-u\cos\theta}\right)\right\}\frac{e^{-in\Omega_1 r/c}}{r},$$

where

$$\Omega_1 = \Omega/[1-(u/c)\cos\theta],$$

and r, θ, ϕ ($= 0$) are the polar coordinates of the point of observation measured from $x = y = 0$, $z = u(t - r/c)$.

6. A point charge executes rectilinear simple harmonic motion. Obtain a Bessel function series for the radiation field and for the time-averaged power per unit solid angle radiated in a direction making an angle θ with the line of motion. Discuss the relative magnitudes of the terms of the series, and show in what sense the results agree with those for an elementary dipole.

REFERENCES

For a general account of Čerenkov radiation and its application to physical phenomena there is the book

J. V. JELLEY, *Čerenkov Radiation and its Applications*, Pergamon (1958).

The theory for a charged particle moving uniformly along an infinite rectilinear path is due to

I. M. FRANK and IG. TAMM, *Dokl. Akad. Nauk, SSSR*, **14**, 109 (1937).

and the case of finite path length to

IG. TAMM, *Zh. fiz. SSSR*, **1**, 439 (1939).

Gyro radiation is fully worked out in

G. A. SCHOTT, *Electromagnetic Radiation and the Mechanical Reactions Arising From It*, Cambridge University Press (1912).

Different versions of, and extensions to, the analysis are to be found in

J. SCHWINGER, *Phys. Rev.*, **75**, 1912 (1949).
K. C. WESTFOLD, *Astrophys. J.*, **130**, 241 (1959).
E. LE ROUX, *Annales d'Astrophysique*, **24**, 71 (1961).

For Bessel functions

G. N. WATSON, *Theory of Bessel Functions*, Cambridge University Press (1944)

is more than adequate for most purposes; but for the asymptotic relation to the Airy functions see

F. J. W. OLVER, *Phil. Trans. Roy. Soc.* A **247**, 328 (1954),

and for information about, and tables of, the latter, see

J. C. P. MILLER, The Airy integral, *Brit. Ass. Math. Tables* Part vol. **B**, Cambridge University Press (1946).

The nature and scope of recent work on radiation from a point charge spiraling in a plasma in a magnetic field can be gauged from the following references:

J. F. MCKENZIE, *Proc. Phys. Soc.*, **84**, 269 (1964).
H. B. LIEMOHN, *Radio Sci.*, **69**, 741 (1965).
V. N. MANSFIELD, *Astrophys. J.*, **147**, 672 (1967).

Accounts of Čerenkov and gyro radiation are given in

J. D. JACKSON, *Classical Electrodynamics*, Wiley (1962).

D. S. JONES, *Theory of Electromagnetism*, Pergamon (1964).

L. D. LANDAU and E. M. LIFSHITZ, *Classical Theory of Fields*, Addison-Wesley (1962).

B. O. LEHNERT, *Dynamics of Charged Particles*, North Holland (1964).

W. K. H. PANOFSKY and M. PHILLIPS, *Classical Electricity and Magnetism* 2nd edn, Addison-Wesley (1962).

CHAPTER 4

DYNAMICAL MOTION OF A POINT CHARGE

4.1 EQUATIONS OF MOTION

In this chapter certain aspects are considered of the motion in a vacuum of a point charge under the influence of an electromagnetic field. The problem that presents itself is that of solving the equations of motion given in relativistic form by (2.187), namely

$$\frac{d}{dt}\left\{\frac{m_0\mathbf{v}}{\sqrt{(1 - v^2/c^2)}}\right\} = e(\mathbf{E} + \mathbf{v} \times \mathbf{B}). \tag{4.1}$$

The problem becomes, in principle, straightforward if it be assumed that **E**, **B** appearing in the right-hand side of (4.1) is a given field; the difficulties are then of the purely technical kind that can also arise in traditional particle mechanics with mechanical forces. The assumption is often a legitimate approximation, and most of the work presented here is in this context. In fact, however, the charge itself also contributes to the electromagnetic field, and its contribution depends on its motion. The inclusion of this " self-force " presents quite fundamental difficulties because of the infinities associated with the concept of a point charge. Attempts to describe the motion of an electron can perhaps be classified in terms of the following alternatives: (a) abandon the idea that the electron is a point charge, and give it some internal structure; (b) introduce into the classical mathematical framework some formalism that succeeds in discarding the infinities; (c) abandon the idea that the electron can be described in any fundamental way by classical theory, and insist that appeal be made to quantum laws. The procedure (b) has been given a good deal of attention, and has led to an equation of motion which at least in certain contexts is acceptable theoretically and is in general agreement with practice. A derivation of this equation is given later in the chapter. Any discussion of the alternatives (a) and (c) is outside the scope of this book.

In attempting to solve (4.1) it may be helpful to make use of the associated energy equation

$$\frac{d}{dt}\left\{\frac{m_0c^2}{\sqrt{(1 - v^2/c^2)}}\right\} = e\mathbf{E}\cdot\mathbf{v}. \tag{4.2}$$

If the electric field is purely static, so that $\mathbf{E} = -\operatorname{grad} \phi$, this gives

$$\frac{m_0 c^2}{\sqrt{(1 - v^2/c^2)}} + e\phi = \text{constant}. \tag{4.3}$$

In the important special case $\mathbf{E} = 0$ evidently the speed v is constant. Since a time-varying magnetic field cannot exist without an associated electric field, this latter case arises only when the charge moves in a magnetostatic field. When v is constant $\dot{\mathbf{v}}$ is along the principal normal to the trajectory and of magnitude v^2/ρ_p, where ρ_p is the principal radius of curvature; hence

$$\rho_p = \frac{m_0 v}{eB \sin\theta \sqrt{(1 - v^2/c^2)}}, \tag{4.4}$$

where θ is the angle between $\mathbf{v}$ and $\mathbf{B}$, commonly called the *pitch angle*.

Often it is not necessary to take account of relativistic effects, and the equations of motion and energy equation can then be taken as

$$m\dot{\mathbf{v}} = e(\mathbf{E} + \mathbf{v} \times \mathbf{B}), \tag{4.5}$$

$$\frac{d}{dt}(\tfrac{1}{2}mv^2) = e\mathbf{E}\cdot\mathbf{v}. \tag{4.6}$$

It may be noted that in the case $\mathbf{E} = 0$, where the field is purely magnetostatic, the relativistic equations are the same as the non-relativistic, except only that m in the latter is replaced by the constant $m_0/\sqrt{(1 - v^2/c^2)}$.

It is well known that the equations of motion can be cast into two alternative forms, familiar in classical mechanics as the Lagrangian and Hamiltonian forms. In the present calculations no extensive use is made of these forms, but it is convenient to state them here. In the non-relativistic treatment the usual form of Lagrange's equations is equivalent to (4.5) if the Lagrangian is taken as

$$\mathscr{L} = \tfrac{1}{2}mv^2 + e\mathbf{v}\cdot\mathbf{A} - e\phi, \tag{4.7}$$

where ϕ and $\mathbf{A}$ are the scalar and vector potentials of the electromagnetic field. Likewise the usual form of Hamilton's equations is equivalent to (4.5) if the Hamiltonian is taken as

$$H = \frac{1}{2m}(\mathbf{p} - e\mathbf{A})^2 + e\phi, \tag{4.8}$$

where

$$\mathbf{p} = m\mathbf{v} + e\mathbf{A}$$

is the generalized momentum.

To recapture the relativistic equation (4.1) the Lagrangian has to be

$$\mathscr{L} = -m_0 c^2 \sqrt{(1 - v^2/c^2)} + e\mathbf{v}\cdot\mathbf{A} - e\phi. \tag{4.9}$$

The corresponding relativistic Hamiltonian is

$$H = c\sqrt{\{m_0^2c^2 + (\mathbf{p} - e\mathbf{A})^2\}} + e\phi,$$

with

$$\mathbf{p} = \frac{m_0\mathbf{v}}{\sqrt{(1 - v^2/c^2)}} + e\mathbf{A}. \tag{4.10}$$

In what follows consideration is given first to exact solutions of (4.5) for some simple special cases, starting with uniform and constant **E**, **B**. Next the situation is treated in which **E** and **B** are arbitrary but "slowly varying." Then some consideration is given to relativistic motions. Finally, the question of the self-force is examined.

4.2 NON-RELATIVISTIC MOTION IN SIMPLE SPECIAL CASES

4.2.1 Uniform static fields

When both **E** and **B** are constant in time and uniform in space, that is, each vector always and everywhere has the same magnitude and same direction, the full solution in the non-relativistic approximation is easily obtained.

First it is noted that in a uniform electrostatic field **E**, with no magnetic field, the particle simply travels with constant acceleration $e\mathbf{E}/m$.

Next, motion in a uniform magnetostatic field **B**, with no electric field, is considered. The speed v is then constant; so is $v_{\|}$, since the force is at right angles to **B**, and so also therefore is the magnitude $v_{\perp}$ of $\mathbf{v}_{\perp}$, where suffices $\|$ and $\perp$ are used to denote components parallel and perpendicular to **B**. Now for motion perpendicular to **B** the force is $ev_{\perp}B$ at right angles to **v**, and since the acceleration at right angles to **v** is $v_{\perp}^2/\rho$, where ρ is the radius of curvature of the projection of the path on to a plane perpendicular to B, it follows that

$$\rho = \frac{mv_{\perp}}{eB}, \tag{4.11}$$

which is constant. Of course, (4.11) is just a special case of the non-relativistic form of (4.4), with $\rho_p \sin^2\theta = \rho$, $v \sin\theta = v_{\perp}$.

The general motion is therefore an arbitrary constant velocity along **B**, superposed on circular motion perpendicular to **B** described with constant angular velocity

$$\Omega = eB/m. \tag{4.12}$$

It is readily checked that a positively charged particle spirals in the sense of a left-handed screw about the magnetic field direction, a negatively charged particle in a right-handed sense.

The expressions Ω and ρ are commonly called the (angular) gyro (or Larmor) frequency and radius. In this non-relativistic approximation Ω is independent of the energy of the particle; for an electron in the earth's magnetic field of, say, 0.5 gauss, it is about 8.8×10^6 rad sec^{-1}. But ρ is proportional to the square root of the energy; for a 10 eV electron in a field of 0.5 gauss it is about 0.21 m.

If **E** and **B** are both present, and each is constant and uniform, the motion can be obtained in the following way. Along **B** there is constant acceleration $eE_{\|}/m$. Perpendicular to **B**, the equation of motion (4.5) is

$$m\dot{\mathbf{v}}_{\perp} = e\mathbf{E}_{\perp} + e\mathbf{v}_{\perp} \times \mathbf{B}. \tag{4.13}$$

It is now observed that (4.13) clearly admits as a *possible* motion that with uniform velocity **u** such that

$$\mathbf{E}_{\perp} + \mathbf{u} \times \mathbf{B} = 0, \tag{4.14}$$

giving, for motion perpendicular to **B**,

$$\mathbf{u} = \frac{\mathbf{E} \times \mathbf{B}}{B^2}. \tag{4.15}$$

In (4.13) it is convenient to single out the velocity **u** explicitly by writing

$$\mathbf{v}_{\perp} = \mathbf{u} + \mathbf{v}'_{\perp}, \tag{4.16}$$

say. The substitution of (4.16) into (4.13) then gives

$$m\dot{\mathbf{v}}'_{\perp} = e\mathbf{v}'_{\perp} \times \mathbf{B},$$

and this equation for $\mathbf{v}'_{\perp}$ is precisely what would obtain for $\mathbf{v}_{\perp}$ in the case $\mathbf{E} = 0$.

The general motion can therefore be visualized simply by observing that the *additional* motion arising from the electric field **E** is a constant acceleration $eE_{\|}/m$ along **B**, and a constant velocity perpendicular to both **E** and **B**, in the sense of a right-handed screw from **E** to **B**, of magnitude $E_{\perp}/B$. The projection of the path of the particle on a plane perpendicular to **B** is a cycloid; it has one of the forms sketched in Fig. 4.1, (a) if $E_{\perp}/B$ exceeds the speed of the circular motion, (b) in the converse case.

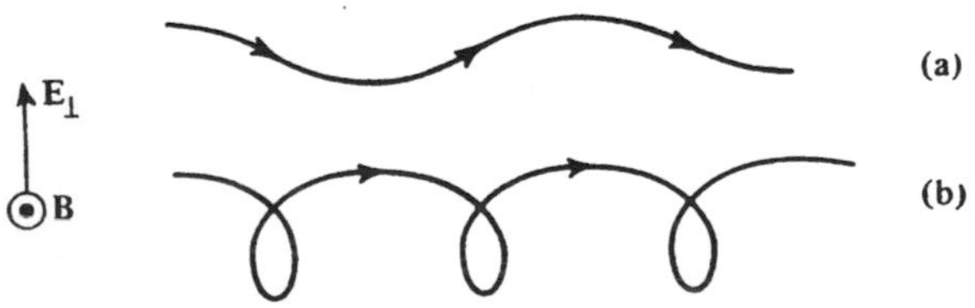

Fig. 4.1. Projection on a plane perpendicular to **B** of the path of a positively charged particle in constant uniform electric and magnetic fields.

In practice, of course, the fields **E** and **B** will vary to some extent. In many situations, however, the departures from constancy and uniformity are relatively quite small; some sort of perturbation theory can then be used, and the local gyro-frequency eB/m and "drift" velocity $\mathbf{E} \times \mathbf{B}/B^2$ naturally feature prominently in the analysis. Before proceeding to a discussion of this sort in §§4.3, 4.4, a few cases are presented in which the fields are not constant and uniform, but are simple enough to yield a straightforward exact analysis. These cases provide concrete examples of the general theory.

4.2.2 Uni-directional magnetostatic field

In this section the case is presented in which there is no electric field, and in which the magnetic field is constant in time and everywhere has the same direction but not necessarily the same magnitude. In rectangular Cartesian coordinates

$$\mathbf{B} = [0, 0, B(x, y)] \tag{4.17}$$

say, where the requirement div $\mathbf{B} = 0$ ensures that B cannot depend on z. Unless **B** is uniform curl **B** does not vanish and there is a current density associated with the field.

Clearly now both v and v_z are constant, and so therefore is

$$v_\perp = \sqrt{(v_x^2 + v_y^2)}.$$

Furthermore, the projection of the path of the particle on the xy-plane has radius of curvature

$$\rho(x, y) = \frac{mv_\perp}{eB}. \tag{4.18}$$

The curvature of the projected path is therefore a constant multiple of the magnetic induction. This fact can lead to a general appreciation of the nature of the path, and could be used for accurate computation.

To investigate the possibility of an analytic solution it is noted that the equations of motion are

$$\dot{v}_x = \Omega v_y, \qquad \dot{v}_y = -\Omega v_x, \tag{4.19}$$

where the local gyro-frequency

$$\Omega = eB/m \tag{4.20}$$

is now a function of x and y, and only motion in the xy-plane need be considered, with $v_\perp = v$.

Since equations (4.19) can be written

$$\frac{dv_x}{dy} = \Omega, \quad \frac{dv_y}{dx} = -\Omega, \tag{4.21}$$

it is clear that a solution by quadratures is possible if Ω depends on only one of x or y. For if it be supposed that B is independent of y, say, the second equation of (4.21) gives

$$v_y = -\int \Omega(x)\,dx, \tag{4.22}$$

so that $v_x = \sqrt{(v^2 - v_y^2)}$ is a known function of x, and t is given in terms of x by a further integration. Alternatively, the path can be determined directly from the fact that

$$\frac{dy}{dx} = \frac{v_y}{v_x} = \frac{v_y}{\sqrt{(v^2 - v_y^2)}} \tag{4.23}$$

is a known function of x.

The same results follow readily from the Lagrangian formulation. The vector potential

$$\mathbf{A} = [0, A(x), 0]$$

corresponds to magnetic induction

$$\mathbf{B} = \left(0, 0, \frac{dA}{dx}\right);$$

and since (4.7) then reads

$$\mathscr{L} = \tfrac{1}{2}m(\dot{x}^2 + \dot{y}^2 + \dot{z}^2) + e\dot{y}A(x),$$

the y-coordinate is ignorable and the corresponding Lagrange equation has first integral

$$\frac{\partial \mathscr{L}}{\partial v_y} = 0,$$

that is

$$mv_y + eA = \text{constant}, \tag{4.24}$$

which is the same as (4.22).

The example leading to the most elementary analytic details seems to be

$$\Omega = \frac{\text{constant}}{x^2}. \tag{4.25}$$

For if $-mu$ and ux_0 are chosen as the respective constants in (4.24) and (4.25), then these equations give

$$v_y = -u(1 - x_0/x), \tag{4.26}$$

from which (4.23) can evidently be integrated in terms of elementary functions. Without pursuing the details, the topology of the path can be ascertained from the fact that v_y^2 cannot exceed the constant v^2. Stated algebraically

$$(v^2 - u^2)\left(x + \frac{ux_0}{v - u}\right)\left(x - \frac{ux_0}{v + u}\right) \geqslant 0. \tag{4.27}$$

Since (4.25) is symmetric in x, and the particle cannot cross the plane $x = 0$, there is no loss of generality in confining attention to positive values of x. Let it also be assumed that $ux_0 > 0$, corresponding to a magnetic field in the positive z-direction. Then if $|u| < v$, (4.27) shows that the path goes to infinity and that

$$x > \frac{ux_0}{v + u}; \tag{4.28}$$

if also $u > 0$ (implying $x_0 > 0$) then (4.28) is less than x_0, the value of x for which v_y vanishes, so that the path is somewhat as sketched in Fig. 4.2(a); whereas if $u < 0$ ($x_0 < 0$) the path is as in Fig. 4.2(b). On the other hand, if $|u| > v$, u must be positive, and the particle is confined to the range

$$\frac{ux_0}{u + v} < x < \frac{ux_0}{u - v}; \tag{4.29}$$

this range contains $x = x_0$, where v_y vanishes, and the nature of the path is indicated in Fig. 4.2(c).

Another comparatively simple example of some interest is that in which the magnetic field varies linearly with x. If, say,

$$\Omega = \frac{2|u|}{x_0^2} x, \tag{4.30}$$

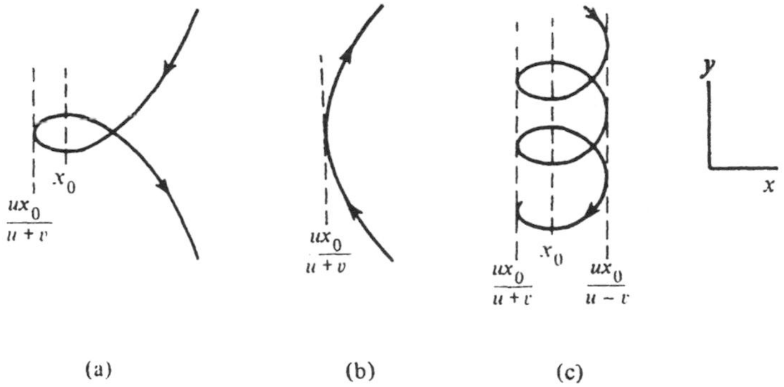

Fig. 4.2. Paths of a charged particle with speed v in the magnetostatic field

$$\mathbf{B} = \left[0, 0, \frac{m}{e}\frac{ux_0}{x^2}\right],$$

where $ux_0 > 0$. (a) $v > u > 0$; (b) $v > -u > 0$; (c) $u > v$.

then (4.22) can be written

$$v_y = u - \frac{|u|}{x_0^2} x^2, \tag{4.31}$$

where u is an arbitrary constant and the constant x_0 can be taken as positive. The solution of (4.23) can be expressed in terms of elliptic integrals, but again the general picture is most easily obtained from the fact that v_y^2 cannot exceed v^2. For this now reads

$$(v^2 - u^2)\left[1 + \frac{|u|\,x^2}{(v-u)x_0^2}\right]\left[1 - \frac{|u|\,x^2}{(v+u)x_0^2}\right] \geqslant 0, \tag{4.32}$$

which shows that three cases can be distinguished as follows. If $|u| < v$, then

$$x^2 \leqslant \frac{v+u}{|u|} x_0^2 \tag{4.33}$$

and the particle is confined between the two planes $x = \pm\sqrt{[(v+u)/|u|]}x_0$; if, also, $u > 0$, the range of x includes $\pm x_0$, the values of x for which v_y vanishes, so that the path is as indicated in Fig. 4.3(a); whereas, if $u < 0$, v_y never vanishes and the type of path is shown in Fig. 4.3(b). On the other hand, $|u| > v$ is only possible when $u > 0$, and (4.32) implies

$$\frac{u+v}{|u|}x_0^2 > x^2 > \frac{u-v}{|u|}x_0^2, \tag{4.34}$$

with the path as in Fig. 4.3(c).

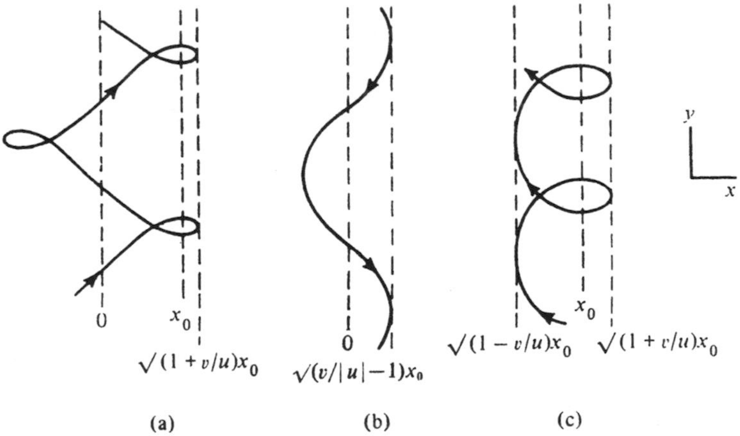

Fig. 4.3. Paths of a charged particle with speed v in the magnetostatic field $\mathbf{B} = \left[0, 0, \frac{m}{e}\frac{2|u|}{x_0^2}x\right]$. (a) $v > u > 0$; (b) $v > -u > 0$; (c) $u > v$.

Suppose, now, that instead of being a function of x only the field (4.17) is a function only of $r = \sqrt{(x^2 + y^2)}$, that is, there is cylindrical symmetry about the z-axis. To conclude this sub-section it is noted that this is a second case where solution by quadratures is possible, and a specific example is considered.

In cylindrical polar coordinates r, θ, z, with

$$\mathbf{B} = [0, 0, B(r)], \tag{4.35}$$

the equation of motion in the θ-direction is

$$\frac{1}{r}\frac{d}{dt}(r^2\dot{\theta}) = -\Omega\dot{r},$$

giving

$$v_\theta = r\dot{\theta} = -\frac{1}{r}\int r\Omega(r)\,dr. \tag{4.36}$$

Then

$$\dot{r}^2 = v_r^2 = v^2 - v_\theta^2$$

is a known function of r, and t is given in terms of r by a further integration. Or, again, the equation of the path is given directly by integrating

$$r\frac{d\theta}{dr} = \frac{v_\theta}{\sqrt{v^2 - v_\theta^2)}}. \tag{4.37}$$

For the Lagrangian formulation the vector potential in cylindrical polar coordinates can be taken as

$$\mathbf{A} = [0, A(r), 0], \tag{4.38}$$

where

$$B = \frac{1}{r}\frac{d}{dr}(rA). \tag{4.39}$$

Then

$$\mathscr{L} = \tfrac{1}{2}m(\dot{r}^2 + r^2\dot{\theta}^2 + \dot{z}^2) + eA(r)r\dot{\theta},$$

so that θ is ignorable and

$$\frac{\partial\mathscr{L}}{\partial\dot{\theta}} = \text{constant},$$

which is readily seen to be identical with (4.36).

An example for which the analytic details are elementary is $A =$ constant. If the constant is written as mu/e, then

$$\Omega = u/r, \tag{4.40}$$

and (4.36) gives

$$v_\theta = -u(1 - r_0/r), \tag{4.41}$$

where u can be taken positive, but r_0 may be positive or negative. The analysis is evidently virtually the same as that in the first example considered in this sub-section, which was based on equation (4.26). The requirement that v_θ^2 cannot exceed the constant v^2 shows that there are three types of path, as indicated in Fig. 4.4. When $u < v$ the paths go to infinity, ultimately approaching equi-angular spirals; cases (a) and (b) correspond to $r_0 > 0$ and $r_0 < 0$ respectively. When $u > v$, r_0 must be positive, and the paths are confined as in (c).

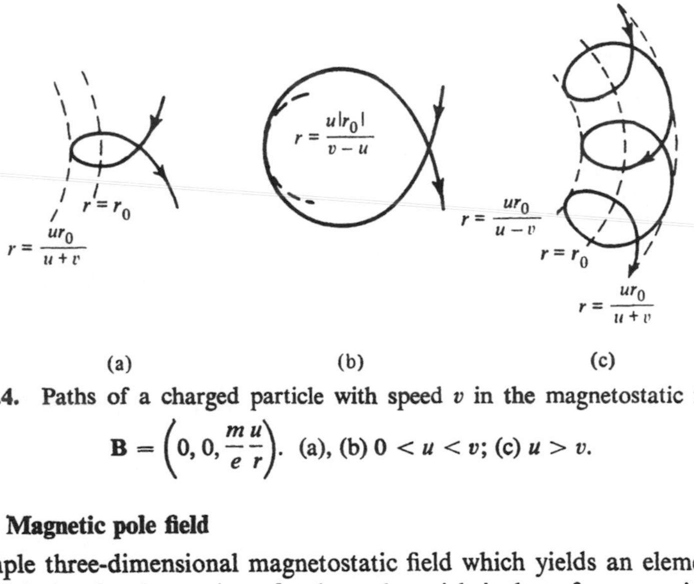

Fig. 4.4. Paths of a charged particle with speed v in the magnetostatic field $\mathbf{B} = \left(0, 0, \frac{m}{e}\frac{u}{r}\right)$. (a), (b) $0 < u < v$; (c) $u > v$.

4.2.3 Magnetic pole field

A simple three-dimensional magnetostatic field which yields an elementary exact solution for the motion of a charged particle is that of a magnetic pole. Suppose that

$$\mathbf{B} = B_0 r_0^2 \frac{\mathbf{r}}{r^3}, \tag{4.42}$$

where $\mathbf{r}$ is the radius vector from the origin, and r_0 and B_0 are positive constants of the dimensions of length and magnetic induction respectively. Then the equation of motion is

$$\dot{\mathbf{v}} = r_0^2 \Omega \frac{\mathbf{v} \times \mathbf{r}}{r^3}, \tag{4.43}$$

where Ω is the constant eB_0/m_0; and it so happens that this form readily yields a vector integral. For if the vector product with $\mathbf{r}$ is taken, and the resulting triple vector product on the right-hand side expanded and $\mathbf{r}\cdot\mathbf{v} = r\dot{r}$ used, it is seen that

$$\begin{aligned}\frac{d}{dt}(\mathbf{r} \times \mathbf{v}) &= r_0^2\Omega\left(\frac{\mathbf{v}}{r} - \frac{\dot{r}}{r^2}\mathbf{r}\right)\\ &= r_0^2\Omega\frac{d}{dt}\left(\frac{\mathbf{r}}{r}\right).\end{aligned}$$

Thus

$$\mathbf{r} \times \mathbf{v} = r_0^2\Omega\left(\frac{\mathbf{r}}{r} - \mathbf{k}\right), \tag{4.44}$$

where $\mathbf{k}$ is a constant vector.

In spherical polar coordinates r, θ, ϕ, with the polar axis taken along $\mathbf{k}$, the radial component of (4.44) gives

$$1 - k\cos\theta = 0,$$

so that θ is a constant, say

$$\theta = \theta_0. \tag{4.45}$$

It is now convenient to exercise the freedom of choice of r_0 and take $k\sin\theta_0 = v/(r_0\Omega)$, where v is the constant speed. The θ-component of (4.44) then takes the form

$$r^2\dot{\phi}\sin\theta_0 = -r_0 v. \tag{4.46}$$

The ϕ-component of (4.44) gives nothing; but

$$\dot{r}^2 + r^2\dot{\phi}^2\sin^2\theta_0 = v^2$$

leads to

$$\dot{r}^2 = v^2(1 - r_0^2/r^2),$$

and if the time origin $t = 0$ is taken at the instant $r = r_0$ this integrates to

$$r^2 = r_0^2 + v^2t^2. \tag{4.47}$$

The particle therefore tracks round the surface of a right circular cone of semi-vertical angle $\tan^{-1}[v/(r_0\Omega)]$, coming in from infinity to reach the minimum value r_0 of r at $t = 0$ and then receding to infinity again. Since, with v constant, the acceleration is along the principal normal to the path, equation (4.53) shows that this direction is normal to the surface of the cone, and the path is therefore a geodesic.

4.2.4 Uniform time-varying magnetic field

When the magnetic field varies with time an electric field must also be present in order that the equation curl $\mathbf{E} = -\dot{\mathbf{B}}$ be satisfied. To consider as simple a case as possible suppose that **B** at every point is parallel to the z-axis and of the same magnitude B, where B varies with time. The electric field is not, of course, uniquely determined by **B**, since it depends also on the current density, which must be present in some form; but it can be taken as

$$\mathbf{E} = (0, -\dot{B}x, 0). \tag{4.48}$$

The equation of motion (4.5) now states

$$m\ddot{x} = eB\dot{y}, \tag{4.49}$$
$$m\ddot{y} = -e\dot{B}x - eB\dot{x}; \tag{4.50}$$

and to the motion thus specified can be added an arbitrary constant velocity in the z-direction. Integration of (4.50) gives

$$\dot{y} = -\Omega x + u,$$

where $\Omega = eB/m$ and u is a constant; and then, from (4.49),

$$\ddot{x} + \Omega^2 x = u\Omega. \tag{4.51}$$

If the particle is moving perpendicular to the y-axis when crossing it, $u = 0$ and

$$\ddot{x} + \Omega^2 x = 0. \tag{4.52}$$

Equation (4.52) is of a form familiar in many branches of mathematical physics, and various exact solutions are available in terms of special functions for certain particular variations of Ω^2 with time. An important example is discussed in §4.4.4, but for the moment we merely note the so-called WKBJ approximate solution, valid when Ω is sufficiently "slowly varying". With appropriate choice of time origin it can be written

$$x = a\sqrt{\left(\frac{\Omega_0}{\Omega}\right)} \cos\left(\int_0^t \Omega\, dt\right), \tag{4.53}$$

where at $t = 0$, $\Omega = \Omega_0$ and $x = a$. Since $\dot{y} = -\Omega x$,

$$y = -a\sqrt{\left(\frac{\Omega_0}{\Omega}\right)} \sin\left(\int_0^t \Omega\, dt\right) \tag{4.54}$$

to the same order of approximation.

4.3 NON-RELATIVISTIC MOTION IN A SLOWLY VARYING FIELD: INTRODUCTION

4.3.1 The guiding center

The action of the magnetic force $e\mathbf{v} \times \mathbf{B}$, which imposes curvature on the trajectory of the particle, is of great importance in many physical situations, both in laboratory experiments and techniques and in naturally occurring phenomena. In the search for methods for the mathematical analysis of motions under the influence of fields other than those few simple geometries that are easily tractable, one approach is to single out for examination the cases in which, broadly speaking, the representative period $2\pi m/(eB)$ and the representative radius of curvature $mv/(eB)$ are very small compared with the overall time and space scales, respectively, of the motion of the particle. Such cases are particularly pertinent in the study of plasmas, for example in the problem of containment by a magnetic field and in the derivation of the hydrodynamic description from consideration of individual particle trajectories. The purpose of this section is to introduce the general concept of "slowly varying" fields and then obtain the approximate working formulas applicable to the situation envisaged.

Consider first a charged particle moving in a purely magnetostatic field. The field is non-uniform, but its space variation is assumed so slight that a picture of the particle proceeding along the field lines and at the same time encircling them at the local gyro-frequency is in some sense a valid approximation. The requisite restriction on the variation of the magnetic field is, in fact, that throughout the duration of a local gyro period the field at the particle shall not have changed appreciably; that is to say, the change of direction of the field, in radians, and the ratio of the change of magnitude to the magnitude, both remain small. The restriction has to be somewhat more severe than simply that the scale of the field variation is large compared with the local gyro radius, in order to accommodate the possibility that the velocity along the field greatly exceeds the gyrational velocity.

If in this picture the oscillatory part of the motion, at the gyro-frequency and at harmonics thereof, is conceived as averaged out, there is left the motion of what is often called the *guiding center*. This somewhat indirect definition of the guiding center is the basis of the mathematical formulation given subsequently, but as an aid to visualization it is worth remarking that if components of the motion involving only the second and higher harmonics of the gyro-frequency are ignored, the guiding center can be defined by the statement that its position vector relative to the particle is

$$\mathbf{v} \times \hat{\mathbf{B}}/\Omega, \tag{4.55}$$

where $\Omega = eB/m$ and $\hat{\mathbf{B}} = \mathbf{B}/B$ are evaluated at the particle; and to the same approximation the motion of the particle can then be described as the local gyro motion superposed on the motion of the guiding center. The objective of the theory, essentially, is to find a direct prescription for the motion of the guiding center.

In a uniform magnetostatic field the guiding center, of course, moves with constant speed along a field line. In a slowly varying magnetostatic field not only may the speed of the guiding center change gradually, but its motion along the field lines may be supplemented by a subsidiary slow motion across the field lines. It is not difficult to get a qualitative impression of the sort of thing that is going to happen in various circumstances. Consider, for example, the effect of a gradual variation of B in a direction perpendicular to $\mathbf{B}$; the gyro spiral tightens up whenever the particle moves into the region of increasing B, and the effect, illustrated in Fig. 4.5, is evidently a slow drift of the guiding center in the direction perpendicular to both $\mathbf{B}$ and grad B. For the special cases worked out in §4.2.2, Figs. 4.2(c), 4.3(c), 4.4(c) show, with exaggerated drift, the type of trajectory predicted by the present argument; whereas, for the trajectories of (a) and (b) of the same figures conditions for the field to be slowly varying are palpably violated.

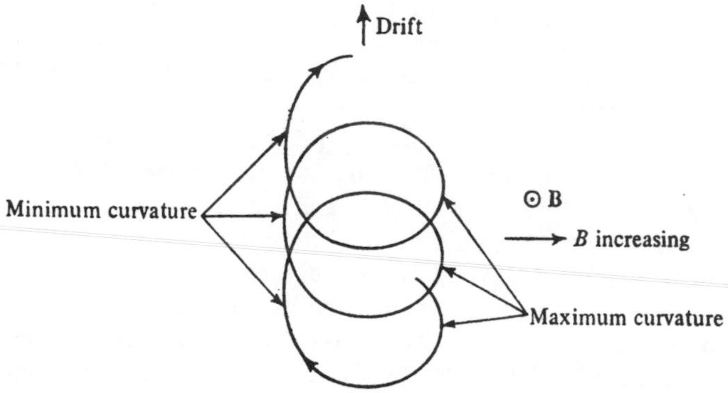

Fig. 4.5. Indicating the origin of the drift of the guiding center in a magnetostatic field $\mathbf{B}$ that has a gradual variation of B in a direction perpendicular to $\mathbf{B}$.

Again, consider the particle in a region in which the magnetic field lines converge gradually, indicating slow increase of B in the direction of $\mathbf{B}$. In the course of its gyration the particle evidently experiences, in addition to the main centrifugal force, also a subsidiary force opposite to the direction in which the field lines converge (see Fig. 4.6). The guiding center of a particle proceeding into the region of convergence is therefore decelerated. This

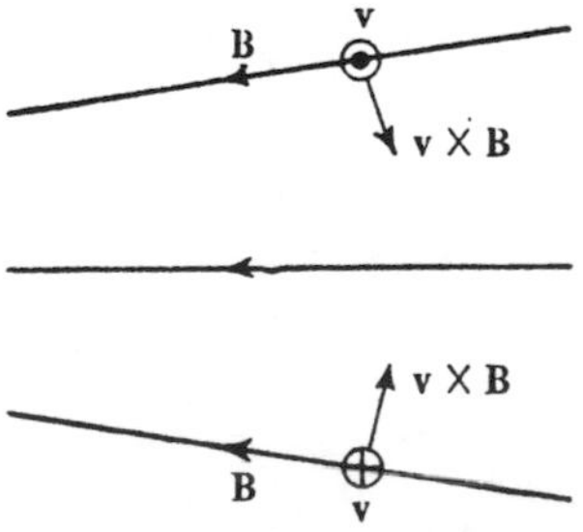

Fig. 4.6. Indicating the origin of the deceleration along $\mathbf{B}$ of the guiding center in a magnetostatic field that has a gradual increase of B in the direction of $\mathbf{B}$.

means that the component of the velocity of the particle parallel to **B** is reduced, and since the speed v is constant this reduction must be at the expense of an increase in the component perpendicular to **B**. The guiding center can be brought to rest for an instant, and subsequently the motion is reversed, with the particle being expelled from the region of convergence. This process is often called "mirroring". The special case of a magnetic pole field, worked out in §4.2.3, illustrates the effect. As the particle spirals in towards the origin along a geodesic on the surface of the cone $\theta = \theta_0$, where θ_0 must be small for the slowly varying approximation to apply, the guiding center correspondingly moves in along $\theta = 0$ with speed $\dot{r}$, which from (4.47) is

$$\frac{t}{\sqrt{(r_0^2/v^2 + t^2)}} v; \tag{4.56}$$

this speed decreases to zero as t increases to zero, and subsequently increases with direction reversed, whilst throughout the motion the component of velocity of the particle perpendicular to **B**, $r\,\dot{\phi} \sin\theta$ is, from (4.46),

$$r_0 v/r. \tag{4.57}$$

The concept of the guiding center is still valid in the more general case in which both electric and magnetic fields are present, and a slow time variation is admitted as well as a slow space variation. The restriction on the variation of the fields, that at the particle they shall not have changed appreciably during a local gyro period, still applies. An obvious effect of an electric field **E** is to contribute to the guiding center the velocity $\mathbf{E} \times \mathbf{B}/B^2$ previously encountered in (4.15) in the discussion of motion in uniform static fields; and this represents a drift across the magnetic field lines which, in contrast to that due to the non-uniformity of **B**, is by no means necessarily slow. Other more subtle effects are left to appear in the detailed mathematical formulation of §4.4.2, but one further point can be emphasized here, namely that because the particle experiences without hindrance the full acceleration due to the component of **E** parallel to **B**, it proves necessary in the analysis to assume that this component is small.

4.3.2 The first adiabatic invariant

The velocity of the guiding center, although it is essentially that of the particle with the gyrational component removed, can nevertheless only be found explicitly with the help of an important piece of information about the gyrational motion itself.

Consider, in the first instance, motion in which any drift of the guiding center across the magnetic field is slow, and denote by $\mathbf{v}_\perp$ the component of the velocity of the particle perpendicular to **B**. To the level of approximation relevant at this stage, $\mathbf{v}_\perp$ also represents the component of velocity of the particle relative to the guiding center perpendicular to **B** evaluated at the guiding center; furthermore it can be regarded as of constant magnitude during a gyro period. In the same spirit, introduce the gyro frequency $\Omega = eB/m$ and the gyro radius $\rho = v_\perp/\Omega$. Then the *magnetic moment* of the gyrating charged particle, defined as the effective current $e\Omega/(2\pi)$ multiplied by the area $\pi\rho^2$ encompassed in one revolution, can be written in the alternative forms

$$M = \tfrac{1}{2}e\Omega\rho^2 = \frac{\tfrac{1}{2}e\mathbf{v}_\perp^2}{\Omega} = \frac{\tfrac{1}{2}m\mathbf{v}_\perp^2}{B}; \tag{4.58}$$

and the required information is commonly summed up in the statement: *the magnetic moment is an adiabatic invariant of the motion.* The remainder of this section is devoted to a brief exposition in explanation and support of this statement; a mathematical proof is given in §4.4.

The term adiabatic invariant was coined in the days of the old quantum theory to describe, in a classical mechanical system subject to some slow perturbation, any quantity which, though not a strict invariant, nevertheless changed more slowly by an order of magnitude than the perturbation itself; the quantity would thus remain unaltered for a finite change of the system, provided the change were executed infinitely slowly. The word adiabatic was used by analogy with the terminology of thermodynamics. In the search for a principle by which to determine the appropriate expressions to quantize in any given problem it was argued that, in order to retain contact with classical mechanics, these expressions must be adiabatic invariants; for since quantized magnitudes could only change by an integral multiple of Planck's constant, a perturbation too gradual to effect such a change must leave the value of the expression unaltered. Within the Hamiltonian structure of classical dynamics adiabatic invariants were to hand in the action integrals

$$\oint p\, dq,$$

where p and q are canonically conjugate momentum and position coordinates which perform a libration in the unperturbed system, the integration being over the libration period.

In the present context the Hamiltonian is given by (4.8), and the unperturbed system is that for which the magnetic field is constant and uniform. In cylindrical polar coordinates r, θ, z, with the z-axis along $\mathbf{B}$, there is periodicity in θ; and with

$$q_\theta = \theta, \qquad p_\theta = \frac{\partial \mathscr{L}}{\partial \dot\theta} = mr^2\dot\theta + erA_\theta,$$

where $\mathbf{A}$ is the vector potential,

$$\oint p_\theta\, dq_\theta = \int_0^{2\pi} (mr^2\dot\theta + erA_\theta)\, d\theta. \tag{4.59}$$

Furthermore, in the notation just used, $r = \rho$, $\dot{\theta} = -\Omega$; and so, making use of the fact that the line integral of **A** round a circuit is the flux of **B** through the circuit, (4.59) is

$$-2\pi m\rho^2\Omega + \pi\rho^2 eB. \tag{4.60}$$

The second term in (4.60) is simply $-\frac{1}{2}$ times the first term, and the whole expression is therefore identical with (4.58) multiplied by $-2\pi m/e$.

These brief remarks have been made in order to show that the demonstration of the adiabatic invariance of the magnetic moment did not appear without any parentage like a rabbit out of a hat. On the other hand it seems desirable to discuss the problem in its specific context, particularly because of its close association with the theory of the motion of the guiding center. Some instructive examples of adiabatic invariance are now mentioned, preparatory and supplementary to the relatively formal analysis of §4.4.

An elementary mechanical system often used by way of illustration is the simple pendulum whose length l is subjected to a gradual change. When l is constant, and the angular displacement remains small, the energy for a bob of unit mass is

$$W = \tfrac{1}{2}l^2\dot{\theta}^2 + \tfrac{1}{2}gl\theta^2,$$

which is, of course, a constant of the motion; and the tension is

$$T = g + l\dot{\theta}^2 - \tfrac{1}{2}g\theta^2,$$

giving

$$\bar{T} = g + \frac{W}{2l},$$

where the bar denotes the average over a period $2\pi\sqrt{(l/g)}$. When l is made to vary the work done on the system is

$$-\int \bar{T}\,dl,$$

and on the understanding that the increment δl in l during time interval $2\pi\sqrt{(l/g)}$ is small this is written

$$-\left(g + \frac{W}{2l}\right)\delta l.$$

The term $-g\delta l$ represents the increase in the mean potential energy, so the other term must represent the increase in the energy of the oscillation. Thus

$$\delta W = -\frac{W}{2l}\,\delta l = W\frac{\delta\omega}{\omega},$$

where $\omega = \sqrt{(g/l)}$ is the angular frequency. Hence $\delta(W/\omega) = 0$, and W/ω is an adiabatic invariant. The fact that the expression is energy divided by frequency is, of course, consistent with Planck's quantum hypothesis.

Much the same sort of argument can be given for the case of a charged particle e in a uniform magnetic field $\mathbf{B}$ which varies slowly with time. With the associated electric field $\mathbf{E}$ being legitimately assumed perpendicular to $\mathbf{B}$, the component of velocity of the particle parallel to $\mathbf{B}$ is constant, but that perpendicular to $\mathbf{B}$, say $\mathbf{v}_\perp$, will vary. If $W = \frac{1}{2}mv_\perp^2$ is the transverse kinetic energy, the small change δW during a single gyration of the particle is given by

$$\delta W = e\oint \mathbf{E}\cdot d\mathbf{s},$$

where the circuit of integration is along the projection of the path of the particle on a plane normal to $\mathbf{B}$. Since $\mathbf{B}$ varies slowly the circuit can be taken to be a circle of radius $\rho = v_\perp/\Omega$, where $\Omega = eB/m$; the line integral of the tangential component of $\mathbf{E}$ is the rate of change of magnetic flux through this circuit, so that

$$\delta W = \pi e \rho^2 \dot{B} = \frac{W}{\Omega}\,\delta\Omega,$$

remembering that $\delta\Omega$ is the small change in Ω in time $2\pi/\Omega$. Thus $\delta(W/\Omega) = 0$, and W/Ω is an adiabatic invariant.

Another way of looking at the same problem is to use the analysis of §4.2.4. Corresponding to the approximate WKBJ solutions (4.53) and (4.54) for the x- and y-coordinates of the particle, the approximate velocity components perpendicular to $\mathbf{B}$ are

$$\dot{x} = -a\sqrt{(\Omega_0\Omega)}\sin\left(\int_0^t \Omega\,dt\right),$$

$$\dot{y} = -a\sqrt{(\Omega_0\Omega)}\cos\left(\int_0^t \Omega\,dt\right);$$

and these give

$$\frac{v_\perp^2}{\Omega} = \frac{\dot{x}^2 + \dot{y}^2}{\Omega} = a^2\Omega_0,$$

which is a constant. If the WKBJ solutions were taken to a higher order of approximation the next term in $v_\perp^2/\Omega$ would have $\dot{\Omega}$ as a factor, so that a finite change in Ω, if achieved infinitely slowly, would result in no change in $v_\perp^2/\Omega$.

With regard to the adiabatic invariance of the magnetic moment in a non-uniform magnetostatic field, problems involving a two dimensional field, such as those worked out in §4.2.2, are of little interest, since throughout the motion $v_\perp^2$ itself remains strictly constant. However, the magnetic pole field discussed in §4.2.3 does bring out the point. Most obviously the particle conforms to the adiabatic invariant theory by remaining on the surface of a tube of magnetic induction, in this case part of the cone $\theta = \theta_0$, where it is recalled that θ_0 must be small for the slowly varying approximation to apply. Likewise, as stated in (4.57), $v_\perp = r_0 v/r$, so that $v_\perp^2$ is proportional to B.

The knowledge that $v_\perp^2/B$ is an adiabatic invariant enables the phenomenon of "mirroring," in which the particle is "reflected" after traveling some way along converging field lines, to be put on a simple quantitative basis. For since v is constant if the field is purely magnetostatic, $\sin^2\theta/B$ is an adiabatic invariant, where θ is the pitch angle that $\mathbf{v}$ makes with $\mathbf{B}$. As the particle progresses along the converging lines, B increases and $\sin^2\theta$ must therefore increase likewise. But $\sin^2\theta$ cannot exceed unity, so for a particle that has pitch angle θ_0 at a point where the magnetic induction is B_0, the "mirror point" is evidently reached when

$$B = B_0/\sin^2\theta_0. \tag{4.61}$$

It is clearly possible for a charged particle to be trapped in a magnetic field in which a "bottle" is formed by virtue of the presence of a mirror point at each end.

Two concluding remarks. If, due to an electric field, the guiding center has a velocity $\mathbf{u}$ whose component $\mathbf{u}_\perp$ perpendicular to $\mathbf{B}$ is not small, then the expression for the adiabatic invariant replacing $v_\perp^2/B$ is, as would be expected,

$$(\mathbf{v}_\perp - \mathbf{u}_\perp)^2/B. \tag{4.62}$$

Secondly, it may happen that the overall field configuration is such that further quasi periodicities are introduced into the motion. In this way there can be up to three adiabatic invariants, the number of degrees of freedom of the particle, rather than just the one, the magnetic moment. The latter is, of course, an inevitable concomitant of a slowly varying field, and is commonly called the first adiabatic invariant; others may or may not exist. Motion involving the so-called second, or longitudinal, adiabatic invariant is typified by that of a particle trapped in a magnetic bottle, the second adiabatic invariant being associated with the oscillation between mirror points. The existence of a third adiabatic invariant is well illustrated by motion in a field which approximates to that of a magnetic dipole, in particular the earth's magnetic field. In this case not only is the gyrating particle reflected at some distance from each pole in turn, but because of the variation of B with distance from the center of the earth, is also subject to a drift which carries it round the magnetic axis. The third adiabatic invariant is associated with this latter precession.

4.4 NON-RELATIVISTIC MOTION IN A SLOWLY VARYING FIELD: ANALYSIS

4.4.1 Motion in a magnetostatic field

In attempting an analysis of the motion of a charged particle e in a slowly varying field it is useful to have an explicit parameter whose magnitude is a measure of the slowness of the field variation. A theoretically convenient procedure is to take $\varepsilon = m/e$ as this parameter, and to express the fact that the field is slowly varying by treating ε as a small quantity in the mathematical sense. This device succeeds because the gyro period and gyro radius vanish with ε, so that for any motion stemming from given initial conditions in a given field an approximation based on the slowly varying character of the field will evidently be the more accurate the smaller is ε. Against the choice of ε is the possible confusion caused by the fact that it is not dimensionless; but in dealing with general fields, to work in terms of any dimensionless parameter is unduly cumbersome.

The next task is to make explicit the distinction between the gyrational motion, which involves the gyro frequency $\Omega = eB/m$ and harmonics thereof, and the motion of the guiding center, which does not. This is achieved by the *ansatz*

$$\mathbf{r} = \mathbf{r}_0 + \mathbf{s} + \mathbf{s}^* \tag{4.63}$$

where

$$\mathbf{s} = \sum_{n=1}^{\infty} \varepsilon^n \, e^{in\theta} \, \mathbf{r}_n, \tag{4.64}$$

$$\theta = \int^t \Omega(\mathbf{r}_0) \, dt. \tag{4.65}$$

A number of points must be made in explanation. First, that $\mathbf{r}$ and $\mathbf{r}_0$ are the position vectors of the particle and guiding center respectively. Secondly, that the position vector of the particle relative to the guiding center, written as $\mathbf{s} + \mathbf{s}^*$ (where the star denotes the complex conjugate), contains only oscillatory terms and vanishes with ε. Thirdly, that the rejection of sines and cosines in favor of complex exponentials exploits a familiar technique which here does indeed make the algebra more compact. Fourthly, that the appearance in the exponents in (4.64) of θ, given by (4.65), rather than Ω itself takes proper account of the continual phase advance in the gyrational motion (for example, compare with the special solution (4.53)). Fifthly, that the complex coefficients $\mathbf{r}_n$ in (4.64), which it should be noted are not lengths but have dimensions length/ε^n, are themselves power series in ε with leading terms independent of ε. And lastly, that in the integrand of (4.65) Ω is evaluated not at the particle, but at the guiding center.

Reference must also be made to two established results which are here accepted without further ado. One is that the solution *can* in fact be represented by means of the expansion (4.64), though the representation is an asymptotic one in the accepted sense that the difference between the solution and the sum of any finite number of terms N of the representative expansion vanishes with ε more rapidly than ε^N. The other is that it is valid to substitute the expansion into the equation of motion and then equate to zero the coefficient of each exp $(in\theta)$ for $n = 0, \pm 1, \pm 2, \ldots$.

This latter procedure is now adopted. The equations for $n = 0$ and $n = 1$ furnish all that is required for establishing both the constancy of the magnetic moment, to order unity, and the formulas for the velocity of the guiding center, to order ε. As a matter of notation, since all functions of position are ultimately expanded about the guiding center, the argument $\mathbf{r}_0$ is left understood; any other argument is noted explicitly. A dot stands for time differentiation following the motion.

The representations of the velocity and acceleration are required. Since, from (4.65),

$$\varepsilon\dot{\theta} = B,$$

a first differentiation of (4.64) gives

$$\dot{\mathbf{s}} = \sum_{n=1}^{\infty} (inB\varepsilon^{n-1}\mathbf{r}_n + \varepsilon^n\dot{\mathbf{r}}_n)\, e^{in\theta}, \tag{4.66}$$

and a second gives

$$\ddot{\mathbf{s}} = \sum_{n=1}^{\infty} [-n^2B^2\varepsilon^{n-2}\mathbf{r}_n + in\varepsilon^{n-1}(\dot{B}\mathbf{r}_n + 2B\dot{\mathbf{r}}_n) + \varepsilon^n\ddot{\mathbf{r}}_n]\, e^{in\theta}. \tag{4.67}$$

Also required is the expansion about the guiding center of the field at the particle. For the magnetic field this is

$$\mathbf{B}(\mathbf{r}) = \mathbf{B} + \varepsilon[(e^{i\theta}\,\mathbf{r}_1 + e^{-i\theta}\,\mathbf{r}_1^*)\cdot\text{grad}]\mathbf{B} + O(\varepsilon^2). \tag{4.68}$$

The immediate discussion is now restricted, for simplicity, to motion in a purely magnetostatic field; the modifications required to treat the general case involve no new technique, and are reserved for §4.4.2. The equation of motion is therefore

$$\varepsilon\dot{\mathbf{v}} = \mathbf{v} \times \mathbf{B}(\mathbf{r}). \tag{4.69}$$

The next step is to substitute from (4.63) and (4.68) into (4.69), making use of (4.66) and (4.67). Equating the terms on either side of (4.69) which do not contain any of the exp $(in\theta)$ as a factor then yields

$$\varepsilon\dot{\mathbf{u}} = \mathbf{u} \times \mathbf{B} + i\varepsilon B[\mathbf{r}_1 \times (\mathbf{r}_1^*\cdot\text{grad})\mathbf{B} - \mathbf{r}_1^* \times (\mathbf{r}_1\cdot\text{grad})\mathbf{B}] + O(\varepsilon^2), \tag{4.70}$$

where

$$\mathbf{u} = \dot{\mathbf{r}}_0$$

is the velocity of the guiding center. And equating the terms on either side of (4.69) which contain the factor exp $(i\theta)$ yields, after a little rearrangement,

$$B(B\mathbf{r}_1 - i\varepsilon\dot{\mathbf{r}}_1) + i(B\mathbf{r}_1 - i\varepsilon\dot{\mathbf{r}}_1) \times \mathbf{B}$$
$$= i\varepsilon[\dot{B}\mathbf{r}_1 + B\dot{\mathbf{r}}_1 + i\mathbf{u} \times (\mathbf{r}_1 \cdot \text{grad})\mathbf{B}] + O(\varepsilon^2), \tag{4.71}$$

the same result being obtained from the terms with factor exp $(-i\theta)$. Equations (4.70) and (4.71) are the basis of the subsequent development.

At this stage it is desirable to introduce at each point in space an orthogonal triad of real unit vectors $\mathbf{e}_1$, $\mathbf{e}_2$, $\mathbf{e}_3$, where $\mathbf{e}_3$ is along $\mathbf{B}$, so

$$\mathbf{B} = B\mathbf{e}_3,$$

and the particular orientation of $\mathbf{e}_1$, $\mathbf{e}_2$ in the plane normal to $\mathbf{B}$ is at the moment left open. We also remark that the $O(\varepsilon)$ part of $\mathbf{r}_1$ does not come into the analysis; it is obviously immaterial to (4.70) and to the $O(1)$ part of (4.71), and as we shall see is eliminated in the treatment of the $O(\varepsilon)$ part of (4.71). From now on, therefore, the symbol $\mathbf{r}_1$ is used for the part independent of ε of the quantity hitherto called $\mathbf{r}_1$.

Consider, first, the $O(1)$ part of (4.71). It can evidently be written

$$\mathbf{r}_1 + i\mathbf{r}_1 \times \mathbf{e}_3 = 0,$$

which implies that

$$\mathbf{r}_1 = \rho(\mathbf{e}_1 + i\mathbf{e}_2), \tag{4.72}$$

where ρ is undetermined. Since, for any ρ, the real and imaginary parts of (4.72) are mutually orthogonal, it is possible to make ρ real by suitable choice of the orientation of $\mathbf{e}_1$, $\mathbf{e}_2$; and this choice is adopted for the sake of convenience. Thus (4.63) reads

$$\mathbf{r} = \mathbf{r}_0 + 2\varepsilon\rho(\mathbf{e}_1 \cos\theta - \mathbf{e}_2 \sin\theta) + O(\varepsilon^2), \tag{4.73}$$

and correspondingly

$$\mathbf{v} = \mathbf{u} - 2B\rho(\mathbf{e}_1 \sin\theta + \mathbf{e}_2 \cos\theta) + O(\varepsilon). \tag{4.74}$$

Equations (4.73), (4.74) evidently accord with the picture of gyration about the guiding center. In particular, the magnetic moment M is defined by

$$\frac{M}{m} = \frac{(\mathbf{v}_\perp - \mathbf{u}_\perp)^2}{2B} = 2B\rho^2 = B\mathbf{r}_1 \cdot \mathbf{r}_1^*, \tag{4.75}$$

where, as usual, the suffix $_\perp$ denotes the component perpendicular to $\mathbf{B}$. Since (4.75) neglects $O(\varepsilon)$ terms, and here $\mathbf{u}_\perp$ is $O(\varepsilon)$, as (4.70) puts in evidence, the inclusion of $\mathbf{u}$ in the definition of M is merely for the sake of explicit adherence to the definition that applies in the general case when $\mathbf{u}_\perp$ is $O(1)$.

Now consider the $O(\varepsilon)$ part of (4.71). If this is to be used to establish the adiabatic invariance of M, the final expression in (4.75) suggests forming the scalar product of (4.71) with $\mathbf{r}_1^*$. It is easily seen that the scalar product of the left-hand side of (4.71) with $\mathbf{r}_1^*$ disappears, since for any vector $\mathbf{a}$,

$$(\mathbf{a} + i\mathbf{a} \times \mathbf{e}_3)\cdot(\mathbf{e}_1 - i\mathbf{e}_2) = 0;$$

there remains

$$\dot{B}\mathbf{r}_1\cdot\mathbf{r}_1^* + B\dot{\mathbf{r}}_1\cdot\mathbf{r}_1^* - i\mathbf{u}\cdot\mathbf{r}_1^* \times (\mathbf{r}_1\cdot\text{grad})\mathbf{B} = 0. \tag{4.76}$$

Now the real part of $-i\mathbf{r}_1^* \times (\mathbf{r}_1\cdot\text{grad})\mathbf{B}$ is

$$\begin{aligned}\rho^2[\mathbf{e}_1 \times (\mathbf{e}_2\cdot\text{grad})\mathbf{B} - \mathbf{e}_2 \times (\mathbf{e}_1\cdot\text{grad})\mathbf{B}]\\ &= \rho^2\left[\left(0, -\frac{\partial B_z}{\partial y}, \frac{\partial B_y}{\partial y}\right) + \left(-\frac{\partial B_z}{\partial x}, 0, \frac{\partial B_x}{\partial x}\right)\right]\\ &= -\rho^2 \,\text{grad}\, B,\end{aligned} \tag{4.77}$$

where the local Cartesian coordinates are defined by the triad $\mathbf{e}_1$, $\mathbf{e}_2$, $\mathbf{e}_3$, and div $\mathbf{B} = 0$ is used. Also $\mathbf{u}\cdot\text{grad}\, B = \dot{B}$. The real part of (4.76) is therefore

$$\dot{M} = 0,$$

and the adiabatic invariance of M is established.

The imaginary part of (4.76) merely gives information about the rate of rotation of $\mathbf{e}_1$, $\mathbf{e}_2$.

With the knowledge that, to order unity, M is constant, (4.70) can be made to yield the guiding center motion. For it is evident from (4.77) that (4.70) can be written, to order ε,

$$\mathbf{u} \times \mathbf{B} = \varepsilon\left(\frac{M}{m}\,\text{grad}\, B + \dot{\mathbf{u}}\right). \tag{4.78}$$

The vector product with $\mathbf{B}$ gives

$$\mathbf{u}_\perp = \frac{\varepsilon}{B^2}\mathbf{B} \times \left(\frac{M}{m}\,\text{grad}\, B + \dot{\mathbf{u}}\right), \tag{4.79}$$

so that $\mathbf{u}_\perp$ is, as previously noted, of order ε, and is determined to order ε if $\mathbf{B} \times \dot{\mathbf{u}}$ is found to order unity. But

$$\begin{aligned}\mathbf{B} \times \dot{\mathbf{u}} &= \mathbf{B} \times (\mathbf{u}\cdot\text{grad})\mathbf{u}\\ &= \frac{u_\parallel^2}{B^2}\mathbf{B} \times (\mathbf{B}\cdot\text{grad})\mathbf{B} + O(\varepsilon),\end{aligned}$$

and (4.79) is therefore equivalent to

$$\mathbf{u}_\perp = \varepsilon\frac{\mathbf{B}}{B^2} \times \left[\frac{M}{m}\operatorname{grad} B + \frac{u_\parallel^2}{B^2}(\mathbf{B}\cdot\operatorname{grad})\mathbf{B}\right]. \tag{4.80}$$

Finally, appeal can be made to the energy equation, which in the present case is just $v^2 = \text{constant}$. Using (4.74), and the fact that $\mathbf{u}_\perp$ is $O(\varepsilon)$, it reads

$$\tfrac{1}{2}u_\parallel^2 + \frac{M}{m}B = \text{constant} + O(\varepsilon). \tag{4.81}$$

In summary, then, to order unity, M is constant and $u_\parallel$ is given by (4.81); and, to order ε, $\mathbf{u}_\perp$ is in turn given by (4.80).

Two further points are worth making. It is not difficult to check that an alternative way of writing (4.80) is

$$\mathbf{u}_\perp = \varepsilon\left(\frac{M}{m} + \frac{u_\parallel^2}{B}\right)\frac{\mathbf{B}}{B^2} \times \operatorname{grad} B + \varepsilon\frac{u_\parallel^2}{B^2}(\operatorname{curl}\mathbf{B})_\perp;$$

and in the important special case $\operatorname{curl}\mathbf{B} = 0$ this reduces to

$$\mathbf{u}_\perp = \varepsilon\left(\frac{M}{m} + \frac{u_\parallel^2}{B}\right)\frac{\mathbf{B}}{B^2} \times \operatorname{grad} B. \tag{4.82}$$

The second point is that the component of (4.78) parallel to $\mathbf{B}$, not used in the above analysis, is

$$\dot{u}_\parallel + \frac{M}{m}(\operatorname{grad} B)_\parallel = O(\varepsilon), \tag{4.83}$$

which is nothing other than the derivative of the form (4.81) of the energy equation. Alternatively, (4.81) and (4.83) could be used to prove $\dot{M} = 0$ without appeal to the $O(\varepsilon)$ part of (4.71).

4.4.2 Motion in a general field

It is an easy matter to generalize the whole of §4.4.1 to take account of the addition of an electric field and a time variation. Clearly nothing needs altering as far as and including (4.68). The equation of motion

$$\varepsilon\dot{\mathbf{v}} = \mathbf{v} \times \mathbf{B}(\mathbf{r}) + \mathbf{E}(\mathbf{r}) \tag{4.84}$$

replaces (4.69), and using for $\mathbf{E}(\mathbf{r})$ the expansion completely analogous to (4.68) it is at once seen that the new equations corresponding to (4.70) and (4.71) differ from the originals only by the addition to the right-hand sides of the terms $\mathbf{E}$ and $-\varepsilon\,(\mathbf{r}_1\cdot\operatorname{grad})\mathbf{E}$ respectively. There is thus no modification required following (4.71) until (4.76) is reached, but (4.76) itself now has on the left-hand side the further term

$$i\mathbf{r}_1^*\cdot(\mathbf{r}_1\cdot\operatorname{grad})\mathbf{E}. \tag{4.85}$$

The real part of (4.85) is

$$\rho^2[\mathbf{e}_2\cdot(\mathbf{e}_1\cdot\text{grad})\mathbf{E} - \mathbf{e}_1\cdot(\mathbf{e}_2\cdot\text{grad})\mathbf{E}]$$
$$= \rho^2\left(\frac{\partial E_y}{\partial x} - \frac{\partial E_x}{\partial y}\right)$$
$$= -\rho^2\frac{\partial B}{\partial t},$$

using curl $\mathbf{E} = -\partial\mathbf{B}/\partial t$ and the fact that $\mathbf{B}$ is along the local z-axis. The analysis, therefore, still leads to the conclusion $\dot{M} = 0$.

Turning now to the velocity of the guiding center, it is evident that (4.78) is replaced by

$$\mathbf{u} \times \mathbf{B} = \varepsilon\left(\frac{M}{m}\,\text{grad}\,B + \dot{\mathbf{u}}\right) - \mathbf{E}, \tag{4.86}$$

so that equation (4.79) acquires on the right-hand side the extra term

$$\mathbf{u}_E = \mathbf{E} \times \mathbf{B}/B^2. \tag{4.87}$$

This term has, of course, already been encountered in the case when the fields are static and uniform. Since it is $O(1)$, in contrast to the rest of $\mathbf{u}_\perp$, which is $O(\varepsilon)$, it makes a contribution to the $O(1)$ part of $\mathbf{B} \times \dot{\mathbf{u}}$. If the $O(1)$ terms of $\mathbf{B} \times \dot{\mathbf{u}}$ are written out formally, by taking $\mathbf{u} = u_\parallel\mathbf{e}_3 + \mathbf{u}_E + O(\varepsilon)$, the expression for $\mathbf{u}_\perp$ generalizing (4.80) is found to be

$$\mathbf{u}_\perp = \mathbf{u}_E + \frac{\varepsilon\mathbf{B}}{B^2} \times \left\{\frac{M}{m}\,\text{grad}\,B + \frac{u_\parallel^2}{B^2}(\mathbf{B}\cdot\text{grad})\mathbf{B}\right.$$
$$+ \frac{u_\parallel}{B}[(\mathbf{B}\cdot\text{grad})\mathbf{u}_E + (\mathbf{u}_E\cdot\text{grad})\mathbf{B}]$$
$$\left. + (\mathbf{u}_E\cdot\text{grad})\mathbf{u}_E + u_\parallel\frac{\partial}{\partial t}\left(\frac{\mathbf{B}}{B}\right) + \frac{\partial\mathbf{u}_E}{\partial t}\right\}. \tag{4.88}$$

The equation giving $u_\parallel$ is the component along $\mathbf{B}$ of (4.86), namely (cf. (4.83))

$$\mathbf{e}_3\cdot\dot{\mathbf{u}} = -\frac{M}{m}(\text{grad}\,B)_\parallel + \frac{1}{\varepsilon}E_\parallel + O(\varepsilon).$$

It is here that the requirement that $E_\parallel$ be $O(\varepsilon)$ shows up; if $E_\parallel$ were $O(1)$, εu would in general be $O(1)$, contrary to the previous assumption. Now

$$\dot{u}_\parallel = \mathbf{e}_3\cdot\dot{\mathbf{u}} + \mathbf{u}\cdot\dot{\mathbf{e}}_3 = \mathbf{e}_3\cdot\dot{\mathbf{u}} + \mathbf{u}_E\cdot\dot{\mathbf{e}}_3 + O(\varepsilon),$$

and

$$\mathbf{u}_E\cdot\dot{\mathbf{e}}_3 = \mathbf{u}_E\cdot\left\{\frac{\partial}{\partial t}\left(\frac{\mathbf{B}}{B}\right) + \left[\left(\frac{u_\parallel}{B}\mathbf{B} + \mathbf{u}_E\right)\cdot\mathrm{grad}\right]\frac{\mathbf{B}}{B}\right\}$$

$$= \frac{\mathbf{u}_E}{B}\cdot\left\{\frac{\partial\mathbf{B}}{\partial t} + \left[\left(\frac{u_\parallel}{B}\mathbf{B} + \mathbf{u}_E\right)\cdot\mathrm{grad}\right]\mathbf{B}\right\},$$

since $\mathbf{u}_E\cdot\mathbf{B} = 0$. Thus, to order unity,

$$\dot{u}_\parallel = -\frac{M}{m}(\mathrm{grad}\ B)_\parallel + \frac{1}{\varepsilon}E_\parallel + \frac{\mathbf{u}_E}{B}\cdot\left\{\frac{\partial\mathbf{B}}{\partial t} + \left[\left(\frac{u_\parallel}{B}\mathbf{B} + \mathbf{u}_E\right)\cdot\mathrm{grad}\right]\mathbf{B}\right\}. \quad (4.89)$$

Consider, finally, the $O(1)$ result derived from the non-oscillatory part of the energy equation

$$\frac{d}{dt}(\tfrac{1}{2}v^2) = \frac{1}{\varepsilon}\mathbf{E}\cdot\mathbf{v}.$$

It is readily seen to be

$$\frac{d}{dt}\left(\frac{1}{2}\mathbf{u}^2 + \frac{M}{m}\mathbf{B}\right) = \frac{1}{\varepsilon}\mathbf{E}\cdot\mathbf{u} + \frac{M}{m}\frac{\partial\mathbf{B}}{\partial t}, \quad (4.90)$$

where curl $\mathbf{E}$ has again been replaced by $-\partial\mathbf{B}/\partial t$. Precisely the same equation, in the form

$$\mathbf{u}\cdot\dot{\mathbf{u}} + \frac{M}{m}\mathbf{u}\cdot\mathrm{grad}\ \mathbf{B} = \frac{1}{\varepsilon}\mathbf{E}\cdot\mathbf{u}, \quad (4.91)$$

is given by taking the scalar product of (4.86) with $\mathbf{u}$. Equations (4.90) and (4.91) could be used to provide an alternative proof that $\dot{M} = 0$.

4.4.3 Further comments on the motion of the guiding center

Equations (4.88) and (4.89) give rather general formulae from which the motion of the guiding center can in principle be determined. Some qualitative indication of the sort of motion to be expected in certain types of field configuration was presented previously, in §4.3, and it is worth emphasizing that similar arguments with a closer attention to detail can in fact be used to interpret, or even obtain, the various terms in (4.88) and (4.89). A brief examination on these lines of the simpler terms now follows; such discussion certainly gives insight into the physical significance of the formal mathematics, but can hardly claim to be independently adequate, particularly because of the difficulty of ensuring that account has been taken of all possible contributions to the guiding center motion.

The appearance in (4.88) of the *electric drift* $\mathbf{u}_E$, given by (4.87), is, of course, expected from the analysis of the simple case of uniform, static fields. A similar term, the *external force drift*

$$\mathbf{u}_F = \frac{1}{e}\frac{\mathbf{F} \times \mathbf{B}}{B^2}, \tag{4.92}$$

would also be present if there were an additional slowly varying force $\mathbf{F}$ acting on the particle, such as gravity. An important distinction between $\mathbf{u}_E$ and $\mathbf{u}_F$ is that the sign of the latter depends on the sign of the charge; hence $\mathbf{u}_F$ tends to set up a current in a neutral distribution of electrons and positive ions, whereas $\mathbf{u}_E$ does not.

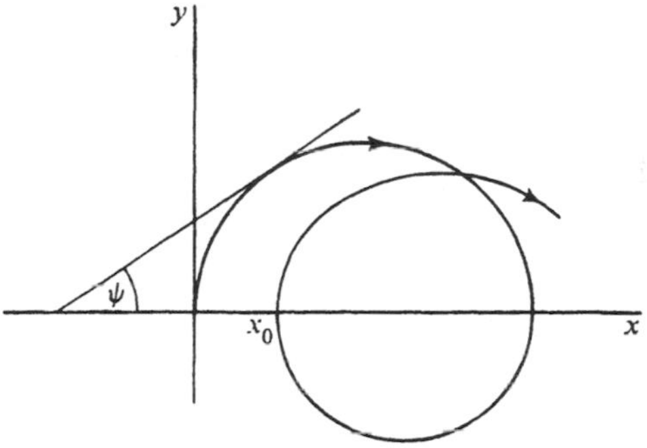

Fig. 4.7. Geometry for the estimation of the gradient drift.

The second term on the right-hand side of (4.88) is the *gradient drift*

$$\varepsilon\frac{M}{mB^2}\mathbf{B} \times \text{grad } B. \tag{4.93}$$

As explained in §4.3.1, it arises when B changes in a direction at right angles to $\mathbf{B}$, because of the resulting variation in the curvature of the path. Suppose that locally $\mathbf{B}$ is in the z-direction, and that the curve in Fig. 4.7 is the projection of the path on the xy-plane. At any point on the curve let ρ, ψ, respectively, be the radius of curvature and angle that the tangent makes with the x-axis. Then the distance traveled in the x-direction in one revolution is

$$x_0 = \int_0^{x_0} dx = \int_{\frac{1}{2}\pi}^{-\frac{3}{2}\pi} \rho \cos\psi \, d\psi; \tag{4.94}$$

and if the path is nearly circular,

$$\rho = \rho_0\left[1 + \left(\frac{\partial\rho}{\partial y}\right)_0 \cos\psi\right]$$

approximately, where the zero suffix denotes quantities evaluated at $x = y = 0$, so that (4.94) gives

$$x_0 = \pi\rho_0\left(\frac{\partial\rho}{\partial y}\right)_0 \tag{4.95}$$

approximately. To get the drift velocity, (4.95) must be divided by the gyro period $2\pi\rho/v_\perp = 2\pi\varepsilon/B$, and when this is done it is easily confirmed that (4.93) is recovered.

The third term on the right-hand side of (4.88) is the *curvature drift*

$$\varepsilon\frac{u_\parallel^2}{B^4}\,\mathbf{B}\times(\mathbf{B}\cdot\text{grad})\mathbf{B}, \tag{4.96}$$

so named because it arises when the lines of magnetic induction are curved. If, at any point P, $\mathbf{e}_3$ is the unit vector along the line of induction l through P, the vector $(\mathbf{e}_3\cdot\text{grad})\mathbf{e}_3$ is along the principal normal to l and has magnitude the curvature of l at P. That, to a first approximation, the guiding center follows the curved line of induction with speed $u_\parallel$ therefore requires a centrifugal force $u_\parallel^2(\mathbf{e}_3\cdot\text{grad})\mathbf{e}_3$. This force can only be accounted for in terms of the appropriate drift velocity perpendicular to $\mathbf{B}$, namely

$$\frac{1}{eB}mu_\parallel^2\mathbf{e}_3\times(\mathbf{e}_3.\text{grad})\mathbf{e}_3,$$

which is indeed (4.96).

Turning now to (4.89) it may be noted that the second term on the right-hand side is self-explanatory, and that the first is the deceleration associated with converging lines of induction, mentioned in §4.3.1. If the guiding center is proceeding into a region of stronger field along an axis, say the z-axis, about which $\mathbf{B}$ has cylindrical symmetry, the retarding force is evidently

$$ev_\perp B\eta, \tag{4.97}$$

where η is the small angle that $\mathbf{B}$ makes with the z-axis at distance ρ, the gyro radius, from it. But at small distance r from the z-axis the component B_r is approximately $r(\partial B_r/\partial r)_0$, where the zero suffix denotes evaluation at $r = 0$; and furthermore

$$\frac{\partial B_r}{\partial r}+\frac{B_r}{r}+\frac{\partial B_z}{\partial z}\equiv \text{div}\,\mathbf{B}=0$$

gives

$$B_r = r\left(\frac{\partial B_r}{\partial r}\right)_0 = -\tfrac{1}{2}r\left(\frac{\partial B}{\partial z}\right)_0.$$

Since, then, η is $-B_r/B_z$ evaluated at $r = \rho$, (4.97) is evidently

$$\tfrac{1}{2}ev_\perp\rho\left(\frac{\partial B}{\partial z}\right)_0 = -M(\text{grad } B)_\parallel\,,$$

which reproduces the first term in the expression (4.89) for $\dot{u}_\parallel$.

Having thus identified the origins of the more notable contributions to the guiding center motion it is also of interest to check the formulas in the simple cases for which exact analysis is available. In §4.2.2 the problem was considered of motion in the magnetostatic field

$$\mathbf{B} = \left(0, 0, \frac{\varepsilon u x_0}{x^2}\right). \tag{4.98}$$

For motion in the xy-plane, with speed $v < u$, the path is illustrated in Fig. 4.2(c). The drift velocity is in the negative y-direction and the exact expression for its magnitude is

$$-\int dy \Big/ \int dt = -\int v_y \, dt \Big/ \int dt, \tag{4.99}$$

where the integrations are taken over one revolution. It is convenient to write

$$v_x = v \cos \gamma, \quad v_y = -v \sin \gamma; \tag{4.100}$$

from which substitution in the equations of motion (4.19) gives

$$\dot{\gamma} = \Omega = u x_0 / x^2. \tag{4.101}$$

Then (4.99) is

$$v \int_0^{2\pi} x^2 \sin \gamma \, d\gamma \Big/ \int_0^{2\pi} x^2 \, d\gamma. \tag{4.102}$$

But the equation (4.26) for v_y is expressible as

$$\frac{x_0}{x} = 1 - \frac{v}{u} \sin \gamma,$$

and since

$$\int_0^{2\pi} \frac{\sin \gamma \, d\gamma}{(1 - k \sin \gamma)^2} = \frac{2\pi k}{(1 - k^2)^{3/2}}, \qquad \int_0^{2\pi} \frac{d\gamma}{(1 - k \sin \gamma)^2} = \frac{2\pi}{(1 - k^2)^{3/2}},$$

the exact drift speed (4.102) is simply

$$v^2/u. \tag{4.103}$$

It is easily confirmed, on the other hand, that the gradient drift (4.93) has only a y-component, which when evaluated at $x = x_0$ is negative with magnitude (4.103). The value x_0 for the x-coordinate of the guiding center is, of course, sufficiently accurate for insertion in (4.93).

A very similar analysis applies to the case when the magnetic field, in cylindrical polar coordinates r, θ, z, is

$$\mathbf{B} = (0, 0, \varepsilon u / r). \tag{4.104}$$

For motion in the xy-plane, with speed $v < u$, the path is illustrated in Fig. 4.4(c). The angular drift velocity is in the negative θ-direction and of magnitude

$$-\int d\theta \Big/ \int dt. \tag{4.105}$$

With

$$\dot{r} = v \cos \gamma, \qquad r\dot{\theta} = -v \sin \gamma,$$

the equations of motion give

$$\dot{\gamma} = \Omega - \frac{v \sin \gamma}{r} = \frac{u}{r}\left(1 - \frac{v}{u} \sin \gamma\right),$$

and the solution (4.41) for $r\dot{\theta}$ can be written

$$\frac{r_0}{r} = 1 - \frac{v}{u} \sin \gamma.$$

Then (4.105) can be evaluated; it turns out to be

$$\frac{u}{r_0}(1 - v^2/u^2)[1 - \sqrt{(1 - v^2/u^2)}]. \tag{4.106}$$

On the other hand, (4.93) has only a θ-component, which when evaluated at $r = r_0$ is negative with magnitude

$$\tfrac{1}{2}v^2/u. \tag{4.107}$$

For the drift approximation to be valid, $v/u \ll 1$ and r_0 times (4.106) then agrees with (4.107).

Whilst reference is being made to the analysis of §4.2.2 the case

$$\mathbf{B} = \left(0, 0, \frac{2\varepsilon\,|\,u\,|}{x_0^2}\, x\right),$$

which is also treated there, may be cited to emphasize the fact that, no matter how slow be the absolute change in **B**, the slowly varying approximation breaks down in the vicinity of any point where $B = 0$. This is well illustrated by Figs. 4.3(a) and (b), where the path crosses $x = 0$. In (a) the drift motion is at least in the same sense, namely y increasing, as that in the slowly varying situation depicted in (c); but in (b) the drift motion is in the opposite sense.

Finally, motion in a magnetic pole field, analysed in §4.2.3, has already been used to illustrate mirroring. Here it only remains to remark that the solution (4.47) for the radial distance r of the particle, and effectively of the guiding center, from the pole, gives $m\ddot{r} = mv^2r_0^2/r^3$; and since $B = B_0r_0^2/r^2$ and $M = \frac{1}{2}mv^2/B_0$ this is indeed $-M$ grad B, in agreement with (4.83).

4.4.4 Further comments on the first adiabatic invariant

The analysis of §§4.4.1 and 4.4.2 showed the magnetic moment M to be a constant of the motion provided terms of order ε were neglected. This means that no matter where the guiding center is transported by its velocity components $u_{\parallel}$ and $\mathbf{u}_E$, which are independent of ε, the change in M can be made arbitrarily small by taking ε small enough. On the other hand, for a prescribed small value of ε, the $O(\varepsilon)$ contributions to the velocity of the guiding center, such as the gradient and curvature drifts, can themselves, during a sufficiently long time interval, carry the guiding center through appreciable variations of the field; and correspondingly there will be appreciable changes in M. A related quantity that is adiabatically invariant to all orders in ε is likely to be of the form $M + \varepsilon M_1 + \varepsilon^2 M_2 + \cdots$, and such does, in fact, exist; even M_1, however, is a comparatively complicated expression and its derivation is beyond the scope of this book. The quantity is not a strict invariant because the power series in ε is an asymptotic but not a convergent representation.

There is a restricted sense in which M itself is adiabatically invariant to all orders in ε. Suppose the problem of the motion of the particle is specified a little less widely than hitherto by requiring that prior to some time t_1 the particle is in a region where the fields $\mathbf{B}_1$, $\mathbf{E}_1$ are static and uniform, and subsequent to some time t_2 the particle is again in a region where the fields $\mathbf{B}_2$, $\mathbf{E}_2$ are static and uniform. Then if the fields at the particle, and all their space and time derivatives, are continuous throughout, the value of M after t_2 is equal to its value before t_1 to a higher order of approximation than any power of ε. This is the case because in the general result that M differs from a constant by terms of order ε, these terms vanish if the field is static and uniform; in spite, therefore, of being constant only to the zeroth order in ε during the transition period from t_1 to t_2, M has the same value to all orders in ε prior to t_1 as subsequent to t_2. If there is a discontinuity in some field derivative the two values of M will only agree to some corresponding order in ε.

Whilst what has just been described may in a sense be said to acknowledge the adiabatic invariance of M to all orders in ε, again it is not strict invariance; for all it implies is that the initial and final values of M differ by a quantity which vanishes with ε more rapidly than any power of ε, and such a quantity is by no means necessarily zero, as the familiar example $\exp(-1/\varepsilon)$ shows. It is possible to confirm in detail the truth of these statements in a variety of cases of the simple but representative problem set out in §4.2.4, and a comparatively general method of doing this is now presented.

In the problem the fields, in rectangular Cartesian coordinates, are

$$\mathbf{B} = (0, 0, B), \qquad \mathbf{E} = (0, -\dot{B}x, 0);$$

and for motion in the xy-plane, with $\dot{y} = 0$ when $x = 0$, x satisfies

$$\ddot{x} + \Omega^2 x = 0, \tag{4.108}$$

where $\Omega = eB/m$.

Suppose, now, that in some time interval τ, Ω swings over smoothly from an effectively constant value Ω_1 to an effectively constant value Ω_2. With appropriate choice of time origin and length scale the final form of x can be taken to be the real part of

$$e^{-i\Omega_2 t}, \tag{4.109}$$

and the initial form will then be the real part of

$$p\, e^{-i\Omega_1 t} + q\, e^{i\Omega_1 t}, \tag{4.110}$$

where p and q are complex constants to be determined; representation in terms of the complex exponential solutions of (4.108) is particularly convenient for the subsequent analysis. Since $\dot{y} = -\Omega x$, the final value of $(\dot{x}^2 + \dot{y}^2)/\Omega$ is Ω_2, and the initial value is

$$\Omega_1[\,|p|^2 + 2\,|pq|\cos(\arg p + \arg q) + |q|^2]. \tag{4.111}$$

The objective, therefore, is to compare (4.111) with Ω_2, under the condition that Ω varies slowly during its transition from Ω_1 to Ω_2. To the WKBJ approximation, quoted in §4.2.4, q would be zero and p would be $\sqrt{(\Omega_2/\Omega_1)}$. Here, an exact solution is obtained by identifying (4.108) with a transformation of the hypergeometric equation, and the manner in which p and q differ from $\sqrt{(\Omega_2/\Omega_1)}$ and zero, respectively, is exposed.

The hypergeometric equation

$$\zeta(1-\zeta)\frac{d^2V}{d\zeta^2} + [\gamma - (\alpha + \beta + 1)\zeta]\frac{dV}{d\zeta} - \alpha\beta V = 0 \tag{4.112}$$

is satisfied by the independent functions

$$F(\alpha, \beta, \gamma; \zeta), \qquad \xi^{1-\gamma}F(\alpha - \gamma + 1, \beta - \gamma + 1, 2 - \gamma; \zeta), \tag{4.113}$$

where

$$F(\alpha, \beta, \gamma; \zeta) = 1 + \frac{\alpha\beta}{1!\gamma}\zeta + \frac{\alpha(\alpha+1)\beta(\beta+1)}{2!\gamma(\gamma+1)}\zeta^2 + \cdots. \tag{4.114}$$

The argument ζ of F can be complex, and α, β, γ are parameters which can take arbitrary complex values. The series (4.114) converges for $|\zeta| < 1$, and F is defined for other values of ζ by analytic continuation. The solutions (4.113) are useful when $|\zeta| \ll 1$; another pair of independent solutions, useful for $|\zeta| \gg 1$, is

$$\begin{aligned} &\xi^{-\alpha}F(\alpha, \alpha - \gamma + 1, \alpha - \beta + 1; 1/\zeta), \\ &\xi^{-\beta}F(\beta, \beta - \gamma + 1, \beta - \alpha + 1; 1/\zeta). \end{aligned} \tag{4.115}$$

Each of (4.115) is, of course, linearly related to the pair (4.113); the connection formula is quoted later, when required.

If now a new function is introduced, defined by

$$U(t) = e^{at/\tau}\,(1 + e^{t/\tau})^d\, V(-e^{t/\tau}), \tag{4.116}$$

it can be checked by straightforward if tedious working that

$$\frac{d^2U}{dt^2} + \Omega^2 U = 0, \tag{4.117}$$

where

$$\Omega^2 = \Omega_1^2 + \frac{e^{t/\tau}[(\Omega_2^2 - \Omega_1^2)(1 + e^{t/\tau}) + \Omega_3^2]}{(1 + e^{t/\tau})^2}, \tag{4.118}$$

provided the relations between α, β, γ and the new parameters a, d, Ω_1, Ω_2, Ω_3 and τ are

$$\begin{gathered} \tfrac{1}{2}(\gamma - 1) = a = -i\Omega_1\tau, \qquad \tfrac{1}{2}(\alpha - \beta) = -i\Omega_2\tau, \\ \tfrac{1}{2}(\alpha + \beta - \gamma + 1) = d = \tfrac{1}{2} - \sqrt{(\tfrac{1}{4} + \tau^2\Omega_3^2)}. \end{gathered} \tag{4.119}$$

For present purposes Ω_1, Ω_2 and τ are taken to be real and positive, and Ω_3^2 is taken to be real.

Formula (4.118) gives a time variation of considerable flexibility. From the value Ω_1 at $t = -\infty$ (attained effectively when $t \ll -\tau$) Ω proceeds to the value Ω_2 at $t = +\infty$ (attained effectively when $t \gg \tau$) either monotonically or with a single stationary value, according as $|\Omega_3^2|$ is less or greater respectively than $|\Omega_1^2 - \Omega_2^2|$. The stationary value is a maximum if $\Omega_3^2 > 0$, a minimum if $\Omega_3^2 < 0$. The solution of the problem originally under discussion is therefore available in terms of hypergeometric functions for a not unduly restricted time variation of the magnetic field.

When $t \gg \tau$ the modulus of the argument of V in (4.116) is large, and the useful forms for $V(\zeta)$ are (4.115). With the further notation (cf. (4.119))

$$b = \tfrac{1}{2}(\alpha - \beta) = -i\Omega_2\tau, \tag{4.120}$$

the corresponding forms for $U(t)$ are seen to be

$$e^{\pm bt/\tau}\,(1 + e^{-t/\tau})^d\, F(a \mp b + d, -a \mp b + d, \mp 2b + 1; -e^{-t/\tau}). \tag{4.121}$$

Evidently (4.121) behaves like $\exp(\mp i\Omega_2 t)$ as $t \to \infty$, and the expression with the upper sign can therefore be identified with the solution for x, which was specified by allocating it the final form (4.109).

When $t \ll -\tau$ the modulus of the argument of V in (4.116) is small, and the useful forms for $V(\zeta)$ are (4.113). The corresponding forms for $U(t)$ are seen to be

$$e^{\pm at/\tau}\,(1 + e^{t/\tau})^d\, F(\pm a + b + d, \pm a - b + d, \pm 2a + 1, -e^{t/\tau}). \tag{4.122}$$

Evidently (4.22) behaves like $\exp(\mp i\Omega_1 t)$ as $t \to -\infty$. The constants p and q of (4.110) are therefore determined by the expression of the function (4.121) with the upper sign as a linear combination of the pair of functions (4.122). Appeal to the known connection formulas for the hypergeometric function gives

$$\begin{aligned} p &= \frac{(-2b)!\,(-2a-1)!}{(-a-b+d-1)!\,(-a-b-d)!}, \\ q &= \frac{(-2b)!\,(2a-1)!}{(a-b+d-1)!\,(a-b-d)!}. \end{aligned} \tag{4.123}$$

These exact formulas can yield explicit information. Take, for simplicity, the case $\Omega_3 = 0$. Then (4.118) can be written

$$\Omega^2 = \tfrac{1}{2}(\Omega_2^2 + \Omega_1^2) + \tfrac{1}{2}(\Omega_2^2 - \Omega_1^2)\tanh\left(\frac{t}{2\tau}\right), \tag{4.124}$$

and the transition from Ω_1 to Ω_2 is monotonic. Furthermore, $\Omega_3 = 0$ implies $d = 0$, from (4.119); and since a and b are pure imaginary, the complex conjugate of p is

$$p^* = \frac{(2b)!\,(2a-1)!}{(a+b-1)!\,(a+b)!}. \tag{4.125}$$

Now, for any η,

$$\eta! = \eta(\eta-1)!, \qquad \eta!\,(-\eta)! = \frac{\pi\eta}{\sin(\pi\eta)}.$$

Hence

$$\begin{aligned} |p|^2 = pp^* &= \frac{b}{a}\,\frac{\sin^2[\pi(a+b)]}{\sin(2\pi a)\sin(2\pi b)} \\ &= \frac{\Omega_2}{\Omega_1}\,\frac{\sinh^2[\pi(\Omega_1+\Omega_2)\tau]}{\sinh(2\pi\Omega_1\tau)\sinh(2\pi\Omega_2\tau)}. \end{aligned} \tag{4.126}$$

Similarly

$$|q|^2 = \frac{\Omega_2}{\Omega_1}\,\frac{\sinh^2[\pi(\Omega_1-\Omega_2)\tau]}{\sinh(2\pi\Omega_1\tau)\sinh(2\pi\Omega_2\tau)}. \tag{4.127}$$

The last step is to observe that the condition that Ω varies slowly means $\Omega_1\tau, \Omega_2\tau \gg 1$. Then, assuming $\Omega_1 > \Omega_2$, $|q/p|$ is of the order of magnitude $\exp(-2\pi\Omega_2\tau)$, and $|p|^2$ differs from Ω_2/Ω_1 by a quantity of the order of magnitude $\exp(-4\pi\Omega_2\tau)$. The initial value (4.111) of $(\dot{x}^2 + \dot{y}^2)/\Omega$ therefore differs from the final value Ω_2 by a quantity of the order of magnitude $\exp(-2\pi\Omega_2\tau)$. Reintroducing the parameter $\varepsilon = m/e$, this can be written $\exp(-2\pi B_2\tau/\varepsilon)$, which vanishes with ε faster than any power of ε.

4.4.5 The second and third adiabatic invariants

At the end of §4.3.2 it was remarked that second and third adiabatic invariants could exist, depending on the field configuration producing the necessary quasi periodicities in the motion.

The second, sometimes called the longitudinal, adiabatic invariant is associated with the motion of the particle parallel to the magnetic field in the case where there are two mirror points. The field variation must be so slow that as the guiding center proceeds back and forth between the mirror points the field traversed in the course of one trip is substantially the same as that traversed in the succeeding trip. This demands, in particular, that the component of the guiding center velocity across the magnetic field be very small, and equation (4.89) for the longitudinal motion is

$$\dot{u}_{\|} = -\frac{M}{m}(\text{grad } B)_{\|} + \frac{1}{\varepsilon}E_{\|} + \text{small terms.} \tag{4.128}$$

As the parallel motion only is in question, the motion can be regarded as one dimensional and expressed in terms of the single distance variable s, say, where $u_{\|} = \dot{s}$. Equation (4.128) is then of the form

$$\dot{u}_{\|} = F(s, \lambda), \tag{4.129}$$

where λ is a parameter which depends slowly on time. The introduction of the parameter λ accounts both for any actual time variation of the field, and also for any apparent time variation of the field as seen by the particle due to the slow drift of the particle across the magnetic field. The field at the particle during a trip back and forth between mirror points can be slightly different from that on the preceding trip due to either of these causes.

It is now shown that the action integral

$$J = \int u_{\|}\, ds \tag{4.130}$$

taken over an "oscillation" from one mirror point to the next and back again, is an adiabatic invariant in the sense that the rate of change of J is second order in $d\lambda/dt$.

To the first order in $d\lambda/dt$, the increment δJ in J from one oscillation to the next is

$$\delta J = \int \delta u_{\|}\, ds, \tag{4.131}$$

where $\delta u_{\|}$ denotes the corresponding increment in $u_{\|}$ for each value of s; the increments in the initial and final values of s, though not in general zero, nevertheless make no contribution to δJ since $u_{\|}$ vanishes at the extremes of the path of integration. Now introduce the quantity

$$W = \tfrac{1}{2}u_{\|}^2 + V(s, \lambda), \tag{4.132}$$

where

$$\frac{\partial V(s, \lambda)}{\partial s} = -F(s, \lambda). \tag{4.133}$$

Hence, to the first order,

$$\delta W = u_{\|} \, \delta u_{\|} + \delta V, \tag{4.134}$$

and (4.131) can be written

$$\delta J = \int_0^T (\delta W - \delta V) \, dt', \tag{4.135}$$

where t' is the time corresponding to position s, so that $ds = u_{\|} \, dt'$, and T is the period of the oscillation.

The objective, then, is to prove that (4.135) vanishes to the first order in $d\lambda/dt$. We note that W is slowly varying, since it would be a constant (the "energy") were λ constant. In fact, from (4.132),

$$\frac{dW}{dt} = u_{\|}\dot{u}_{\|} + u_{\|} V_s + V_\lambda \frac{d\lambda}{dt},$$

where V_s, V_λ are the partial derivatives of $V(s, \lambda)$; and from (4.129), (4.133) this gives simply

$$\frac{dW}{dt} = V_\lambda \frac{d\lambda}{dt}. \tag{4.136}$$

It is convenient to introduce, as a small quantity to indicate magnitudes, a representative value η of $d\lambda/dt$. Then, to order η, δW is independent of t' and from (4.136) is

$$\delta W = \eta \int_0^T V_\lambda[s(t), \lambda(t)] \, dt. \tag{4.137}$$

Also, evidently, to order η,

$$\delta V = \eta T V_\lambda[s(t'), \lambda(t')]. \tag{4.138}$$

It follows at once that (4.135) vanishes to order η, which is the required result.

The way in which the second adiabatic invariant can be used is typified by the situation of a particle in a magnetic bottle in which the two mirror points approach or recede from each other. If the distance between the mirror points along a line of magnetic induction is d, the constancy of J implies that the average value of $u_{\|}$ is proportional to $1/d$. The longitudinal kinetic energy $\frac{1}{2}mu_{\|}^2$ is therefore proportional to $1/d^2$. For a collection of particles in the bottle the number per unit volume, neglecting any variation of lateral spread, is sensibly proportional to $1/d$, so that the parallel pressure

$p_{\|}$ is proportional to $1/d^3$. That $p_{\|}$ times the cube of the effective volume is constant is the adiabatic gas relation with an equivalent specific heat ratio of three.

Another context is that of a particle trapped in an approximate dipole field, the familiar example being the geomagnetic field. If the field were static and strictly axially symmetric J would remain strictly constant; the situation could be described by saying that the line of induction along which mirroring takes place simply precesses uniformly round the axis, the precession being due, of course, to the gradient drift (4.93). But if the field departed slightly from axial symmetry, or had a slow time variation, the adiabatic invariance of J would yield significant information. For the case of the static imperfect dipole field the motion of the particle is prescribed completely by its initial position and the values of M (the magnetic moment) and J; in particular, the field line along which mirroring takes place reverts to the initial line after each circumambulation of the axis.

The approximate dipole field with time variation slow even compared with the period of the precession just mentioned is the standard peg on which to hang the discussion of the third, sometimes called the flux, adiabatic invariant Φ. As has just been indicated, if the time variation of the field were halted at any stage, the section of the field line along which mirroring takes place would sweep out a surface akin to the curved surface of a barrel, albeit distorted in some way; and Φ is the flux of magnetic induction through this open ended barrel. When the field *is* slowly time varying the barrel-like surface does not quite close up, and the definition of Φ is reminiscent of the description of the first adiabatic invariant in terms of the flux through the curve formed by the projection of the particle's motion on a plane perpendicular to the magnetic field.

An outline only of one proof of the adiabatic invariance of Φ will be given here. The field is represented in terms of coordinates α, β, both constant along any line of magnetic induction, such that the vector potential is

$$\mathbf{A} = \alpha \operatorname{grad} \beta; \tag{4.139}$$

it follows that

$$\mathbf{B} = (\operatorname{grad} \alpha) \times (\operatorname{grad} \beta), \tag{4.140}$$

and

$$\mathbf{E} = -\operatorname{grad}(\phi + \psi) + \frac{\partial \beta}{\partial t} \operatorname{grad} \alpha - \frac{\partial \alpha}{\partial t} \operatorname{grad} \beta, \tag{4.141}$$

where ϕ is the scalar potential and $\psi = \alpha\, \partial\beta/\partial t$. Then from

$$\dot{\alpha} = \frac{\partial \alpha}{\partial t} + \mathbf{u}_{\perp} \cdot \operatorname{grad} \alpha, \tag{4.142}$$

a corresponding expression for β, and

$$\mathbf{u}_\perp = \mathbf{u}_E + \varepsilon\frac{\mathbf{B}}{B} \times \left\{\frac{M}{m}\operatorname{grad} B + \frac{u_\parallel^2}{B^2}(\mathbf{B}\cdot\operatorname{grad})\mathbf{B}\right\}, \tag{4.143}$$

which is (4.88) with the $\mathbf{u}_E$ and $\partial/\partial t$ terms inside the curly bracket dropped, it can be shown that

$$e\bar{\dot{\alpha}} = -\partial K/\partial\beta, \qquad e\bar{\dot{\beta}} = \partial K/\partial\alpha, \tag{4.144}$$

where the bars denote the average value over a mirror period, and

$$K = \tfrac{1}{2}u_\parallel^2 + \frac{M}{m}B + \frac{1}{\varepsilon}(\phi + \psi) \tag{4.145}$$

is regarded as a function of α, β, M, J and a parameter which is a slowly varying function of time; it is observed that the zero-order energy equation (4.90), with the $\mathbf{u}_E$ and $\partial/\partial t$ terms dropped, is $K =$ constant. The Hamiltonian form of (4.144) implies the adiabatic invariance of

$$\int \alpha\, d\beta = \int \alpha(\operatorname{grad}\beta)\cdot d\mathbf{s} = \int \mathbf{A}\cdot d\mathbf{s} = \Phi.$$

4.5 SOME EXAMPLES OF RELATIVISTIC MOTION

4.5.1 Motion in uniform static fields

Charged particles can well acquire such high energies that the relativistic equation of motion (4.1) must be used. One simple but important general feature is noted immediately. When $\mathbf{E}$ is everywhere zero, so that the field is purely magnetostatic, v is constant and the only modification of any result derived from the non-relativistic equation is in the replacement of the rest mass m_0 by $m_0/\sqrt{(1 - v^2/c^2)}$. In a uniform magnetostatic field $\mathbf{B}_0$, for example, the motion consists of an arbitrary constant velocity parallel to $\mathbf{B}_0$ together with uniform circular motion perpendicular to $\mathbf{B}_0$ at angular frequency

$$\frac{eB_0}{m_0}\sqrt{(1 - v^2/c^2)}. \tag{4.146}$$

The gyro frequency is thus inversely proportional to the relativistic energy.

If an electric field is present, v varies during the motion, and no such simple general result is available. In this section an elementary analysis is given of the motion, first in a uniform electrostatic field and then in mutually perpendicular electrostatic and magnetostatic fields.

Suppose there is a uniform electrostatic field $(E, 0, 0)$. If the motion is confined to the x-axis the equation of motion integrates at once to

$$\frac{m_0 v}{\sqrt{(1 - v^2/c^2)}} = eEt, \tag{4.147}$$

with appropriate choice of time zero. The equation can easily be solved for v in terms of t and integrated again. But it is even quicker to use also the energy equation (4.3), which here is

$$\frac{m_0c^2}{\sqrt{(1-v^2/c^2)}} = eEx, \tag{4.148}$$

with appropriate choice of origin; for the elimination of v between (4.147) and (4.148) gives immediately

$$x = c\sqrt{\left(\frac{m_0^2c^2}{e^2E^2} + t^2\right)}. \tag{4.149}$$

In the motion specified by (4.149) the speed tends to the limit c as $t \to \pm\infty$. As t comes up to zero the particle moves towards the origin along the positive x-axis with a deceleration which brings it to rest at $t = 0$; it then accelerates back on the return journey with a speed that approaches c asymptotically. The motion is, in fact, that sometimes called *hyperbolic*, in which the space part of the relativistic four-acceleration is constant. Hyperbolic motion is of theoretical interest in connection with the force of radiation reaction, and is considered further in §4.6.

The general motion in the electrostatic field $(E, 0, 0)$ can be found in much the same way. The particle evidently remains in a plane, say $z = 0$. Then, with appropriate choice of time zero, the x- and y-components of the equation of motion integrate to

$$\frac{m_0\dot{x}}{\sqrt{(1-v^2/c^2)}} = eEt \tag{4.150}$$

$$\frac{m_0\dot{y}}{\sqrt{(1-v^2/c^2)}} = p_0, \tag{4.151}$$

where the constant p_0 is the initial value of the relativistic momentum, and $v^2 = \dot{x}^2 + \dot{y}^2$. By squaring and adding

$$\frac{m_0^2v^2}{1-v^2/c^2} = e^2E^2t^2 + p_0^2, \tag{4.152}$$

and the elimination of v^2 between (4.152) and the energy equation (4.148) gives

$$x = c\sqrt{\left(\frac{p_0^2 + m_0^2c^2}{e^2E^2} + t^2\right)}. \tag{4.153}$$

Also, from (4.151) and (4.148),

$$\dot{y} = \frac{c^2p_0}{eE}x;$$

and substitution for x from (4.153) followed by integration results in

$$y = \frac{cp_0}{eE}\sinh^{-1}\left\{\frac{eEt}{\sqrt{(p_0^2 + m_0^2c^2)}}\right\}. \tag{4.154}$$

The equation of the path is obtained by eliminating t between (4.153) and (4.154). It can be written

$$x = \frac{c\sqrt{(p_0^2 + m_0^2c^2)}}{eE}\cosh\left(\frac{eEy}{cp_0}\right). \tag{4.155}$$

It should perhaps be remarked that the motion (4.153), (4.154) is *not* hyperbolic. General hyperbolic motion, in a plane rather than along a line, could be derived from (4.149) by a Lorentz transformation; and such motion would be that in which the particle was acted on by the field that was the transformation of $(E, 0, 0)$, a field which would, in fact, be partly magnetic.

Consider, now, mutually perpendicular uniform electrostatic and magnetostatic fields, say $(E, 0, 0)$ and $(0, 0, B)$. It is recalled that in the non-relativistic approximation, apart from an arbitrary constant velocity in the z-direction, the motion consists of the constant velocity E/B in the y-direction superposed on uniform circular motion in the xy-plane. Clearly such motion is inadmissible when $E > cB$, and this case can only be handled relativistically; in fact, as will be seen, the motion then takes on something of the character of the case $B = 0$, in being no longer bounded in the x-coordinate.

It is convenient to take $t = 0$ when $\dot{x} = 0$, and also then $x = a$, $y = z = 0$; the oscillation of x for $E < cB$ may be located between a and $-a$. With these initial conditions the once integrated components of the equation of motion are

$$\frac{m_0\dot{x}}{\sqrt{(1 - v^2/c^2)}} = eEt + eBy, \tag{4.156}$$

$$\frac{m_0\dot{y}}{\sqrt{(1 - v^2/c^2)}} = -eB(x - a) - p_2, \tag{4.157}$$

$$\frac{m_0\dot{z}}{\sqrt{(1 - v^2/c^2)}} = p_3, \tag{4.158}$$

where the constants $-p_2, p_3$ are the initial values of the y- and z-components of the relativistic momentum; and the energy equation is

$$\frac{m_0c^2}{\sqrt{(1 - v^2/c^2)}} = eE(x - a) + \mathscr{E}, \tag{4.159}$$

where $\mathscr{E}$ is the initial value of the relativistic energy given in terms of p_2 and p_3 by the relation

$$\mathscr{E}^2/c^2 = p_2^2 + p_3^3 + m_0^2c^2. \tag{4.160}$$

Squaring and adding (4.156), (4.157) and (4.158), and then substituting for v^2 on the left-hand side from (4.159), gives

$$e^2(Et+By)^2+[eB(x-a)+p_2]^2+p_3^2 = \frac{1}{c^2}[eE(x-a)+\mathscr{E}]^2 - m_0^2c^2; \quad (4.161)$$

from which, again with the help of (4.159), equation (4.156) can be converted into a first order differential equation for x, namely

$$\frac{1}{c^2}[eE(x-a)+\mathscr{E}]\dot{x}$$
$$= \pm\sqrt{\left\{\frac{1}{c^2}[eE(x-a)+\mathscr{E}]^2-[eB(x-a)+p_2]^2-p_3^2-m_0^2c^2\right\}}. \quad (4.162)$$

The coefficient of e^2x^2 in the quadratic expression under the radical sign on the right-hand side of (4.162) is $E^2/c^2 - B^2$. It is therefore positive or negative according as E is greater or less than cB, and the implication is evidently that for $E > cB$ the x-coordinate ranges from a to infinity, whereas for $E < cB$ the x-coordinate oscillates between a and some lesser value. This latter case is now examined in a little more detail.

It is convenient, by choice of origin, to make $-a$ the other value of x where $\dot{x} = 0$. This puts (4.162) in the form

$$\frac{E/c}{\sqrt{(c^2B^2 - E^2)}}\dot{x} = \pm\frac{\sqrt{(a^2 - x^2)}}{x - a + \mathscr{E}/(eE)}, \quad (4.163)$$

with

$$a = \frac{Bp_2 - E\mathscr{E}/c^2}{e(B^2 - E^2/c^2)}, \quad (4.164)$$

which can also be written

$$a = \frac{m_0}{eB\sqrt{(1 - u^2/c^2)}}\frac{u_2 - E/B}{1 - E^2/(c^2B^2)}, \quad (4.165)$$

where $-u_2$ and u are the initial values of $\dot{y}$ and v respectively. As a check it may be noted that (4.165) reduces correctly to the relativistic gyro radius when $E = 0$, and also has the non-relativistic approximation

$$m_0(u_2 - E/B)/(eB),$$

which is again seen to be the gyro radius when allowance is made for the superposition of the drift velocity E/B on the gyro motion.

Integration of (4.163) gives

$$ct = \frac{E/(cB)}{\sqrt{[1 - E^2/(c^2B^2)]}}\left\{\sqrt{(a^2 - x^2)} + \left(\frac{\mathscr{E}}{eE} - a\right)\cos^{-1}\left(\frac{x}{a}\right)\right\}. \quad (4.166)$$

The period of libration of the x-coordinate is therefore

$$T = \frac{2\pi\mathscr{E}}{eBc^2} \frac{1 - eEa/\mathscr{E}}{\sqrt{[1 - E^2/(c^2B^2)]}}. \tag{4.167}$$

The y-coordinate can be obtained from (4.161) in the form

$$y = -\frac{E}{B}t \mp \sqrt{[1 - E^2/(c^2B^2)]}\sqrt{(a^2 - x^2)}. \tag{4.168}$$

In the period T, y decreases by $-ET/B$, so the drift velocity is still E/B in the relativistic treatment.

If (4.166) is written as

$$x = a \cos \Phi, \tag{4.169}$$

where

$$\Phi = \frac{\sqrt{[1 - E^2/(c^2B^2)]}}{\dfrac{E}{cB}\left(\dfrac{\mathscr{E}}{eE} - a\right)}\left\{ct - \frac{E/(cB)}{\sqrt{[1 - E^2/(c^2B^2)]}}\sqrt{(a^2 - x^2)}\right\}, \tag{4.170}$$

it is easy to confirm that

$$\frac{d\Phi}{dt} = \frac{eB}{m_0}\sqrt{(1 - v^2/c^2)}\sqrt{[1 - E^2/(c^2B^2)]}. \tag{4.171}$$

The expression (4.170) is an "instantaneous" frequency, which indicates how the relativistic gyro frequency is modified by the electric field.

A rather more formal analysis of the problem of relativistic motion in uniform electrostatic and magnetostatic fields proceeds from the four-dimensional equations (2.174). These are linear equations for ct, x, y, z in terms of the interval s as the independent variable, and can be solved as such. The solutions are obtained in the form typified by (4.169) with Φ specified by (4.171). From this starting point the investigation of adiabatic invariants and drifts in slowly varying fields has been carried through relativistically.

4.5.2 Motion in the field of a time-harmonic plane wave

The motion of a charged particle in a time-harmonic wave field is of interest in the study of the interaction of plasmas and electromagnetic waves, and in other contexts. Linearized non-relativistic theory is the standard basis for many results, such as the scattering cross-section calculation of §4.6.2 and the whole of the development of Chapter 5. It is possible, however, with quite elementary analysis, to treat exactly the relativistic motion of a particle in the field of a time-harmonic plane wave. This is now done, first for the case in which there is no field other than that of the plane wave, and subsequently for the case in which there is also a uniform magnetostatic field parallel to the direction of propagation of the plane wave. The latter case is of particular interest because of the resonance which occurs when the wave frequency has a certain value closely related to the gyro-frequency.

For a homogeneous progressive plane wave traveling in the direction of the unit vector **n** the electric and magnetic vectors are perpendicular to **n** and are related by

$$\mathbf{B} = \frac{1}{c}\mathbf{n} \times \mathbf{E}. \tag{4.172}$$

Substituting for **B** into the right-hand side of the equation of motion (4.1) gives

$$\frac{d}{dt}\left\{\frac{\mathbf{v}}{\sqrt{(1 - v^2/c^2)}}\right\} = \frac{e}{m_0}(1 - \mathbf{n}\cdot\mathbf{v}/c)\mathbf{E} + \frac{e}{m_0 c}(\mathbf{E}\cdot\mathbf{v})\mathbf{n}. \tag{4.173}$$

It is first observed that the scalar product of (4.173) with **n**, namely

$$\frac{d}{dt}\left\{\frac{\mathbf{n}\cdot\mathbf{v}}{\sqrt{(1 - v^2/c^2)}}\right\} = \frac{e}{m_0 c}\mathbf{E}\cdot\mathbf{v}, \tag{4.174}$$

together with the energy equation (4.2), establishes that

$$\lambda \equiv \frac{1 - \mathbf{n}\cdot\mathbf{v}/c}{\sqrt{(1 - v^2/c^2)}} \tag{4.175}$$

is a constant.

Now discard t in favor of the phase variable

$$\phi = \omega(t - \mathbf{n}\cdot\mathbf{r}/c), \tag{4.176}$$

where ω is the angular frequency of the wave and **r** is the position vector of the particle. Then

$$\mathbf{v} = \frac{d\mathbf{r}}{dt} = \omega(1 - \mathbf{n}\cdot\mathbf{v}/c)\mathbf{r}', \tag{4.177}$$

where the dash denotes differentiation with respect to ϕ; and, using (4.175),

$$\frac{d}{dt}\left\{\frac{\mathbf{v}}{\sqrt{(1 - v^2/c^2)}}\right\} = \lambda\frac{d}{dt}\left(\frac{\mathbf{v}}{1 - \mathbf{n}\cdot\mathbf{v}/c}\right) = \omega^2\lambda(1 - \mathbf{n}\cdot\mathbf{v}/c)\mathbf{r}''. \tag{4.178}$$

The substitution of (4.177) and (4.178) into (4.173) enables the factor $1 - \mathbf{n}\cdot\mathbf{v}/c$ to be cleared, and leaves

$$\mathbf{r}'' = \frac{e}{\lambda m_0 \omega^2}\left[\mathbf{E} + \frac{\omega}{c}(\mathbf{E}\cdot\mathbf{r}')\mathbf{n}\right], \tag{4.179}$$

which is a second-order linear differential equation for r as a function of ϕ.

As an example, take the case of a linearly polarized wave, with

$$\mathbf{n} = (0, 0, 1), \qquad \mathbf{E} = E\cos\phi(1, 0, 0), \qquad \mathbf{B} = \frac{E}{c}\cos\phi(0, 1, 0),$$

where E is a constant and $\phi = \omega(t - z/c)$. Then the components of (4.179) are

$$x'' = \frac{eE}{\lambda m_0 \omega^2} \cos\phi, \qquad y'' = 0, \qquad z'' = \frac{eE}{c\lambda m_0 \omega} x' \cos\phi. \tag{4.180}$$

The solution of (4.180) when all non-oscillatory terms are discarded is

$$x = -\frac{eE}{\lambda m_0 \omega^2} \cos\phi, \qquad y = 0, \qquad z = -\frac{e^2E^2}{8c\lambda^2 m_0^2 \omega^3} \sin(2\phi), \tag{4.181}$$

which gives a trajectory in the xz-plane somewhat as sketched in Fig. 4.8.

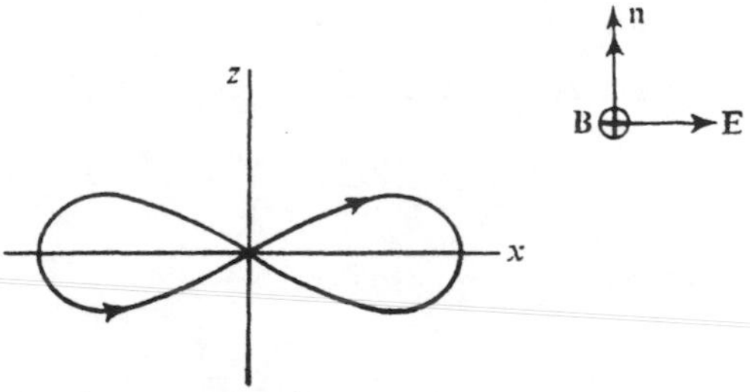

Fig. 4.8. Oscillatory part of the trajectory of a charged particle in the field of a time harmonic linearly polarized plane wave.

In the usual approximation to the problem of motion in the field of an electromagnetic wave the point is made straight away that the magnetic force on the particle is smaller by a factor v/c than the electric force, and can therefore be neglected in a non-relativistic treatment. The approximate solution corresponding to (4.181) is then simply

$$x = -\frac{eE}{m_0 \omega^2} \cos(\omega t), \qquad y = z = 0. \tag{4.182}$$

It is in fact apparent in (4.181) that the amplitude of the z-excursion is less than that of x by a factor of the magnitude of v/c.

Consider now the original problem of an arbitrarily polarized plane wave traveling in the direction $\mathbf{n}$, but with the superposition of a uniform magnetostatic field $\mathbf{B}_0 = B_0\mathbf{n}$. The equation of motion (4.173) acquires the additional term $(e/m_0)\mathbf{v} \times \mathbf{B}_0$ on the right-hand side. This evidently leaves unaltered (4.174) and the conclusion that (4.175) is a constant. There is, however, a corresponding term to add to the right-hand side of (4.179), which with this modification reads

$$\mathbf{r}'' = \frac{e}{\lambda m_0 \omega^2}\left[\mathbf{E} + \frac{\omega}{c}(\mathbf{E}\cdot\mathbf{r}')\mathbf{n}\right] + \frac{\Omega}{\omega}\mathbf{r}' \times \mathbf{n}, \tag{4.183}$$

where the constant

$$\Omega = eB_0/(\lambda m_0) \tag{4.184}$$

is the relativistic gyro-frequency associated with the field $\mathbf{B}_0$, divided by the factor $1 - \mathbf{n}\cdot\mathbf{v}/c$.

As an example, take the case of a circularly polarized wave, with $\mathbf{n} = (0, 0, 1)$ and

$$\mathbf{E} = E(\cos\phi, -\sin\phi, 0), \qquad \mathbf{B} = \frac{E}{c}(\sin\phi, \cos\phi, 0). \tag{4.185}$$

The components of (4.183) are then

$$x'' = \frac{eE}{\lambda m_0\omega^2}\cos\phi + \frac{\Omega}{\omega}y', \tag{4.186}$$

$$y'' = -\frac{eE}{\lambda m_0\omega^2}\sin\phi - \frac{\Omega}{\omega}x', \tag{4.187}$$

$$z'' = \frac{eE}{c\lambda m_0\omega}(x'\cos\phi - y'\sin\phi). \tag{4.188}$$

With $\xi = x + iy$, equations (4.186) and (4.187) combine into

$$\xi'' + i\frac{\Omega}{\omega}\xi' = \frac{eE}{\lambda m_0\omega^2}e^{-i\phi}, \tag{4.189}$$

which puts in evidence the fact that there is resonance when $\omega = \Omega$. If e and B_0 are positive, so is Ω, and the resonance arises by virtue of the fact that the rotation of the field vectors in (4.185) is in the sense of a left-handed screw about $\mathbf{B}_0$, the same as that of the gyration of the particle were it only under the influence of $\mathbf{B}_0$.

The solution of (4.189) gives expressions for x and y, and these are fed back into (4.188), which in turn is solved for z. The general solution in the resonance case $\omega = \Omega$ is readily shown to be

$$x = \alpha_1 + \alpha_2\cos\phi - \alpha_3\sin\phi + \frac{E}{B_0\omega}\phi\sin\phi, \tag{4.190}$$

$$y = \alpha_4 + \alpha_3\cos\phi + \alpha_2\sin\phi + \frac{E}{B_0\omega}\phi\cos\phi, \tag{4.191}$$

$$z = \alpha_5 + \alpha_6\phi + \frac{E\alpha_3}{2cB_0}\phi^2 + \frac{E^2}{6cB_0^2\omega}\phi^3, \tag{4.192}$$

where $\alpha_1, \alpha_2, \ldots, \alpha_6$ are constants. It is not difficult to confirm mathematically that (4.192) gives just one real value for z as a function of t. Furthermore, remembering that $\phi = \omega(t - z/c)$, it is clear that, as $t \to \infty$,

$$\phi \sim \left(\frac{6c^2B_0^2}{E^2}\omega t\right)^{1/3}, \qquad z \sim ct - \frac{c}{\omega}\left(\frac{6c^2B_0^2}{E^2}\omega t\right)^{1/3}. \tag{4.193}$$

Ultimately, therefore, the particle velocity in the z-direction approaches c, and the projection on the xy-plane of the position vector of the particle becomes parallel to the magnetic induction of the wave, by virtue of the dominance of the final terms in (4.190) and (4.191).

4.6 THE SELF-FORCE

4.6.1 Derivation

The analysis so far presented in this chapter has been made on the premise that the field vectors **E**, **B** at the particle, which determine the Lorentz force, are suitably well-behaved functions of position and time; and where their behavior was quoted explicitly the assumption was tacit that this represented a field that was independent of the particle. In fact, of course, the particle itself gives rise to a field, and this field must certainly, in principle at least, be superposed on the "external" field and taken into account. For the sake of a label, the part of the Lorentz force contributed by the field due to the particle may be called the *self-force*.

When the self-force is in question a dichotomy arises in thinking of the electron in classical terms; for the tendency has been to regard it as a fundamental "point" particle, whereas a calculation of the self-force that avoids infinities and self-propulsion appears to require some detailed knowledge of its structure. A number of different theories have been proposed, as mentioned in §4.1, but as yet there are no grounds for granting any of them complete acceptance. This section presents some discussion of what is perhaps the simplest and most standard theoretical formulation, a formulation which can be used to good effect in practical calculations for the case of quasi periodic motion when the self-force is small compared with the external force.

The theory treats, by the classical electrodynamics of Chapter 2, a model electron in which the charge distribution is assumed to be rigid in the sense that, at any time t, there is a Lorentz frame in which each element of the charge is instantaneously at rest. To keep the algebra from getting unwieldy the calculation is performed in the rest frame, but the main result is subsequently generalized by the appropriate relativistic transformation. In the rest frame the self-force is the volume integral

$$\int \rho \mathbf{E}\, d\tau \tag{4.194}$$

taken over the charge distribution, naturally assumed to be spherically symmetrical, where, at the volume element $d\tau$, ρ is the charge density, and **E** the field generated solely by the charge itself.

The complication in the analysis arises from the fact that, since **E** has to be calculated at internal points, the difference in the retarded times associated with the contributions to **E** made by different volume elements is significant. However, on the understanding that the dimensions of the charge are certainly small, it is natural to represent retarded quantities as expansions about the instantaneous time t at which the force (4.194) is evaluated. The idea is similar to that indicated by (2.68) in the discussion of the dipole approximation to the field radiated by a charge distribution, but has here to be carried through to considerably higher accuracy. Within this general technique alternative formulations are available for determining **E**. One is to use the integrals (2.61), (2.62) for the retarded potentials, or equivalently (2.64) for **E**, in which the integrands represent the contributions to **E** from each fixed volume element of space; the other is to use the expression (2.122) to obtain the contribution from each moving element of charge. The latter method is given here.

In (2.122) $\mathbf{r} = r\hat{\mathbf{r}}$ is the vector from the charge element to the field point, and the objective is to represent the entire retarded quantity as an expansion in ascending powers of $r(t)$. The lowest power in the expansion is evidently r^{-3}, and it is necessary to find the terms explicitly at least up to that independent of r; those involving positive powers of r vanish for a point electron.

Heavy square brackets signify, as in Chapter 2, evaluation at the retarded time $t - [r]/c$, and quantities not so bracketed are understood to be evaluated at t. Then the Taylor expansion of $[r]$ to the requisite number of terms, remembering that $\mathbf{v} = -\dot{\mathbf{r}} = 0$, is

$$[\mathbf{r}] = \mathbf{r} - \frac{[r^2]}{2c^2}\dot{\mathbf{v}} + \frac{[r^3]}{6c^3}\ddot{\mathbf{v}} + \cdots \tag{4.195}$$

On squaring,

$$[r^2] = r^2\left(1 + \frac{\mathbf{r}\cdot\dot{\mathbf{v}}}{c^2}\right)^{-1}\left(1 + \frac{\mathbf{r}\cdot\ddot{\mathbf{v}}}{3c^3}r + \frac{\dot{\mathbf{v}}^2}{4c^4}r^2 + \cdots\right), \tag{4.196}$$

from which $[r^2]$ in the second term on the right-hand side of (4.195) can be replaced by $r^2(1 - \mathbf{r}\cdot\dot{\mathbf{v}}/c^2)$, and $[r^3]$ in the third term by r^3, to give

$$[\mathbf{r}] = \mathbf{r} - \frac{\dot{\mathbf{v}}}{2c^2}r^2 + \frac{(\mathbf{r}\cdot\dot{\mathbf{v}})\dot{\mathbf{v}}}{2c^4}r^2 + \frac{\ddot{\mathbf{v}}}{6c^3}r^3 + O(r^4). \tag{4.197}$$

Similarly

$$[\mathbf{v}] = -\frac{[r]}{c}\dot{\mathbf{v}} + \frac{[r^2]}{2c^2}\ddot{\mathbf{v}} + \cdots$$

gives

$$[\mathbf{v}] = -\frac{r}{c}\left(\dot{\mathbf{v}} - \frac{\mathbf{r}\cdot\dot{\mathbf{v}}}{2c^2}\dot{\mathbf{v}} - \frac{\ddot{\mathbf{v}}}{2c}r\right) + O(r^3); \tag{4.198}$$

and again, at once,

$$[\dot{\mathbf{v}}] = \dot{\mathbf{v}} - \frac{\ddot{\mathbf{v}}}{c} r + O(r^2). \tag{4.199}$$

In both (4.198) and (4.199) just the number of explicit terms subsequently required have been retained.

Now from (2.122) the contribution to $\mathbf{E}$ from the charge element de' is

$$d\mathbf{E} = \frac{de'}{4\pi\varepsilon_0}(\mathbf{K}_1 + \mathbf{K}_2), \tag{4.200}$$

where

$$\mathbf{K}_1 = \left[\frac{1}{\kappa^3 r^3}(1 - v^2/c^2)\left(\mathbf{r} - \frac{\mathbf{v}}{c} r\right)\right], \tag{4.201}$$

$$\mathbf{K}_2 = -\frac{1}{c^2}\left[\frac{1}{\kappa^3 r^3}\mathbf{r} \times \left\{\dot{\mathbf{v}} \times \left(\mathbf{r} - \frac{\mathbf{v}}{c} r\right)\right\}\right]. \tag{4.202}$$

For the evaluation of $\mathbf{K}_1$ it is seen that $1/[\kappa^3 r^3]$ must be developed up to the r^{-1} terms, and $\mathbf{r} - \mathbf{v}r/c$ up to the r^3 terms; for $\mathbf{K}_2$, fewer terms are required. Since

$$[\kappa r] = [r] - [\mathbf{r}\cdot\mathbf{v}]/c, \tag{4.203}$$

and from (4.196)

$$[r] = r\left\{1 - \frac{\mathbf{r}\cdot\dot{\mathbf{v}}}{2c^2} + \frac{3(\mathbf{r}\cdot\dot{\mathbf{v}})^2}{8c^4} + \frac{\dot{\mathbf{v}}^2}{8c^4} r^2 + \frac{\mathbf{r}\cdot\ddot{\mathbf{v}}}{6c^3} r\right\} + O(r^4), \tag{4.204}$$

the subtraction from (4.204) of $1/c$ times the scalar product of (4.197) and (4.198) yields

$$[\kappa r] = r\left\{1 + \frac{\mathbf{r}\cdot\dot{\mathbf{v}}}{2c^2} - \frac{(\mathbf{r}\cdot\dot{\mathbf{v}})^2}{8c^4} - \frac{3\dot{\mathbf{v}}^2}{8c^4} r^2 - \frac{\mathbf{r}\cdot\ddot{\mathbf{v}}}{3c^3} r\right\} + O(r^4), \tag{4.205}$$

which in turn gives

$$\frac{1}{[\kappa^3 r^3]} = \frac{1}{r^3}\left\{1 - \frac{3\mathbf{r}\cdot\dot{\mathbf{v}}}{2c^2} + \frac{15(\mathbf{r}\cdot\dot{\mathbf{v}})^2}{8c^4} + \frac{9\dot{\mathbf{v}}^2}{8c^4} r^2 + \frac{\mathbf{r}\cdot\ddot{\mathbf{v}}}{c^3} r\right\} + O(1). \tag{4.206}$$

Also, from (4.197), (4.198) and the first two terms of (4.204),

$$\left[\mathbf{r} - \frac{\mathbf{v}}{c} r\right] = \mathbf{r} + \frac{\dot{\mathbf{v}}}{2c^2} r^2 - \frac{(\mathbf{r}\cdot\dot{\mathbf{v}})\dot{\mathbf{v}}}{2c^4} r^2 - \frac{\ddot{\mathbf{v}}}{3c^3} r^3 + O(r^4); \tag{4.207}$$

and

$$[1 - v^2/c^2] = 1 - \frac{\dot{\mathbf{v}}^2}{c^4} r^2 + O(r^3). \tag{4.208}$$

All the material is now to hand for writing down $\mathbf{K}_1$ and $\mathbf{K}_2$. The complete expressions are rather lengthy, but (4.194), which now, with $\rho\, d\tau = de$, appears as

$$\frac{1}{4\pi\varepsilon_0}\iint(\mathbf{K}_1 + \mathbf{K}_2)\, de\, de', \tag{4.209}$$

is integrated over a spherically symmetric distribution, so that the terms in $\mathbf{K}_1$ and $\mathbf{K}_2$ that contain odd powers of the vector $\mathbf{r}$ make no contribution to the self-force. If $\mathbf{K}_1'$ and $\mathbf{K}_2'$ stand for the corresponding expressions with the odd powers of $\mathbf{r}$ omitted it is found that

$$\mathbf{K}_1' = -\frac{3(\mathbf{r}\cdot\dot{\mathbf{v}})\mathbf{r}}{2c^2r^3} + \frac{\dot{\mathbf{v}}}{2c^2r} - \frac{\ddot{\mathbf{v}}}{3c^3} + \frac{(\mathbf{r}\cdot\ddot{\mathbf{v}})\mathbf{r}}{c^3r^2} + O(r), \tag{4.210}$$

$$\mathbf{K}_2' = \frac{(\mathbf{r}\cdot\dot{\mathbf{v}})\mathbf{r}}{c^2r^3} - \frac{\dot{\mathbf{v}}}{c^2r} + \frac{\ddot{\mathbf{v}}}{c^3} - \frac{(\mathbf{r}\cdot\ddot{\mathbf{v}})\mathbf{r}}{c^3r^2} + O(r), \tag{4.211}$$

so that

$$\mathbf{K}_1' + \mathbf{K}_2' = -\frac{(\mathbf{r}\cdot\dot{\mathbf{v}})\mathbf{r}}{2c^2r^3} - \frac{\dot{\mathbf{v}}}{2c^2r} + \frac{2\ddot{\mathbf{v}}}{3c^3} + O(r). \tag{4.212}$$

When (4.212) is substituted into (4.209) and the integration performed, the contribution from the first term is, from the spherical symmetry, precisely one third that of the second term. The final expression for the self-force is therefore

$$\mathbf{F}_s = -\frac{4U}{3c^2}\dot{\mathbf{v}} + \frac{e^2}{6\pi\varepsilon_0c^3}\ddot{\mathbf{v}} + O(r_e), \tag{4.213}$$

where $-e$ is the total charge (negative for an electron), r_e the radius, and

$$U = \frac{1}{8\pi\varepsilon_0}\iint\frac{de\, de'}{r} \tag{4.214}$$

the electrostatic energy, of the charge distribution.

4.6.2 Interpretation

On the basis of the calculation just presented it is presumed that the non-relativistic equation of motion of an electron is

$$m'\dot{\mathbf{v}} = \mathbf{F}_e + \mathbf{F}_s, \tag{4.215}$$

where $\mathbf{F}_e$ is the external force and $\mathbf{F}_s$ is given by (4.213). The "inertial mass" is for the moment written m' rather than m_0, because there is no way, by observation, of distinguishing its effect from that of the "electromagnetic mass"

$$\frac{4U}{3c^2} \tag{4.216}$$

implicit in the term in $\mathbf{F}_s$ proportional to $\dot{\mathbf{v}}$. In the early days of electron theory it was indeed hoped that all the mass could be accounted for electromagnetically, but this idea, at least in the simple formulation presented here, fails because the energy and momentum of the field of a moving charge do not constitute a 4-vector as must the energy and momentum of a particle. The discrepancy is shown up in (4.216) by the factor $\frac{4}{3}$ in the mass to energy ratio. However, the obvious procedure is to combine (4.216) and m' into the observed mass m_0. If also the $O(r_e)$ terms in $\mathbf{F}_s$ are dropped on the supposition that they are negligible, or even perhaps zero were the aesthetically attractive idea of a "point" electron tenable, the equation of motion becomes

$$\dot{\mathbf{v}} = \frac{1}{m_0}\mathbf{F}_e + \frac{e^2}{6\pi\varepsilon_0 m_0 c^3}\ddot{\mathbf{v}}. \tag{4.217}$$

Equation (4.217) is the non-relativistic statement of the standard formulation previously mentioned as being, in certain circumstances, adequate for practical calculations. Before discussing the equation the questionable aspects of its derivation should be stressed. The "point" electron admittedly removes the $O(r_e)$ terms in $\mathbf{F}_s$, but at the same time makes (4.216) infinite. On the other hand, the electron with finite structure puts limits on the validity of the approximations implicit in the expansions used in the derivation of $\mathbf{F}_s$. The only order of magnitude value of a non-zero r_e that might be used is the "classical electron radius"

$$r_e = \frac{e^2}{4\pi\varepsilon_0 m_0 c^2} = 2.8 \times 10^{-15}\ \text{m}, \tag{4.218}$$

suggested theoretically both by the identification of the electrostatic energy with m_0c^2 and from the scattering cross-section (see §4.6.3). Then a rough criterion for the expansions to converge rapidly is that (cf. (2.69)) the time scale of variations in the motion of the electron is much greater than

$$\tau = \frac{2r_e}{3c} = \frac{e^2}{6\pi\varepsilon_0 m_0 c^3} = 6.3 \times 10^{-24}\ \text{sec}, \tag{4.219}$$

which is seen to be the coefficient of $\ddot{\mathbf{v}}$ in (4.217), the factor $\frac{2}{3}$ in the definition having been inserted purely for the convenience of precise identification. Since frequencies of the order of 10^{23} c/s demand quantum treatment it would appear reasonable to introduce r_e and to regard (4.217) as an approximate equation. On the other hand, mention should be made of a rather attractive "point" electron theory in which, by insisting on the inadmissibility of the concept of an isolated charge in otherwise empty infinite space, the analysis presented here is modified in such a way that cancellation removes the divergent $\dot{\mathbf{v}}$ term in $\mathbf{F}_s$ altogether, and (4.217) is established as exact, at least without mathematical impropriety.

Consider, now, some of the implications of (4.217). First it is easily shown that the $\dddot{\mathbf{v}}$ term maintains the energy balance, on average for certain motions, when allowance is made for radiation. For if the scalar product of (4.217) with $\mathbf{v}$ is integrated over some time interval, and the second term on the right-hand side integrated by parts, the result can be written

$$\int_{t_1}^{t_2} \mathbf{F}_e \cdot \mathbf{v}\, dt = m_0\left[\tfrac{1}{2}\mathbf{v}^2 - \tau \mathbf{v}\cdot\dot{\mathbf{v}}\right]_{t_1}^{t_2} + \int_{t_1}^{t_2} P\, dt, \tag{4.220}$$

where

$$P = \frac{e^2\dot{\mathbf{v}}^2}{6\pi\varepsilon_0 c^3} \tag{4.221}$$

is the rate at which the electron puts energy into the radiation field (see 2.106). Provided, therefore, that $\mathbf{v}\cdot\dot{\mathbf{v}}$ has the same value at t_1 and t_2, and not otherwise, the work done by the external force is equal to the sum of the energy radiated and the increase in kinetic energy. The $\ddot{\mathbf{v}}$ term in (4.217) is sometimes called the *radiation reaction.*

That (4.217) implies energy balance only under certain restrictions is forcibly demonstrated by considering motion for which $\dot{\mathbf{v}}$ is constant. The rate of radiation of energy (4.221) is then constant, but the equation is simply $m_0\dot{\mathbf{v}} = \mathbf{F}_e$ and takes no account of it whatsoever. As is shown a little later, the relativistic equation exhibits the same feature for "hyperbolic" motion.

A more common criticism of equation (4.217) is of its prediction of a "runaway" electron. With $\mathbf{F}_e = 0$ the general solution is

$$\mathbf{v} = \mathbf{v}_0 + \mathbf{v}_1 e^{t/\tau}, \tag{4.222}$$

where $\mathbf{v}_0$ and $\mathbf{v}_1$ are constants, and τ is given by (4.219). The criticism is valid if (4.217) claims to be exact; but if the classical electron radius is introduced and the equation regarded as an approximation, the motion (4.222) violates the condition for its validity, namely that the time scale must greatly exceed τ. Indeed this condition implies in general that the radiation reaction gives rise only to a small correction to the motion, which is commonly the case in practical application.

To obtain the relativistic form of (4.217),

$$\frac{d}{dt}\left\{\frac{m_0\mathbf{v}}{\sqrt{(1 - v^2/c^2)}}\right\} = \mathbf{F}_e + \mathbf{F} \tag{4.223}$$

say, it is only necessary, as indicated by the theory of §2.3.3, to find a 4-vector

$$\mathscr{F}^i = \left\{\frac{\mathbf{v}\cdot\mathbf{F}}{\sqrt{(1 - v^2/c^2)}}, \frac{\mathbf{F}}{\sqrt{(1 - v^2/c^2)}}\right\}, \tag{4.224}$$

the space components of which give $e^2\ddot{\mathbf{v}}/(6\pi\varepsilon_0 c^3)$ in the frame in which the charge is at rest. Clearly the 4-vector

$$\mathscr{F}^i = \frac{e^2}{6\pi\varepsilon_0 m_0 c^3}\left\{c^2\frac{d^2p^i}{ds^2} + \lambda p^i\right\}, \tag{4.225}$$

where p^i is the 4-momentum (2.184), has the correct space components when $\mathbf{v} = 0$, for any value of the constant λ; and the correct relation between space and time components can be achieved by choosing λ so that $p_i \mathscr{F}^i = 0$. Since $p_i p^i = m_0^2 c^2$, this evidently implies

$$\lambda = -\frac{1}{m_0^2} p_i \frac{d^2 p^i}{ds^2};$$

or, using $d^2(p_i p^i)/ds^2 = 0$,

$$\lambda = \frac{1}{m_0^2} \frac{dp_i}{ds} \frac{dp^i}{ds}. \tag{4.226}$$

For the explicit expression of $\mathbf{F}$ in terms of $\mathbf{v}$ and its derivatives it is recalled from §2.3.4 that (4.226) is $-6\pi\varepsilon_0 c/e^2$ times the radiated power (2.129), so that

$$\lambda = -\frac{1}{c^2(1 - v^2/c^2)^2}\left\{\dot{\mathbf{v}}^2 + \frac{(\mathbf{v}\cdot\dot{\mathbf{v}}/c)^2}{1 - v^2/c^2}\right\}. \tag{4.227}$$

Furthermore it is easily verified that

$$\frac{c}{m_0}\frac{dp^i}{ds} = \left\{\frac{\mathbf{v}\cdot\dot{\mathbf{v}}/c}{1 - v^2/c^2}, \frac{\dot{\mathbf{v}}}{1 - v^2/c^2} + \frac{\mathbf{v}\cdot\dot{\mathbf{v}}/c^2}{(1 - v^2/c^2)^2}\mathbf{v}\right\}, \tag{4.228}$$

$$\begin{aligned}\frac{c^2}{m_0}\frac{d^2p^i}{ds^2} = \Bigg\{&\frac{1}{c(1 - v^2/c^2)^{5/2}}\left(\frac{4(\mathbf{v}\cdot\dot{\mathbf{v}}/c)^2}{1 - v^2/c^2} + \dot{\mathbf{v}}^2 + \mathbf{v}\cdot\ddot{\mathbf{v}}\right),\\ &\mathbf{K} + \frac{\mathbf{v}}{c^2(1 - v^2/c^2)^{5/2}}\left(\frac{4(\mathbf{v}\cdot\dot{\mathbf{v}}/c)^2}{1 - v^2/c^2} + \dot{\mathbf{v}}^2 + \mathbf{v}\cdot\ddot{\mathbf{v}}\right)\Bigg\},\end{aligned} \tag{4.229}$$

where

$$\mathbf{K} = \frac{\ddot{\mathbf{v}}}{(1 - v^2/c^2)^{3/2}} + \frac{3\mathbf{v}\cdot\dot{\mathbf{v}}/c^2}{(1 - v^2/c^2)^{5/2}}\dot{\mathbf{v}} = \frac{d}{dt}\left\{\frac{\dot{\mathbf{v}}}{(1 - v^2/c^2)^{3/2}}\right\}. \tag{4.230}$$

Identifying the space parts of (4.224) and (4.225), using (4.227) and (4.229), gives

$$\mathbf{F} = \frac{e^2}{6\pi\varepsilon_0 c^3}\sqrt{(1 - v^2/c^2)}\left\{\mathbf{K} + \frac{\mathbf{K}\cdot\mathbf{v}/c^2}{1 - v^2/c^2}\mathbf{v}\right\}. \tag{4.231}$$

This can be alternatively written

$$\mathbf{F} = \frac{e^2}{6\pi\varepsilon_0 c^3}\sqrt{(1 - v^2/c^2)}\left\{\mathbf{K}_\perp + \frac{\mathbf{K}_\parallel}{1 - v^2/c^2}\right\}, \tag{4.232}$$

where $\mathbf{K}_\perp$ and $\mathbf{K}_\parallel$ are the components of $\mathbf{K}$ perpendicular and parallel to $\mathbf{v}$.

It is evident from (4.232) that $\mathbf{F}$ vanishes if and only if $\mathbf{K} = 0$, that is

$$\frac{\dot{\mathbf{v}}}{(1 - v^2/c^2)^{3/2}} = \text{a constant vector.} \tag{4.233}$$

Equation (4.233) gives a hyperbolic trajectory and is, in fact, the so-called hyperbolic motion, mentioned in §4.5.1, for which the rate of change of acceleration, measured in the frame S' in which the particle is instantaneously at rest, is zero. For the 4-vector given by (4.229) in the frame S, in which the particle has velocity $\mathbf{v}$, is $(\dot{\mathbf{v}}'^2/c, \ddot{\mathbf{v}}')$ in S'; and a Lorentz transformation from S to S' shows, with a little algebra, that

$$\begin{aligned} \ddot{\mathbf{v}}' &= \mathbf{K} + \left(\frac{1}{\sqrt{(1 - v^2/c^2)}} - 1\right)\frac{\mathbf{K}\cdot\mathbf{v}}{v^2}\mathbf{v} \\ &= \mathbf{K}_\perp + \frac{\mathbf{K}_\parallel}{\sqrt{(1 - v^2/c^2)}}. \end{aligned} \tag{4.234}$$

Hence $\ddot{\mathbf{v}}' = 0$ if and only if $\mathbf{K} = 0$.

It was shown in §4.5.1 that rectilinear motion in a uniform electrostatic field is hyperbolic motion. In this case, then, the radiation reaction $\mathbf{F}$ in the relativistic equation (4.223) is zero, and the predicted motion is (4.149) as calculated without regard to radiation. On the other hand the expression (2.149) for the radiated power is a non-zero constant.

4.6.3 Applications

The inclusion of the radiation reaction term as a small perturbation in what would otherwise be periodic motion leads to sensible predictions about the way in which the motion is affected. The discussion of the self-force is completed in the present section by outlining three applications of this kind.

Consider, first, the classical model of a bound electron that executes rectilinear oscillations of angular frequency ω_0. The non-relativistic equation of motion is

$$\ddot{x} + \omega_0^2 x = \tau \dddot{x}, \tag{4.235}$$

where the term on the right-hand side is the radiation reaction per unit mass, τ being given by (4.219). Solutions are $\exp(i\Omega t)$, with

$$\Omega^2 = \omega_0^2 + i\tau\Omega^3. \tag{4.236}$$

On the assumption that $\omega_0 \ll 1/\tau$ the values of Ω of interest are those close to $\pm\omega_0$; the radiation reaction term is small for these values, and to an adequate approximation they are

$$\Omega = \pm\omega_0 + \tfrac{1}{2}i\tau\omega_0^2. \tag{4.237}$$

The third value of Ω given by (4.236) is close to $-i/\tau$, and must be discarded as a further example of the appearance of an unacceptable prediction when the radiation reaction term is large.

The complex representation of x can therefore be taken as

$$e^{-\frac{1}{2}\tau\omega_0^2 t}\, e^{i\omega_0 t} \tag{4.238}$$

for $t > 0$, and zero for $t < 0$. The radiation field is proportional to $\ddot{x}$, which is approximately $-\omega_0^2 x$, and its complex frequency spectrum is proportional to

$$\int_0^\infty e^{-\frac{1}{2}\tau\omega_0^2 t}\, e^{i\omega_0 t}\, e^{-i\omega t}\, dt = \frac{1}{i(\omega - \omega_0) + \frac{1}{2}\tau\omega_0^2}. \tag{4.239}$$

The corresponding dependence of intensity (square of field amplitude) on frequency is given by

$$\frac{1}{(\omega - \omega_0)^2 + \frac{1}{4}\tau^2\omega_0^4}. \tag{4.240}$$

The frequency spread between the half-intensity points on either side of ω_0 is

$$\Delta\omega = \tau\omega_0^2, \tag{4.241}$$

and this is a measure of the spectral line broadening due to radiation damping. The associated spread in wavelength $\lambda = 2\pi c/\omega$ is

$$\Delta\lambda = \frac{2\pi c}{\omega_0^2}\Delta\omega = \tfrac{4}{3}\pi r_e = 1.18 \times 10^{-14} \text{ m}. \tag{4.242}$$

As a second example consider the classical model for the scattering of an electromagnetic wave by an electron. Suppose the incident wave has electric vector $\mathbf{E}$ and angular frequency ω. The non-relativistic equation of motion is

$$\ddot{\mathbf{x}} + \omega_0^2\mathbf{x} = \tau\dddot{\mathbf{x}} - \frac{e}{m_0}\mathbf{E}, \tag{4.243}$$

which differs from (4.235) only in the inclusion of the forcing term. Here, however, it is the forced oscillation alone which is of interest; this is represented by the complex solution

$$\mathbf{x} = \frac{e/m_0}{\omega^2 - \omega_0^2 - i\tau\omega^3}\mathbf{E}, \tag{4.244}$$

where it is assumed that the amplitude of the oscillation is small compared with the wavelength $2\pi c/\omega$, so that the space dependence of $\mathbf{E}$ in (4.243) can be ignored and the equation is linear. Now the power radiated by the electron is $e^2/(6\pi\varepsilon_0 c^3)$ times the square of its acceleration. The time-averaged radiated power is therefore

$$\frac{e^2}{6\pi\varepsilon_0 c^3}\frac{\frac{1}{2}(e/m_0)^2\omega^4}{(\omega^2 - \omega_0^2)^2 + \tau^2\omega^6}E_0^2, \tag{4.245}$$

where E_0 is the amplitude of $\mathbf{E}$. The ratio of (4.245) to the time-averaged power in the incident wave crossing unit area normal to the direction of propagation, which latter quantity is $\frac{1}{2}\varepsilon_0 c E_0^2$, gives the *scattering cross-section* σ. Thus

$$\sigma = \tfrac{8}{3}\pi \frac{\omega^4}{(\omega^2 - \omega_0^2)^2 + \tau^2\omega^6} r_e^2. \tag{4.246}$$

For a free electron $\omega_0 = 0$, the radiation damping is unimportant if $\omega \ll 1/\tau$, and then

$$\sigma = \tfrac{8}{3}\pi r_e^2, \tag{4.247}$$

a well-known result which gives an interpretation of r_e. For a strongly bound electron with $\omega_0 \gg \omega$ the radiation damping is unimportant if $\omega \ll (\omega_0/\omega)^2/\tau$, and then

$$\sigma = \tfrac{8}{3}\pi \left(\frac{\omega}{\omega_0}\right)^4 r_e^2; \tag{4.248}$$

this expression exhibits fourth power dependence on frequency in conformity with the Rayleigh law of scattering. Evidently, in general, radiation damping can be important when ω is close to ω_0; in this case an approximation is

$$\sigma = \tfrac{2}{3}\pi \frac{\omega^2}{(\omega - \omega_0)^2 + \frac{1}{4}\tau^2\omega^4} r_e^2. \tag{4.249}$$

As a final example consider the case treated in §3.3 of an electron executing, apart from radiation damping, uniform circular motion in a magnetostatic field of induction B. In this case the radiated power does not oscillate, and the degradation of the energy of the motion is easily analysed, relativistically, by a direct appeal to energy balance. The rate of decrease of the energy $\mathscr{E} = m_0c^2/\sqrt{(1 - v^2/c^2)}$ of the particle is equated to the radiated power

$$\frac{e^4B^2}{6\pi\varepsilon_0 m_0^2 c} \frac{v^2/c^2}{1 - v^2/c^2}, \tag{4.250}$$

which is just (3.110) with Ω replaced by (4.146). With (4.250) expressed in terms of $\mathscr{E}$ rather than v there results the differential equation

$$\frac{d\mathscr{E}}{dt} = -\frac{e^4B^2}{6\pi\varepsilon_0 m_0^4 c^5}(\mathscr{E}^2 - m_0^2c^4). \tag{4.251}$$

If $\mathscr{E} = \mathscr{E}_0$ when $t = 0$, this integrates to

$$\mathscr{E} = m_0c^2 \frac{\mathscr{E}_0 + m_0c^2 + (\mathscr{E}_0 - m_0c^2)\, e^{-\gamma t}}{\mathscr{E}_0 + m_0c^2 - (\mathscr{E}_0 - m_0c^2)\, e^{-\gamma t}}, \tag{4.252}$$

where

$$\gamma = \frac{e^4B^2}{3\pi\varepsilon_0 m_0^2 c^3} = 2\left(\frac{eB}{m_0}\right)^2 \tau, \tag{4.253}$$

with τ given by (4.219). Since eB/m_0 is the angular gyro frequency, and τ is of the order of 10^{-23} sec, it is apparent that the decay time is very long unless the magnetic field is exceptionally strong.

PROBLEMS

(Only in Problem 11 is the radiation reaction force to be included in the equations of motion.)

1. A point charge e, of mass m, moves in electrostatic and magnetostatic fields whose components in cylindrical polar coordinates (r, θ, z) are $\mathbf{E} = (aE_0/r, 0, 0)$, $\mathbf{B} = (0, 0, B)$, where a and B are independent of position. If the charge is released from rest at $r = a$, show that, in a non-relativistic analysis, its radial velocity is subsequently given by
$$\dot{r}^2 = \frac{aeB}{m}\left[\frac{2E_0}{B}\log(r/a) - \frac{aeB}{4m}\left(\frac{r}{a} - \frac{a}{r}\right)^2\right].$$
Estimate the maximum radial excursion of the charge for the case $a = 1$ cm, $B = 100$ gauss, $E_0 = 5 \times 10^4$ volt m^{-1}.
2. A charged particle moves in a magnetic pole field, as described in §4.2.3. Find the total excursion of the angle ϕ. What is the ultimate velocity of the particle?
3. A charged particle moves in the field of a magnetic dipole of constant moment, and the field at the particle can be treated as slowly varying. The particle mirrors at $\theta = \alpha$, and when it is in the equatorial plane of the dipole ($\theta = \frac{1}{2}\pi$) its pitch angle is χ. Plot the graph of α against χ.
4. Assuming that the earth's magnetic field is that of a dipole of moment 8.1×10^{22} amp m^2, calculate the drift velocity at the magnetic equator, distance 2×10^4 km from the earth's center, of a 10^4 eV electron with pitch angle 45°.
5. Write down: (a) the power P_e radiated by an electron in rectilinear motion under the action of a uniform electrostatic field $\mathbf{E}$, (b) the power P_m radiated by an electron of relativistic energy $\mathscr{E}$ in circular motion under the action of a uniform magnetostatic field $\mathbf{B}$. Show that, if $\mathscr{E} \gg m_0c^2$,
$$P_m/P_e = \left(\frac{\mathscr{E}}{m_0c^2}\right)^4 \frac{c^2B^2}{E^2}.$$
What value of E makes this ratio unity when $\mathscr{E}$ is 10 meV and B is 1 gauss?
6. Show that, if Problem 1 is generalized by allowing the field B to be an arbitrary function of r, and by adopting a relativistic treatment, then
$$\frac{\dot{r}^2}{c^2} = 1 - \frac{1 + \left(\dfrac{e\Phi}{2\pi m_0 cr}\right)^2}{\left(1 + \dfrac{aE_0}{m_0c^2}\log\dfrac{r}{a}\right)^2},$$
where Φ is the flux of magnetic induction between circles of radii a and r.
Confirm that, when B is independent of r, the result of Problem 1 is recovered in the non-relativistic approximation.

7. A betatron is a machine designed to accelerate electrons to relativistic energies by keeping them in a circular orbit of constant radius. Suppose the magnetic and electric fields in cylindrical polar coordinates (r, θ, z) are given by

$$\mathbf{B} = (0, 0, B(r, t)), \qquad \mathbf{E} = (0, E(r, t), 0).$$

Show that an electron can gyrate in a fixed circular orbit of radius a if, at each time t, $B(a, t)$ is half the mean value of B within the orbit.

8. Show that, in the betatron action described in Problem 7, the energy of the electron is

$$m_0c^2\sqrt{[1 + (aeB/m_0c)^2]}.$$

Calculate this when $B = 10^4$ gauss, $a = 1$ m, and find the electron's gain in energy in one revolution if $dB/dt = 10^6$ gauss sec^{-1}.

Now estimate the loss of energy per revolution due to radiation, and comment on the result. In what part of the frequency spectrum is most of the power radiated?

9. Show that in a relativistic treatment of Problem 8 of Chapter 2 the total energy radiated is

$$\frac{e^2k}{6\pi\varepsilon_0}\left[\frac{V/c}{1 - V^2/c^2} - \tfrac{1}{2}\log\left(\frac{1 + V/c}{1 - V/c}\right)\right].$$

10. Consider the problem discussed in §4.5.2, with the generalization that the wave and particle are in a medium of refractive index μ, rather than in a vacuum. Show that there is a constant of the motion, λ, analogous to (4.175), and that the equation of motion analogous to (4.183) reads

$$(\Gamma\mathbf{r}')' = \frac{e}{\lambda m_0\omega^2}\left[\mathbf{E} + \frac{\mu\omega}{c}(\mathbf{E}\cdot\mathbf{r}')\mathbf{n}\right] + \frac{\Omega}{\omega}\mathbf{r}' \times \mathbf{n},$$

where

$$\Gamma = (1 - \mu\mathbf{n}\cdot\mathbf{v}/c)/(\mu - \mathbf{n}\cdot\mathbf{v}/c),$$

and the dash denotes differentiation with respect to $\phi = \omega(t - \mu\mathbf{n}\cdot\mathbf{r}/c)$.

11. Show that if an external force acting on a charged particle keeps its speed constant, then, at each instant, the rate at which the external force does work is equal to the rate of radiation of energy.

Obtain the relativistic form of the "energy" equation (4.220).

REFERENCES

Attention is drawn to three books in particular.

T. G. NORTHROP, *The Adiabatic Motion of Charged Particles* (Interscience Tracts on Physics and Astronomy, No. 21), Interscience (1963).

This book is devoted to an account of the motion of a charged particle in slowly varying fields, and the treatment is closely followed in the present chapter. (See also A. BAÑOS, *J. Plasma Physics* **1**, 305 (1967).)

F. ROHRLICH, *Classical Charged Particles: Foundations of their Theory*, Addison-Wesley (1965).

This, as the author writes in the preface, "presents a consistent and coherent classical theory of point charges," and will help the reader interested in a deeper investigation of the self-force problem.

C. STÖRMER, *The Polar Aurora*, Oxford University Press (1955).

This includes a connected account of the author's well known work (not described here) on the exact analysis of the motion of a point charge in the field of a magnetic dipole.

Many other books contain accounts of charged particle dynamics, including

H. ALFVÉN and C. G. FÄLTHAMMAR, *Cosmical Electrodynamics* 2nd edn, Clarendon Press (1963).

S. CHANDRASEKHAR, *Plasma Physics* (notes compiled by S. K. Trehan), University of Chicago Press (1960).

J. L. DELCROIX, *Plasma Physics*, Wiley (1965).

V. C. A. FERRARO and C. PLUMPTON, *An Introduction to Magneto-Fluid Mechanics* 2nd edn, Oxford University Press (1966).

S. GARTENHAUS, *Elements of Plasma Physics*, Holt, Rinehart and Winston (1964).

E. H. HOLT and R. E. HASKELL, *Foundations of Plasma Physics*, Macmillan (1965).

J. D. JACKSON, *Classical Electrodynamics*, Wiley (1962).

W. B. KUNKEL (Editor), *Plasma Physics in Theory and Application*, McGraw-Hill (1966).

B. O. LEHNERT, *Dynamics of Charged Particles*, North-Holland (1964).

J. G. LINHART, *Plasma Physics*, North-Holland (1960).

C. L. LONGMIRE, *Elementary Plasma Physics*, Interscience (1963).

W. K. H. PANOFSKY and M. PHILLIPS, *Classical Electricity and Magnetism*, Addison-Wesley (1962).

G. SCHMIDT, *Physics of High Temperature Plasmas: An Introduction*, Academic Press (1966).

G. W. SUTTON and A. SHERMAN, *Engineering Magnetohydrodynamics*, McGraw-Hill (1965).

W. B. THOMPSON, *An introduction to Plasma Physics*, 2nd (revised) impression, Pergamon (1964).

A treatment of relativistic motion in slowly varying fields is given by

P. O. VANDERVOORT, *Ann. Phys.* **10**, 401 (1960).

Many papers have discussed the effect of perturbation fields in violating the adiabatic variants. For example

G. HAERENDEL, *J. Geophys. Res.* **71**, 1857 (1966).

For work on corrections to the second adiabatic invariant see

T. G. Northrop, C. S. Lin and M. D. Kruskal, *Phys. Fluids* **9**, 1503 (1966).

The method used in §4.5.2 is due to

A. A. Kolomenskii and A. N. Lebedev, *Sov. Phys. Doklady*, **7**, 745 (1963); *Sov. Phys. JETP*, **17**, 179 (1963).

For this problem see also

M. J. Laird and F. B. Knox, *Phys. Fluids*, **8**, 755 (1965).

CHAPTER 5

WAVES IN AN IONIZED GAS: MAGNETO-IONIC THEORY

5.1 MACROSCOPIC DESCRIPTION OF PLASMA

5.1.1 The hydrodynamic formalism

In the last chapter a description was given of the motion of a charged particle under the influence of electric and magnetic fields. A collection of such particles is now considered, forming part or all of an ionized gas, the so-called plasma introduced in Chapter 1. The distribution and motion of the charged particles in the gas produce electric and magnetic fields, and these in turn influence the motion of the particles. The situation is broadly similar to the classical model of the electromagnetic behavior of a medium, for example, that of a dielectric in optics; in the present case, though, the charged particles are "free" in the sense that they are not subject to any quasi-elastic force binding them to essentially fixed atoms.

In a rigorous discussion of this type of model the particle motion is initially described in terms of the actual field acting on the particle, and macroscopic properties of the field and plasma are subsequently derived by some averaging process. Such an ambitious programme leads, however, to considerable complication, and is not taken up until later. The relatively elementary approach is to decide, first, what is in macroscopic terms the "effective" field acting on a charged particle, and then to use this "effective" field in the equations describing the motion. In the present chapter we follow this procedure, and make the further assumption (see §1.3) that the effective field acting on each particle is the same as the "average" field which would be obtained by experiment, and which appears in the macroscopic version of Maxwell's equations.

There remains a broad division, in possible theoretical developments, between the methods of hydrodynamics and those of kinetic theory. In the former, to which the present chapter is devoted, the thermal velocity spread is ignored and the plasma (as well as the field) is treated as a macroscopic continuum. In the latter, the starting point is the appropriate form of Boltzmann's equation for the distribution function in coordinate and velocity space. The kinetic treatment is certainly required for the deepest understanding of the problem, and would be expected to be capable of yielding the hydrodynamic model through suitable approximating and averaging procedures. Such matters are considered later on.

In comparing the theories of neutral and fully ionized gases in Chapter 1, we stressed that the bases for the recovery of equations of the hydrodynamic type from the equations of kinetic theory are different in the two cases. For the plasma they are in some respects less clear cut; for example, the phenomenon of Landau damping predicted by the kinetic treatment (Chapter 8) is lost in the hydrodynamic approximation.

Whilst it must therefore be recognized that the hydrodynamic approach has its limitations, it is undoubtedly of great importance in the general development and application of the theory. Often problems are inherently so complicated, and the links between theory and experiment so fragile, that it is only prudent to abjure anything but the simplest mathematical formalism. Even that, however, can all too readily lead to results of disheartening complexity, and the programme of this chapter has been drawn up partly with the purpose of proceeding relatively gently from the less to the more elaborate situations.

5.1.2 The linear approximation

One of the earliest and most extensive applications of the theory of electromagnetic waves in ionized gases was to the propagation of radio waves in the ionosphere. The accepted mathematical formulation in that context is of long standing and is known as *magneto-ionic theory*. In what follows, the specific approach of magneto-ionic theory is adopted, and developments going outside the original boundaries of magneto-ionic theory are in a sense regarded as extensions to it. The material could certainly be presented in other equally valid ways. The one adopted emphasizes the individual contributions of the separate constituent gases of the plasma, which are here confined to three, namely electron, positive ion and neutral.

The equations of magneto-ionic theory are those of Maxwell together with equations of motion of the hydrodynamic type. In addition, these latter are, in their standard form, linearized, with the electromagnetic field and associated fluid motion that are to be investigated regarded as perturbations on a steady state. In many applications the steady state is one in which the gas is at rest and the field, if any, is purely magnetostatic; attention is confined to this case in the present chapter, and some consideration given to streaming plasmas in the next. The existence of a steady state demands local electrical neutrality; that is, equal concentrations at each point of electrons and positive ions, assuming the latter to be singly ionized. Furthermore, the steady state of the plasma is taken to be uniform throughout all space; the interesting problems posed by inhomogeneity, to which a great deal of attention has been given, are not really relevant to a description of the basic characteristics of a plasma, and are outside the objective of this book.

The linear approximation is a very familiar mathematical technique. It consists simply in neglecting, in the equations, quadratic and higher order terms in the perturbation variables, and the relative simplification it introduces need hardly be stressed. Moreover, it is often valid in practical plasma problems, although there are also important situations in which it is inadequate.

The analysis now proceeds from the assumption of a perturbation which is harmonic both in time and space. In other words, monochromatic plane wave solutions are sought. The refractive index and polarization characteristics of these waves, in particular, exemplify the way in which fluid and electromagnetic field can interact. The expression for the refractive index is implicit in the dispersion relation between the angular frequency ω and the wave number k (2π divided by the wavelength), and much of the discussion is concerned with examination of the dispersion relation in different contexts.

The bulk of the chapter is devoted to collision-free plasmas. In this case neutral particles, of course, play no part, and the treatment is confined to the two-constituent plasma of electrons and singly ionized positive ions. The main complications in the theory are then introduced by the presence of an ambient magnetostatic field, and by the inclusion of the partial pressure terms for the respective constituent gases. The former influences the particle motion in a way which accounts for many of the distinctive features of magneto-ionic theory; the latter admits the possibility of compressional disturbances. If the pressure terms are excluded, the equations of motion for monochromatic but not necessarily spatially harmonic fields give a linear relation between the complex representations of the current and the electric field; the corresponding expression for the conductivity, tensor or scalar according to whether there is or is not a pervading magnetostatic field, is a function of the angular frequency ω. There is dispersion in the usual sense. If the pressure terms are included, the relation between the complex representations of current and electric field involves the gradient of the latter; the simple concept of conductivity is then only valid for fields that are harmonic in space as well as in time, and the expression for it given directly by the equations of motion appears as a function of both ω and k explicitly. The explicit dependence on k represents what is sometimes known as *spatial dispersion*. Spatial dispersion arises in various contexts, and is a distinctive feature of a physical model. However, from the standpoint of the mathematical development just outlined it evidently presents no instrinsically new problem.

Since the positive ions are at least 1837 times as massive as the electrons their motion will often be relatively unimportant, and some algebraic simplification can be achieved by treating them as stationary. This is indeed done in the common form of magneto-ionic theory, and likewise here, after a look at the simplest case of a collision-free plasma with no magnetic field and with pressure terms excluded. Then come extensions to the theory in the shape of the inclusion of positive ion motions and pressure terms. Finally, the effect of collisions is investigated with the use of the "frictional force" approximation of magneto-ionic theory.

5.2 COLLISIONLESS PLASMA WITHOUT MAGNETIC FIELD

5.2.1 The equations for an electron gas

In this section the simplest model of a plasma interacting with an electromagnetic field is considered. The mathematics is quite trivial, but the problem does illustrate some of the important physical principles which appear in more elaborate guises in more complicated models.

The problem is that of an unbounded electron gas, which in the unperturbed state is uniform and at rest. The gas is assumed to be "cold"; that is, in the language of kinetic theory, the thermal velocities are zero, or in the language of hydrodynamics there is no pressure force. Also, to begin with, the perturbation motion of the electrons only is considered, with the positive ions simply forming a stationary background of charge which neutralizes the unperturbed gas at each point.

First, the equation of motion of an electron is invoked. If, in the Eulerian representation of hydrodynamics, $\mathbf{v}$ is written for its perturbation velocity, this is just

$$m\frac{\partial \mathbf{v}}{\partial t} = -e\mathbf{E}, \tag{5.1}$$

where the constants $-e$, m are respectively the electronic charge and mass, and $\mathbf{E}$ is taken to be the "average" field on the grounds already explained. Linearization excludes from the left-hand side of (5.1) the relativistic mass correction factor and the contribution $m(\mathbf{v}\cdot\text{grad})\mathbf{v}$ from the convective part of the acceleration, and from the right-hand side the magnetic force $-e\mathbf{v}\times\mathbf{B}$. It may be noted that the magnitudes of the excluded terms are smaller than those retained by factors of the order $(v/c)^2$, v/c' and again v/c', respectively, where c is the vacuum speed of light and c' is the phase speed of the particular wave motion ultimately considered.

With the assumption of a harmonic time variation at angular frequency ω (5.1) gives, in the complex representation,

$$i\omega m\mathbf{v} = -e\mathbf{E}. \tag{5.2}$$

Furthermore, the current density is

$$\mathbf{j} = -Ne\mathbf{v}, \tag{5.3}$$

where, by virtue of the linear approximation, N is the unperturbed number density of the electrons. The combination of (5.2) and (5.3) can then be written

$$\mathbf{j} = \sigma\mathbf{E}, \tag{5.4}$$

where

$$\sigma = -i\frac{Ne^2}{\omega m}. \tag{5.5}$$

In other words, in respect of its interaction with an electromagnetic field, the electron gas behaves as a medium with scalar conductivity given by (5.5). Since this expression is purely imaginary the medium is in fact more aptly described as a lossless dielectric. Substitution of (5.4) into the right-hand side of the Maxwell equation

$$\text{curl}\,\mathbf{H} = i\omega\varepsilon_0\mathbf{E} + \mathbf{j} \tag{5.6}$$

yields the dielectric constant of the medium in the form

$$1 - \frac{Ne^2}{\varepsilon_0 m\omega^2}. \tag{5.7}$$

The quantity $Ne^2/(\varepsilon_0 m)$ has the dimensions of the square of an angular frequency; this angular frequency is known, for reasons which will shortly be apparent, as the *plasma frequency*, ω_p say. Then (5.7) is

$$1 - \omega_p^2/\omega^2. \tag{5.8}$$

If $\omega \neq \omega_p$, in essence only the standard theory of transverse plane waves in a dielectric medium is involved. The details are presented in a moment, but consideration is first given to the interesting special case $\omega = \omega_p$.

5.2.2 Plasma oscillations

For $\omega = \omega_p$ (5.6) states that $\text{curl}\,\mathbf{H} = 0$. In conjunction with the other Maxwell equation

$$\text{curl}\,\mathbf{E} = -i\omega\mu_0\mathbf{H} \tag{5.9}$$

the implication is that, in any *source-free* field that is *finite at infinity*, **H** must be identically zero and **E** therefore a potential field. For $\text{curl}\,\text{curl}\,\mathbf{E} = 0$ implies $\text{curl}\,\mathbf{E} = -\text{grad}\,\psi$, which in turn implies $\nabla^2\psi = 0$; and the only solution of Laplace's equation without any singularity, and finite at infinity, is $\psi = \text{constant}$, that is, $\text{grad}\,\psi = 0$. It is this type of field which is studied in the present section. Fields with $\omega = \omega_p$ for which **H** is non-zero can only exist in association with sources in the medium. It is pointed out at the end of §5.2.3 that they really belong to the same genus as the transverse waves discussed there.

At the plasma frequency, then, natural oscillations of a purely electrostatic character are possible, described by the relations

$$\mathbf{H} = 0, \tag{5.10}$$

$$\mathbf{E} = -\text{grad}\,\phi, \tag{5.11}$$

and the universal equation

$$\operatorname{div} \mathbf{E} = \frac{\rho}{\varepsilon_0}, \tag{5.12}$$

where ρ is the charge density resulting from the perturbation of the electron distribution. These are sometimes known as plasma oscillations, in contrast to those associated with transverse electromagnetic waves. In the latter $\operatorname{div} \mathbf{E} = 0$ and there is no charge density.

Plasma oscillations are readily envisaged in physical terms. Any initial small displacement, with arbitrary space variation, of the uniform electron distribution results in an electrostatic restoring force determined by (5.12) which merely gives rise to an oscillation of the amplitude of the space variation with angular frequency ω_p. No traveling disturbance is involved.

That any charge density must oscillate at the plasma frequency can also be seen directly from the substitution of (5.1) and (5.12) into the linear form

$$\operatorname{div} (Ne\mathbf{v}) = \dot{\rho} \tag{5.13}$$

of the charge conservation relation

$$\operatorname{div} \mathbf{j} + \dot{\rho} = 0. \tag{5.14}$$

For these substitutions give

$$\ddot{\rho} + \frac{Ne^2}{\varepsilon_0 m} \rho = 0. \tag{5.15}$$

Another particularly simple situation which evidences natural oscillations at the plasma frequency is worth noting. Although strictly it falls outside the purview of the previous theory, because the unperturbed charged particle distribution is not uniform throughout all space, it does illustrate clearly the principles involved. The problem concerns a one-dimensional model in which the unperturbed electron distribution is confined to a slab bounded by two parallel plane faces, the uniform distribution within the slab being neutralized by an identical positive ion distribution. If a perturbation is now set up by a bodily displacement of the slab of electrons through a small distance x in a direction normal to the faces of the slab, the positive ions being assumed to remain at rest, effective surface charge densities $\pm Nex$ appear on the faces, with an associated uniform electric field Nex/ε_0 between and normal to them. The electric field thus provides each electron with a restoring force per unit mass equal to $Ne^2/(\varepsilon_0 m)$ times its displacement, and the result is simple harmonic motion at the plasma frequency.

The mechanism of plasma oscillations is a facet of the continual effort made by a plasma to preserve its electrical neutrality. Replacement of (5.12) by the assumption that the number densities of electrons and positive ions remains everywhere equal is a useful simplifying approximation which is valid, in other contexts, in the analysis of sufficiently slowly varying disturbances. For the insertion of the numerical values of e, m and ε_0 gives (see §1.2)

$$\omega_p^2 = \frac{Ne^2}{\varepsilon_0 m} = 3.2 \times 10^3 N, \tag{5.16}$$

from which it is apparent that the plasma frequency is likely to be high. Referring to Table 3 in §1.2 we see that for a laboratory plasma the number of electrons per cubic meter is between 10^{18} and 10^{24}, giving $\omega_p/(2\pi)$ between about 10^4 Mc/s (centimeter radio waves) and 10^7 Mc/s (infrared); for the ionosphere at the F-layer peak with $N \simeq 10^{12}$, say, $\omega_p/(2\pi) \simeq 10$ Mc/s; and even for the tenuous outer reaches of the exosphere, with N perhaps 10^8, it is still, at 100 kc/s, comfortably within the broadcasting spectrum.

5.2.3 Transverse waves

The theory of §5.2.1 is now developed for the case $\omega \neq \omega_p$. In this case the medium behaves as a lossless dielectric, with a frequency-dependent dielectric constant which is less than unity and which becomes negative at frequencies below the plasma frequency.

For the traveling wave solution in which the space variation is specified by the factor

$$e^{-ikx} \tag{5.17}$$

the dispersion relation is

$$c^2k^2 = \omega^2 - \omega_p^2, \tag{5.18}$$

and the refractive index μ $(=ck/\omega)$ is given by

$$\mu^2 = 1 - \omega_p^2/\omega^2. \tag{5.19}$$

The field vectors $\mathbf{E}$ and $\mathbf{H}$ lie in the yz-plane, and without loss of generality it may be taken that their only non-zero components are

$$E_y = e^{-ikx}, \tag{5.20}$$

$$H_z = Y_0 \mu e^{-ikx}, \tag{5.21}$$

where Y_0 is the vacuum admittance.

If $\omega > \omega_p$, then $0 < \mu^2 < 1$ and (5.20) and (5.21) represent an unattenuated wave traveling with phase speed $c/\mu > c$. If a number of such waves, having different frequencies near ω, form a group, then the group speed $\partial\omega/\partial k$ is $c\mu$. The product of the phase and group speed is c^2.

The complex Poynting vector in the single traveling wave has an x-component only, of magnitude

$$E_y H_z^* = Y_0 \mu, \tag{5.22}$$

and the time-averaged rate of flow of energy across unit area normal to the x-axis is

$$\tfrac{1}{2}\mathrm{Re}\, E_y H_z^* = \tfrac{1}{2} Y_0 \mu. \tag{5.23}$$

If $\omega < \omega_p$ then μ is purely imaginary and (5.20) and (5.21) represent an *evanescent* wave in which each field vector oscillates everywhere in phase, with an amplitude which decays exponentially with the exponent proportional to x. The disturbance is not propagated. Evidently $\mathrm{Re}\, E_y H_z^* = 0$, which confirms that on a time average there is no energy transport. Physically, of course, a single evanescent wave can exist in at most a half-space.

The dependence of the refractive index on frequency is often regarded from a slightly different standpoint, with emphasis on the fact that whether or not a wave of prescribed frequency can be propagated through the plasma depends on the number density of electrons. When N exceeds $\varepsilon_0 m\omega^2/e^2$ the wave cannot be propagated, and then the medium is said to be *overdense*.

The nature of the distinction between propagated and evanescent waves is one of the most significant features of fields in plasmas, and is, implicitly at least, at the root of much of the detailed development of the subject. Deductions from the basic equations are commonly depicted in a graphical form in which the distinction is immediately apparent. There is, however, a rather bewildering variety of ways of doing this, and no one method of presentation seems generally favored. Three are mentioned here.

One way is to demonstrate the behavior of the refractive index, taking μ^2 as ordinate—so that positive values indicate propagation and negative values evanescence—and ω as abscissa. The curve is then specified by (5.19) and is shown in Fig. (5.1).

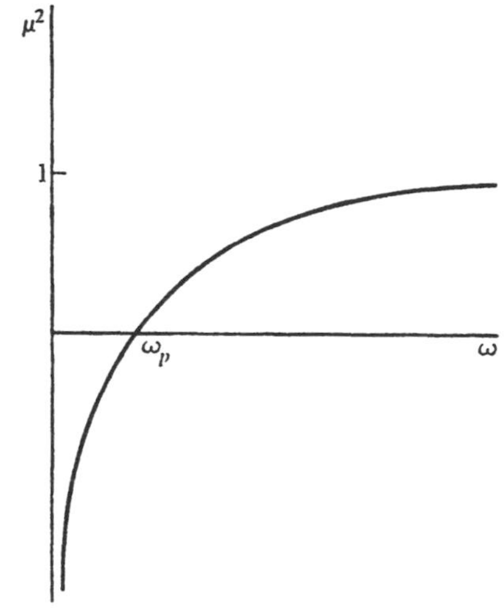

Figure 5.1

Another plot often used is that of ck against ω (with variants including omission of the factor c and interchange of ordinate and abscissa). This is a direct representation of the dispersion relation, particularly useful in displaying the group velocity $\partial\omega/\partial k$. The curve given by (5.18) is depicted in Fig. (5.2).

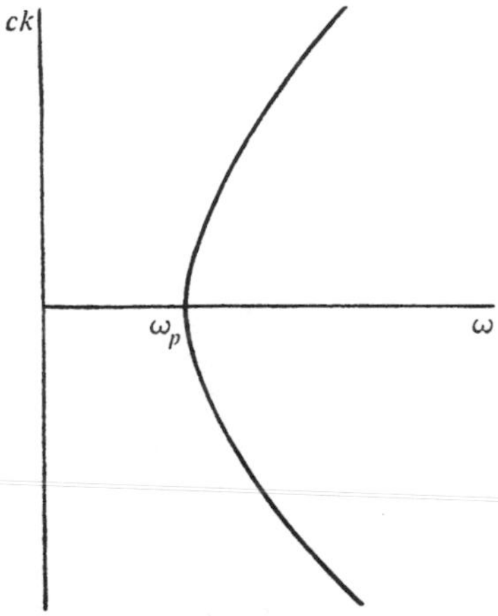

Figure 5.2

Yet another representation is adopted in magneto-ionic theory. The symbol X is used for the dimensionless parameter ω_p^2/ω^2, and (5.19) is written

$$\mu^2 = 1 - X. \tag{5.24}$$

For a given ω, X is proportional to N; and it is customary to take μ^2 as ordinate with X as abscissa, because of the importance in ionospheric work of the variation of μ^2 with N. The result is shown in Fig. (5.3).

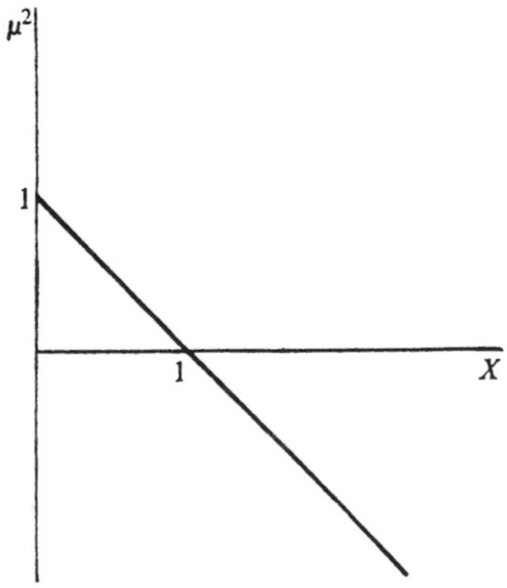

Figure 5.3

The situation described so far is particularly simple. But when the complications of an ambient magnetostatic field and the pressure force are included, the multiplicity of diagrams required for a comprehensive description is evidently excessive for inclusion in this book and only selected examples are offered. Before proceeding to these matters, two further points are considered. The first, on non-electrostatic oscillations when $\omega = \omega_p$, immediately; the second, on the motion of the positive ions, in §5.2.4.

Near the beginning of §5.2.2 it was stated that, when $\omega = \omega_p$, fields for which **H** is not zero can only exist in association with sources, and further comment was promised. The point is simply this. Of the same genus as the transverse waves considered in the present section are inhomogeneous plane waves which have an exponential amplitude variation in the direction normal to that of their phase propagation. In brief, for a mathematical solution the space variation can be specified by the factor

$$e^{-ik_0(lx + my)}, \tag{5.25}$$

where $k_0 = \omega/c$, provided only that

$$l^2 + m^2 = \mu^2. \tag{5.26}$$

At the plasma frequency this condition is

$$l^2 + m^2 = 0, \tag{5.27}$$

so that if l is real m is purely imaginary, and the solution tends to infinity either as y tends to infinity or as y tends to minus infinity. Physically such a wave can exist throughout at most a half space and must therefore be associated with a source. Conversely, a given current distribution oscillating at the plasma frequency generates a non-radiated field which can be synthesized from such waves.

5.2.4 The motion of positive ions

The inclusion of the motion of the positive ions in all the preceding analysis is easily effected. In addition to (5.1) for the electrons there is a corresponding equation of motion for the ions. Thus

$$m_e \frac{\partial \mathbf{v}_e}{\partial t} = -e\mathbf{E}, \tag{5.28}$$

$$m_i \frac{\partial \mathbf{v}_i}{\partial t} = e\mathbf{E}, \tag{5.29}$$

where the suffixes e and i characterize quantities referring respectively to electrons and positive ions.

Evidently (5.3) is replaced by

$$\mathbf{j} = Ne(\mathbf{v}_i - \mathbf{v}_e), \tag{5.30}$$

which gives the conductivity

$$\sigma = -i\frac{Ne^2}{\omega}\left(\frac{1}{m_e} + \frac{1}{m_i}\right) \tag{5.31}$$

It follows at once that the plasma frequency is ω_p, where

$$\omega_p^2 = \omega_{pe}^2 + \omega_{pi}^2 = \frac{Ne^2}{\varepsilon_0}\left(\frac{1}{m_e} + \frac{1}{m_i}\right), \tag{5.32}$$

and the refractive index μ, where

$$\mu^2 = 1 - \frac{\omega_p^2}{\omega^2} = 1 - X = 1 - X_e - X_i, \tag{5.33}$$

with the notation

$$\omega_{pe}^2 = \frac{Ne^2}{\varepsilon_0 m_e}, \qquad \omega_{pi}^2 = \frac{Ne^2}{\varepsilon_0 m_i}, \tag{5.34}$$

$$X_e = \frac{\omega_{pe}^2}{\omega^2}, \qquad X_i = \frac{\omega_{pi}^2}{\omega^2}. \tag{5.35}$$

Sometimes ω_{pe} is called the electron plasma frequency and ω_{pi} the ion plasma frequency.

In these formulas the contribution of the positive ion motion is smaller than that of the electron motion by the factor m_e/m_i, and in this context, therefore, the approximation of treating the positive ions as fixed is clearly justified. Later on, however, it is seen that there are more complicated situations in which the motion of the positive ions can be most important.

5.3 COLLISIONLESS MAGNETO-IONIC THEORY

5.3.1 The conductivity tensor

A plasma model of much wider theoretical interest arises when it is supposed that there is a pervading magnetostatic field. This case is also of great practical importance; magnetic fields are likely to be present in many astrophysical problems; the ionosphere and exosphere are situated in the earth's magnetic field; and magnetic fields are commonly employed in the control and exploration of laboratory plasmas.

The main general effect produced by a magnetic field is to compel the ionized gas to act in its response to electromagnetic waves as an anisotropic medium. That is to say, as in analogous optical phenomena, the refractive index of a plane wave depends on the direction in which the wave is traveling, and for each direction has two possible values; moreover, for a given direction, each value of the refractive index is associated with a quite specific polarization. From the mathematical standpoint these features arise from the fact that the linear relation between the current density and the electric field is now a tensor rather than a scalar relation.

Apart from the supposed absence of collisions, an analysis on the lines of that in §5.2.1, with the neglect of the motion of the positive ions but the inclusion of a uniform magnetostatic field $\mathbf{B}_0$, constitutes the basis of the conventional form of magneto-ionic theory and is developed in the present section. The first step is to derive the tensor conductivity.

The equation of motion of an electron is

$$m\frac{\partial \mathbf{v}}{\partial t} = -e(\mathbf{E} + \mathbf{v} \times \mathbf{B}_0), \tag{5.36}$$

where the magnetic force has been added to the right-hand side of (5.1); that is, at angular frequency ω, in the complex representation,

$$i\omega m\mathbf{v} = -e(\mathbf{E} + \mathbf{v} \times \mathbf{B}_0). \tag{5.37}$$

To arrive at the conductivity tensor it is only necessary to solve (5.37) for $\mathbf{v}$ in terms of $\mathbf{E}$. The algebra can be done in a variety of ways. The procedure adopted here is to start with the simplification achieved by taking the direction of the z-axis along the magnetostatic field, which is often a convenient choice in specific problems, and subsequently to write down the result for arbitrarily rotated axes by inspection of the tensor form.

When

$$\mathbf{B}_0 = (0, 0, B_0), \tag{5.38}$$

the three components of (5.37) are

$$i\omega m V_x = -eE_x - eB_0 v_y, \tag{5.39}$$

$$i\omega m V_y = -eE_y + eB_0 v_x, \tag{5.40}$$

$$i\omega m V_z = -eE_z, \tag{5.41}$$

which are readily rewritten

$$v_x = \frac{ie}{\omega m}\frac{1}{1 - \Omega^2/\omega^2}\left(E_x + i\frac{\Omega}{\omega}E_y\right), \tag{5.42}$$

$$v_y = \frac{ie}{\omega m}\frac{1}{1 - \Omega^2/\omega^2}\left(-i\frac{\Omega}{\omega}E_x + E_y\right), \tag{5.43}$$

$$v_z = \frac{ie}{\omega m}E_z, \tag{5.44}$$

where the angular gyro frequency $\Omega = eB_0/m$ has been introduced. Then $\mathbf{j} = -Ne\mathbf{v}$ can be expressed in suffix notation

$$j_i = \sigma_{ij}E_j, \tag{5.45}$$

where the conductivity tensor is given by

$$\sigma_{ij} = -i\frac{Ne^2}{\omega m}\frac{1}{1-\Omega^2/\omega^2}\begin{pmatrix} 1 & i\Omega/\omega & 0 \\ -i\Omega/\omega & 1 & 0 \\ 0 & 0 & 1-\Omega^2/\omega^2 \end{pmatrix}. \quad (5.46)$$

Again it is equally possible, and perhaps more appropriate, to make the mathematical statement in terms of a tensor relation between **D** and **E**. Thus

$$D_i = \varepsilon_0 \kappa_{ij} E_j, \quad (5.47)$$

where the dielectric tensor is

$$\kappa_{ij} = \delta_{ij} - i\frac{\sigma_{ij}}{\varepsilon_0 \omega} = \delta_{ij} - \frac{\omega_p^2}{\omega^2-\Omega^2}\begin{pmatrix} 1 & i\Omega/\omega & 0 \\ -i\Omega/\omega & 1 & 0 \\ 0 & 0 & 1-\Omega^2/\omega^2 \end{pmatrix}; \quad (5.48)$$

or

$$\kappa_{ij} = \begin{pmatrix} 1-\dfrac{\omega_p^2}{\omega^2-\Omega^2} & -i\dfrac{\omega_p^2\Omega}{\omega(\omega^2-\Omega^2)} & 0 \\ i\dfrac{\omega_p^2\Omega}{\omega(\omega^2-\Omega^2)} & 1-\dfrac{\omega_p^2}{\omega^2-\Omega^2} & 0 \\ 0 & 0 & 1-\dfrac{\omega_p^2}{\omega^2} \end{pmatrix}. \quad (5.49)$$

Four features, certainly, of (5.49) could have been anticipated. First, that the elements in the third row or column have the same values as in the case of no magnetostatic field; this because if **E** were parallel to $\mathbf{B}_0$ the electrons would move only parallel to $\mathbf{B}_0$, experiencing no magnetic force. Secondly, that $\kappa_{11} = \kappa_{22}$, $\kappa_{12} = -\kappa_{21}$; this because of the nature of the symmetry about the direction of $\mathbf{B}_0$. Thirdly, that $\kappa_{ij}(-\Omega) = \kappa_{ji}(\Omega)$; this because a reversal of the direction of the magnetostatic field only alters the sense of the gyration of the electrons. Fourthly, that the tensor is Hermitian, that is $\kappa_{ij} = \kappa_{ji}^*$; this because, in the absence of collisions, there is no loss mechanism.

The expression for the dielectric tensor when the direction of $\mathbf{B}_0$ with respect to the axes is arbitrary is easily found by writing (5.48) in general tensor form. This is evidently

$$\kappa_{ij} = \delta_{ij} - \frac{\omega_p^2}{\omega^2-\Omega^2}(\delta_{ij} - \Omega_i\Omega_j/\omega^2 + i\varepsilon_{ijk}\Omega_k/\omega), \quad (5.50)$$

where

$$\boldsymbol{\Omega} = e\mathbf{B}_0/m. \quad (5.51)$$

In terms of the dimensionless quantities $X = \omega_p^2/\omega^2$, $\mathbf{Y} = \boldsymbol{\Omega}/\omega$,

$$\kappa_{ij} = \delta_{ij} - \frac{X}{1-Y^2}(\delta_{ij} - Y_iY_j + i\varepsilon_{ijk}Y_k); \quad (5.52)$$

and displayed in full,

$$\kappa_{ij} = \delta_{ij} - \frac{X}{1-Y^2}\begin{pmatrix} 1-Y_1^2 & iY_3-Y_1Y_2 & -iY_2-Y_1Y_3 \\ -iY_3-Y_2Y_1 & 1-Y_2^2 & iY_1-Y_2Y_3 \\ iY_2-Y_3Y_1 & -iY_1-Y_3Y_2 & 1-Y_3^2 \end{pmatrix}. \quad (5.53)$$

5.3.2 The refractive index

Maxwell's equations for a time-harmonic electromagnetic field in the plasma are

$$\text{curl}\,\mathbf{E} = -i\omega\mu_0\mathbf{H}, \quad (5.54)$$

$$\text{curl}\,\mathbf{H} = i\omega\varepsilon_0\boldsymbol{\kappa}\mathbf{E}, \quad (5.55)$$

where $\boldsymbol{\kappa}$ is the tensor (5.53).

If the field is that of a plane wave with space variation specified by the factor

$$e^{-i\mathbf{k}\cdot\mathbf{r}} \quad (5.56)$$

then the equations are

$$-i\mathbf{k}\times\mathbf{E} = -i\omega\mu_0\mathbf{H}, \quad (5.57)$$

$$-i\mathbf{k}\times\mathbf{H} = i\omega\varepsilon_0\boldsymbol{\kappa}\mathbf{E}, \quad (5.58)$$

from which the elimination of $\mathbf{H}$ gives

$$-\mu^2\hat{\mathbf{k}}\times(\hat{\mathbf{k}}\times\mathbf{E}) = \boldsymbol{\kappa}\mathbf{E}, \quad (5.59)$$

where $\hat{\mathbf{k}}$ is the unit vector along $\mathbf{k}$, so that $\mathbf{k} = k\hat{\mathbf{k}}$, and $\mu = ck/\omega$ is the refractive index.

In suffix notation (5.59) is

$$\{\kappa_{ij} - \mu^2(\delta_{ij} - \hat{k}_i\hat{k}_j)\}E_j = 0, \quad (5.60)$$

and the characteristics of the plane wave solutions are implicit in this equation. A non-zero field exists only if the determinant of the tensor multiplying $\mathbf{E}$ vanishes. This condition is the dispersion relation, which gives the possible values of μ; and the corresponding ratios of the components of $\mathbf{E}$, which specify the polarization, are then determinate.

One way of simplifying the algebra is to take the z-axis along $\mathbf{B}_0$, so that from (5.49),

$$\kappa_{ij} = \begin{pmatrix} \kappa_1 & -i\kappa_2 & 0 \\ i\kappa_2 & \kappa_1 & 0 \\ 0 & 0 & \kappa_0 \end{pmatrix}, \quad (5.61)$$

where

$$\kappa_0 = 1 - X, \qquad \kappa_1 = 1 - \frac{X}{1-Y^2}, \qquad \kappa_2 = \frac{XY}{1-Y^2}. \quad (5.62)$$

And then it is also convenient to orientate the axes to make $\hat{\mathbf{k}}$ lie in a coordinate plane; that is, say,

$$\hat{\mathbf{k}} = (0, \sin\theta, \cos\theta), \tag{5.63}$$

where θ is the angle which the direction of propagation makes with the magnetostatic field.

With this choice of axes the dispersion relation appears in the form

$$\begin{vmatrix} \kappa_1 - \mu^2 & -i\kappa_2 & 0 \\ i\kappa_2 & \kappa_1 - \mu^2\cos^2\theta & \mu^2\sin\theta\cos\theta \\ 0 & \mu^2\sin\theta\cos\theta & \kappa_0 - \mu^2\sin^2\theta \end{vmatrix} = 0; \tag{5.64}$$

that is

$$(\kappa_0\cos^2\theta + \kappa_1\sin^2\theta)\mu^4 - [\kappa_0\kappa_1(1+\cos^2\theta) + (\kappa_1^2 - \kappa_2^2)\sin^2\theta]\mu^2 + \kappa_0(\kappa_1^2 - \kappa_2^2) = 0. \tag{5.65}$$

Equation (5.65) is a quadratic for μ^2, yielding two values which may be regarded, for prescribed gyro and plasma frequencies, as functions of ω and θ.

The cases $\theta = 0$ (or equivalently $\theta = \pi$) and $\theta = \frac{1}{2}\pi$ are particularly simple.

For $\theta = 0$, provided $\kappa_0 \neq 0$, (5.65) gives at once,

$$\mu^2 = \kappa_1 \pm \kappa_2; \tag{5.66}$$

that is, from (5.62),

$$\mu^2 = 1 - \frac{X}{1 \pm Y} = 1 - \frac{\omega_p^2}{\omega(\omega \pm \Omega)}\,. \tag{5.67}$$

These are the squares of the refractive indices for plane waves that, if propagated, travel in the direction of the magnetostatic field, and if evanescent have maximum space rate of amplitude decay in that direction.

Alternatively, for $\theta = 0$, (5.65) is satisfied if $\kappa_0 = 0$, which implies oscillations at the plasma frequency ω_p. Equations (5.57) and (5.58) show that in this case all components of the wave field are zero except for $\mathbf{E}_z$, and that μ and E_z are arbitrary. Such oscillations are simply the plasma oscillations discussed in §5.2.2, but with the restriction here that the motion of the electrons is parallel to the magnetostatic field and consequently unaffected by it.

For $\theta = \frac{1}{2}\pi$, (5.65) gives

$$\mu^2 = \kappa_0,\ (\kappa_1^2 - \kappa_2^2)/\kappa_1; \tag{5.68}$$

that is, either

$$\mu^2 = 1 - X = 1 - \omega_p^2/\omega^2, \tag{5.69}$$

or

$$\mu^2 = 1 - \frac{X(1-X)}{1-X-Y^2} = 1 - \frac{\omega_p^2(\omega^2 - \omega_p^2)}{\omega^2(\omega^2 - \omega_p^2 - \Omega^2)}. \tag{5.70}$$

These are the squares of the refractive indices for plane waves traveling or evanescent in a direction perpendicular to that of the magnetostatic field. That one of the values, namely (5.69), does not involve the magnetostatic field, and is identical with (5.24), is to be expected, since a purely transverse plane wave, linearly polarized with the electric vector along $\mathbf{B}_0$ so that the electron motion is unaffected by $\mathbf{B}_0$, is clearly a possible field.

When the angle θ between the direction of propagation and $\mathbf{B}_0$ is arbitrary, both the variation of μ with θ for a given ω and the variation of μ with ω for a given θ are of interest. Rather than attempt to write down directly the solutions of (5.65) for μ^2 it is simpler and more informative to begin by solving for θ. In fact, immediately,

$$\mu^2 \cos^2\theta = -\frac{(\mu^2 - \kappa_0)(\kappa_1\mu^2 - \kappa_1^2 + \kappa_2^2)}{(\kappa_0 - \kappa_1)\mu^2 - \kappa_0\kappa_1 + \kappa_1^2 - \kappa_2^2}, \tag{5.71}$$

$$\mu^2 \sin^2\theta = \frac{\kappa_0(\mu^2 - \kappa_1 - \kappa_2)(\mu^2 - \kappa_1 + \kappa_2)}{(\kappa_0 - \kappa_1)\mu^2 - \kappa_0\kappa_1 + \kappa_1^2 - \kappa_2^2}, \tag{5.72}$$

where the right-hand side of (5.72) of course vanishes for the values of μ^2 displayed in (5.66), and that of (5.71) for the values (5.68).

Now from (5.62),

$$\kappa_0\kappa_1 - \kappa_1^2 + \kappa_2^2 = \frac{XY^2}{1-Y^2} = \kappa_0 - \kappa_1. \tag{5.73}$$

If, then, the refractive index is thought of as the vector $c\mathbf{k}/\omega$, with components parallel and perpendicular to B_0 of magnitudes

$$\mu_\parallel = \mu\cos\theta, \qquad \mu_\perp = \mu\sin\theta, \tag{5.74}$$

and if the further notation

$$\lambda = \frac{\mu^2 - \kappa_0}{\mu^2 - 1} = 1 - \frac{X}{1-\mu^2} \tag{5.75}$$

is introduced, (5.71) and (5.72) appear in the form

$$\mu_\parallel^2 = \frac{\lambda(\varepsilon - \lambda)}{\varepsilon(\lambda - 1)}, \tag{5.76}$$

$$\mu_\perp^2 = \frac{\lambda^2 - Y^2}{\varepsilon(\lambda - 1)}, \tag{5.77}$$

where

$$\varepsilon = \frac{\kappa_0 - \kappa_1}{\kappa_1 - \kappa_1^2 + \kappa_2^2} = \frac{Y^2}{1-X}. \tag{5.78}$$

The relations (5.76) and (5.77) are a most useful restatement of (5.65). On the one hand they can be treated as a simple parametric form of the equation for μ as a function of θ for a given ω; values accorded to λ between one of the limits 0, ε and one of the limits $\pm Y$ give corresponding values of $\mu_{\parallel}$ and $\mu_{\perp}$, so that the polar representation of μ as a function of θ is easily obtained by plotting the equivalent Cartesian representation. On the other hand, either one of the relations (5.76), (5.77), when the left-hand side is expressed in terms of λ and θ from (5.74) and (5.75), gives just

$$\lambda^2 - \varepsilon \sin^2 \theta \lambda - Y^2 \cos^2 \theta = 0; \tag{5.79}$$

whence

$$\lambda = \tfrac{1}{2}\varepsilon \sin^2 \theta \mp \sqrt{\{(\tfrac{1}{2}\varepsilon \sin^2 \theta)^2 + Y^2 \cos^2 \theta\}}, \tag{5.80}$$

and so, from (5.75),

$$\mu^2 = 1 - \frac{X}{1 - \dfrac{\frac{1}{2}Y^2 \sin^2 \theta}{1 - X} \pm \sqrt{\left\{\left(\dfrac{\frac{1}{2}Y^2 \sin^2 \theta}{1 - X}\right)^2 + Y^2 \cos^2 \theta\right\}}}. \tag{5.81}$$

The expression (5.81) is the Appleton–Hartree formula when collisions are not taken into account. For a specific θ it gives the dispersion relation between k $(= \omega\mu/c)$ and ω in the form

$$c^2k^2 = \omega^2 - \frac{\omega_p^2\omega(\omega^2 - \omega_p^2)}{\omega(\omega^2 - \omega_p^2) - \frac{1}{2}\Omega^2\omega \sin^2 \theta \pm \sqrt{\{(\frac{1}{2}\Omega^2\omega \sin^2 \theta)^2 + (\omega^2 - \omega_p^2)^2 \Omega^2 \cos^2 \theta\}}}. \tag{5.82}$$

Comparatively simple formulas are obtained only when one of the two terms in the radical is much smaller than the other. The so-called quasi-longitudinal approximation

$$\mu^2 = 1 - \frac{X}{1 \pm Y \cos \theta} \tag{5.83}$$

is commonly used when

$$\left|\frac{2(1 - X) \cos \theta}{Y \sin^2 \theta}\right| \gg 1,$$

and a quasi-transverse approximation when the inequality is reversed.

In the next two sub-sections some account is given of the information provided by the different formulations of equation (5.65).

5.3.3 The variation of refractive index with frequency

In attempting to give some overall picture of the general behavior of the refractive index it is convenient to keep distinct its specific variation with θ.

It is supposed for the moment that θ is fixed, and consideration is given to some of the possible ways of displaying the dependence of the refractive index on the remaining parameters.

Three particular graphical representations are treated, which are analogous to those discussed in §5.2.3 for the case of no magnetic field. In the first two ω_p and Ω are regarded as fixed, and μ^2 and ck respectively are plotted as ordinates against ω as abscissa; in the third, which is standard in magneto-ionic theory, Y is regarded as fixed, and μ^2 as ordinate is plotted against X as abscissa.

The general nature of the curves for any fixed value of θ, and indeed the way in which the curves change as the value of θ changes, is most readily understood by making use of the important fact that, for θ alone varying, the variation of μ^2 with θ, in the range $0 \leqslant \theta \leqslant \frac{1}{2}\pi$, is monotonic. This is proved by noting that, from (5.75) and (5.80),

$$\frac{\partial \mu^2}{\partial \theta} = \frac{d\mu^2}{d\lambda}\frac{\partial \lambda}{\partial \theta}$$

$$= -(1-\mu^2)^2 \frac{\varepsilon}{X} \sin\theta\cos\theta \left\{1 \pm \frac{\frac{1}{2}\varepsilon^2 \sin^2\theta - Y^2}{\sqrt{[(\frac{1}{2}\varepsilon^2 \sin^2\theta - Y^2)^2 - Y^2(Y^2 - \varepsilon^2)]}}\right\},$$

which does not change sign as θ progresses from 0 to $\frac{1}{2}\pi$. Indeed, between the limits $\theta = 0$ and $\theta = \frac{1}{2}\pi$, manifestly $\partial\mu^2/\partial\theta$ can vanish only when $\mu^2 = 1$, or when $Y^2 = \varepsilon^2$, or when $X = 1$ (ε infinite, with the factor in curly brackets vanishing like a multiple of $1/\varepsilon^2$); all of which conditions are independent of θ.

Thus it is known that, for a given θ, the curve of, for example, μ^2 against ω lies between the corresponding curves for the two cases $\theta = 0$, $\theta = \frac{1}{2}\pi$; and moreover that, for a fixed value of ω, μ^2 proceeds for increasing θ monotonically from its value at $\theta = 0$ to its value at $\theta = \frac{1}{2}\pi$.

For $\theta = 0$, μ^2 as a function of ω is given by (5.67). It is, of course, to be expected that only positive values of ω need be considered, and this is ensured by the alternative signs in (5.67). For $\omega \geqslant 0$, it is seen that $\mu^2 \to \infty$ both as $\omega \to \Omega$ and as $\omega \to 0$, but that μ^2 is otherwise regular; μ^2 vanishes at

$$\omega = \pm\tfrac{1}{2}\Omega + \sqrt{(\tfrac{1}{4}\Omega^2 + \omega_p^2)}. \tag{5.84}$$

In addition to (5.67) the case $\theta = 0$ admits the possibility $\omega = \omega_p$, and this must be included in the representation.

For $\theta = \frac{1}{2}\pi$, μ^2 as a function of ω is given either by (5.69) or by (5.70). In the former, $\mu^2 \to -\infty$ as $\omega \to 0$, but is otherwise regular, and vanishes at $\omega = \omega_p$. In the latter, $|\mu^2| \to \infty$ as $\omega \to \sqrt{(\Omega^2 + \omega_p^2)}$, $\mu^2 \to -\infty$ as $\omega \to 0$, and μ^2 vanishes at the values of ω given by (5.84).

The relation (5.84) is, of course, just $\kappa_1 = \pm\kappa_2$, shown explicitly in (5.66) and (5.68) to give zeros of μ^2 for both $\theta = 0$ and $\theta = \frac{1}{2}\pi$. An equivalent statement to (5.84) is

$$X = 1 \mp Y, \tag{5.85}$$

and this is just the condition $Y^2 = \varepsilon^2$ under which it was seen that, for one branch, $\partial\mu^2/\partial\theta$ vanished for all values of θ. The only other possibilities for $\partial\mu^2/\partial\theta$ to be zero occur either when $\mu^2 = 1$, or when $\omega = \omega_p$ (ε infinite); and in fact, from (5.82), the former implies the latter.

The picture is now clear. In a plot of μ^2 as ordinate against ω as abscissa, the curves for $\theta = 0$ and $\theta = \frac{1}{2}\pi$ intersect at four points; one where $\mu^2 = 1$, with $\omega = \omega_p$, and the other three where $\mu^2 = 0$, with $\omega = \omega_p$ or with ω given by (5.84). All the curves corresponding to other values of θ also go through these four points, but meet each other nowhere else.

The situation is illustrated in Fig. 5.4, where the dashed curve is that for $\theta = 0$, the dotted curve that for $\theta = \frac{1}{2}\pi$, and the full line curve that for an intermediate value of θ. The general way in which the dotted curve goes over to the dashed curve is seen at a glance, as is the rôle played by the straight line $\omega = \omega_p$. The figure is purely diagrammatic, in the sense that the overall topology is correct without corresponding values of ω and μ^2 being quanti-

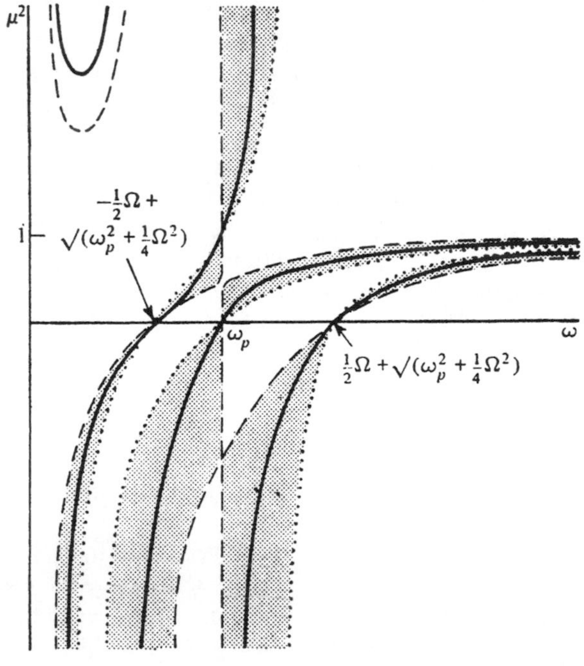

Figure 5.4

tatively represented on any scale. The asymptotes to the full-line curve are, apart from $\mu^2 = 1$,

$$\omega^2 = \tfrac{1}{2}(\Omega^2 + \omega_p^2) \pm \sqrt{[\tfrac{1}{4}(\Omega^2 + \omega_p^2)^2 - \omega_p^2\Omega^2 \cos^2\theta]}. \tag{5.86}$$

As θ moves towards $\frac{1}{2}\pi$ the left-hand one of these closes on $\omega = 0$ and squeezes out the branch of the curve between it and $\omega = 0$.

Figure 5.4 is drawn for the case $\omega_p > \Omega\sqrt{2}$. Topological variants arise when $\Omega < \omega_p < \Omega\sqrt{2}$ and when $\Omega > \omega_p$. If $\omega_p > \Omega\sqrt{2}$ the asymptote to the dashed ($\theta = 0$) curve lies to the left of the first zero of μ^2, that at (5.84) with the lower sign. The frequency band

$$\Omega < \omega < -\tfrac{1}{2}\Omega + \sqrt{(\tfrac{1}{4}\Omega^2 + \omega_p^2)} \tag{5.87}$$

therefore admits no propagating mode for any value of θ. If $\Omega < \omega_p < \Omega\sqrt{2}$ the topology is the same as in Fig. 5.4, except that for sufficiently small values of θ the branch in the upper left-hand corner overlaps the first zero of μ^2. But if $\Omega > \omega_p$ the $\theta = 0$ curve has a further division at $\omega = \omega_p$, $\mu^2 = \Omega/(\Omega - \omega_p)$, with the part of $\omega = \omega_p$ above $\mu^2 = \Omega/(\Omega - \omega_p)$ linking to the part of $\mu^2 = 1 - \omega/(\omega - \Omega)$ to the left of $\omega = \omega_p$.

The topology of the alternative plot in which ck is ordinate and ω is

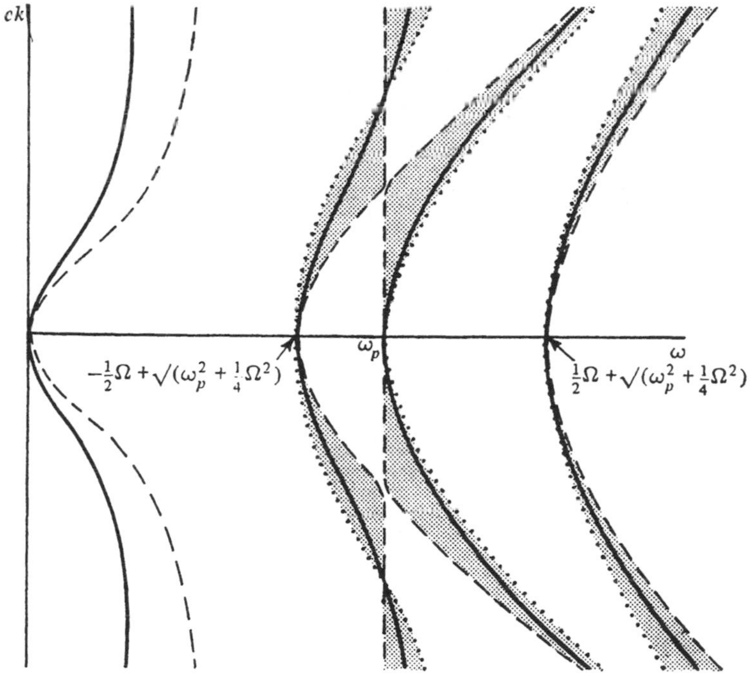

Figure 5.5

abscissa is readily obtained from the features already noted. The plot when $\omega_p > \Omega\sqrt{2}$ is shown diagrammatically in Fig. 5.5; again the dashed curve is for $\theta = 0$, the dotted curve is for $\theta = \frac{1}{2}\pi$, and the full-line curve for an intermediate value of θ.

Finally, the plot frequently used in magneto-ionic theory, with μ^2 as ordinate and X as abscissa, is shown diagrammatically in Figs. 5.6, 5.7 and 5.8. Here Y is regarded as fixed, so that what is being represented is essenti-

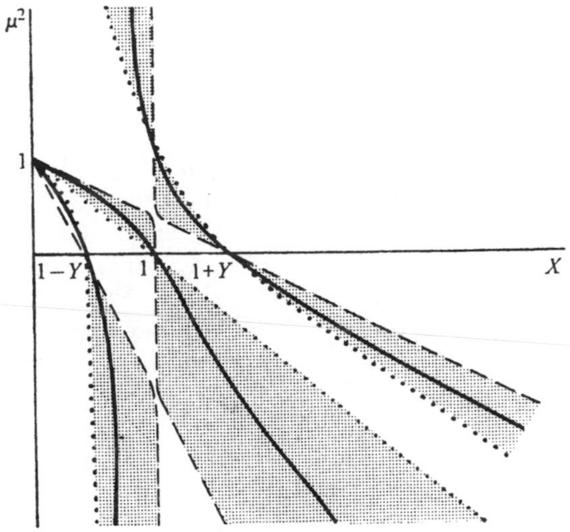

Figure 5.6

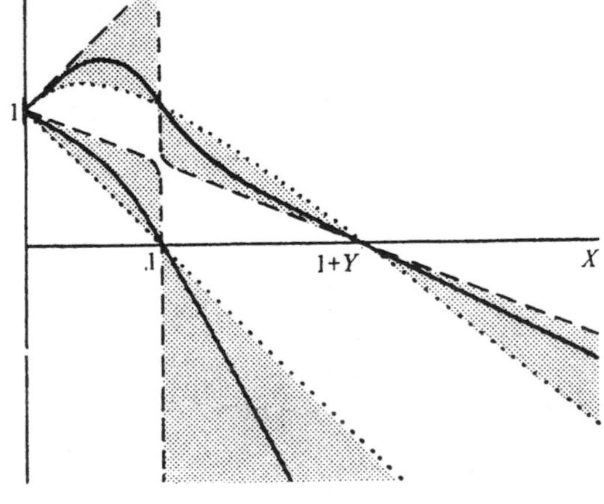

Figure 5.7

ally the variation of μ^2 with electron density for a given frequency and a given magnetostatic field. The figures are for the respective cases $Y < 1$, $\sec\theta > Y > 1$, $Y > \sec\theta$. The basic difference between the two latter is in the existence or not of an upper branch of the curve with asymptotes

$$X = \frac{Y^2 - 1}{Y^2\cos^2\theta - 1}, \qquad \mu^2 = \frac{X}{Y\cos\theta - 1}. \tag{5.88}$$

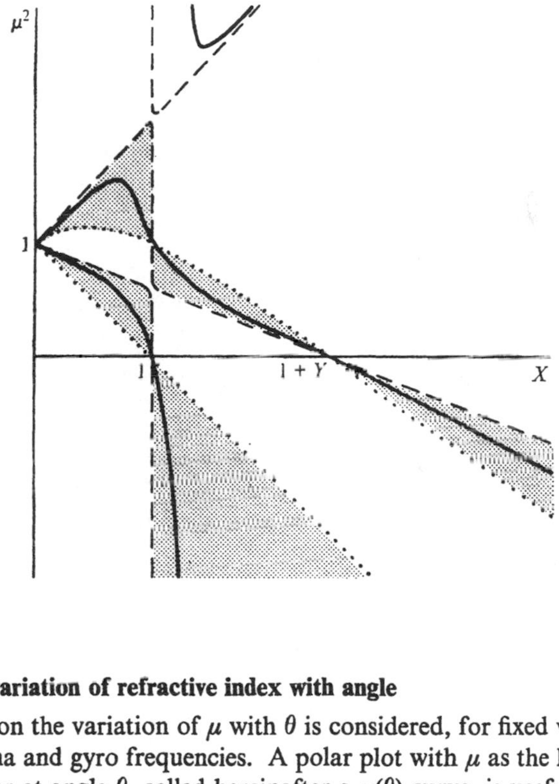

Figure 5.8

5.3.4 The variation of refractive index with angle

In this section the variation of μ with θ is considered, for fixed values of the wave, plasma and gyro frequencies. A polar plot with μ as the length of the radius vector at angle θ, called hereinafter a $\mu(\theta)$ curve, is useful for several purposes; in particular, because the curve has the property that for any point P on it a plane wave with phase propagation in the direction OP is associated with energy flow in the direction normal to the curve at P.

The $\mu(\theta)$ curves can be quite easily drawn with the help of the Cartesian parametric representation (5.76), (5.77). By symmetry it is, of course, only necessary to consider the quadrant in which $\mu_\parallel$ and $\mu_\perp$ are both positive. The type of curve obtained depends on the values of Y and ε. If λ can run from one of its terminal values at $\theta = 0$ $(\pm Y)$ to one of those at $\theta = \frac{1}{2}\pi$ $(0, \varepsilon)$ without encountering the value 1, then propagation is possible at any angle to the magnetostatic field. On the other hand, if λ can run from one terminal value to 1 without encountering another terminal value, then the $\mu(\theta)$ curve has an asymptote $\theta = \theta_0$, where

$$\tan^2\theta_0 = \frac{1 - Y^2}{\varepsilon - 1} = \frac{XY^2}{X + Y^2 - 1} - 1; \tag{5.89}$$

and this asymptote defines a cone about the direction of the magnetostatic field which delineates the range of angles within which propagation is possible.

The general feature that μ varies monotonically with θ has already been stressed. Another general feature, which is made evident, for example, in Fig. 5.4, is that on each individual $\mu(\theta)$ curve either $\mu^2 > 1$ or $\mu^2 < 1$; and furthermore the two $\mu(\theta)$ curves associated with any one given pair of values of Y, ε do not meet. The only exceptions to these statements arise in the special limiting cases $X = 0, 1$.

Distinguishing characteristics of the $\mu(\theta)$ curves are whether or not μ at $\theta = 0$ exceeds μ at $\theta = \frac{1}{2}\pi$, and whether or not points of inflection exist.

A convenient way of displaying the dependence of the shapes of the curves on the values of the wave, plasma and gyro frequencies is to think of the plane of the dimensionless parameters X, Y divided into regions, the boundaries of the regions being specified by certain relations between X and Y which represent limiting cases between two types of $\mu(\theta)$ curve. Since there are, for the most part, two curves associated with each pair of values X, Y it is simpler to show two such demarcations of the X, Y plane, as in Fig. 5.9.

The X, Y equations of the boundaries of the regions are given explicitly

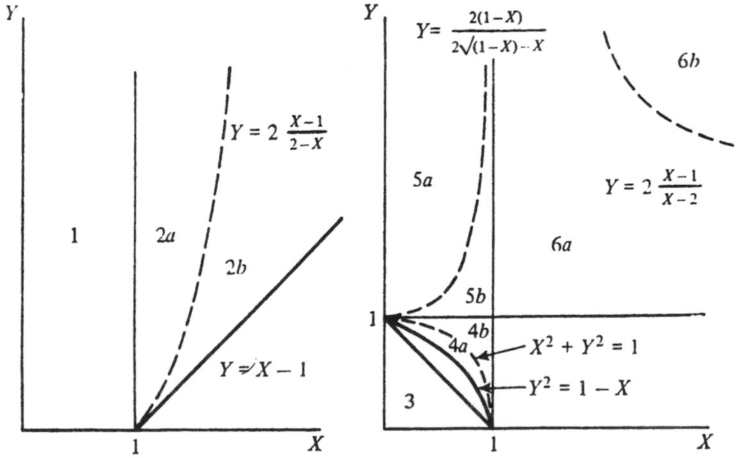

Figure 5.9

in Fig. 5.9. The range of values of the parameter λ in each region, and the corresponding range of values of μ are as follows:

Region 1 $$0 \leqslant \theta \leqslant \tfrac{1}{2}\pi, \qquad -Y \leqslant \lambda \leqslant 0,$$

$$\sqrt{\left(1 - \frac{X}{1+Y}\right)} \geqslant \mu \geqslant \sqrt{(1-X)}$$

Region 2 $$0 \leqslant \theta \leqslant \tfrac{1}{2}\pi, \qquad -Y \geqslant \lambda \geqslant \varepsilon,$$

$$\sqrt{\left(1 - \frac{X}{1+Y}\right)} \leqslant \mu \leqslant \sqrt{\left[\frac{Y^2 - (X-1)^2}{Y^2 + X - 1}\right]}$$

Region 3 $$0 \leqslant \theta \leqslant \tfrac{1}{2}\pi, \qquad Y \geqslant \lambda \geqslant \varepsilon,$$

$$\sqrt{\left(1 - \frac{X}{1-Y}\right)} \leqslant \mu \leqslant \sqrt{\left[\frac{(1-X)^2 - Y^2}{1 - X - Y^2}\right]}$$

Region 4 $$\theta_0 < \theta \leqslant \tfrac{1}{2}\pi, \qquad 1 < \lambda \leqslant \varepsilon,$$

$$\infty > \mu \geqslant \sqrt{\left[\frac{Y^2 - (1-X)^2}{Y^2 - 1 + X}\right]}$$

Region 5 $$0 \leqslant \theta \leqslant \tfrac{1}{2}\pi, \qquad Y \leqslant \lambda < \varepsilon,$$

$$\sqrt{\left(1 + \frac{X}{Y-1}\right)} \geqslant \mu \geqslant \sqrt{\left[\frac{Y^2 - (1-X)^2}{Y^2 - 1 + X}\right]}$$

Region 6 $$0 \leqslant \theta < \theta_0, \qquad Y \geqslant \lambda > 1,$$

$$\sqrt{\left(1 + \frac{X}{Y-1}\right)} \leqslant \mu < \infty.$$

The angle θ_0 in Regions 4 and 6 is given by (5.89).

The dashed curves in Fig. 5.9 form divisions of regions 2, 4, 5, and 6 which separate $\mu(\theta)$ curves with inflections from those without inflections. The shapes of the $\mu(\theta)$ curves are shown diagrammatically in Fig. 5.10.

The $\mu(\theta)$ curves in sub-regions 2a, 4b and 6b have (in $0 \leqslant \theta \leqslant \tfrac{1}{2}\pi$) just one point of inflection. In 2a and 6b the point of inflection is associated with a point (other than $\theta = 0$) where the tangent to the $\mu(\theta)$ curve is perpendicular to the magnetostatic field, and in 4b with a point (other than $\theta = \tfrac{1}{2}\pi$) where the tangent to the $\mu(\theta)$ curve is parallel to the magnetostatic field. The equations of the dividing lines in regions 2, 4, and 6, shown dashed in Fig. 5.9, are therefore easily found from consideration of the slope of the $\mu(\theta)$ curve; for this, from (5.76) and (5.77), is

$$-\cot\theta \, \frac{\lambda^2 - 2\lambda + Y^2}{\lambda^2 - 2\lambda + \varepsilon}, \tag{5.90}$$

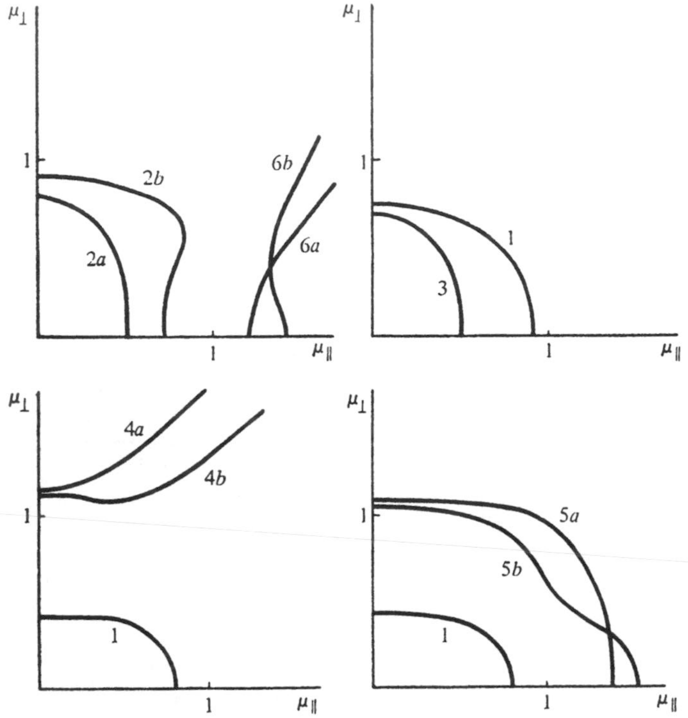

Figure 5.10

and it is only necessary to seek the condition for the denominator or numerator to vanish in the range covered by λ in the region in question.

The equation of the dividing line between sub-regions 5a and 5b is more difficult to come by. In 5b there are two points of inflection, and the limiting case is that for which these coalesce. Now it can be verified that if the curvature has a zero at $\lambda = \sqrt{\varepsilon}$, then it is a double zero; the required X, Y relation is therefore specified by the vanishing of the curvature at $\lambda = \sqrt{\varepsilon}$.

The arrangement of Figs. 5.9 and 5.10 is such that for a given pair of values X, Y it can be seen at a glance for what directions there are two, one or no plane wave modes of propagation. It should be remarked that there are, of course, many other similar ways, differing in detail, of presenting much the same information. The XY-plane of Fig. 5.9 might be replaced, for example, by the plane of the two dimensionless parameters ω/ω_p, Ω/ω_p; this would probably be more useful if the main interest were in a fixed electron density under conditions of changing frequency and changing magnetostatic field. Again, it is not uncommon to find, rather than μ, the speed of phase propagation c/μ, plotted radially against angle.

This section concludes with the observation that the idealized physical problem has, even at this stage, given rise to a great variety of situations, as depicted, for example, in Fig. 5.10. In addition to noting the curious shapes of some of the curves, and stressing again the importance of the distinction between evanescent and propagated plane wave modes, attention should also be drawn to the fact that μ can be very large. It is possible in this way for electromagnetic waves to travel with speeds of lower order of magnitude than the vacuum speed of light. In §5.4 the inclusion of the motion of positive ions is considered, and particular interest then centers on low-frequency slow waves whose characteristics are very different from those traditionally associated with electromagnetic waves.

5.3.5 Further remarks

One of the main applications of the preceding theory is to the propagation of radio waves in natural plasmas, in particular the earth's ionosphere and exosphere. In this context the polarization of a wave, as well as its refractive index, is an important characteristic. The relative amplitudes and phases of the components of the electric vector are given by (5.60); and in general, of course, the polarization is elliptical and the electric vector has a component in the direction of phase propagation as well as components transverse to it. No details are given here, except for the observation that in the special case $\theta = 0$, with the magnetostatic field and the direction of phase propagation along the z-axis and the refractive index given by (5.67), then

$$E_z = 0, \qquad E_x = \pm i E_y; \tag{5.91}$$

so here the electric vector is transverse to the direction of phase propagation, and the two modes have circular polarization with opposite senses of rotation. On the other hand, for $\theta = \frac{1}{2}\pi$, the mode (5.69), in which the magnetostatic field plays no part, has the expected linear polarization; whereas the mode (5.70), with the magnetostatic field along the z-axis and the direction of phase propagation along the y-axis, has

$$E_z = 0, \qquad E_x = i\,\frac{1 - X - Y^2}{XY}\,E_y. \tag{5.92}$$

In the earth's ionosphere $\Omega/(2\pi) \simeq 1$ Mc/s, and $\omega_p/(2\pi)$ increases with height up to the F-layer peak, where it has a maximum $\omega_{pm}/(2\pi) \simeq 10$ Mc/s. On a simple ray theory, and with the effect of collisions ignored, it is apparent from Figs. 5.6, 5.7 and 5.8 that for $\Omega < \omega < \omega_{pm}$ the two characteristic waves do not penetrate beyond the respective heights at which $X = 1 - Y$ and $X = 1$; whilst for $\omega < \Omega$ the heights beyond which there is no penetration are specified by $X = 1$, $X = 1 + Y$. Thus reliable radio communication via the ionosphere can be carried out with frequencies below about 10 Mc/s; and waves at frequencies down in the kc/s range are for the most part confined to and beneath the lower edge of the ionosphere. The modes of propagation involved are those associated with regions 1, 3 ($\omega < \Omega$) and 1, 5 and 2 ($\Omega < \omega < \omega_{pm}$) of the XY-plane, in the notation of Fig. 5.9.

Another interesting possibility is that of waves at frequencies in the vicinity of 1 kc/s traveling through the exosphere in the mode of propagation associated with Region 6 of the XY-plane Such waves can be generated by a lightning stroke, and in practice the values both of μ and of the group refractive index $c\partial k/\partial\omega$ can be as great as several hundred. The energy is funneled along magnetic field lines, and can therefore return to earth in the neighborhood of the point magnetically conjugate to the initiating lightning stroke after a journey taking several seconds. On reception these "whistlers" commonly produce a falling tone, lasting about a second, due to the dispersion arising from the frequency dependence of the refractive index.

5.4 THE INCLUSION OF THE MOTION OF POSITIVE IONS

5.4.1 The refractive index

When a collisionless electron–ion gas is pervaded by a magnetostatic field the motion of the positive ions can in certain circumstances be important. This is in contrast with the case of no magnetic field, for which, as shown in §5.2.4, the positive ion motion can always be ignored. The difference arises from the influential rôle played by the gyro frequencies. It is to be expected that the magneto-ionic theory described in §5.3 is an adequate approximation only if the wave frequency is appreciably higher than the positive ion gyro frequency, and that when the wave frequency becomes comparable with or less than the ion gyro frequency the characteristics of the electromagnetic field may be quite different from those so far encountered.

The detailed investigation can proceed in a way entirely similar to that given in §5.3. It is only necessary to augment the conductivity tensor by the positive-ion contribution, and this clearly leaves the structure of the tensor unaltered, though the expressions for its elements become more complicated.

The equation of motion of a positive ion differs from that of an electron, namely (5.36), only in that $-e$ is replaced by e and the electron mass m_e (written m in (5.36)) by the ion mass m_i. Clearly, therefore, the ion contribution to the conductivity tensor differs from the electron contribution (5.46) only in that Ω_e (written Ω in (5.46)) is replaced by $-\Omega_i$, where

$$\Omega_i = eB_0/m_i. \tag{5.93}$$

And this in turn implies that a plane-wave analysis developed as in §5.3.2 leads to equation (5.65) for the refractive index μ, with now

$$\kappa_0 = 1 - X_e - X_i, \qquad \kappa_1 = 1 - \frac{X_e}{1 - Y_e^2} - \frac{X_i}{1 - Y_i^2},$$

$$\kappa_2 = \frac{X_e Y_e}{1 - Y_e^2} - \frac{X_i Y_i}{1 - Y_i^2}, \tag{5.94}$$

where

$$X_e = \frac{Ne^2}{\varepsilon_0 m_e \omega^2}, \qquad X_i = \frac{Ne^2}{\varepsilon_0 m_i \omega^2}, \qquad Y_e = \frac{eB_0}{m_e \omega}, \qquad Y_i = \frac{eB_0}{m_i \omega}. \tag{5.95}$$

If a symbol for the ratio of the electron mass to ion mass is introduced, say

$$\eta = m_e/m_i, \tag{5.96}$$

then $X_i = \eta X_e$, $Y_i = \eta Y_e$, and (5.94) can be written

$$\kappa_0 = 1 - X, \qquad \kappa_1 = 1 - \frac{X(1-\eta Y^2)}{(1-Y^2)(1-\eta^2 Y^2)},$$

$$\kappa_2 = \frac{(1-\eta)XY}{(1-Y^2)(1-\eta^2 Y^2)}, \tag{5.97}$$

where $X = X_e + X_i$, as in §5.2.4, and $Y = Y_e$, as in §5.3.

The investigation of equation (5.65) can proceed as in §5.3.2 in that (5.71) and (5.72) still apply. Now, however, from (5.97),

$$\kappa_0\kappa_1 - \kappa_1^2 + \kappa_2^2 = \frac{XY^2(\Gamma - \eta X)}{(1-Y^2)(1-\eta^2 Y^2)} = \left(1 - \frac{\eta X}{\Gamma}\right)(\kappa_0 - \kappa_1), \tag{5.98}$$

where

$$\Gamma = 1 - \eta + \eta^2(1 - Y^2); \tag{5.99}$$

so the introduction of the notation

$$\lambda = \frac{\mu^2 - \kappa_0}{\mu^2 - (\kappa_0\kappa_1 - \kappa_1^2 + \kappa_2^2)/(\kappa_0 - \kappa_1)} = 1 - \frac{(1 - \eta/\Gamma)X}{1 - \eta X/\Gamma - \mu^2} \tag{5.100}$$

enables (5.71) and (5.72) to be written

$$\mu_{\parallel}^2 = \frac{\lambda\left[\dfrac{\Gamma(\Gamma - \eta X)}{(1-\eta)^2}\varepsilon - \lambda\right]}{\left(\dfrac{\Gamma}{1-\eta}\right)^2 \varepsilon(\lambda - 1)}, \tag{5.101}$$

$$\mu_{\perp}^2 = \frac{\lambda^2 - \left(\dfrac{\Gamma}{1-\eta}\right)^2 Y^2}{\left(\dfrac{\Gamma}{1-\eta}\right)^2 \varepsilon(\lambda - 1)}, \tag{5.102}$$

which clearly reduce to (5.76) and (5.77) respectively if η is set equal to zero.

To get an explicit expression for μ^2 it need only be noted that, from (5.100),

$$\mu^2 = 1 - \frac{\eta X}{\Gamma} - \frac{\left(1 - \dfrac{\eta}{\Gamma}\right)X}{1 - \lambda}. \tag{5.103}$$

The substitution from (5.103) into the left-hand sides of either (5.101) or (5.102) then gives, easily,

$$\lambda^2 - \frac{\Gamma(\Gamma - \eta X)}{(1-\eta)^2}\varepsilon \sin^2\theta\lambda - \left(\frac{\Gamma Y}{1-\eta}\right)^2 \cos^2\theta = 0, \qquad (5.104)$$

which is the generalization of (5.79). And finally the two solutions for λ from (5.104) when substituted into (5.103) give an explicit expression for μ^2 similar in form to the Appleton–Hartree formula (5.81), which can be regained by setting η equal to zero.

The various ways of displaying the functional behavior of μ^2 considered in §§5.3.3 and 5.3.4 can be adapted without any further difficulty of principle to take account of positive ion motions. To set out the equivalent results in the same detail would, however, take up an unjustifiable amount of space. The discussion is therefore restricted to exemplifying the situation without attempting to be comprehensive. Before dealing with the variation of μ^2 throughout the entire frequency range it is of particular interest to make an approximate examination of what happens at frequencies well below the ion gyro frequency.

5.4.2 The low-frequency approximation

The condition for the elements of the dielectric tensor given by (5.97) to have approximately the same values as when positive ion motions are ignored is $\eta Y^2 \ll 1$, that is

$$\omega^2 \gg \Omega_e \Omega_i. \qquad (5.105)$$

Departures from the theory of §5.3 may therefore be expected when the angular frequency is comparable with or less than the geometric mean of the electron and ion gyro frequencies.

Particularly simple and instructive are the results in the low-frequency limit for which

$$\omega \ll \Omega_i. \qquad (5.106)$$

For if $\eta Y \gg 1$, (5.97) gives approximately

$$\kappa_0 = 1 - X, \qquad \kappa_1 = 1 + \frac{X}{\eta Y^2}, \qquad \kappa_2 = \frac{X}{\eta^2 Y^3}, \qquad (5.107)$$

and then

$$\kappa_1^2 \gg \kappa_2^2. \qquad (5.108)$$

Now the inequality (5.108) introduces a great simplification into (5.65), the basic equation for μ^2. It implies that κ_2^2 can be neglected, the left-hand side of the equation factorizes, and the two values of μ^2 are

$$\kappa_1, \qquad \frac{\kappa_0 \kappa_1}{\kappa_0 \cos^2\theta + \kappa_1 \sin^2\theta}, \qquad (5.109)$$

With the approximations (5.107) the first of these values may be written

$$1 + \frac{\omega_p^2}{\eta\Omega^2}, \tag{5.110}$$

which is independent of both frequency and angle; whereas the second is

$$\left(1 + \frac{\omega_p^2}{\eta\Omega^2}\right)\frac{\omega_p^2 - \omega^2}{\omega_p^2 \cos^2\theta - \left(1 + \dfrac{\omega_p^2}{\eta\Omega^2}\sin^2\theta\right)\omega^2}. \tag{5.111}$$

The expressions (5.110) and (5.111) representing μ^2 as a function of ω are plotted diagrammatically in Fig. 5.11, where the dashed and dotted curves are for the respective limits $\theta = 0$, $\theta = \frac{1}{2}\pi$. It must be remembered that the representation is an approximation valid only for values of ω much less than Ω_i.

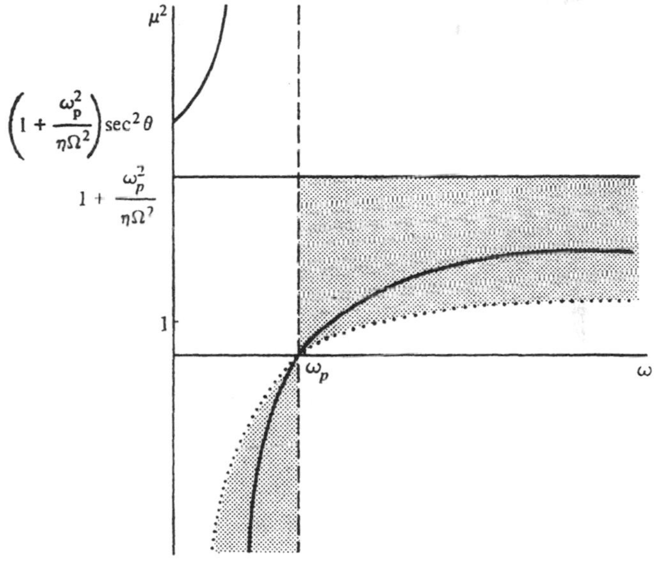

Figure 5.11

The $\mu(\theta)$ curves given by (5.109) are the circle $\mu^2 = \kappa_1$ and the conic

$$\frac{\mu_{\parallel}^2}{\kappa_1} + \frac{\mu_{\perp}^2}{\kappa_0} = 1. \tag{5.112}$$

In the present context it is likely that $\omega < \omega_p$, in which case (5.112) is a hyperbola with asymptotes making angle α with the magnetostatic field, where

$$\tan\alpha = \sqrt{(-\kappa_0/\kappa_1)}. \tag{5.113}$$

Waves in this mode can propagate only in directions making angles less than α with the magnetic field.

The commonly experienced case $\omega_p^2 \gg \eta\Omega^2 = \Omega_e\Omega_i$ gives rise to further simplification, since then, approximately

$$\kappa_0 = -\frac{\omega_p^2}{\omega^2}, \qquad \kappa_1 = \frac{\omega_p^2}{\eta\Omega^2}, \tag{5.114}$$

and $-\kappa_0 \gg \kappa_1$. Except for values of θ close to $\frac{1}{2}\pi$ the two values of μ^2 are therefore

$$\frac{\omega_p^2}{\eta\Omega^2}, \qquad \frac{\omega_p^2}{\eta\Omega^2\cos^2\theta}, \tag{5.115}$$

corresponding to wave speeds

$$\frac{\sqrt{(\Omega_e\Omega_i)}}{\omega_p}c, \qquad \frac{\sqrt{(\Omega_e\Omega_i)}}{\omega_p}c\cos\theta. \tag{5.116}$$

It may be noted that

$$\frac{\sqrt{(\Omega_e\Omega_i)}}{\omega_p}c = \frac{B_0}{\sqrt{(\mu_0\rho)}} \tag{5.117}$$

where μ_0 is the vacuum permeability and

$$\rho = N(m_e + m_i) \tag{5.118}$$

is the mass density of the electron–ion gas.

The expression

$$V_A = \frac{B_0}{\sqrt{(\mu_0\rho)}} \tag{5.119}$$

is often called the Alfvén speed. Its magnitude for an electron–proton gas, with, for illustration, $B_0 = 1$ gauss and $N = 10^{12}$ m^{-3}, is about 2×10^6 m sec^{-1}. It can only be a valid approximation when it is much less than the vacuum speed of light; the more accurate expression from the value (5.110) for μ^2 may be written

$$\frac{V_A}{\sqrt{(1 + V_A^2/c^2)}}. \tag{5.120}$$

The very low-frequency waves under discussion are variously called magnetohydrodynamic (mhd), hydromagnetic, or Alfvén waves. They have the property that, in the associated motion of the charged particles, the velocity component perpendicular to the magnetostatic field is approximately the same for the electrons and ions, the common value being

$$\mathbf{E} \times \mathbf{B}_0/B_0^2. \tag{5.121}$$

This may be seen by reference to equations (5.42) and (5.43). In its motion across the magnetostatic field the gas therefore moves as a whole.

The polarizations of the waves for which the values of μ^2 are those in (5.109) are easily found from (5.60) and (5.64). In the first wave $E_y = E_z = 0$, so that the electric vector is perpendicular both to the direction of propagation and to the magnetostatic field; whereas in the second,

$$E_x = 0, \qquad E_y/E_z = -(\kappa_0/\kappa_1)\cot\theta,$$

so that the electric vector is in the plane containing the directions of propagation and of the magnetostatic field, and has, in particular, a component along the former.

5.4.3 The variation of refractive index with frequency and angle

In §5.3.3 it was seen that $\partial\mu^2/\partial\theta$ could only vanish under one of the special conditions $Y = \pm\varepsilon$, $X = 1$, conditions independent of θ. A quite similar result holds when positive ion motions are included. From the results of §5.4.1 it readily appears that the corresponding conditions are

$$Y = \pm\frac{\Gamma - \eta X}{1 - \eta}\varepsilon, \qquad X = 1, \tag{5.122}$$

where Γ is given by (5.99).

The first condition in (5.122) reduces to either $\eta Y = 1 - \eta$, that is

$$\omega = \frac{\Omega_i}{1 - \eta}, \tag{5.123}$$

or $X = (1 \mp Y)(1 \pm \eta Y)$, that is

$$\omega = \pm\tfrac{1}{2}(\Omega_e - \Omega_i) + \sqrt{[\tfrac{1}{4}(\Omega_e + \Omega_i)^2 + \omega_p^2]}. \tag{5.124}$$

In a plot of μ^2 as ordinate against ω as abscissa $\partial\mu^2/\partial\theta$ therefore vanishes at five points. Three are where $\mu^2 = 0$, with ω equal to ω_p or to either of the expressions (5.124); a fourth is where $\omega = \omega_p$, $\mu^2 = 1 - \eta/\Gamma$; and a fifth is where $\omega = \Omega_i/(1 - \eta)$, $\mu^2 = 1 - (1 - \eta)^2\omega_p^2/\Omega_i^2$. These points each have their obvious counterpart in the case when positive ion motions are neglected, except for the last named, for which $\mu^2 \to -\infty$ as $\Omega_i \to 0$.

Now for $\theta = 0$, either $\omega = \omega_p$ or $\mu^2 = \kappa_1 + \kappa_2$, the latter giving, from (5.97),

$$\mu^2 = 1 - \frac{X}{(1 - Y)(1 + \eta Y)} = 1 - \frac{\omega_p^2}{(\omega - \Omega_e)(\omega + \Omega_i)}, \tag{5.125}$$

$$\mu^2 = 1 - \frac{X}{(1 + Y)(1 - \eta Y)} = 1 - \frac{\omega_p^2}{(\omega + \Omega_e)(\omega - \Omega_i)}. \tag{5.126}$$

And for $\theta = \frac{1}{2}\pi$, $\mu^2 = \kappa_0$ or $\mu^2 = (\kappa_1^2 - \kappa_2^2)/\kappa_1$, giving

$$\mu^2 = 1 - X = 1 - \omega_p^2/\omega^2, \tag{5.127}$$

$$\mu^2 = 1 - \frac{X(1-X-\eta Y^2)}{(1-Y^2)(1-\eta^2 Y^2) - X(1-\eta Y^2)}$$
$$= 1 - \frac{\omega_p^2(\omega^2 - \omega_p^2 - \Omega_e\Omega_i)}{(\omega^2-\Omega_e^2)(\omega^2-\Omega_i^2) - \omega_p^2(\omega^2-\Omega_e\Omega_i)}. \tag{5.128}$$

It follows that $\omega = \omega_p$ intersects (5.127) where $\mu = 0$, and (5.128) where

$$\mu^2 = 1 - \frac{\omega_p^2\Omega_e\Omega_i}{\omega_p^2(\Omega_e^2 + \Omega_i^2 - \Omega_e\Omega_i) - \Omega_e^2 - \Omega_i^2}, \tag{5.129}$$

the expression (5.129) being just $1 - \eta/\Gamma$ with $\omega = \omega_p$. Also (5.125) and (5.126) intersect (5.128) where $\mu = 0$, with ω given respectively by the two values (5.124). Again (5.126) intersects (5.127) at

$$\omega = \Omega_i/(1 - \eta), \qquad \mu^2 = 1 - (1 - \eta)^2 p^2/\Omega_i^2.$$

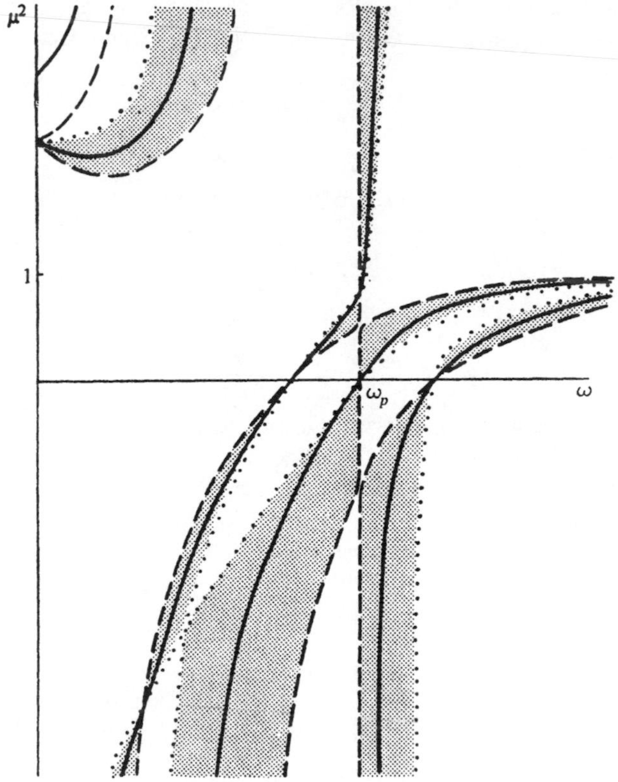

Figure 5.12

It is thus established that the curve of μ^2 as ordinate against ω (or X) as abscissa for any value of θ passes through the five points of intersection of the $\theta = 0$ curve with the $\theta = \frac{1}{2}\pi$ curve; and that, for any fixed value of ω (or X), μ^2 progresses monotonically as θ passes from 0 to $\frac{1}{2}\pi$. As previously, therefore, the broad features of curves for general values of θ can be ascertained from those for the special cases $\theta = 0$, $\theta = \frac{1}{2}\pi$.

A plot of μ^2 against ω is shown diagrammatically in Fig. 5.12, for the case in which the two values of ω in (5.124), specifying zeros of μ^2, exceed Ω_e and lie either side of ω_p. The condition for this to happen is

$$\omega_p > \sqrt{[2\Omega_e(\Omega_e - \Omega_i)]}, \tag{5.130}$$

and comparison should be made with Fig. 5.4, which represents the corresponding case when Ω_i is set equal to zero. For a given value of η ($\ll 1$) there are, in fact, six topologically different cases.

Figure 5.13 is a diagrammatic plot of ck against ω under the same condition (5.130). Comparison should be made with Fig. 5.5.

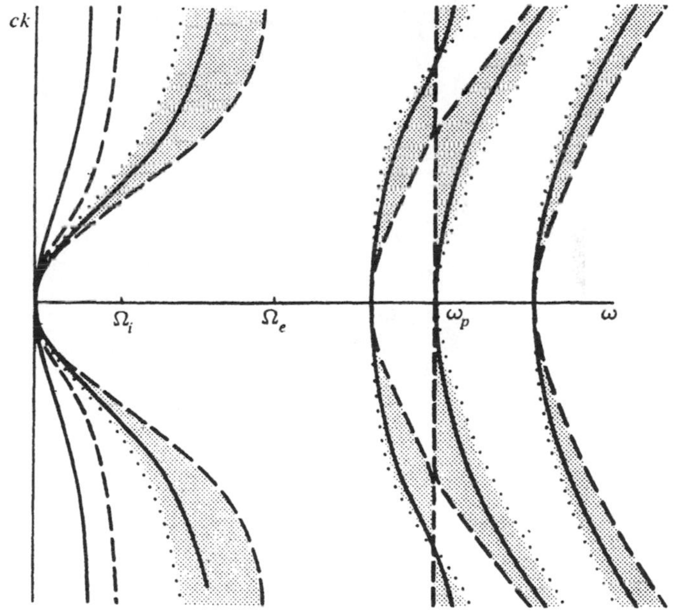

Figure 5.13

The plot of μ^2 against X, for a fixed value of Y, is comparatively simple analytically, since the $\theta = 0$ equations (5.125), (5.126) represent straight lines, and the $\theta = \frac{1}{2}\pi$ equations (5.127) and (5.128) represent a straight line and a hyperbola respectively. Again there are some half a dozen different

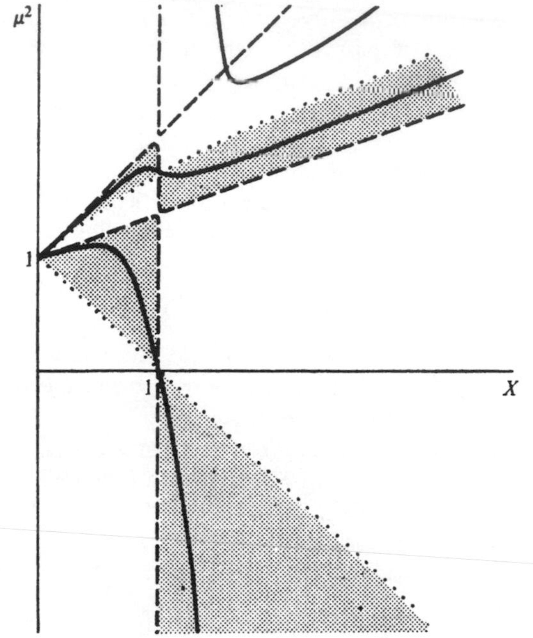

Figure 5.14

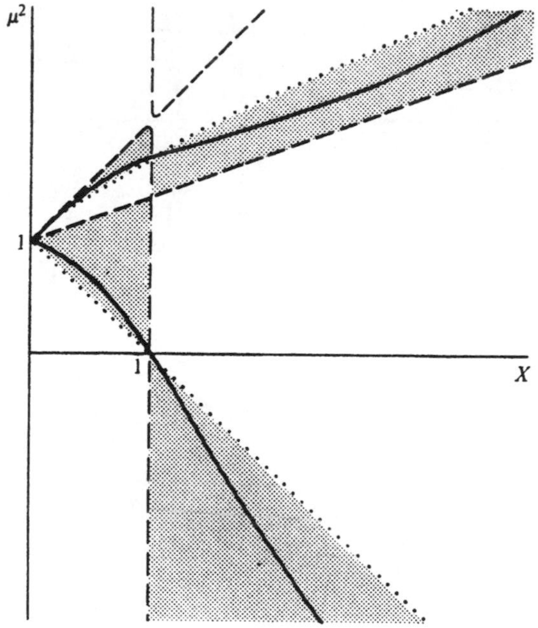

Figure 5.15

topologies. The case $\eta Y > 1$, that is $\omega < \Omega_i$, is shown diagrammatically in Figs. 5.14 and 5.15, the former or latter according to whether $\tan^2 \theta$ is less than or greater than $(Y^2 - 1)(\eta^2 Y^2 - 1)/(\eta Y^2 - 1)$.

5.5 THE INCLUSION OF PARTIAL PRESSURES

5.5.1 Pressure and temperature

Thus far in the analysis of wave motion in an electron–ion gas the pressure forces, which play an essential part in the hydrodynamical theory of neutral gases, have not been included. The treatment has been for what is often termed a "cold" plasma, since the pressure forces owe their existence to the thermal velocities of the gas particles, and would be strictly zero only at zero temperature. The introduction of pressure forces means that the basis of the theory of sound waves is built into the analysis, and broadly speaking it is to be expected that the results already developed remain valid under the generalization in so far as the speeds of the waves are sufficiently in excess of the relevant speed of sound. The implication is that electromagnetic waves with speeds comparable with that of light in vacuum are only slightly modified, whereas the electrostatic plasma waves and the extremely low-frequency magnetohydrodynamic waves may be subject to important modifications. Those particular waves in which there are no density fluctuations, such as the transverse waves in a plasma without external magnetostatic field, are, of course, completely unaffected.

In a gas, the pressure exerts a force per unit volume equal to the negative of its gradient. For a gas that in equilibrium is uniform this is

$$-\gamma KT \operatorname{grad} n, \tag{5.131}$$

on the working hypothesis that fluctuations of pressure and density take place adiabatically; here γ is the specific heat ratio, K Boltzmann's constant, T the absolute temperature and n the excess number density of particles over the equilibrium number density N.

Now the equation of continuity is

$$\frac{\partial n}{\partial t} + \operatorname{div}\,[(N + n)\mathbf{v}] = 0 \tag{5.132}$$

where v is the gas velocity. When it is linearized, and the time variation taken as $\exp(i\omega t)$, it becomes

$$i\omega n + N \operatorname{div} \mathbf{v} = 0, \tag{5.133}$$

and this enables (5.131) to be written

$$-i\gamma \frac{KTN}{\omega} \operatorname{grad} \operatorname{div} \mathbf{v}. \tag{5.134}$$

Separate forces (5.134), corresponding to the partial pressures of the electron and ion constituents of the plasma, are, then, to be included in their respective equations of motion. The equilibrium number density must have the same value N for both electrons and ions, but the values of the specific heat ratio and the temperature can differ between them, and are distinguished in the analysis by writing γ_e, T_e for the electrons, and γ_i, T_i for the ions. The simplest case, that of a plasma with no external magnetostatic field, is treated first, and then some consideration is given to the rather complicated situation that arises when a magnetostatic field is present.

5.5.2 The electron gas

With the inclusion of pressure forces the motion of positive ions can be important even in the absence of an external magnetostatic field, but it remains true that at sufficiently high frequencies a good approximation is obtained by taking them to be at rest. For the sake of algebraic simplicity the nature of the effect of pressure forces is introduced in this section by considering electron motions only, and an important approximate dispersion relation is established. Positive ion motions are included in the next section, and the extent to which they can give rise to further waves is elucidated.

The linearized electron equation of motion is

$$i\omega m\mathbf{v} = -e\mathbf{E} - i\frac{\gamma KT}{\omega}\,\text{grad div}\,\mathbf{v}, \tag{5.135}$$

where in this section the subscript e is omitted, as in §5.2.1. With the space variation specified by the factor $\exp(-ikx)$ the components of this equation are

$$i\omega m v_x = -eE_x + i\frac{\gamma KT}{\omega}k^2 v_x, \tag{5.136}$$

$$i\omega m v_y = -eE_y, \tag{5.137}$$

$$i\omega m v_z = -eE_z. \tag{5.138}$$

Two points may be noted. First, that the equations for the vector components perpendicular to the direction of propagation are unaffected by the pressure term in (5.135); the transverse wave described in §5.2.3 is therefore unmodified, as has already been mentioned. Secondly, that the conductivity or dielectric tensor, which describe the relationship between $\mathbf{v}$ and $\mathbf{E}$, now explicitly contain the wavelength $2\pi/k$ as well as the frequency ω. This is an example of what is sometimes called *spatial dispersion*.

Evidently the separation of transverse electromagnetic waves from longitudinal "electrostatic" or "plasma" waves can be made just as in §5.2. The former need no further comment. In the latter, E_x is the only non-zero component of the vectors **E**, **H**, and the linearized current density also has only an x-component

$$j_x = -Nev_x. \tag{5.139}$$

Substitution from (5.136) gives

$$j_x = -i\frac{Ne^2}{m\omega}\frac{1}{1 - a^2k^2/\omega^2}E_x, \tag{5.140}$$

where

$$a = \sqrt{(\gamma KT/m)} \tag{5.141}$$

is the "speed of sound" in the electron gas. Apart from a numerical factor of the order of unity, the expression for a is, of course, the same as that for the average thermal speed of the electrons; at $T = 1000°K$ it is of the order of 10^5 m sec^{-1}. Equation (5.140) shows that the influence of the pressure term is small when the wave speed ω/k is much greater than a.

Now from Maxwell's equations, in the present context,

$$i\omega\varepsilon_0 E_x + j_x = 0; \tag{5.142}$$

and together (5.140), (5.142) lead to the dispersion relation

$$a^2k^2 = \omega^2 - \omega_p^2, \tag{5.143}$$

where ω_p is the electron plasma frequency.

Equation (5.143) is an important result. It can be written as an expression for the square of the refractive index, thus

$$\mu^2 = \left(\frac{c}{a}\right)^2\left(1 - \frac{\omega_p^2}{\omega^2}\right); \tag{5.144}$$

and in another useful form

$$\omega^2 = \omega_p^2(1 + \gamma h^2k^2), \tag{5.145}$$

where

$$h = \frac{a}{\omega_p\sqrt{\gamma}} = \sqrt{\left(\frac{\varepsilon_0 KT}{Ne^2}\right)} \tag{5.146}$$

is the Debye length.

The dispersion relation states that waves are propagated, as opposed to being evanescent, only if the frequency exceeds the plasma frequency; that to any frequency above the plasma frequency there corresponds a definite wavelength; and that as the frequency increases much beyond the plasma frequency the speed of propagation tends downwards to the speed of sound in the electron gas. On the other hand, if the temperature is put equal to zero the dispersion relation reverts to $\omega = \omega_p$, demanding oscillations only at the plasma frequency, but without restriction on the wavelength; and it may be asked to what extent this is a valid approximation. It is evident from the form (5.145) that ω is close to ω_p if $h^2k^2 \ll 1$, that is, if the wavelength is much greater than 2π times the Debye length. Now the kinetic treatment given in Chapter 8 shows in fact that unless this condition holds the wave is damped, and that the damping becomes increasingly severe as the wavelength decreases to and below the Debye length. This so-called "Landau damping" is quite different from the collisional damping discussed in §5.6; but the inability of the plasma to sustain undamped oscillations in the face of thermal agitation uninhibited by particle encounters is, perhaps, to be expected if it is realized that at a frequency comparable with the plasma frequency and at a wavelength comparable with the Debye length an electron with average thermal velocity traverses a wavelength in a time comparable with the period of oscillation. The dispersion relation derived here is therefore only valid for values of ω/ω_p not much greater than unity, and it is in this sense that $\omega = \omega_p$ represents an acceptable approximation. The corresponding values of the wavelength range from infinity, at $\omega = \omega_p$, to a value many times the Debye length; but in practice the restriction is not severe, since the Debye length is not likely to exceed a few centimeters, and is commonly much less. Finally it may be remarked that the phase velocity is

$$\frac{\omega}{k} = \frac{\sqrt{(1 + \gamma h^2k^2)}}{\sqrt{\gamma}hk}a, \tag{5.147}$$

which is much greater than a in the permitted wavelength range.

What is the value to be accorded to γ in these formulas? It is known that

$$\gamma = 1 + 2/s, \tag{5.148}$$

where s is the number of degrees of freedom; and though normally for electrons $s = 3$, in the present case the motion considered is one dimensional and there are no collisions to randomize the velocities during an oscillation. Effectively, therefore, $s = 1$. Hence $\gamma = 3$, and the dispersion relation is

$$\omega^2 = \omega_p^2 + \frac{3KT}{m}k^2. \tag{5.149}$$

As is seen later, this is agreement with the corresponding result obtained on the basis of kinetic theory.

5.5.3 The electron–ion gas

If the motion of positive ions is included it obviously remains the case that the wave field can be separated into a transverse field and a longitudinal field, with the former not involving the pressure term. The current density

$(j_x, 0, 0)$ for the longitudinal field is obtained simply by adding the electron and positive ion contributions, and the relation corresponding to (5.140) is

$$j_x = -i\frac{Ne^2}{\omega}\left(\frac{1/m_e}{1 - a_e^2k^2/\omega^2} + \frac{1/m_i}{1 - a_i^2k^2/\omega^2}\right)E_x, \tag{5.150}$$

where

$$a_e = \sqrt{(\gamma_e KT_e/m_e)}, \qquad a_i = \sqrt{(\gamma_i KT_i/m_i)} \tag{5.151}$$

are the velocities of sound for the electron gas and for the ion gas respectively.

The combination of equations (5.142) and (5.150) gives the dispersion relation

$$\frac{\omega_{pe}^2}{\omega^2 - a_e^2k^2} + \frac{\omega_{pi}^2}{\omega^2 - a_i^2k^2} = 1, \tag{5.152}$$

with the notation of (5.34).

The consequences of this relation are perhaps most readily envisaged from a plot of k^2 against ω^2. The curve is a hyperbola. When both ω^2 and k^2 tend to infinity one of the terms on the left-hand side of (5.152) tends to zero; the other term must then tend to unity, so that the asymptotes are

$$\omega^2 = a_e^2k^2 + \omega_{pe}^2, \qquad \omega^2 = a_i^2k^2 + \omega_{pi}^2. \tag{5.153}$$

For one branch of the hyperbola $\omega^2 = \omega_{pe}^2 + \omega_{pi}^2 = \omega_p^2$ when $k^2 = 0$, and $k^2 = -(\omega_{pe}^2/a_e^2 + \omega_{pi}^2/a_i^2)$ when $\omega^2 = 0$. The other branch of the hyperbola goes through the origin, and near the origin

$$\omega^2/k^2 = (a_e^2\omega_{pi}^2 + a_i^2\omega_{pe}^2)/\omega_p^2 = (\gamma_e KT_e + \gamma_i KT_i)/(m_i + m_e). \tag{5.154}$$

The plot is shown diagrammatically in Fig. 5.16. It is of practical interest to consider $T_e \geqslant T_i$, which implies $\omega_{pi}^2/a_i^2 \geqslant \omega_{pe}^2/a_e^2$.

As it stands, the plot predicts two modes of propagation for $\omega > \omega_p$, and one for $\omega < \omega_p$. Again, however, the picture has to be revised in the light of the kinetic treatment; the only modes free from appreciable damping are those for which k^2 is much less than $\omega_{pe}^2/(\gamma a_e^2)$, and with this restriction on k^2 Fig. 5.16 predicts one mode of propagation with ω^2 greater than but close to ω_p^2, and one mode approximated by (5.154). More precisely the approximated dispersion equation of the former is easily seen from (5.152) to be

$$\frac{a_e^2\omega_{pe}^2 + a_i^2\omega_{pi}^2}{\omega_p^2}k^2 = \omega^2 - \omega_p^2, \tag{5.155}$$

which, since $\omega_{pe} \gg \omega_{pi}$ and $a_e \gg a_i$, is virtually identical with (5.143). The new mode introduced by considering the motion of the positive ions is therefore described by (5.154). If $\gamma_e = \gamma_i = \gamma$ and $T_e = T_i$, then (5.154) is effectively

$$\omega^2/k^2 = 2a_i^2 = 2\gamma KT/m_i \tag{5.156}$$

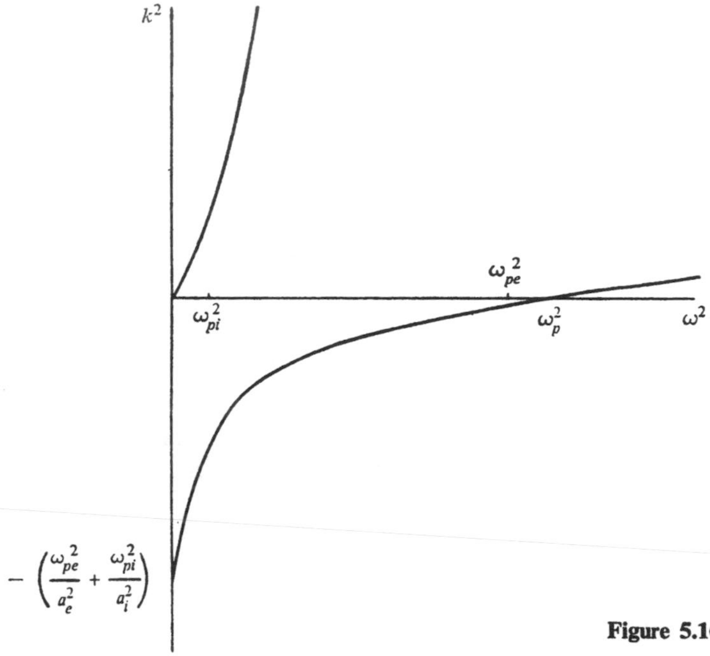

Figure 5.16

whereas if $T_e \gg T_i$ the corresponding result is

$$\omega^2/k^2 = \gamma_e KT_e/m_i. \tag{5.157}$$

At this low-frequency end of the spectrum the speeds of propagation are therefore, in the first case, $\sqrt{2}$ times the speed of sound in the ion gas, and in the second case a hybrid sound speed in which the electron temperature is associated with the ion mass.

5.5.4 The electron gas with magnetostatic field

If the plasma is in a magnetostatic field the inclusion of the partial pressure terms increases the algebraic complication of the analysis to such an extent that a complete treatment in this text would be unjustifiably long. The magneto-ionic waves of §5.4 and the two plasma-sonic waves of §5.5 must be represented, with the result that the dispersion equation is of the fourth degree in k. It factors explicitly only in special cases. If the direction of propagation is along the magnetostatic field $\mathbf{B}_0$, the two circularly polarized transverse magneto-ionic waves, which have refractive indices given by (5.125) and (5.126), are unaffected by the pressure term; and conversely, the longitudinal plasma-sonic waves, with dispersion relation (5.152), are unaffected by $\mathbf{B}_0$. If the direction of propagation is perpendicular to $\mathbf{B}_0$, clearly the wave with refractive index (5.127), which is linearly polarized with electric vector along $\mathbf{B}_0$, is unaffected either by $\mathbf{B}_0$ or by the pressure terms; the removal of the associated factor from the dispersion relation then leaves a cubic.

In place of a full treatment, an account is given, by approximate methods, of certain cases which broadly illustrate the new features resulting from the generalization of the previous development in this chapter. In the present section the electron gas only is considered, with the positive ions assumed to be stationary. The dispersion relation is then a cubic, since for $\mathbf{B}_0 = 0$ there is only one plasma-sonic wave; and propagation perpendicular to $\mathbf{B}_0$ essentially only involves a quadratic. Because of "coupling" between the plasma wave and the magneto-ionic waves new features might be expected at frequencies of the order of the plasma frequency; and when, as is usual in practice, the plasma frequency is relatively high, the result should not be significantly in error through the neglect of positive ion motions. In the next section, the low-frequency approximation for $\omega \to 0$ is investigated; this is of particular interest because it involves the coupling of magnetohydrodynamic waves and sound waves.

The electron equation of motion (cf. equations (5.37) and (5.135)) is now

$$i\omega m\mathbf{v} = -e\mathbf{E} - e\mathbf{v} \times \mathbf{B}_0 - \frac{i\gamma KT}{\omega} \text{ grad div } \mathbf{v}. \tag{5.158}$$

To study propagation perpendicular to the magnetostatic field the space dependence is taken to be exp $(-iky)$, with $\mathbf{B}_0 = (0, 0, B_0)$. Then the components of (5.158) are

$$i\omega m v_x = -eE_x - eB_0 v_y, \tag{5.159}$$

$$i\omega m v_y = -eE_y + eB_0 v_x + \frac{i\gamma KT}{\omega} k^2 v_y, \tag{5.160}$$

$$i\omega m v_z = -eE_z. \tag{5.161}$$

Equation (5.161) is concerned only with the transverse wave which is affected neither by $\mathbf{B}_0$ nor by the pressure term; it is of no further concern here. Equation (5.159) and (5.160) can be written

$$i\omega j_x + \Omega j_y = \omega_p^2 \varepsilon_0 E_x, \tag{5.162}$$

$$\Omega j_x - i\omega(1 - a^2k^2/\omega^2) j_y = -\omega_p^2 \varepsilon_0 E_y. \tag{5.163}$$

From Maxwell's equations

$$j_y = -i\omega\varepsilon_0 E_y, \qquad j_x = -i\omega(1 - c^2k^2/\omega^2)\varepsilon_0 E_x, \tag{5.164}$$

and substitution gives

$$(c^2k^2 - \omega^2 + \omega_p^2)E_x + i\omega\Omega E_y = 0, \tag{5.165}$$

$$i\Omega(c^2k^2 - \omega^2)E_x + \omega(a^2k^2 - \omega^2 + \omega_p^2)E_y = 0. \tag{5.166}$$

The dispersion relation is therefore

$$(c^2k^2 - \omega^2 + \omega_p^2)(a^2k^2 - \omega^2 + \omega_p^2) + \Omega^2(c^2k^2 - \omega^2) = 0, \tag{5.167}$$

which can alternatively be written

$$(c^2k^2 - \omega^2 + \omega_p^2)(a^2k^2 - \omega^2 + \omega_p^2 + \Omega^2) = \Omega^2\omega_p^2. \tag{5.168}$$

To examine (5.168) we consider a plot of c^2k^2 against ω^2. The curve is a hyperbola with asymptotes given by equating the left-hand side of (5.168) to zero. An example is shown diagrammatically in Fig. 5.17 by the full-line curve. The dotted curve represents the case when a is set zero; it cuts the ω^2-axis in the same two points as the full line curve. The dashed lines represent the case $\Omega = 0$.

The corresponding ways in which the graphical representations of §5.3 are modified by the pressure terms are easily seen. In the plot of μ^2 against ω (Fig. 5.4), for example, the left-hand branch of the dotted curve representing

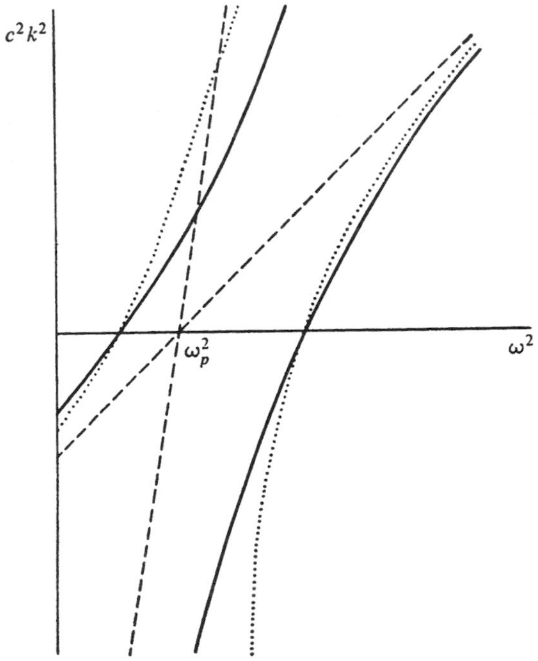

Figure 5.17

the case $\theta = \frac{1}{2}\pi$ bends round to approach $\mu^2 = c^2/a^2$ as ω tends to infinity, whereas the right-hand branch crosses over $\omega^2 = \Omega^2 + \omega_p^2$ at a negative value of μ^2 and has $\omega^2 = 0$ as asymptote.

In the light of the kinetic treatment (see §9.3) these results must be viewed with caution. For propagation perpendicular to $\mathbf{B}_0$ there is no Landau damping, but there are resonances at all harmonics of the gyro frequency, which make the true picture much more complicated.

If the analysis of this section is carried through for a general angle θ between the direction of propagation and $\mathbf{B}_0$, the resulting dispersion relation, written as an equation for the square of the refractive index in terms of the dimensionless parameters X and Y, is the cubic

$$A_0\mu^6 + A_1\mu^4 + A_2\mu^2 + A_3 = 0, \tag{5.169}$$

where, with $\beta = a/c$,

$$A_0 = \beta^2(1 - Y^2\cos^2\theta), \tag{5.170}$$

$$A_1 = -1 + X + Y^2 - XY^2\cos^2\theta - 2\beta^2(1 - X - Y^2\cos^2\theta), \tag{5.171}$$

$$A_2 = 2(1 - X)(1 - X - Y^2) - XY^2\sin^2\theta + \beta^2[(1 - X)^2 - Y^2\cos^2\theta], \tag{5.172}$$

$$A_3 = -(1 - X)[(1 - X)^2 - Y^2]. \tag{5.173}$$

When $\theta = 0$, equation (5.169) is

$$[(1 - Y)\mu^2 - 1 + X + Y][(1 + Y)\mu^2 - 1 + X - Y](\beta^2\mu^2 - 1 + X) = 0; \tag{5.174}$$

and when $\theta = \frac{1}{2}\pi$ it is

$$(\mu^2 - 1 + X)\{\beta^2\mu^4 - [1 - X - Y^2 + \beta^2(1 - X)]\mu^2 + (1 - X)^2 - Y^2\} = 0 \tag{5.175}$$

in which the vanishing of the second factor on the left-hand side is equivalent to (5.168).

5.5.5 The electron–ion gas with magnetostatic field in the low-frequency approximation

In this section the motion of the positive ions is taken into account. The frequencies at which these motions are important are those comparable with or less than the ion gyro frequency. Relatively simple results are obtained when $\omega \to 0$, and the mathematical treatment is developed for this limiting case; the results should then be applicable for frequencies well below the ion gyro and ion plasma frequencies.

At such low frequencies there is coupling between the two magneto-hydrodynamic (mhd) waves described in §5.4.2 and the "sound" wave of

§5.5.3, whose speed is given by (5.154); and in each of these three waves the speed ω/k tends to a finite limit as $\omega \to 0$. In the present problem it is therefore to be expected that the assumption that ω/k remains finite as $\omega \to 0$ will yield three propagated characteristic waves.

The second of the squares of the mhd wave refractive indices in (5.109) undergoes a sharp transition as the angle θ between the direction of propagation and the magnetostatic field approaches $\frac{1}{2}\pi$. For when ω is very small, $|\kappa_0| \simeq \omega_p^2/\omega^2$ and is much greater than $\kappa_1 \simeq 1 + \omega_p^2/(\Omega_e\Omega_i)$; so $\mu^2 \cos^2\theta \simeq 1 + \omega_p^2/(\Omega_e\Omega_i)$ if θ is not too close to $\frac{1}{2}\pi$, whereas $\mu^2 \simeq -\omega_p^2/\omega^2$ if $\theta = \frac{1}{2}\pi$. Put mathematically, as $\omega \to 0$ μ^2 tends to a limit *non-uniformly* in the range $\theta \leqslant \frac{1}{2}\pi$. For this reason it is now particularly desirable to consider a general angle θ.

With space dependence $\exp(-iky)$ and $\mathbf{B}_0 = (0, B_0\cos\theta, B_0\sin\theta)$ the electron equations of motion give

$$i\omega j_{ex} + \Omega_e \sin\theta\, j_{ey} - \Omega_e\cos\theta\, j_{ez} = \omega_{pe}^2\varepsilon_0 E_x, \quad (5.176)$$

$$-\Omega_e\sin\theta\, j_{ex} + i\omega\left(1 - \frac{a_e^2k^2}{\omega^2}\right) j_{ey} = \omega_{pe}^2\varepsilon_0 E_y, \quad (5.177)$$

$$\Omega_e\cos\theta\, j_{ex} + i\omega j_{ez} = \omega_{pe}^2\varepsilon_0 E_z, \quad (5.178)$$

where $\mathbf{j}_e = -Ne\mathbf{v}_e$ is the electron contribution to the current density. These correspond to the form (5.162), (5.163), to which (5.176), (5.177) reduce when $\theta = \frac{1}{2}\pi$. Similarly, from the positive ion equations of motion

$$i\omega\, j_{ix} - \Omega_i\sin\theta\, j_{iy} + \Omega_i\cos\theta\, j_{iz} = \omega_{pi}^2\varepsilon_0 E_x, \quad (5.179)$$

$$\Omega_i\sin\theta\, j_{ix} + i\omega\left(1 - \frac{a_i^2k^2}{\omega^2}\right) j_{iy} = \omega_{pi}^2\varepsilon_0 E_y, \quad (5.180)$$

$$-\Omega_i\cos\theta\, j_{ix} + i\omega j_{iz} = \omega_{pi}^2\varepsilon_0 E_z. \quad (5.181)$$

And, finally, from Maxwell's equations

$$j_x = -i\omega(1 - c^2k^2/\omega^2)\varepsilon_0 E_x, \quad (5.182)$$

$$j_y = -i\omega\varepsilon_0 E_y, \quad (5.183)$$

$$j_z = -i\omega(1 - c^2k^2/\omega^2)\varepsilon_0 E_z, \quad (5.184)$$

where $\mathbf{j} = \mathbf{j}_e + \mathbf{j}_i$ is the current density.

The self-consistency of the nine equations (5.176)–(5.184) yields the general dispersion relation. The limiting case $\omega \to 0$ can be treated by successive approximation. If ω is set zero, and k/ω is assumed to remain finite, the equations give

$$\mathbf{j}_e = -\mathbf{j}_i, \quad (5.185)$$

$$j_{ex} = -\frac{\omega_p^2}{\Omega\sin\theta}E_y = \frac{\omega_p^2}{\Omega\cos\theta}E_z, \quad (5.186)$$

$$\sin\theta\, j_{ey} - \cos\theta\, j_{ez} = \frac{\omega_p^2}{\Omega} E_x, \tag{5.187}$$

where $\omega_p^2 = \omega_{pe}^2 + \omega_{pi}^2$, as before, and it also proves convenient in this section to write $\Omega = \Omega_e + \Omega_i$, so that

$$\omega_{pe}^2/\Omega_e = \omega_{pi}^2/\Omega_i = \omega_p^2/\Omega. \tag{5.188}$$

The next approximation is obtained by substituting from these $\omega = 0$ relations into the small terms of equations (5.176)–(5.184) (those with a factor ω). The result may be written

$$\sin\theta\, j_{ey} - \cos\theta\, j_{ez} = \frac{\omega_p^2}{\Omega}\,\varepsilon_0\left[E_x + \frac{i\omega}{\Omega_e \sin\theta}\,E_y\right], \tag{5.189}$$

$$\sin\theta\, j_{ex} - \frac{i\omega}{\Omega_e}\left(1 - \frac{a_e^2k^2}{\omega^2}\right) j_{ey} = -\frac{\omega_p^2}{\Omega}\,\varepsilon_0 E_y, \tag{5.190}$$

$$\cos\theta\, j_{ex} + \frac{i\omega}{\Omega_e} j_{ez} = \frac{\omega_p^2}{\Omega}\,\varepsilon_0 E_z, \tag{5.191}$$

$$\sin\theta\, j_{iy} - \cos\theta\, j_{iz} = -\frac{\omega_p^2}{\Omega}\,\varepsilon_0\left[E_x - \frac{i\omega}{\Omega_i \sin\theta} E_y\right], \tag{5.192}$$

$$\sin\theta\, j_{ix} - \frac{i\omega}{\Omega_i}\left(1 - \frac{a_i^2k^2}{\omega^2}\right) j_{ey} = \frac{\omega_p^2}{\Omega}\,\varepsilon_0 E_y, \tag{5.193}$$

$$\cos\theta\, j_{ix} + \frac{i\omega}{\Omega_i} j_{ez} = -\frac{\omega_p^2}{\Omega}\,\varepsilon_0 E_z, \tag{5.194}$$

$$j_x = -i\omega\varepsilon_0(1 - c^2k^2/\omega^2)E_x, \tag{5.195}$$

$$j_y = -i\omega\varepsilon_0 E_y \tag{5.196}$$

$$j_z = i\omega\,\varepsilon_0 \cot\theta\,(1 - c^2k^2/\omega^2)E_y. \tag{5.197}$$

It is now observed that the elimination of E_x from (5.189) and (5.192) gives an expression for E_y in terms of j_y, j_z which together with (5.196) and (5.197) yields the dispersion relation

$$\mu^2\cos^2\theta = 1 + \frac{\omega_p^2}{\Omega_e\Omega_i}. \tag{5.198}$$

This relation must hold when the $\omega = 0$ value of E_y, which from (5.186) is the same as the $\omega = 0$ value of $-\tan\theta E_z$, is not zero. It gives the refractive index of one of the three propagated characteristic waves.

For the other two waves the $\omega = 0$ values of E_y and E_z are zero. With the neglect of terms of order ω^2, (5.196) and (5.197) then imply $j_y = j_z = 0$, and consequently (5.189) and (5.192) both read

$$\sin\theta\, j_{ez} - \cos\theta\, j_{ex} = \frac{\omega_p^2}{\Omega} E_x. \qquad (5.199)$$

Elimination of E_x between (5.199) and (5.195) gives an expression for j_x in terms of j_{ey} and j_{ez}; the addition of (5.190) and (5.193) gives a second such expression; and addition of (5.191) and (5.194) a third. A little algebra shows that the identity of these three expressions demands the relation

$$a^2\cos^2\theta\,\mu^4 - \left(\frac{a^2\omega_p^2}{\Omega_e\Omega_i} + c^2 + a^2\cos^2\theta\right)\mu^2 + c^2\left(1 + \frac{\omega_p^2}{\Omega_e\Omega_i}\right) = 0, \qquad (5.200)$$

where now

$$a^2 = \frac{a_e^2\omega_{pi}^2 + a_i^2\omega_{pe}^2}{\omega_p^2} = \frac{\gamma_e KT_e + \gamma_i KT_i}{m_e + m_i} \qquad (5.201)$$

is the square of the composite "sound" speed previously met in (5.154).

The implications of the simple relations (5.198) and (5.200) can be put most vividly in the common case for which the Alfvén speed (see (5.119))

$$V_A = \frac{\sqrt{(\Omega_e\Omega_i)}}{\omega_p} c = \frac{B}{\sqrt{(\mu_0\rho)}} \qquad (5.202)$$

is much less than c. For then (5.198) gives a phase speed

$$c/\mu = V_A\cos\theta, \qquad (5.203)$$

and (5.200) gives

$$c^2/\mu^2 = \tfrac{1}{2}\{a^2 + V_A^2 \pm \sqrt{[(a^2 + V_A^2)^2 - 4a^2V_A^2\cos^2\theta]}\}. \qquad (5.204)$$

The wave characterized by (5.198) is evidently none other than one of the mhd waves of §5.4.2, which remains unaffected by the pressure terms. The waves characterized by (5.200) are sometimes called magnetosonic waves: "fast" for the upper sign in (5.204), for which the speed exceeds both a and V_A, and "slow" for the lower sign, for which the speed is less than both a and V_A.

To check the case $\theta = 0$ it need only be observed that (5.200) is then

$$(a^2\mu^2 - c^2)\left(\mu^2 - 1 - \frac{\omega_p^2}{\Omega_e\Omega_i}\right) = 0. \qquad (5.205)$$

Thus one of the magnetosonic waves itself reverts to a pure mhd wave, and the other is the "sound" wave unaffected by $\mathbf{B}_0$.

For $\theta = \frac{1}{2}\pi$, on the other hand, it must be recognized that (5.198) is invalid. In this case the three z-component equations of the set (5.176)–(5.184) involve only j_{ez}, j_{iz} and E_z, and give the dispersion relation

$$\mu^2 = 1 - \omega_p^2/\omega^2, \qquad (5.206)$$

which replaces (5.198) in the way already explained. The squares of the phase speeds of the magnetosonic waves are now, from (5.200),

$$0, \qquad a^2 + \frac{1 - \dfrac{a^2}{c^2}\sin^2\theta}{1 + V_A^2/c^2} V_A^2; \tag{5.207}$$

or, in the approximation (5.204), simply

$$0, \qquad a^2 + V_A^2. \tag{5.208}$$

We end this sub-section with the remark that the hydrodynamic treatment of pressure given here relies on the existence of a small number of collisions in order to maintain an isotropic pressure. With insufficient collisions, and a strong magnetic field, there are pressures $p_{\|}$ and $p_{\perp}$, along and normal to $\mathbf{B}_0$ respectively, which satisfy new adiabatic equations (see §§9.3.3, 11.4.4, 11.7.1).

5.6 THE INCLUSION OF COLLISIONS

5.6.1 The effective frictional force

Throughout the treatment so far the idealization of a collisionless plasma has been adopted. In practice, however, each particle will suffer random fluctuations of velocity due to "encounters" with other particles. For a neutral particle such an encounter is a relatively well defined event; it is a collision in the ordinary sense of the word. For two charged particles a long range electromagnetic interaction between them is always present, and the concept of a collision is less precise. It is, nevertheless, still a useful concept since only the smoothed out effect of electromagnetic forces has been represented in the previous analysis, and no account has been taken of local effects which give rise to random fluctuations of velocity. At an elementary level the consequences of encounters between charged particles can be regarded as broadly similar to those of collisions in which one partner is neutral, in spite of the fact that with the former it is the accumulation of the comparatively frequent small velocity changes that is important, rather than the abrupt large changes arising from close collisions.

It is possible to give a relatively simple approximate treatment of the effect of collisions by the inclusion in the equation of motion of each plasma constituent an appropriate "frictional" force. The idea is essentially that introduced in the early classical theory of the flow of current in a conductor. Consider in the first instance a cold, weakly ionized gas, consisting of electrons, positive ions and a much greater number of neutral particles, under the sole influence of an electrostatic field $\mathbf{E}$. A steady current flows, due sensibly to the "drift" velocity of the electrons. Each electron is always accelerated by a force $-e\mathbf{E}$, but from time to time it has collisions with stationary neutral particles which deflect its course in a random way, and the result is an average rate of progression in the direction of $\mathbf{E}$. A rough quantitative calculation proceeds as follows.

Since the velocity after a collision has no preferred direction it may be taken as zero. The directed distance traveled by an electron between successive collisions is then

$$\mathbf{r} = -\frac{e\mathbf{E}}{2m}\tau^2, \tag{5.209}$$

where τ is the corresponding time interval. Now τ can be assumed to have a probability density

$$\nu e^{-\nu\tau} \tag{5.210}$$

with average value

$$\bar{\tau} = \nu\int_0^\infty \tau e^{-\nu\tau}\,d\tau = 1/\nu; \tag{5.211}$$

so that the *average* directed distance traveled between collisions is

$$\bar{\mathbf{r}} = -\frac{e\mathbf{E}}{2m}\overline{\tau^2} = -\frac{e\mathbf{E}}{m\nu^2}. \tag{5.212}$$

Thus in a time interval T long compared with $1/\nu$ there are approximately νT collisions and the electron has traveled a directed distance approximately $-e\mathbf{E}\nu T/(m\nu^2)$. The drift velocity is therefore

$$\mathbf{v} = -\frac{e\mathbf{E}}{m\nu}; \tag{5.213}$$

and the current density $-Ne\mathbf{v}$, where N is the number of electrons per unit volume, corresponds to a conductivity

$$\frac{Ne^2}{m\nu}. \tag{5.214}$$

The expression of this result in terms of the electron equations of motion is simply a matter of recognizing that the electric force must be counterbalanced by an effective frictional force to produce a steady velocity. Explicitly,

$$0 = -e\mathbf{E} - m\nu\mathbf{v}, \tag{5.215}$$

which is merely a restatement of (5.213). In this context the second term on the right-hand side of (5.215) may be labeled the rate of destruction of momentum by collisions.

This analysis is applied to plasma theory by introducing quite freely the force $-m\nu\mathbf{v}$, or its equivalent for collisions between other types of particle, in the diverse time-varying situations of the sort already considered. At the elementary level, the collision frequency ν is treated as a parameter in the theory to which experiment can assign a numerical value by comparison of wave absorption, say, with theoretical prediction. Various refinements of the theory are available, but are not pursued here. Instead an indication is given of some of the consequences of adopting the simple approach.

Collisions give rise to absorption. The propagated waves derived in previous sections of this chapter no longer proceed without attenuation. If, for example, the problem just discussed is developed for a harmonic time variation, (5.215) is replaced by

$$i\omega m\mathbf{v} = -e\mathbf{E} - m\nu\mathbf{v}, \tag{5.216}$$

and the current density is

$$\mathbf{j} = -\frac{i\,Ne^2}{m(\omega - i\nu)}\mathbf{E}. \tag{5.217}$$

The medium therefore behaves as a lossy dielectric, with dielectric constant

$$1 - \frac{Ne^2}{\varepsilon_0 m(\omega^2 + \nu^2)} \tag{5.218}$$

and conductivity

$$\frac{Ne^2\nu}{m(\omega^2 + \nu^2)}\,. \tag{5.219}$$

It may be noted that when $\nu = 0$ (5.218) reduces to (5.7), and that when $\omega = 0$ (5.219) reduces to (5.214). If $\nu \ll \omega$ the dielectric constant is negligibly affected by collisions and the conductivity is approximately $Ne^2\nu/(m\omega^2)$.

For plane wave propagation (5.217) leads to a complex refractive index μ, where

$$\mu^2 = 1 - \frac{Ne^2}{\varepsilon_0 m\omega(\omega - i\nu)} = 1 - \frac{\omega_p^2}{\omega(\omega - i\nu)}\,. \tag{5.220}$$

In magneto-ionic terminology

$$Z = \nu/\omega, \tag{5.221}$$

and

$$\mu^2 = 1 - \frac{X}{1 - iZ}\,. \tag{5.222}$$

The attenuation factor for a transverse plane wave propagated along Ox is $\exp(-\omega\chi/c)$, where χ is minus the imaginary part of μ.

The ionosphere is a weakly ionized gas and collisions often play an important part in modifying the propagation of radio waves through it. The collision frequency ν varies greatly with height and in different circumstances can be greater than, or comparable with, or less than ω. For this reason the magneto-ionic theory of §5.2 is commonly developed with the inclusion of the collision term $-m\nu\mathbf{v}$ on the right-hand side of (5.37). As regards the conductivity tensor this evidently means no more than replacing ω by $\omega - i\nu$ in (5.46). It turns out that the Appleton–Hartree refractive index is given by

$$\mu^2 = 1 - \frac{X}{1 - iZ + \dfrac{\frac{1}{2}Y^2\sin^2\theta}{1-X-iZ} \pm \sqrt{\left\{\left(\dfrac{\frac{1}{2}Y^2\sin^2\theta}{1-X-iZ}\right)^2 + Y^2\cos^2\theta\right\}}} \tag{5.223}$$

in place of (5.81).

In problems where the frequency is sufficiently high for only electron motions to be important, electron collisions with positive ions need not be differentiated from those with neutral particles; in both cases the much heavier particle remains virtually stationary. In the lower part of the ionosphere, for example, the value of Z in (5.223) is determined by collisions with neutral molecules, whereas in the F-region and beyond collisions with positive ions become increasingly dominant.

On the other hand it has been seen that the positive ion motions are certainly important at frequencies comparable with or below their gyro frequency, and then collisions between positive ions and neutral particles may mean that the motion of the latter must also be taken into account. In this context the frictional force in the electron equation of motion is

$$-m_e\nu_e(\mathbf{v}_e - \mathbf{v}_n) - m_e\nu_{ei}(\mathbf{v}_e - \mathbf{v}_i), \tag{5.224}$$

where ν_e, ν_{ei} are respectively the electron–neutral and electron–ion collision frequencies, and $\mathbf{v}_e$, $\mathbf{v}_i$, $\mathbf{v}_n$ are respectively the electron, ion and neutral particle drift velocities. The corresponding force in the ion equation of motion is

$$-m_i\nu_i(\mathbf{v}_i - \mathbf{v}_n) + m_e\nu_{ei}(\mathbf{v}_e - \mathbf{v}_i), \tag{5.225}$$

where ν_i is half the ion–neutral collision frequency, on the assumption that m_i is approximately equal to the neutral particle mass m_n, and the second term is, of course, just the negative of the second term in (5.224). The force in the neutral particle equation of motion is simply the negative of the sum of (5.224) and (5.225).

In the next section it is shown how, in a partially ionized gas, mhd waves at the low-frequency limit involve the motion of the neutral particles.

5.6.2 The low-frequency limit

A cold, partially ionized gas pervaded by a magnetostatic field $\mathbf{B} = (0, 0, B_0)$ is considered. If electron–ion collisions are neglected the electron, positive ion and neutral particle equations of motion are respectively

$$i\omega m_e \mathbf{v}_e = -e\mathbf{E} - e\mathbf{v}_e \times \mathbf{B}_0 - m_e \nu_e(\mathbf{v}_e - \mathbf{v}_n), \tag{5.226}$$

$$i\omega m_i \mathbf{v}_i = e\mathbf{E} + e\mathbf{v}_i \times \mathbf{B}_0 - m_i \nu_i(\mathbf{v}_i - \mathbf{v}_n), \tag{5.227}$$

$$i\omega q m_n \mathbf{v}_n = -m_e \nu_e(\mathbf{v}_n - \mathbf{v}_e) - m_i \nu_i(\mathbf{v}_n - \mathbf{v}_i), \tag{5.228}$$

where q is the ratio of the unperturbed number density of neutral particles to the unperturbed number density of ions (or electrons).

From (5.228)

$$\mathbf{v}_n = \frac{m_e \nu_e \mathbf{v}_e + m_i \nu_i \mathbf{v}_i}{i\omega q m_n + m_e \nu_e + m_i \nu_i}, \tag{5.229}$$

and substitution into (5.226), (5.227) yields two equations for $\mathbf{v}_e$ and $\mathbf{v}_i$ in terms of $\mathbf{E}$, which lead in turn to a dielectric tensor

$$\begin{pmatrix} \kappa_1 & -i\kappa_2 & 0 \\ i\kappa_2 & \kappa_1 & 0 \\ 0 & 0 & \kappa_0 \end{pmatrix}. \tag{5.230}$$

The form of (5.230) is the same as that of the dielectric tensors of §§5.2 and 5.3, but the κ's now have imaginary parts. Moreover, the full algebra produces most unwieldy expressions. However, in the specific case to be examined, namely $\omega \ll \Omega_i$, the κ's have quite simple approximations which can be derived in the following way.

In the first place it may be noted that κ_0 is independent of $\mathbf{B}_0$, and must therefore to a good approximation be the same as (5.220), since in the case of no magnetic field approximate identification of the total current with the electron current clearly remains valid when neutral particles are present.

To obtain κ_1 and κ_2 it is recalled, as found in §5.4.2, that $\mathbf{v}_e \simeq \mathbf{v}_i$ for $\omega \ll \Omega_i$. It is a natural assumption, which is soon confirmed *a posteriori*, that this result continues to hold here. Furthermore $m_e \nu_e \ll m_i \nu_i$, since ν_i/ν_e has a value perhaps of the order of 30. With $m_n \simeq m_i$ a good approximation to (5.229) is therefore

$$\mathbf{v}_n = \frac{\nu_i}{i\omega q + \nu_i} \mathbf{v}_i = \frac{\nu_i}{i\omega q + \nu_i} \mathbf{v}_e. \tag{5.231}$$

Substitution of (5.231) into (5.226) and (5.227) then shows that the equations for $\mathbf{v}_e$ and $\mathbf{v}_i$ are the same as in the absence of neutral particles, save that the masses m_e and m_i appearing in the latter are replaced by "apparent" masses, both of which contain the extra factor

$$1+\frac{q\nu_i}{i\omega q+\nu_i}. \tag{5.232}$$

Since the ratio of the apparent masses is the same as that of the actual masses the results of §5.4.2 can at once be translated into the new context. In particular, one of the values of μ^2 is

$$\kappa_1 = 1+\frac{\omega_p^2}{\Omega_e\Omega_i}\left(1+\frac{q\nu_i}{i\omega q+\nu_i}\right). \tag{5.233}$$

Two limiting cases can be distinguished. If $\omega \gg \nu_i/q$ (remembering also that $\omega \ll \Omega_i$) the expression (5.233) is approximately

$$\kappa_1 = 1+\frac{\omega_p^2}{\Omega_e\Omega_i}\left(1-i\frac{\nu_i}{\omega}\right), \tag{5.234}$$

and (5.231) shows that $\mathbf{v}_n \ll \mathbf{v}_i$. Thus the motion of the neutral particles is negligible, but their collisions with positive ions give rise to attenuation. For low attenuation the phase speed of the mhd wave is

$$\frac{V_A}{\sqrt{(1+V_A^2/c^2)}}, \tag{5.235}$$

where

$$V_A = \frac{B_0}{\sqrt{(\mu_0\rho)}}, \tag{5.236}$$

with $\rho = N(m_e+m_i)$ representing the mass density of the charged constituents, just as in §5.4.2.

On the other hand, if $\omega \ll \nu_i/q$, (5.233) is approximately

$$\kappa_1 = 1+\frac{\omega_p}{\Omega_e\Omega_i}(1+q), \tag{5.237}$$

and $\mathbf{v}_n \simeq \mathbf{v}_i$. The phase speed is now

$$\frac{V_A'}{\sqrt{(1+V_A'^2/c^2)}}, \tag{5.238}$$

where

$$V_A' = \frac{B_0}{\sqrt{(\mu_0\rho_n)}}, \tag{5.239}$$

with $\rho_n = N(1+q)(m_e+m_i)$ representing the total mass density of the gas. For a weakly ionized gas with $q \gg 1$ the speed (5.238) is much less than (5.235).

PROBLEMS

1. Uniform electron and positive ion distributions are confined to an infinitely long circular cylindrical region, the whole being neutral. The cylinder of electrons is now displaced bodily through a small distance ξ perpendicular to its axis, the ions remaining fixed. Find the restoring force on each electron.

   Do a similar calculation for a spherical distribution.

2. A *charge sheet* is an infinite, plane, uniform distribution of surface charge density. Consider a one dimensional plasma model consisting of parallel charge sheets free to move in a uniform neutralizing background of stationary charge. Show that, provided the charge sheets never cross one another, the motion of each is (without approximation) a simple harmonic oscillation at the plasma frequency.

3. For plane wave propagation in a plasma as described in section 5.2.3, calculate the time average of the combined energy density of the field and the electrons. Verify that the product of this quantity with the group velocity is indeed identical with the expression (5.23) for the time-averaged power flux density given by the Poynting vector.

4. Show that for a monochromatic field of angular frequency ω in a non-magnetic medium characterized by a (real) permittivity $\varepsilon(\omega)$ the time-averaged electric energy density is

$$\frac{1}{4}\frac{\partial}{\partial\omega}(\omega\varepsilon)\mathbf{E}\cdot\mathbf{E}^*.$$

   [Hint: integrate the expression $\mathbf{E}\cdot\dot{\mathbf{D}}$ for the rate of change of electric energy density, and ensure convergence at the lower limit $t = -\infty$ by crediting ω with a small negative imaginary part.]

   Verify that the expression does give the combined energy density of the field and the electrons calculated in Problem 3.

5. A monochromatic homogeneous plane wave in vacuum is incident normally on the plane boundary of a semi-infinite uniform plasma. Find the reflected and transmitted waves, considering frequencies both above and below the plasma frequency.

   For what value of the frequency is there a $\frac{1}{2}\pi$ change of phase on reflection?

6. Solve (5.37) by vector algebra to obtain

$$\mathbf{v} = \frac{i\omega\varepsilon_0}{Ne}\frac{X}{1 - Y^2}[\mathbf{E} + iY\mathbf{E}\times\mathbf{b} - Y^2(\mathbf{E}\cdot\mathbf{b})\mathbf{b}],$$

   where $\mathbf{b}$ is the unit vector along $\mathbf{B}_0$. Confirm that this result is in agreement with (5.53).

7. Consider a plasma immersed in a very strong magnetic field $(0, 0, B_0)$. Write down the dielectric tensor in the limit $B_0 \to \infty$, and obtain the corresponding refractive indices.

   Show that for any plane wave in the medium either $E_z = 0$ or $H_z = 0$.

8. Obtain the quasi-transverse approximation to the refractive indices analogous to (5.83).

9. A plane wave with propagation vector **k** travels in a magneto-ionic medium, as described in §5.3. Let E_a and E_b be the components of **E** transverse to **k** in the respective directions $\mathbf{k} \times \mathbf{B}_0$ and $\mathbf{k} \times (\mathbf{k} \times \mathbf{B}_0)$. Show that

$$\frac{E_a}{E_b} = \frac{i\lambda}{Y \cos \theta}$$

where λ is given by (5.80). What is this ratio in the quasi-longitudinal approximation?

10. Consider Faraday rotation in a uniform magneto-ionic medium by taking the linearly polarized field formed by the superposition of two characteristic plane waves of equal amplitude propagated along the magnetic field $\mathbf{B}_0$. Specifically, find the angle through which the direction of **E** is seen to rotate as the observer moves a distance z along $\mathbf{B}_0$.

    How is the result affected if first, propagation is not necessarily along $\mathbf{B}_0$, but the quasi-longitudinal approximation is valid; and secondly, the electron density is allowed to vary slightly with position? Show that if everywhere $X \ll 1$, then the Faraday rotation is a measure of the total number of electrons in a cylinder whose generators are parallel to the direction of propagation.

11. Sketch roughly the variation with height above the earth's surface (on a logarithmic scale) of the Alfvén speed in the ionosphere at the magnetic equator. (The table gives approximate values of N, the number of electrons per cubic meter, at different heights

height

km	300	400	500	750	1000	2000	5000	10,000	25,000
N	6×10^{11}	10^{12}	10^{12}	2×10^{11}	6×10^{10}	3×10^{9}	10^{9}	6×10^{8}	4×10^{8}

Assume that the average ionic molecular weight falls linearly from 16 to 15 between 300 and 1000 km, and then down to 1 at (and beyond) 2500 km.)

12. When a charged particle moves with uniform velocity **v** along the magnetic field $\mathbf{B}_0$ in a magneto-ionic medium it radiates (Čerenkov-wise) at the frequencies and angles for which the coherence condition

$$v \cos \theta = c/\mu$$

holds, where μ is the refractive index for propagation at angle θ to $\mathbf{B}_0$. Find roughly the sort of frequencies at which this condition can be satisfied. (Include the effect of positive ions by working, for example, from Fig. 5.13.)

13. For plane wave propagation $\exp[i(\omega t - \mathbf{k}\cdot\mathbf{r})]$ establish the relations

$$\omega\mathbf{D}\cdot\mathbf{E}^* = \mathbf{k}\cdot(\mathbf{E}^* \times \mathbf{H}), \qquad \omega\mathbf{B}\cdot\mathbf{H}^* = \mathbf{k}\cdot(\mathbf{E} \times \mathbf{H}^*).$$

Now envisage small changes in each component k_i of **k**, entailing corresponding changes in ω and the field vectors. By differentiating the relations partially with respect to k_i show that

$$S_i = \frac{1}{4}\frac{\partial}{\partial k_i}(\omega\varepsilon_{lm})E_l^* E_m + \frac{1}{4}\frac{\partial}{\partial k_i}(\omega\mu_{lm})H_l^* H_m,$$

where

$$\mathbf{S} \equiv \mathrm{Re}\,\tfrac{1}{2}\mathbf{E} \times \mathbf{H}^*,$$

and $D_l = \varepsilon_{lm}E_m$, $B_l = \mu_{lm}H_m$, with $\varepsilon_{lm} = \overset{*}{\varepsilon}_{ml}$, $\mu_{lm} = \overset{*}{\mu}_{ml}$.

Deduce that if ε_{lm} and μ_{lm} depend on $\mathbf{k}$ only through dependence on ω (no spatial dispersion), then $\mathbf{S}$ is equal to the product of the group velocity $\partial\omega/\partial k_i$ and the time-averaged energy density (cf. Problem 4).

14. Show that the electron gas with pressure term included, as in §5.5.2, has dielectric tensor

$$\kappa_{ij} = \left(1 - \frac{\omega_p^2}{\omega^2}\right)\delta_{ij} - \frac{\omega_p^2}{\omega^2}\frac{a^2}{\omega^2 - k^2a^2}k_ik_j.$$

Consider a longitudinal plasma wave propagated in the x-direction, and deduce from the equation for S_i displayed in Problem 13 that $\partial\kappa_{11}/\partial k_i = 0$. Verify this directly from the expression for κ_{ij}.

15. Consider conventional magneto-ionic theory with collisions included as in §5.6.1. Show that, when $Z \neq 0$,
    (a) the dielectric tensor is not Hermitian;
    (b) the refractive index μ cannot be zero;
    (c) when $X = 1$, there is an angle θ at which the two values of μ are the same, being given by

$$\mu^2 = 1 - \frac{1}{1 - i\sqrt{(Y^2 + Z^2)}}.$$

REFERENCES

For the standard version of magneto-ionic theory in the notation conventional to ionospheric work, and for references to the early papers, see

J. A. Ratcliffe, *The Magneto-Ionic Theory and Its Applications to the Ionosphere*, Cambridge University Press (1959).

Other books which include accounts of "cold" plasma theory are

W. P. Allis, S. J. Buchsbaum and A. Bers, *Waves in Anisotropic Plasmas*, M.I.T. Press (1963).

J. F. Denisse and J. L. Delcroix, *Plasma Waves*, Interscience (1963).

M. A. Heald and C. B. Wharton, *Plasma Diagnostics with Microwaves*, Wiley (1965).

E. H. Holt and R. E. Haskell, *Foundations of Plasma Dynamics*, Macmillan (1965).

T. H. Stix, *The Theory of Plasma Waves*, McGraw–Hill (1962).

For comprehensive treatments of wave propagation in *inhomogeneous* cold plasmas (not discussed here), see

K. G. Budden, *Radio Waves in the Ionosphere*, Cambridge University Press (1961).

V. L. Ginzburg, *Propagation of Electromagnetic Waves in Plasmas* (translated), Pergamon (1964).

The existence of magnetohydrodynamic waves was first reported in

H. Alfvén, *Nature* **150**, 405 (1942).

Magnetohydrodynamic waves are described in fluid mechanical terms in many books, for example

H. Alfvén and C. G. Fälthammar, *Cosmical Electrodynamics*, 2nd edn, Clarendon Press (1963).

T. G. Cowling, *Magnetohydrodynamics* (Interscience Tracts on Physics and Astronomy No. 4), Interscience (1957).

V. C. A. Ferraro and C. Plumpton, *An introduction to Magneto-Fluid Mechanics*, 2nd edn, Oxford University Press (1966).

W. F. Hughes and F. J. Young, *The Electrodynamics of Fluids*, Wiley (1966).

G. W. Sutton and A. Sherman, *Engineering Magnetohydrodynamics*, McGraw–Hill (1965).

For the link between magnetohydrodynamic waves and magneto-ionic theory, see

E. Åström, *Ark. Fiz.* **2**, 443 (1950).

J. A. Fejer, *J. Atmos. Terr. Phys.* **18**, 135 (1960).

C. O. Hines, *Proc. Camb. Phil. Soc.* **49**, 299 (1953).

CHAPTER 6

PLASMA STREAMS

6.1 INSTABILITY AND GROWTH

6.1.1 The rôle of stream velocity

In the previous chapter, the plasmas considered were, in the unperturbed state, at rest. It is, however, of theoretical and practical interest to consider also plasmas in bulk motion, and in this chapter the nature of the characteristic waves supported by plasma streams is investigated. The idealizations of the previous chapter are maintained; in particular the thermal velocity spread is ignored, and the basic situation treated is that in which one unbounded, homogeneous stream of identical charged particles moves with uniform velocity through another. Both the total charge density and the total current density in the unperturbed state are taken to be everywhere zero. Particle collisions are neglected throughout the discussion, though they could in principle be taken into account in the way indicated in §5.6. The two-stream problem contains fundamentally new features, whereas that of a single stream, say an electron–ion gas with a common unperturbed velocity for each constituent, is simply equivalent to the uniform motion of the observer past a stationary plasma.

Plasma streams appear in practice in a number of different contexts. They are basic to various laboratory devices typified by the so-called traveling wave tube or amplifier; and in nature an example which has recently attracted attention is the "solar wind", the name given to what is surmised to be a stream of charged particles continuously emanating from the sun and affecting the outer reaches of the earth's environment.

The interesting new feature encountered in waves associated with plasma streams is the possibility of instability or growth. These terms are defined more precisely a little later on. For the moment it may be observed that whereas for a stationary plasma with a harmonic plane wave perturbation the time-averaged work done by the electric field on a charged particle is zero, for a streaming plasma this is no longer necessarily the case. It is not easy to envisage the various possibilities without setting out the analysis, but a hint of what to expect may be obtained from considerations of the following sort. It can happen in a number of circumstances, as already seen, that a wave can have a phase speed appreciably less than the vacuum speed of light; the charged particles of a stream can therefore move with a speed close to that of the wave, and are in consequence under the action of an electric field whose phase at the particle scarcely varies. This can result in the steady conversion of the macroscopic energy of the stream to the field energy of the wave, or vice versa. In the former case the wave is amplified as it progresses. Such traveling wave amplification is the essence of the mode of operation of the laboratory devices already mentioned, and is of interest in the interpretation of natural radio noise because of the possibility of sources of disturbance being built up in this way. An allied but somewhat different sequence of events takes place when particles initially accelerated away from some region by the electric field of the wave are returned to it by the streaming motion with a different density; instability can occur if the bunching is appropriately phased, and then the mechanism is essentially that of an oscillator.

The theoretical approach adopted in this chapter follows much the same lines as that already made familiar, in that the equations are linearized and we seek solutions that are harmonic in space and time. In the case of instability this procedure can only lead to the prediction of its existence and an indication of its nature; what happens after the initial build up is quite another question to which no answer is attempted here. The equations of motion and Maxwell's equations are used as before, the former being modified by the existence of the stream velocity. The opportunity evidently presents itself of introducing the Lorentz transformation of special relativity to extend to moving plasmas results originally obtained for stationary plasmas; and although only limited use is made of this technique, it is convenient to keep the equations of motion relativistic for the purpose of comparison. It may be recalled that the equations of Chapter 5 for a stationary plasma, for example (5.2), are relativistic to the linear approximation; but now terms involving the ratio of the stream velocity to the vacuum velocity of light are involved. A relativistic treatment is also of value in its own right, since it may be of practical interest to consider streams moving at relativistic speeds.

The apparent indication of some form of instability or growth by the linearized plane wave analysis requires rather more by way of interpretation than might at first sight be supposed, and the following two sub-sections are devoted to elucidating the different situations that can arise.

6.1.2 The interpretation of instabilities

The preliminary point is made that, for the sake of simplicity, it will be assumed in the discussion that all coefficients in the dispersion equation are real; they are, in fact, real when collisions are absent, as may be confirmed in the various cases treated in Chapter 5. As a consequence, complex values of ω corresponding to a real value of k occur, if at all, in conjugate pairs; and likewise for complex values of k corresponding to a real value of ω. With this proviso it may assuredly be asserted that the existence of complex values of ω for some range of real values of k implies an instability; for the amplitude of the dominant part of a disturbance that has space variation corresponding to any such value of k will increase exponentially with time. Often, however, interest lies in the possibility of amplification, in the sense of a wave of given (real) frequency growing as it travels; the problem being described in much the same terms, for example, as that of a wave of given frequency traveling through a conducting medium, but with growth replacing attenuation. From this point of view it is at first sight tempting to suppose that the existence of complex values of k for some range of real values of ω implies amplification. But a forewarning against too glib an acceptance of the latter interpretation is apparent in the familiar case of evanescent waves in an overdense medium; for example, the dispersion equation $c^2k^2 = \omega^2 - \omega_p^2$ yields a conjugate pair of purely imaginary values of k for every real value of ω less than ω_p, but there is here, of course, no expectation whatever of wave amplification. It has already been explained that with streaming plasmas there is a presumption of the possibility of wave amplification, and a criterion must be sought which distinguishes between amplification and evanescence. Furthermore, such considerations pose the question of whether there is not some analogous distinction to be drawn between types of wave exhibiting growth in time.

For an adequate investigation of these matters it is necessary to discuss fields less idealized than that of a single harmonic plane wave. An evanescent wave, for instance, is a non-physical concept in uniform unbounded space. We therefore introduce what is often called a wave packet or wave group. These terms designate the field formed by a superposition of a number of harmonic plane waves, the frequencies or wavelengths of which are grouped around a central frequency or wavelength. For a vacuum this results in a field which is in some sense limited both in time and space, the manner of this limitation depending, of course, on the amplitudes and phases of the superposed plane waves. For a medium with a given dispersion relation the behavior of a wave packet in time and space indicates the physical nature of the disturbance associated with complex k for real ω, or complex ω for real k.

In the first place, let the dispersion equation be regarded as giving k as a function of ω (which function may, of course, be many-valued), and suppose that for some finite range of real values of ω the corresponding values of one branch $k(\omega)$ are complex. Suppose, in particular, that $k(\omega_0)$ is complex. Now form a wave packet with central frequency ω_0 by the superposition of plane waves of different frequencies, thus

$$\int_{-\infty}^{\infty} f(\omega)\, e^{-ik(\omega)x}\, e^{i\omega t}\, d\omega. \tag{6.1}$$

In (6.1) the function $f(\omega)$ determines the amplitude and phase of each plane wave, and is to be thought of as dying away rapidly as ω departs from ω_0. The restriction of the analysis to one space coordinate is not a limitation; the investigation pertains to a particular direction of wave propagation, the influence of which appears in the dependence on direction of the function $k(\omega)$. On the assumption that $k(\omega)$ and $f(\omega)$ are sufficiently well-behaved functions of ω it is known that, for any given value of x, (6.1) tends to zero as $t \to \pm\infty$. This result is the Riemann–Lebesgue lemma; it follows in essence from an integration by parts in the form

$$-\frac{i}{t}\left[f(\omega)\, e^{-ik(\omega)x}\, e^{i\omega t}\right]_{-\infty}^{\infty} + \frac{i}{t}\int_{-\infty}^{\infty} \frac{d}{d\omega}\left[f(\omega)\, e^{-ik(\omega)x}\right] e^{i\omega t}\, d\omega,$$

and evidently depends on the fact that the path of integration is confined to the real axis. The wave packet is therefore localized in time at any given point. Is it also localized in space at any given time? If it is, then it must be a confined pulse which grows as it travels, and the complex value of $k(\omega_0)$ represents true traveling wave amplification. If not, then the packet is not an acceptable physical idealization and the complex value of $k(\omega_0)$ corresponds to evanescence.

In order to see whether or not (6.1) is localized in space an attempt is made to reverse the rôles of ω and k. If it is permissible, by the rules governing integration in the complex plane, to distort the real axis in (6.1) into a line that is a map in the ω-plane of the real k-axis, then (6.1) can be written

$$\int_{-\infty}^{\infty} F(k)\, e^{i\omega(k)t}\, e^{-ikx}\, dk, \tag{6.2}$$

and this tends to zero as $x \to \pm\infty$. If it is not permissible to distort the path into one on which k is real, then in general the expression tends to infinity at one of the extremes $x \to \pm\infty$.

Whether a complex value $k(\omega_0)$ corresponding to a real frequency represents amplification or evanescence is therefore indicated by whether or not the path of integration along the real axis in the ω-plane can be legitimately distorted into a map of the real k-axis.

Now let the dispersion equation be regarded as giving ω as a function of k. An instability resulting from a complex value $\omega(k_0)$ corresponding to a real wave number k_0 can be investigated by constructing a wave packet, with central wavelength $2\pi/k_0$, in the form (6.2), where $F(k)$ determines the amplitude and phase of each plane wave, and is to be thought of as dying away rapidly as k departs from k_0. This packet is localized in space at any given time. Is it also localized in time at any given point? If it is, then it must be a confined pulse which travels as it grows, and the complex value of $\omega(k_0)$ represents *convected* instability; otherwise the instability is *non-convected*, and is sometimes called *absolute*. Following an argument similar to that already presented, whether the instability is convected or non-convected is indicated by whether or not the path of integration along the real axis in the k-plane can be legitimately distorted into a map of the real ω-axis.

Convected instability and traveling wave amplification are thus different ways of describing the same phenomenon. Non-convected instability, on the other hand, is typified by the oscillator rather than the amplifier.

In practice the diagnosis of the nature of the growth or instability can often be achieved directly from an inspection of the plot of k against ω. Take the illustrative case in which k is a two-valued function of ω, and ω is a two-valued function of k, so that there is not more than a single range of real values of ω for which k is complex, and not more than a single range of real values of k for which ω is complex. Then the possible topologies of the plot of k against ω are as shown in Figs. 6.1, 6.2a, 6.3a and 6.4a.

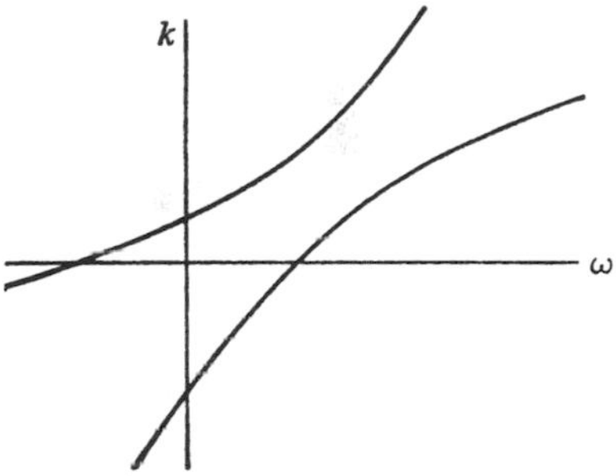

Fig. 6.1 No instability or evanescent wave.

In Fig. 6.1, k is real for all values of ω, and ω is real for all real values of k. There is no growth or instability.

In Fig. 6.2a, ω is real for all real values of k, so there is no instability. But k is complex for $\omega_1 < \omega < \omega_2$; does this represent amplification or evanescence? Figure 6.2b shows the map in the complex ω-plane of the real k-axis; the arrows indicate the path traced out as k proceeds through real values from $-\infty$ to ∞. Clearly it is not possible to bridge the gap between

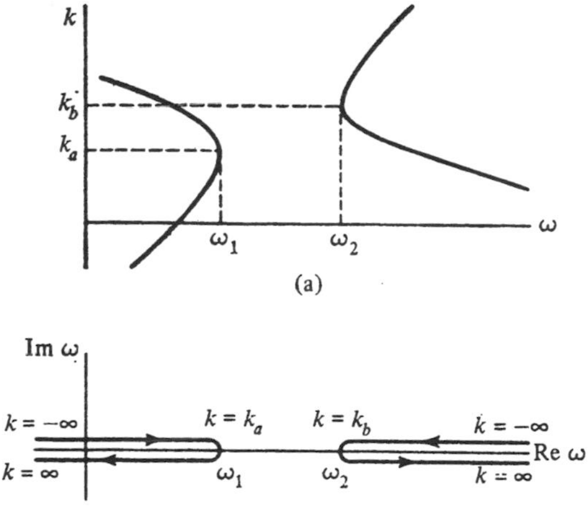

Fig. 6.2 Evanescent waves for $\omega_1 < \omega < \omega_2$.

ω_1 and ω_2 with any path along which k is real, because by hypothesis k is complex for real values of ω between ω_1 and ω_2, and when k is real ω is real. Hence the real ω-axis cannot legitimately be distorted into a map of the real k-axis, and the complex values of k for $\omega_1 < \omega < \omega_2$ therefore represent evanescent waves.

In Fig. 6.3a, k is real for all real values of ω, but ω is complex for $k_1 < k < k_2$. By an argument analogous to that just given the real k-axis

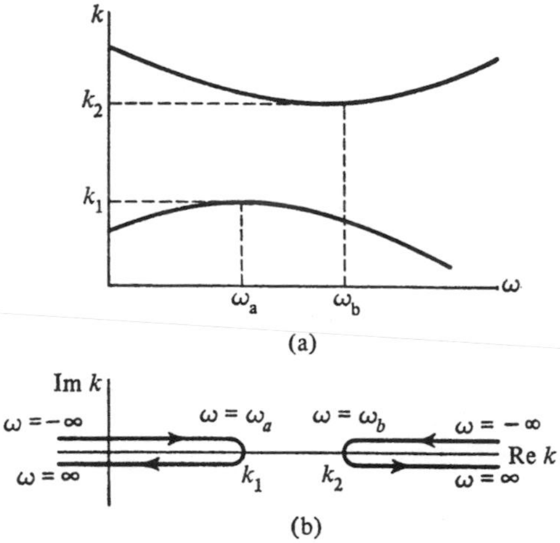

Fig. 6.3 Non-convected instability for $k_1 < k < k_2$.

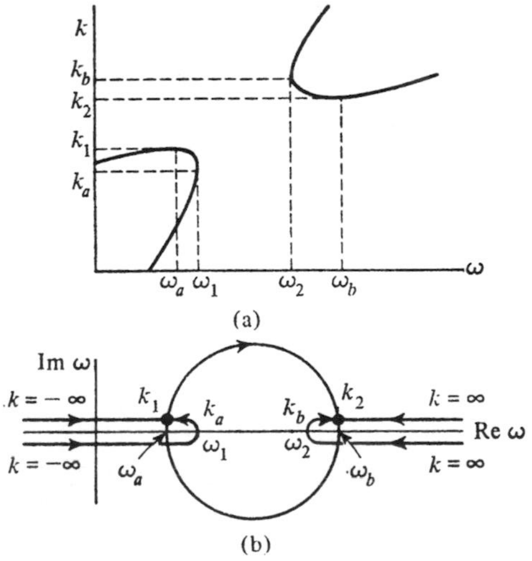

Fig. 6.4 Convected instability for $k_1 < k < k_2$; amplifying wave for $\omega_1 < \omega < \omega_2$.

cannot legitimately be distorted into a map of the real ω-axis, and the instability is therefore non-convected.

Finally, in Fig. 6.4a, k is complex for $\omega_1 < \omega < \omega_2$ and ω is complex for $k_1 < k < k_2$. Figure 6.4b shows the map of the real k-axis in the complex ω-plane. As k increases through real values from $-\infty$ to k_1, ω approaches ω_a along the real axis by a double path, one branch of which has a turning value at ω_1. A further increase in k to take it from k_1 to k_2 must imply some route for ω away from the real axis, since by hypothesis ω is complex for $k_1 < k < k_2$. The topology is therefore as shown in Fig. 6.4b, and the real ω-axis can be distorted into a map of the real k-axis. The complex values of k for $\omega_1 < \omega < \omega_2$ therefore represent amplifying waves; and, similarly, the complex values of ω for $k_1 < k < k_2$ represent convected instability.

In more general cases, local regions in the plot of k against ω can often be identified with one of the topologies in Figs. (6.1)–(6.4), and interpreted accordingly.

6.1.3 Source excitation

The nature of the disturbances associated with solutions of the dispersion equation can also be understood by treating the case in which a localized source is switched on at time $t = 0$, the disturbance being everywhere zero for $t < 0$. If we continue to work with the one-dimensional model, and suppose for convenience that the source function is proportional to $\delta(x)\,\delta(t)$, the disturbance can be written in the form

$$\int_{-\infty-i\gamma}^{\infty-i\gamma}\int_{-\infty}^{\infty}\frac{e^{i(\omega t-kx)}}{D(\omega, k)}\,dk\,d\omega, \tag{6.3}$$

where $D(\omega, k) = 0$ is the dispersion relation. For a vacuum (6.3) is the representation analogous to (2.94). The path for the k-integration is along the real axis. The path for the ω-integration is parallel to the real axis, and, in order to ensure that (6.3) is zero for $t < 0$, lies below the poles in the complex ω-plane, given by $D(\omega, k) = 0$, for all real values of k.

Useful information can be obtained from each of two procedures: one, to carry out the k-integration in (6.3); the other, to carry out the ω-integration in (6.3). In considering these in turn it is helpful first to draw attention to two specific possibilities for the ω-path of integration.

We denote the solutions of $D(\omega, k) = 0$, regarded as an equation for ω in terms of k, by $\omega(k)$, and distinguish between two cases: (a) all $\omega(k)$ lie on or above the real axis for all real values of k, (b) some $\omega(k)$ lie below the real axis for some finite range of real values of k. In case (a), the ω-path of integration in (6.3) can be taken indefinitely close to the real axis ($\gamma \to 0+$). In case (b), its approach to the real axis is barred by the segments in the lower half plane of the lines traced by the $\omega(k)$ as k runs through real values from

$-\infty$ to ∞; there is thus some curved path, C say, defined by the prescription that it lies as close as possible to the real axis.

Now consider the k-integration in (6.3). The path is closed by an infinite semi-circle, below the real axis for $x > 0$, and the residue theorem applied. Thus (6.3) is

$$-2\pi i \int \sum \frac{e^{i[\omega t - k(\omega)x]}}{(\partial D/\partial k)_{k=k(\omega)}} \, d\omega, \tag{6.4}$$

where $k(\omega)$ denotes the solutions of $D(\omega, k) = 0$ regarded as an equation for k in terms of ω, and the summation is over those $k(\omega)$ that lie in the lower half of the complex k-plane when ω takes the values specified by the ω-path of integration. Let us examine this condition on $k(\omega)$ in more detail.

In case (a), the condition is conveniently stated in the form that the $k(\omega)$ have negative imaginary part when ω assumes an infinitesimal negative imaginary part. For $k(\omega)$ that are real for all real values of ω, the implication is that those contributing to the solution (6.4) (for $x > 0$) have $dk(\omega)/d\omega > 0$ for all real values of ω; the waves are thus ordinary propagated waves, with group velócity in the direction of increasing x. For $k(\omega)$ that are complex for some range of real values of ω, the implication is that those contributing to (6.4) have negative imaginary part for the real values of ω in question; the waves are thus evanescent.

In case (b), the condition is that the $k(\omega)$ have negative imaginary part for those values of ω on path C, or on an equivalent path. By taking the path as in (6.3), with γ sufficiently large, the condition can be stated in the form that the $k(\omega)$ have negative imaginary part as $|\omega| \to \infty$ with $\text{Im}\,\omega < 0$.

It is also useful to note that, since (6.4) is zero for $t < 0$, the integrand must behave as $|\omega| \to \infty$ in such a way as to allow the path to be closed by an infinite semi-circle below the real axis, and must be free of poles and branch points below the path C; for the truth of which statement it should perhaps be emphasized that it is essential, in forming the integrand, to include the summation over relevant $k(\omega)$.

We now revert to (6.3) and carry out the ω-integration. For $t > 0$ the path is closed by an infinite semi-circle above the real axis, and (6.3) is

$$2\pi i \int_{-\infty}^{\infty} \sum \frac{e^{i[\omega(k)t - kx]}}{(\partial D/\partial \omega)_{\omega=\omega(k)}} \, dk, \tag{6.5}$$

where the summation is over all $\omega(k)$.

In case (a), by hypothesis, all $\text{Im}\,\omega(k) \geqslant 0$ for all real values of k, and (6.5) is a superposition of ordinary propagated waves and waves damped in time.

In case (b), since some $\text{Im}\,\omega(k) < 0$ for some range of real values of k, there are instabilities. Are they convected or non-convected? Suppose, for the sake of definiteness, that there is just one $\omega(k)$ exhibiting $\operatorname{Im}\omega(k) < 0$, and that for $k_1 < k < k_2$. Then the part of (6.5) contributing to the instability is

$$2\pi i \int_{k_1}^{k_2} \frac{e^{i[\omega(k)t - kx]}}{(\partial D/\partial\omega)_{\omega=\omega(k)}}\, dk = -2\pi i \int_{C'} \frac{e^{i[\omega t - k(\omega)x]}}{(\partial D/\partial k)_{k=k(\omega)}}\, d\omega, \qquad (6.6)$$

where C' is the path traced by $\omega(k)$ as k goes through real values from k_1 to k_2. Evidently C' is a segment of the path C previously defined, and the representation on the right of (6.6) looks to be nothing but a reversion to (6.4). The method of arriving at (6.6) has, however, yielded the further useful information that $k(\omega)$ is specifically the inverse function to that $\omega(k)$ which has negative imaginary part for some real k. The object of returning to an ω-integral is to ascertain, for any fixed x, the behavior of the disturbance as $t \to \infty$. As discussed in §6.1.2, the outcome hinges on whether or not C' can legitimately be flattened on to the real ω-axis. If it can, (6.6) tends to zero as $t \to \infty$; if not, to infinity. The instability is therefore convected if $k(\omega)$ and $1/(\partial D/\partial k)_{k=k(\omega)}$ are free of poles and branch points inside the loop, L say, formed by C' and the portion of the real axis between $\omega(k_1)$ and $\omega(k_2)$.

Slight variants on the condition just stated are available. Only in unimportant, exceptional circumstances can $1/(\partial D/\partial k)_{k=k(\omega)}$ have poles or branch points differing from those of $dk(\omega)/d\omega$, so the question to settle is simply whether $k(\omega)$ is free of such singularities in L. Again, it has already been observed that such singularities cannot exist below C; their absence in L therefore implies their absence throughout the lower half-plane, which in simple cases can perhaps be immediately checked by inspection. Where the detection of singularities requires computational methods it is useful to know that the search area need not extend outside L.

If $k(\omega)$ has poles or branch points inside L the instability is non-convected. It is emphasized, however, that when more than one branch $\omega(k)$ is involved, account must be taken of summation over branches. The proviso is important because it is possible to have mutually canceling contributions from each of two branches that have a common branch point.

It is easy to confirm that the predictions of the analytic procedure outlined in this sub-section, when applied to the special cases of Figs. (6.1)–(6.4), are in agreement with the results previously arrived at.

6.2 A SINGLE STREAM

Th general aim, as in Chapter 5, is first to derive the conductivity tensor from the dynamics of the charged particles, and then to introduce it into Maxwell's equations to find the refractive indices and polarizations of the characteristic plane waves. Each charged particle stream, thought of as comprising all identical particles with identical unperturbed uniform velocities, makes a contribution to the conductivity tensor; and the conductivity tensor for the whole system is simply obtained by the addition of the separate contributions from each stream. In this section, therefore, the calculations are carried out only for a single stream of electrons; it is assumed that both the charge and the current densities of the electron stream in the unperturbed state are neutralized by those of a positive ion stream, but the perturbation motion of the positive ions is ignored. The main object is to find the conductivity tensor, but the opportunity is also taken of deriving the dispersion relation for a single stream, so that an immediate check is afforded by identifying it with that obtained from the dispersion relation of a stationary electron gas by a Lorentz transformation.

The space and time dependence of the wave perturbation is taken as

$$e^{i(\omega t - kx)}$$

and the uniform velocity of the electron stream as

$$\mathbf{U} = U(\cos\alpha, \sin\alpha, 0), \tag{6.7}$$

so that the direction of wave propagation makes an angle α with the stream. If $\mathbf{v}$ is the perturbation velocity of the electrons, and if n is the excess number of electrons per unit volume above the unperturbed number N, then the current density is

$$\mathbf{j} = -e(N + n)(\mathbf{U} + \mathbf{v}) + eN\mathbf{U}, \tag{6.8}$$

or, when linearized,

$$\mathbf{j} = -eN\mathbf{v} - e\mathbf{U}n. \tag{6.9}$$

The second term on the right-hand side of (6.9) is the contribution from the convection of charge density. The elimination of n in favor of $\mathbf{v}$, which is required since the particle dynamics leads to the expression of $\mathbf{v}$ in terms of $\mathbf{E}$, is achieved by means of the charge conservation relation

$$\operatorname{div}\mathbf{j} - e\frac{\partial n}{\partial t} = 0, \tag{6.10}$$

which with harmonic space and time dependence reads

$$kj_x + e\omega n = 0. \tag{6.11}$$

Substitution for n from (6.11) into (6.9) gives the component equations

$$j_x = -eN\frac{\omega}{\omega - kU\cos\alpha}v_x, \tag{6.12}$$

$$j_y = -eN\left(\frac{kU\sin\alpha}{\omega - kU\cos\alpha}v_x + v_y\right), \tag{6.13}$$

$$j_z = -eNv_z. \tag{6.14}$$

It is assumed that $\omega \neq kU\cos\alpha$; equality is a special case which is dealt with at the end of the section.

The expression of **v** in terms of **E** is determined by the equation of motion of the electrons. Its relativistic form is

$$\frac{D}{Dt}\left\{\frac{\mathbf{U}+\mathbf{v}}{\sqrt{[1-(\mathbf{U}+\mathbf{v})^2/c^2]}}\right\} = -\frac{e}{m}[\mathbf{E}+(\mathbf{U}+\mathbf{v})\times\mathbf{B}], \tag{6.15}$$

where it should be emphasized that **B** is the perturbation magnetic induction, there being no external magnetostatic field. The operator on the left-hand side is

$$\frac{D}{Dt} \equiv \frac{\partial}{\partial t} + (\mathbf{U}+\mathbf{v})\cdot\text{grad},$$

so that when the non-linear terms are dropped (6.15) reads

$$i(\omega - kU\cos\alpha)\left[\frac{\mathbf{U}\cdot\mathbf{v}/c^2}{1-U^2/c^2}\mathbf{U}+\mathbf{v}\right] = -\frac{e}{m}\sqrt{(1-U^2/c^2)}(\mathbf{E}+\mathbf{U}\times\mathbf{B}). \tag{6.16}$$

Furthermore, **B** is expressed directly in terms of **E** through the Maxwell equation

$$\text{curl}\,\mathbf{E} = -i\omega\mathbf{B};$$

explicitly

$$B_x = 0, \qquad B_y = -\frac{k}{\omega}E_z, \qquad B_z = \frac{k}{\omega}E_y. \tag{6.17}$$

The components of (6.16) are therefore

$$\left(1-\frac{U^2}{c^2}\sin^2\alpha\right)v_x + \frac{U^2}{c^2}\sin\alpha\cos\alpha\, v_y = i\frac{e}{m}\frac{(1-U^2/c^2)^{3/2}}{\omega - kU\cos\alpha}\left(E_x + \frac{kU}{\omega}\sin\alpha\, E_y\right), \tag{6.18}$$

$$\frac{U^2}{c^2}\sin\alpha\cos\alpha\, v_x + \left(1-\frac{U^2}{c^2}\cos^2\alpha\right)v_y = i\frac{e}{m}\frac{(1-U^2/c^2)^{3/2}}{\omega}E_y, \tag{6.19}$$

$$v_z = i\frac{e}{m}\frac{\sqrt{(1-U^2/c^2)}}{\omega}E_z. \tag{6.20}$$

Equations (6.18) and (6.19) give

$$v_x = i\frac{e}{m}\frac{\sqrt{(1-U^2/c^2)}}{\omega - kU\cos\alpha}\left\{\left(1-\frac{U^2}{c^2}\cos^2\alpha\right)E_x + \sin\alpha\left(\frac{kU}{\omega}-\frac{U^2}{c^2}\cos\alpha\right)E_y\right\}, \tag{6.21}$$

$$v_y = i\frac{e}{m}\frac{\sqrt{(1-U^2/c^2)}}{\omega - kU\cos\alpha}\left\{-\frac{U^2}{c^2}\sin\alpha\cos\alpha\, E_x + \left(1-\frac{kU}{\omega}\cos\alpha - \frac{U^2}{c^2}\sin^2\alpha\, E_y\right)\right\}, \tag{6.22}$$

and the substitution of (6.21), (6.22) and (6.20) into (6.12), (6.13) and (6.14) yields relations between the components of **j** and **E** in the form

$$j_i = \sigma_{ij} E_j. \tag{6.23}$$

The conductivity tensor can be written explicitly

$$\sigma_{ij} = -i\omega\varepsilon_0 \frac{\omega_p^2}{\omega^2}$$
$$\times \begin{bmatrix} \dfrac{1-(U^2/c^2)\cos^2\alpha}{[1-\mu(U/c)\cos\alpha]^2} & \dfrac{\mu-(U/c)\cos\alpha}{[1-\mu(U/c)\cos\alpha]^2}\dfrac{U}{c}\sin\alpha & 0 \\ \dfrac{\mu-(U/c)\cos\alpha}{[1-\mu(U/c)\cos\alpha]^2}\dfrac{U}{c}\sin\alpha & 1-\dfrac{1-\mu^2}{[1-\mu(U/c)\cos\alpha]^2}\dfrac{U^2}{c^2}\sin^2\alpha & 0 \\ 0 & 0 & 1 \end{bmatrix}, \tag{6.24}$$

where μ is the refractive index ck/ω, and

$$\omega_p^2 = \frac{Ne^2}{\varepsilon_0 m}\sqrt{\left(1-\frac{U^2}{c^2}\right)} \tag{6.25}$$

is the square of the proper plasma frequency of the electron gas, that is, the plasma frequency corresponding to the proper number density measured in a frame moving with the unperturbed stream.

If the equivalent dielectric tensor

$$\kappa_{ij} = \delta_{ij} - \frac{i\sigma_{ij}}{\omega\varepsilon_0} \tag{6.26}$$

is introduced (cf. (5.48)), and written

$$\boldsymbol{\kappa} = \begin{pmatrix} \kappa_1 & \kappa_3 & 0 \\ \kappa_3 & \kappa_2 & 0 \\ 0 & 0 & \kappa_0 \end{pmatrix}, \tag{6.27}$$

then the elimination of **B** from Maxwell's equations, that is (5.54), (5.55) with harmonic space and time dependence, leads to

$$\mu^2(0, E_y, E_z) = \boldsymbol{\kappa}\cdot\mathbf{E}. \tag{6.28}$$

The dispersion relation is therefore

$$\begin{vmatrix} \kappa_1 & \kappa_3 & 0 \\ \kappa_3 & \kappa_2-\mu^2 & 0 \\ 0 & 0 & \kappa_0-\mu^2 \end{vmatrix} = 0, \tag{6.29}$$

and the refractive indices of the characteristic waves are specified either by

$$\mu^2 = \kappa_0, \tag{6.30}$$

or by

$$\kappa_1(\kappa_2 - \mu^2) = \kappa_3^2, \tag{6.31}$$

where it must be remembered that κ_1, κ_2 and κ_3 are themselves functions of μ as well as ω.

Expressions for the κ's are given by (6.26) and (6.24). Evidently (6.30) is simply

$$\mu^2 = 1 - \omega_p^2/\omega^2. \tag{6.32}$$

The explicit form of (6.31), on the other hand, appears at first sight rather complicated, being

$$\left[1-\frac{\omega_p^2}{\omega^2}\frac{1-(U^2/c^2)\cos^2\alpha}{[1-\mu(U/c)\cos\alpha]^2}\right]\left\{1-\frac{\omega_p^2}{\omega^2}\left[1-\frac{1-\mu^2}{[1-\mu(U/c)\cos\alpha]^2}\frac{U^2}{c^2}\sin^2\alpha\right]-\mu^2\right\}$$
$$= \frac{\omega_p^4}{\omega^4}\frac{[\mu-(U/c)\cos\alpha]^2}{[1-\mu(U/c)\cos\alpha]^4}\frac{U^2}{c^2}\sin^2\alpha. \tag{6.33}$$

However, a little algebra shows that it simplifies to

$$\left[1 - \frac{\omega_p^2}{\omega^2}\frac{1 - U^2/c^2}{[1 - \mu(U/c)\cos\alpha]^2}\right]\left(1 - \frac{\omega_p^2}{\omega^2} - \mu^2\right) = 0. \tag{6.34}$$

There are thus two characteristic waves, each of which can travel in either the positive or negative x-direction, with a common value of refractive index

$$\mu = \sqrt{(1 - \omega_p^2/\omega^2)}; \tag{6.35}$$

and two others with respective refractive indices given by

$$\mu\frac{U}{c}\cos\alpha = 1 \pm \frac{\omega_p}{\omega}\sqrt{(1 - U^2/c^2)}. \tag{6.36}$$

Reference to (6.28) shows that one of the waves (6.35) can be purely transverse, with E_z, B_y, j_z, v_z the only non-zero components of the electromagnetic and velocity vectors. For $\alpha \neq 0$, all the other waves are neither purely transverse nor purely longitudinal, having in particular, non-zero components E_x, E_y. When $\alpha = 0$, (6.35) pertains to purely transverse waves, and (6.36) to purely longitudinal (plasma) waves.

It is easy to confirm that (6.35) and (6.36) agree with results obtained from a Lorentz transformation of the dispersion relation for a stationary electron gas. For suppose the frame of reference S' moves with the stream, that is, with velocity (6.7) relative to the frame S in which ω and k are measured; and suppose the corresponding measurements in S' are ω', k'. Then the dispersion relations in S' are

$$c^2k'^2 = \omega'^2 - \omega_p^2 \tag{6.37}$$

for transverse oscillations, remembering that ω_p is the proper plasma frequency; and

$$\omega'^2 = \omega_p^2 \tag{6.38}$$

for longitudinal oscillations. But components (ω, $ck \cos \alpha$, $ck \sin \alpha$, 0) in the frame S specify a four-vector; the square of its length is the invariant

$$\omega^2 - c^2k^2 = \omega'^2 - c^2k'^2, \tag{6.39}$$

and the transformation of the time-like component is

$$\omega' = \frac{\omega - kU \cos \alpha}{\sqrt{(1 - U^2/c^2)}}. \tag{6.40}$$

Substitution from (6.39) into (6.37) gives

$$c^2k^2 = \omega^2 - \omega_p^2, \tag{6.41}$$

which is identical with (6.35); and the substitution of (6.40) in (6.38) gives

$$\omega - kU \cos \alpha = \pm\omega_p\sqrt{(1 - U^2/c^2)}, \tag{6.42}$$

which is identical with (6.36). The components of the field vectors in the characteristic waves for the streaming gas could, of course, likewise be obtained from those for the stationary gas by the relativistic transformations (2.172).

It is also seen from (6.40) that the special case $\omega = kU \cos \alpha$, deferred earlier on, corresponds in S' to the trivial result $\omega' = 0$. If $\omega' = 0$, then

$$\mathbf{E}' = 0, \qquad B_x' = 0$$

$$-eN\sqrt{(1 - U^2/c^2)}v_y' = j_y' = ik'H_z',$$

$$-eN\sqrt{(1 - U^2/c^2)}v_z' = j_z' = -ik'H_y',$$

which just represents steady current flow in planes perpendicular to the x-axis, the value of k' being arbitrary. It is simply the convection of this stationary pattern that is seen in S.

Having confirmed in detail the correctness of the deductions from the conductivity tensor (6.24) it is now possible to proceed with confidence to the typical problem in which a basically new phenomenon is exhibited, the problem of two interpenetrating streams.

6.3 TWO STREAMS

6.3.1 Propagation along parallel streams

If there are two unbounded, uniform, interpenetrating streams of charged particles, the conductivity tensor is obtained by adding the two expressions obtained from (6.24) by inserting the respective values of U, α and ω_p for each stream. The composite tensor of course retains the structure (6.27), and the dispersion relations are still formally given by (6.30) and (6.31). The explicit form of equation (6.30) may be written

$$\mu^2 = 1 - p^2/\omega^2, \tag{6.43}$$

where

$$p^2 = \omega_{p1}^2 + \omega_{p2}^2, \tag{6.44}$$

with ω_{p1}, ω_{p2} standing for the proper plasma frequencies of the respective streams; and the characteristic wave specified by (6.43) is purely transverse. The explicit form of (6.31) is, however, too complicated to be easily manageable. In order to keep the algebra simple we restrict the discussion by taking the streams to be parallel to one another and either parallel or perpendicular to the direction of wave propagation. The former case is treated here, the latter in §6.3.2.

It is assumed then that for each stream $\alpha = 0$ or $\alpha = \pi$. In consequence, $\kappa_3 = 0$, $\kappa_2 = \kappa_0$; and it is seen at once that the only characteristic wave not described by (6.43) is a purely longitudinal (plasma) wave, with dispersion relation $\kappa_1 = 0$, that is

$$\frac{q_1^2}{(\omega - kU_1)^2} + \frac{q_2^2}{(\omega - kU_2)^2} = 1, \tag{6.45}$$

where

$$q_1^2 = (1 - U_1^2/c^2)\omega_{p1}^2, \qquad q_2^2 = (1 - U_2^2/c^2)\omega_{p2}^2. \tag{6.46}$$

In (6.46) U_1 and U_2 are the respective values of the stream velocities, and may be positive or negative.

Perhaps the most immediate way to see that (6.45) predicts an instability is to write it as

$$F(w) \equiv \frac{q_1^2}{(w - U_1)^2} + \frac{q_2^2}{(w - U_2)^2} = k^2,$$

with $w = \omega/k$, and plot $F(w)$ as a function of w. Taking $U_1 > U_2$ it is easily confirmed that the curve must be of the form sketched in Fig. 6.5, so clearly

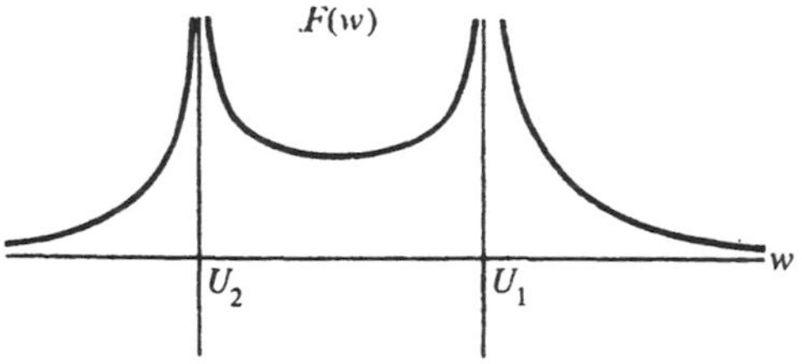

Fig. 6.5 Demonstrating the two-stream instability.

there is a pair of conjugate complex values of ω for all real values of k for which k^2 is less than the minimum value, $(q_1^{2/3} + q_2^{2/3})^3/(U_1 - U_2)^2$, reached by $F(w)$ between $w = U_1$ and $w = U_2$.

A plot of k against ω will also display the salient features of (6.45), and can be used for its interpretation in the way described in §6.1. As regards the general nature of the curves there are two cases to be distinguished according to whether the streams travel in the same or opposite directions. These are considered in turn.

If the streams travel in the same direction it may be supposed that $U_1 > U_2 > 0$. The plot of k against ω then appears as shown diagrammatically in Fig. 6.6. For large values of $|\omega|$ and $|k|$ the asymptotes are obviously the four lines $\omega - kU_1 = \pm q_1$, $\omega - kU_2 = \pm q_2$, and the diagram actually represents the situation when $q_1 > q_2$ and $q_2/U_2 > q_1/U_1$, though these inequalities are of no special significance. Mathematically there are, of course, four values of ω for each real value of k, and vice versa. But the diagram makes it clear that there is a range of real values of k, say that between $\pm k_1$, for which ω is complex, and a range of real values of ω, say that between $\pm\, \omega_1$, for which k is complex. From the nature of the curves it can be inferred that there is convected instability for wavelengths greater than $2\pi/k_1$, and amplification for angular frequencies less than ω_1.

If the streams travel in opposite directions it may be supposed that $U_1 > -U_2 > 0$, and the plot is then as shown in Fig. 6.7. Here one pair of branches gives complex ω for $-k_1 < k < k_1$, but real k for real ω; and the other pair of branches gives complex k for $-\omega_1 < \omega < \omega_1$, but real ω for real k. The inference from the nature of the curves is that the complex values of ω correspond to non-convected instability, and complex values of k to evanescent waves.

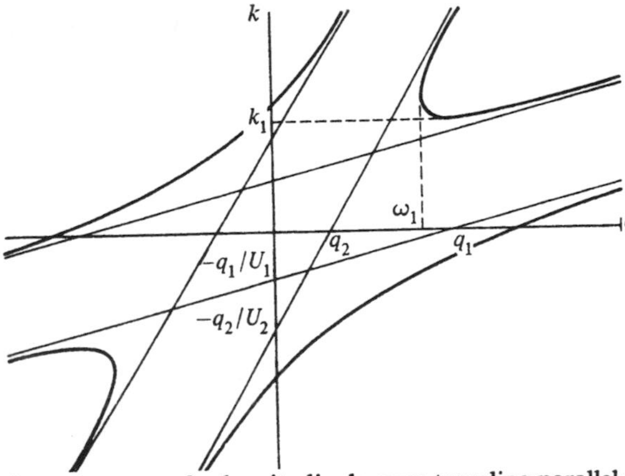

Fig. 6.6 A dispersion curve for longitudinal waves traveling parallel to two streams that flow in the same direction.

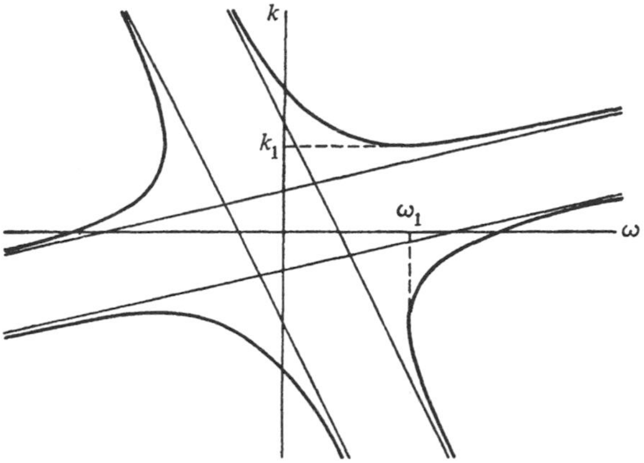

Fig. 6.7 A dispersion curve for longitudinal waves traveling parallel to two streams that flow in opposite directions.

It appears, therefore, that a traveling wave amplifier could be constructed from a device in which the charged particle streams travel in the same direction, and an oscillator from a device in which the streams are oppositely directed. The analysis in the latter case has given rise to the conjecture, of some interest in astrophysics, that colliding plasmas would rapidly trade their bulk motion for random motion associated with the instability; the order of magnitude estimate of the minimum distance over which the plasmas must interact to produce instability is $2\pi/k_1$, which will in general be something like the relative velocity of the plasmas divided by the plasma frequency.

A more detailed investigation of (6.45) is particularly easy in the quite practical case $q_1 = q_2$. Apart from the relativistic corrections, which would usually be negligible, this equality implies equal values of Ne^2/m for the two streams; two electron streams, for example, must have the same density.

To keep a common notation for this section and §6.3.2 we introduce

$$q^2 = q_1^2 + q_2^2, \tag{6.47}$$

so that for the present $q_1^2 = q_2^2 = \frac{1}{2}q^2$. If also

$$U_1 = U + V, \qquad U_2 = U - V, \tag{6.48}$$

where U_1 and U_2 are now arbitrary, it is readily confirmed that (6.45) can be written

$$(\omega - kU)^4 - 2(\tfrac{1}{2}q^2 + k^2V^2)(\omega - kU)^2 + k^2V^2(k^2V^2 - q^2) = 0. \tag{6.49}$$

This can be regarded as a quadratic for $(\omega - kU)^2$ in terms of k, and the explicit solution for ω in terms of k is

$$\omega = kU \pm [\tfrac{1}{2}q^2 + k^2V^2 \pm \sqrt{(\tfrac{1}{4}q^4 + 2q^2k^2V^2)}]^{1/2}, \tag{6.50}$$

where the four values come from the various combinations of the alternative signs.

The value below which positive values of k give complex values of ω is at once seen from (6.49) to be

$$k_1 = q/V, \tag{6.51}$$

in agreement with the general result previously obtained from a consideration of Fig. 6.5. Furthermore, differentiation with respect to k^2 of the quantity in square brackets on the right-hand side of (6.50) shows that the imaginary part of ω is a maximum when

$$k = \tfrac{1}{2}\sqrt{(3/2)}q/V, \tag{6.52}$$

and its value is then $q/(2\sqrt{2})$, giving a growth rate described by

$$e^{qt/2\sqrt{2}}. \tag{6.53}$$

Another instructive case is that in which one of the streams is highly rarefied. For the sake of definiteness we reimpose the condition $U_1 > U_2 > 0$ and assume that the density of the first stream is so low that q_1 can be treated mathematically as a small parameter. Then (6.45) is conveniently rewritten

$$(\omega - kU_2 - q_2)(\omega - kU_2 + q_2)[(\omega - kU_1)^2 - q_1^2] = q_1^2 q_2^2, \tag{6.54}$$

and with the notation (6.48) this is

$$(\Delta + 2kV - q_2)(\Delta + 2kV + q_2)(\Delta^2 - q_1^2) = q_1^2 q_2^2, \tag{6.55}$$

where

$$\Delta = \omega - kU_1. \tag{6.56}$$

If $2kV \neq q_2$, approximations to Δ are evidently either $-2kV \pm q_2$ or the values given by

$$\Delta^2 - q_1^2 = q_1^2 q_2^2/(4k^2V^2 - q_2^2), \tag{6.57}$$

since in the latter the assumption that Δ is of the order of magnitude of the small parameter q_1 is verified *a posteriori*. That is, approximately, either

$$\omega = kU_2 \pm q_2, \tag{6.58}$$

or

$$\omega = kU_1 \pm \frac{2kV}{\sqrt{(4k^2V^2 - q_2^2)}}\, q_1. \tag{6.59}$$

For real positive values of k greater than $q_2/(2V)$, the values (6.59) of ω are real; and for real positive values of k less than $q_2/(2V)$ the values (6.59) of ω are complex with imaginary part proportional to q_1.

If on the other hand, $2kV = q_2$, the approximations to Δ from (6.55) are evidently $-2kV - q_2$ and the values given by

$$\Delta(\Delta^2 - q_1^2) = \tfrac{1}{2}q_2q_1^2; \tag{6.60}$$

and (6.60) in turn implies approximately that

$$\Delta^3 = \tfrac{1}{2}q_2q_1^2, \tag{6.61}$$

since the assumption that Δ is of larger order than q_1 is verified *a posteriori.* Now (6.61) states that

$$\omega = kU_1 + 2^{-1/3}q_2^{1/3}q_1^{2/3}\delta, \tag{6.62}$$

where δ stands for any one of the cube roots of unity, and the factor multiplying δ is real. For $\delta = \exp(-2i\pi/3)$, (6.62) corresponds to convective instability or amplification. Note that the imaginary part of (6.62) is proportional to $q_1^{2/3}$.

What happens in the case of a rarefied stream can therefore be elucidated as follows. The characteristic waves of the dense stream are but little affected by the rarefied stream, and to a zero order approximation have the dispersion relation $(\omega - kU_2)^2 = q_2^2$. However, for those waves whose speed is close to the speed U_1 of the rarefied stream there is the opportunity for a persistent transfer of energy from the bulk motion of the stream to the wave perturbation. The state of maximum growth rate is thus intimately associated with the relations $\omega = kU_1$, $\omega - kU_2 = q_2$. These relations do indeed jointly imply $2kV = q_2$, from which (6.62) was derived. Moreover, in the light of (6.62) we can make the more precise statement that, at a given (real) frequency, maximum amplification occurs when the stream speed U_1 is slightly greater than the phase speed of the characteristic wave.

6.3.2 Propagation across parallel streams

The two-stream instability that has just been discussed is the very familiar one associated with purely longitudinal oscillations. It is only one example out of many of various types and various degrees of complication. As an illustration of an instability associated with a partly transverse electromagnetic disturbance we now consider the almost equally simple case in which the direction of spatial dependence is perpendicular to the parallel streams, which may flow in the same or opposite senses.

The conductivity tensor (6.24) for a single stream with $\alpha = \frac{1}{2}\pi$ is

$$\sigma_{ij} = -i\omega\varepsilon_0\frac{\omega_p^2}{\omega^2}\begin{pmatrix} 1 & kU/\omega & 0 \\ kU/\omega & 1 - U^2/c^2 + k^2U^2/\omega^2 & 0 \\ 0 & 0 & 1 \end{pmatrix}. \tag{6.63}$$

The stream flows in the positive or negative y-direction according to whether U is positive or negative respectively, and the space–time dependence of the disturbance is specified by exp $[i(\omega t - kx)]$. The dielectric tensor for two streams therefore takes the form (6.27), where now

$$\kappa_0 = \kappa_1 = 1 - p^2/\omega^2, \tag{6.64}$$

$$\kappa_2 = 1 - q^2/\omega^2 - (p^2 - q^2)c^2k^2/\omega^4, \tag{6.65}$$

$$\kappa_3 = -(U_1\omega_{p1}^2 + U_2\omega_{p2}^2)k/\omega^3, \tag{6.66}$$

with ω_{p1}, ω_{p2} again representing the proper plasma frequencies of the respective streams, and

$$p^2 = \omega_{p1}^2 + \omega_{p2}^2, \qquad q^2 = (1 - U_1^2/c^2)\omega_{p1}^2 + (1 - U_2^2/c^2)\omega_{p2}^2,$$

as in (6.44), (6.46) and (6.47).

The dispersion relations for the characteristic waves are (6.30) and (6.31). The former has already been seen to be (6.43), which holds for any value of α and pertains to a purely transverse wave without instability; whereas the latter, which pertains to a partly transverse wave with $E_z = 0$ but E_x and E_y not in general zero, now takes the form

$$\left(1 - \frac{p^2}{\omega^2}\right)\left[1 - \frac{q^2}{\omega^2} - \left(1 + \frac{p^2 - q^2}{\omega^2}\right)\frac{c^2k^2}{\omega^2}\right] = \left(U_1\omega_{p1}^2 + U_2\omega_{p2}^2\right)^2\frac{k^2}{\omega^6}, \tag{6.67}$$

and is shown to exhibit non-convected instability.

Since the coefficient of c^2k^2/ω^6 in (6.67) is

$$p^2(p^2 - q^2) - (U_1\omega_{p1}^2 + U_2\omega_{p2}^2)^2/c^2 = \omega_{p1}^2\omega_{p2}^2(U_1 - U_2)^2/c^2,$$

the equation can be written

$$\left(1 - \frac{p^2}{\omega^2}\right)\left(1 - \frac{q^2}{\omega^2}\right) = \left(1 - \frac{q^2}{\omega^2} - \frac{(U_1 - U_2)^2}{c^2}\frac{\omega_{p1}^2\omega_{p2}^2}{\omega^4}\right)\frac{c^2k^2}{\omega^2};$$

or

$$c^2k^2 = \frac{\omega^2(\omega^2 - p^2)(\omega^2 - q^2)}{(\omega^2 + \omega_1^2)(\omega^2 - \omega_2^2)}, \tag{6.68}$$

where

$$\left.\begin{matrix}\omega_1^2\\ \omega_2^2\end{matrix}\right\} = \tfrac{1}{2}\{\sqrt{[q^4 + 4\omega_{p1}^2\omega_{p2}^2(U_1 - U_2)^2/c^2]} \mp q^2\}. \tag{6.69}$$

It is easy to see that, assuming $(U_1 - U_2)^2/c^2$ to be small, ω_2 lies in the narrow range between p and q, whereas

$$\omega_1 \simeq \frac{|U_1 - U_2|}{c}\frac{\omega_{p1}\omega_{p2}}{q}. \tag{6.70}$$

The easiest way to envisage (6.68) is perhaps through the plot of c^2k^2 against ω^2, that of k against ω being less instructive for a reason which will be apparent in a moment. The former plot is evidently somewhat as sketched in Fig. 6.8. The dashed line is $c^2k^2 = \omega^2 - p^2$, and apart from the relativistic dependence of p^2 on U_1 and U_2 represents the limiting case which obtains when the stream velocities are equal so that ω_2 coincides with q and ω_1 goes to zero; it gives a close approximation to the full curve when $\omega^2 > p^2$. The portion of the curve for which ω^2 is negative when k^2 is positive represents an instability that is pure growth in time, without propagation; it is seen to exist no matter what real value k takes, which is why the k versus ω plot is relatively uninformative. The growth of the instability is described by exp (γt), where the coefficient γ is not greater than ω_1; it appears from (6.70) that the growth coefficient is less than that for the longitudinal instability of §6.3.1 by a factor of the order of the ratio of the relative speed of the streams to the vacuum speed of light.

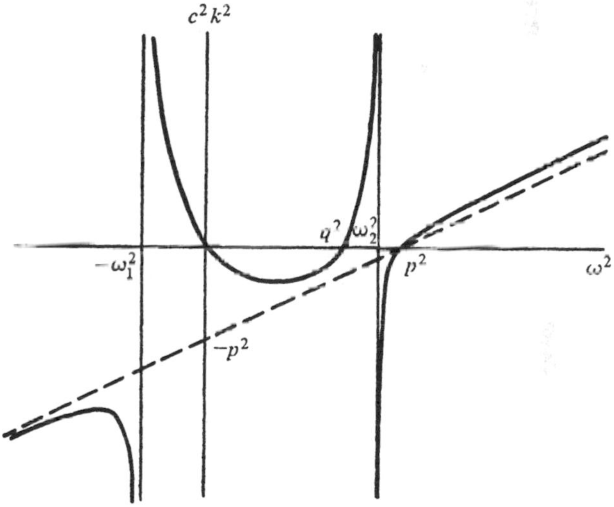

Fig. 6.8 A dispersion curve for waves traveling across two parallel streams.

The special case in which the streams travel with the same speed in opposite directions, but are otherwise identical, is particularly simple. Then $U_1 = -U_2 = U$ say, $\omega_{p1}^2 = \omega_{p2}^2 = \frac{1}{2}p^2$ and $q^2 = (1 - U^2/c^2)p^2$, so that

$$\kappa_2 = 1 - (1 - U^2/c^2)p^2/\omega^2 - k^2U^2p^2/\omega^4, \qquad \kappa_3 = 0. \tag{6.71}$$

The dispersion relation (6.31) separates into the pair of independent relations

$$\kappa_1 = 0, \qquad \mu^2 = \kappa_2. \tag{6.72}$$

The former describes a purely longitudinal oscillation in which k is arbitrary and E_x is the only non-zero field component. The latter describes a purely transverse wave in which E_y and B_z are the only non-zero field components, and is

$$c^2k^2 = \frac{\omega^2[\omega^2 - (1 - U^2/c^2)p^2]}{\omega^2 + p^2U^2/c^2}, \tag{6.73}$$

in agreement with (6.68) when the common factor $\omega^2 - p^2$ is removed from numerator and denominator on the right-hand side. Equation (6.73) is a quadratic for ω^2, and its explicit solution shows that one of the values of ω^2 is negative for all positive values of k^2.

In this last case we can credit the dynamical origin of the instability entirely to the Lorentz force $\pm UB_z$ acting on the streams. The streams are deflected in the x-direction, in opposite senses, and the pattern of deflections has wavelength $2\pi/k$. There is a corresponding pattern of current flow in the y-direction, these currents being the source of B_z. The result is a pinch effect; the plasma fragments into a series of plane pinches, and there is no space charge and no pressure to oppose the collapse. Comparison can be drawn with the ordinary plasma oscillation: the mixture has objects of opposite signs (the members of the two streams) and is "neutral" in the uniform state (no currents); but here like objects attract, unlike repel, and when there is some separation the force set up tends to enhance it rather than restore equilibrium, leading to growth rather than oscillation; the comparatively slow growth rate is accounted for by the fact that the magnetic force between charges with relative velocity u is of the order of u^2/c^2 times the electrostatic force. The situation is similar, if not completely analogous, to the Jeans instability in stellar dynamics.

6.3.3 The inclusion of partial pressures

The analysis so far has been for "cold" streams. Some hint of the effect of temperature can be obtained from the inclusion of hydrodynamic pressure forces, as in §5.5.

To treat the simplest case it is supposed, as for the most part in §6.3.1, that any stream velocity is parallel to the direction of wave motion, and the latter is taken to be the positive x-direction. The pressure force per unit volume in a particular stream is given in non-relativistic form by (5.131), and since, from (6.9) and (6.10),

$$n = N\frac{k}{\omega - kU}v_x, \tag{6.74}$$

the force per charged particle is

$$\left(i\gamma KT\frac{k^2}{\omega - kU}v_x, 0, 0\right). \tag{6.75}$$

Transverse oscillations are unaffected by the pressure, and need no further consideration. For longitudinal oscillations the non-relativistic equation of motion is now

$$i(\omega - kU)v_x = -\frac{e}{m}E_x + i\frac{a^2k^2}{\omega - kU}v_x\,, \tag{6.76}$$

where $a = \sqrt{(\gamma KT/m)}$, as in (5.141), and is the "speed of sound" for the stream constituent.

From (6.76)

$$v_x = i\frac{e}{m}\frac{\omega - kU}{(\omega - kU)^2 - a^2k^2}E_x\,, \tag{6.77}$$

and substitution into (6.10) gives

$$j_x = -i\frac{Ne^2}{m}\frac{\omega}{(\omega - kU)^2 - a^2k^2}E_x. \tag{6.78}$$

The formal amendment to the two-stream theory of §6.3.1 is now obvious. The dispersion relation to replace (6.45) is

$$\frac{\omega_{p1}^2}{(\omega - kU_1)^2 - a_1^2k^2} + \frac{\omega_{p2}^2}{(\omega - kU_2)^2 - a_2^2k^2} = 1, \tag{6.79}$$

where ω_{p1}, ω_{p2} are here stream plasma frequencies without any relativistic factors. The non-relativistic form of (6.45) is recovered on putting $a_1 = a_2 = 0$, and it may also be noted that (5.152) is recovered on putting $U_1 = U_2 = 0$.

Evidently when $a_1 \ll U_1$ and $a_2 \ll U_2$ (6.79) is much the same as (6.45) and instability is present. To see that the theory predicts inhibition of the instability at sufficiently great sound speeds it is perhaps easiest to look at the counterpart of Fig. 6.5. That is, we write $w = \omega/k$ and consider

$$G(w) \equiv \frac{\omega_{p1}^2}{(w - U_1)^2 - a_1^2} + \frac{\omega_{p2}^2}{(w - U_2)^2 - a_2^2} = k^2.$$

We can assume without loss of generality that

$$U_1 + a_1 > U_2 + a_2,$$

and then the plot of $G(w)$ against w takes one of two forms according to whether $U_1 - a_1$ is greater or less than $U_2 + a_2$. If $U_1 - a_1 > U_2 + a_2$, the situation is as indicated in Fig. 6.9: of the three stationary values of $G(w)$ the minimum must be positive, because $G(w)$ is positive when w lies between $U_2 + a_2$ and $U_1 - a_1$, and neither of the maxima can exceed the minimum, because this would imply a quartic equation having more than four roots; hence there is a range of positive values of k^2 for which two of the values of ω are a conjugate complex pair. If $U_1 - a_1 < U_2 + a_2$, the situation is quite different, as indicated in Fig. 6.10 (it being immaterial to the argument whether or not $U_1 - a_1$ exceeds $U_2 - a_2$): evidently for any positive value of k^2 all four values of ω are real. Inhibition of the instability is therefore predicted when $a_1 + a_2 > U_1 - U_2$.

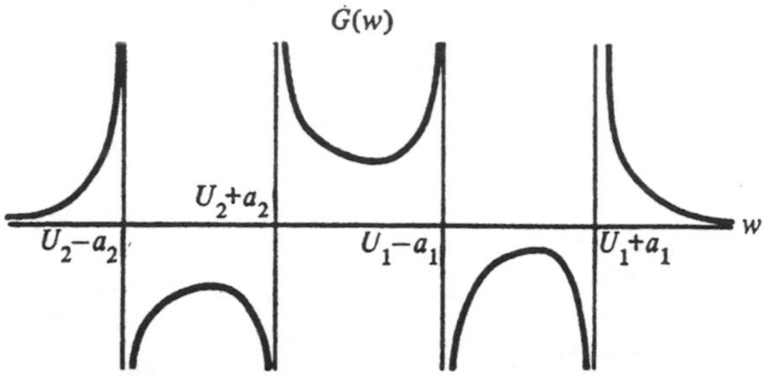

Fig. 6.9 Demonstrating the maintenance of the two-stream instability with the inclusion of partial pressures, when $a_1 + a_2 < U_1 - U_2$.

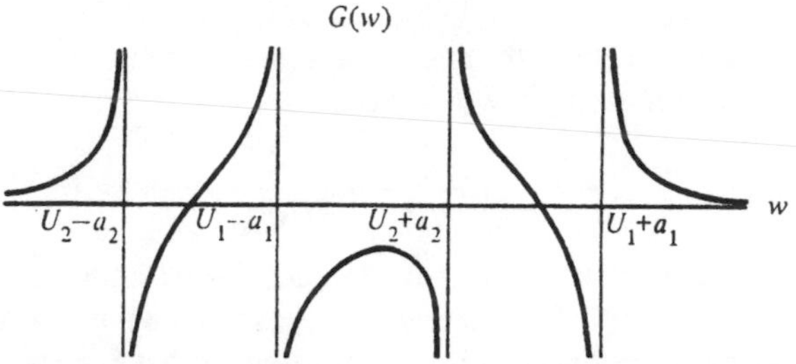

Fig. 6.10 Demonstrating the inhibition of the two-stream instability, by the inclusion of partial pressures, when $a_1 + a_2 > U_1 - U_2$.

PROBLEMS

(The plasmas are taken to be homogeneous and unbounded. The analysis is linear in perturbation quantities, and except in Problem 7 the perturbation motion of the positive ions is neglected.)

1. Confirm the results of §6.1.2 pertaining to Figs. 6.1–6.4 by working through the argument of §6.1.3 for the case where the dispersion relation takes the form

$$ak^2 + 2bk\omega + c\omega^2 = d,$$

where a, b, c, d are independent of k and ω.

2. Derive the conductivity tensor (6.24) for a streaming plasma by considering a coordinate frame S' moving with the stream, and transforming the relation $j'_i = \sigma'_{ij}E'_j$, in which σ'_{ij} is the known conductivity tensor (at the appropriate frequency) of a stationary plasma.

3. A uniform plasma stream travels with speed U through a stationary plasma. Show that the dispersion relation for a longitudinal wave propagated along the stream can be written

$$kU = \omega[1 \pm \sqrt{\nu\omega}/\sqrt{(\omega^2 - \omega_p^2)}],$$

where ω_p is the plasma frequency of the stationary plasma and, for $U^2/c^2 \ll 1$, ν is the ratio of the stream density to the stationary plasma density.

Deduce that, if

$$\nu \ll | (kU/\omega_p)^2 - 1 |,$$

then

$$\omega = kU[1 \mp \sqrt{\nu\omega_p}/\sqrt{(k^2U^2 - \omega_p^2)}].$$

4. Generalize the two-stream, longitudinal-wave, dispersion relation (6.45) to the case of an arbitrary number of parallel streams. Indicate qualitatively the way in which the number of solutions corresponding to instability depends on the magnitude of k.
5. A plasma stream travels with speed U through a homogeneous, isotropic medium of dielectric constant κ. Consider a plane wave propagated at an angle α to the stream. Show that the dispersion relation is either

$$c^2k^2 - \kappa\omega^2 + \omega_p^2 = 0,$$

in which case the wave is purely transverse, or

$$(c^2k^2-\kappa\omega^2+\omega_p^2)[\kappa(\omega-kU\cos\alpha)^2 - (1-U^2/c^2)\omega_p^2] + (\kappa-1)\omega_p^2k^2U^2\sin^2\alpha = 0,$$

where ω_p is the proper plasma frequency of the stream.
6. A plasma stream with speed U travels through a stationary plasma in the direction of a uniform magnetostatic field $\mathbf{B}_0$. Show that for propagation parallel to $\mathbf{B}_0$ the wave is either longitudinal or transverse, and that the dispersion relation in the latter case is

$$c^2k^2 = \omega^2 - \omega_p^2\frac{\omega}{\omega \pm \Omega} - \omega_{ps}^2\frac{\omega - kU}{\omega - kU \pm \sqrt{(1 - U^2/c^2)}\Omega},$$

where ω_p, ω_{ps} are the plasma frequencies of the stationary and streaming plasmas, respectively, and $\Omega = eB_0/m$.

What is the dispersion relation for the longitudinal wave?
7. Show how the dispersion relation in Problem 6 is modified when the perturbation motion of the positive ions is taken into account.

REFERENCES

Part of the account here on the interpretation of instabilities follows that given by

P. Sturrock, *Phys. Rev.* **112,** 1488 (1958); **117,** 1426 (1960).

For a somewhat different approach, see

A. Bers and R. J. Briggs, *Bull. Am. Phys. Soc.* **9**, 304 (1964)

and the book

R. J. Briggs, *Electron Stream Interaction with Plasmas* (Research Monograph No. 29), M.I.T. Press (1964).

Another discussion is given by

R. N. Sudan, *Phys. Fluids* **8**, 1899 (1965).

The earliest papers on the two-stream instability are

J. R. PIERCE, *J. App. Phys.* **19**, 231 (1948).

A. V. HAEFF, *Phys. Rev.* **74**, 1532 (1948).

See also

O. BUNEMAN, *Phys. Rev. Letters* **1,** 8 (1958).

In more recent work the case in which a superposed magnetic field is present has naturally been considered. See, for example,

J. NEUFELD and H. WRIGHT, *Phys. Rev.* **129,** 1489 (1963); **131,** 1395 (1963).

C. F. KNOX, *J. Plasma Phys.* **1,** 1 (1967).

CHAPTER 7

BOLTZMANN'S EQUATION

Up to now, we have considered only the electrodynamics of a single particle, or of a plasma in the approximation in which each species may be regarded as a continuous (and usually pressureless) charged fluid. For the remainder of the book we investigate those properties of plasmas for which the thermal distribution of velocities must be taken into account. We begin by setting up the means of describing this velocity distribution and an equation for calculating its approximate development in time. In doing so, we shall proceed rather intuitively, postponing the detailed discussion of the approximation involved. Later, in Chapter 12, we shall reconsider this question in a much more fundamental way. This order of presentation is not, therefore, the most direct, but it has the advantage that both the physical ideas and the mathematical methods developed in the meantime will themselves be found to be very helpful in the more complex fundamental discussion.

7.1 PHASE SPACE AND LIOUVILLE'S THEOREM

We must, of course, reject the possibility that we could solve simultaneously the equations of motion of a large number of interacting particles; and even if we could the results would be of no practical value. A statistical description, involving a reduction in the amount of information to be handled, is required. When discussing general principles in statistical mechanics one makes use of phase space. This is a space with coordinates $q_1 \ldots q_k, p_1 \ldots p_k,$ where the q's are the coordinates for all the degrees of freedom which the gas has, and the p's are the corresponding momenta. If, for instance, the gas consists of N identical molecules each with s degrees of freedom, $k = Ns$. This space is often called Γ-space, and a point in it corresponds to a single microscopic state for the whole gas. Instead of constructing an orbit in phase space, corresponding to an exact solution of all the equations of motion, one considers a probability density $\rho(q_1 \ldots q_k, p_1 \ldots p_k, t)$, such that $\rho \, dq_1 \ldots dq_k \, dp_1 \ldots dp_k$ is the probability of finding the gas in states represented by the region $(q_1, q_1 + dq_1)$, etc., at time t. This probability density clearly satisfies a conservation equation

$$\frac{\partial \rho}{\partial t} + \sum_{i=1}^{k} \frac{\partial}{\partial q_i}(\rho \dot{q}_i) + \sum_{i=1}^{k} \frac{\partial}{\partial p_i}(\rho \dot{p}_i) = 0, \tag{7.1}$$

by analogy with the equation of continuity in fluid mechanics. Using the Hamiltonian equations

$$\dot{q}_i = \frac{\partial H}{\partial p_i}, \qquad \dot{p}_i = -\frac{\partial H}{\partial q_i},$$

where $H(q_1 \ldots q_k, p_1 \ldots p_k, t)$ is the Hamiltonian for the whole system, we find

$$\frac{\partial \dot{q}_i}{\partial q_i} = \frac{\partial^2 H}{\partial p_i \partial q_i} = -\frac{\partial \dot{p}_i}{\partial p_i},$$

so that (7.1) may be written more simply

$$\frac{\partial \rho}{\partial t} + \sum_i \dot{q}_i \frac{\partial \rho}{\partial q_i} + \sum_i \dot{p}_i \frac{\partial \rho}{\partial p_i} = 0. \tag{7.2}$$

This is also often written

$$\frac{D\rho}{Dt} = 0, \tag{7.2a}$$

or

$$\frac{\partial \rho}{\partial t} + [\rho, H] = 0. \tag{7.2b}$$

In the first of these, D/Dt is the mobile operator in Γ-space, again analogous to that used in fluid mechanics:

$$\frac{D}{Dt} \equiv \frac{\partial}{\partial t} + \sum_i \dot{q}_i \frac{\partial}{\partial q_i} + \sum_i \dot{p}_i \frac{\partial}{\partial p_i}, \tag{7.3}$$

and (7.2a) merely states that regarded as a "fluid" in Γ-space, the probability density is incompressible, or equivalently that ρ is invariant on the orbits of the system in Γ-space. In (7.2b), $[\rho, H]$ is the classical Poisson bracket of ρ and H, and this form provides the natural comparison with the corresponding results for systems obeying quantum mechanics.

Equation (7.2), in any of its forms, is a statement of *Liouville's Theorem*. In the statistical mechanics of systems in thermal equilibrium, its chief importance is in showing that one can without obvious inconsistency ascribe equal *a priori* probability to equal volumes of phase space. For non-equilibrium phenomena it is less immediately useful, for then we are concerned with the development in time of macroscopic quantities. Any attempt actually to solve (7.2) is equivalent to constructing the orbits in Γ-space, as is obvious from (7.2a); this would involve the complete solution of the mechanics of all the particles, and this we have already rejected.

If solutions could be found, and $\rho(q_1 \ldots q_k, p_1 \ldots p_k, t)$ used as a weighting function for calculating the mean values of macroscopic quantities, the process could be regarded as averaging over a collection of replicas of the whole system with a variety of initial conditions. Such a collection is called an "ensemble".

However, Liouville's equation is *in principle* the starting point for all work on the kinetic theory of gases. The methods used in practice are merely devices for approximating to it. This will be made explicit in Chapter 12. A further fundamental point is that it is in the approximation procedures that the phenomenon of macroscopic irreversibility enters.

7.2 THE BOLTZMANN–VLASOV EQUATIONS

A simpler way to describe a gas of many identical point masses is to use the phase space for a *single* particle, often called μ-space, as opposed to the phase space for the whole gas (Γ-space). This is the method normally used in plasma physics, and as Cartesian components are used for most purposes, it has become customary to use the components of velocity in place of the components of momentum. So we introduce a function $f(\mathbf{x}, \mathbf{v}, t)$, where $\mathbf{x}$ is the position and $\mathbf{v}$ the velocity of a particle, such that

$f\, d^3x\, d^3v =$ the number of particles in the range

$$(\mathbf{x}, \mathbf{x} + d\mathbf{x}), (\mathbf{v}, \mathbf{v} + d\mathbf{v}). \tag{7.4}$$

This also obeys a conservation equation analogous to (7.1):

$$\frac{\partial f}{\partial t} + \frac{\partial}{\partial x_i}(v_i f) + \frac{\partial}{\partial v_i}(a_i f) = 0. \tag{7.5}$$

Here the suffixes denote Cartesian components, and summation over $i = 1, 2, 3$, is assumed. $\mathbf{a}(\mathbf{x}, \mathbf{v}, t)$ is the acceleration of a particle which is at $\mathbf{x}$ and has velocity $\mathbf{v}$ at time t. We shall usually have

$$\mathbf{a} = \frac{e}{m}[\mathbf{E}(\mathbf{x}, t) + \mathbf{v} \times \mathbf{B}(\mathbf{x}, t)] \tag{7.6}$$

for a particle of mass m and charge e, in electric field $\mathbf{E}$ and magnetic field $\mathbf{B}$; additional forces such as a gravitational field can readily be included. Since $\partial a_i/\partial v_i \equiv 0$ for any acceleration given by (7.6), we can again simplify (7.5) in the form

$$\frac{\partial f}{\partial t} + v_i \frac{\partial f}{\partial x_i} + a_i \frac{\partial f}{\partial v_i} = 0. \tag{7.7}$$

We shall call this equation the "Boltzmann" equation.

In an exact description, our function $f(\mathbf{x}, \mathbf{v}, t)$ would become a collection of delta-functions, one for each particle, and $\mathbf{a}$ at any particle would (after deducting the self-field of the particle) be that due to all the other particles together with any external apparatus. In this form, (7.7) is merely a restatement of the problem of calculating the orbits of the particles. The statistical aspects of our problem are introduced when we replace this f by a smooth (in fact, differentiable) function, $\bar{f}$. For the present, $\bar{f}$ may be regarded as meaningful only when applied to a volume $d^3x\, d^3v$ of phase space big enough to contain a large number of particles; (7.4) will apply to such a volume, but must now be regarded as a statement that the *mean* or *expected* number of particles (rather than the exact or actual number) is $\bar{f}\, d^3x\, d^3v$. The question then arises whether an equation such as (7.7) is satisfied by $\bar{f}$, and if so what value of $\mathbf{a}$ should be used. Clearly the simplest procedure is to replace the electromagnetic field by similarly smoothed functions $\bar{\mathbf{E}}(\mathbf{x}, t)$ and $\bar{\mathbf{B}}(\mathbf{x}, t)$, and to suppose that (7.6) and (7.7) then apply, so that

$$\frac{\partial \bar{f}}{\partial t} + v_i \frac{\partial \bar{f}}{\partial x_i} + \bar{a}_i \frac{\partial \bar{f}}{\partial v_i} = 0, \tag{7.8}$$

where

$$\bar{\mathbf{a}} = \frac{e}{m}[\bar{\mathbf{E}}(\mathbf{x}, t) + \mathbf{v} \times \bar{\mathbf{B}}(\mathbf{x}, t)]. \tag{7.9}$$

When several species of charged particles are present, we use a distribution function $\bar{f}$ for each species, with a suitable suffix, and there is an equation of the form (7.8) for each.

To make $\bar{\mathbf{E}}$ and $\bar{\mathbf{B}}$ consistent with the sources generated by the plasma itself we must write down Maxwell's equations:

$$\begin{aligned} \operatorname{div} \bar{\mathbf{B}} &= 0 \\ \operatorname{curl} \bar{\mathbf{E}} + \dot{\bar{\mathbf{B}}} &= 0 \\ \varepsilon_0 \operatorname{div} \bar{\mathbf{E}} &= \rho_{\text{ext}} + \sum_s \bar{\rho}_s \\ \frac{1}{\mu_0} \operatorname{curl} \bar{\mathbf{B}} - \varepsilon_0 \dot{\bar{\mathbf{E}}} &= \mathbf{j}_{\text{ext}} + \sum_s \bar{\mathbf{j}}_s. \end{aligned} \tag{7.10}$$

Here ρ_{ext} and $\mathbf{j}_{\text{ext}}$ are the charge and current due to external apparatus (if any); s enumerates the species, and

$$\begin{aligned} \bar{\rho}_s &= e_s \int \bar{f}_s\, d^3v \\ \bar{\mathbf{j}}_s &= e_s \int \mathbf{v} \bar{f}_s\, d^3v, \end{aligned} \tag{7.11}$$

in obvious notation. The use of $\bar{\mathbf{E}}$ in (7.9) without any correction due to polarization effects has already been discussed in §1.3.

We now have a closed system of equations, (7.8)–(7.11), often known collectively as the "Vlasov equations", for treating the particle motions and the electromagnetic field in a self-consistent way. We must, however, recognize that in applying equation (7.8) to our "smoothed-out" distribution functions and components of electromagnetic field, an approximation is being made. The force acting on any one particle is assumed to be a continuous and slowly varying function of space, representing the effect of all the other particles. The vast majority of all these particles may indeed be so represented, but there will always be a few, close to the one under observation, whose fields are varying rather rapidly in the relevant region of physical space. Events involving such very close particles are of course collisions, and equation (7.8) is often called the "collisionless Boltzmann equation". We may use it with good approximation only if the collective effect of the distant particles is much greater than any systematic effect arising from the collisions involving one, or a small number, of the nearby particles. Encouraged by the rough calculations made earlier (§1.2) we shall assume that this procedure is indeed valid, leaving the detailed investigation until Chapter 12. From now on we shall write $\bar{f}$, $\bar{\mathbf{E}}$, $\bar{\mathbf{B}}$, etc., as f, $\mathbf{E}$, $\mathbf{B}$, etc., and regard them as continuous and definite functions of their arguments, so that the statistical aspects disappear. Physically, this is equivalent (see §12.4.3) to subdividing (or "pulverizing") the particles into many smaller ones, in such a way that e/m remains fixed, and regarding f as the density of a continuous fluid in phase space.

To improve on this approximation, we may write formally

$$\frac{\partial f}{\partial t} + v_i \frac{\partial f}{\partial x_i} + a_i \frac{\partial f}{\partial v_i} = \left(\frac{\partial f}{\partial t}\right)_c. \tag{7.12}$$

Here, the right-hand side represents any residual contribution, due to collisions, to the rate of change of f. Several expressions for $(\partial f/\partial t)_c$ suitable for plasmas have been proposed, and will be mentioned later; meanwhile we shall concentrate on developing the collision-free theory, using (7.8).

It should be emphasized that for gases consisting only of *neutral* molecules, the situation is quite different. In that case it is the influence of the distant particles which may be neglected in comparison with that of nearby particles, to such an extent that no mutual interactions need be included in $\mathbf{a}$. In practice, the only contribution to $\mathbf{a}$ is the external gravitational field, if any. Further, only binary collisions need to be included in $(\partial f/\partial t)_c$. These facts all follow from the short-range nature of the van der Waals forces between pairs of neutral molecules. Thus the orbit of each particle may be regarded as a succession of free orbits punctuated by very brief "encounters", each involving only one other particle; the velocity changes abruptly at each encounter. An integral form for $(\partial f/\partial t)_c$ can then be set up. To do this, the dynamics of binary collisions must be worked out from the law of interaction, and an assumption about the statistics of the collisions (hypothesis of molecular chaos) must be made. Except in very special circumstances (such as the interior of shock waves), the term $(\partial f/\partial t)_c$ dominates in (7.12), so that f differs only slightly from the solution of $(\partial f/\partial t)_c = 0$, namely the Maxwellian distribution. Equation (7.12) may thus be solved by a technique of successive iteration (see §11.5, and Chapman and Cowling, 1952).

All the remarks made in the previous paragraph become untrue when particles interact with an inverse square law. Binary collisions are not then the dominant effect, and if one uses the equations and methods developed for neutral gases, the results which emerge contain divergent integrals, and so appear to be meaningless. Actually, results of the right order of magnitude can often be obtained by "cutting off" these divergent integrals, but this is really quite arbitrary. On the other hand, there are several phenomena, not present in collision-dominated gases, arising from the collective effect of the many distant particles; here the collision-free theory is quite adequate.

7.3 PLASMA IN THERMAL EQUILIBRIUM

Consider a uniform infinite plasma, in which the distribution functions f_i, f_e for ions (singly charged) and electrons are each independent of position in physical space, and of time, and with no electromagnetic field, so that $\mathbf{E} = \mathbf{B} = 0$. Maxwell's equations are completely satisfied provided the charge and current densities for the two species are equal and opposite. Thus

$$\left.\begin{aligned} \int f_i\, d^3v &= \int f_e\, d^3v = N, \text{ say,} \\ \text{and} \qquad \int \mathbf{v} f_i d^3v &= \int \mathbf{v} f_e\, d^3v = N\mathbf{u}, \text{ say,} \end{aligned}\right\} \tag{7.13}$$

so that N is the density, $\mathbf{u}$ the drift velocity, for both species. The collision-free Boltzmann equations are trivially satisfied. Thus any distribution functions satisfying (7.13) give a possible steady state.

Within the collision-free approximation there is no basis for choosing, from among all these equilibrium states, one which is to be regarded as representing "thermal equilibrium" (except that some of these equilibria will later (Chapter 10) be found to be unstable; however, there is a large class of stable states). We shall nevertheless refer to a particular distribution as the state of "thermal equilibrium". This is, of course, the Maxwellian

$$f(\mathbf{x}, \mathbf{v}, t) = N\left[\frac{m}{2\pi KT}\right]^{3/2} \exp\left[-m(\mathbf{v} - \mathbf{u})^2/2KT\right], \tag{7.14}$$

where T is the temperature, K is Boltzmann's constant, and N and $\mathbf{u}$ have the same meaning as in (7.13). That this *is* the distribution appropriate for thermal equilibrium can be proved by means of the corrections to Boltzmann's equation to account for collisions, or by considering the Gibbs distribution in Γ-space. The elementary method of the most probable state would, however, lead one to expect this result. Meanwhile, the particular attention paid to the Maxwellian distribution may, if desired, be regarded merely as a way of providing convenient examples.

We are ignoring the possibility that some of the ions and electrons may recombine to form neutral atoms or molecules. A more complete discussion of thermal equilibrium would include consideration of the chemical equilibrium between the ionized and neutral gas, described by the Saha equation (see, for instance, Fay, 1965). Here we take the plasma to be fully and permanently ionized.

As a generalization of these results, a source-free magnetic field $\mathbf{B}(\mathbf{x})$ may be included, and (7.14), with $\mathbf{u} = 0$, continues to be the equilibrium distribution. The proof of this is left to the reader.

7.4 HYDRODYNAMIC VARIABLES AND EQUATIONS

We give here an introductory account of the relation between Boltzmann's equation and the hydrodynamic description of a plasma, so that the terminology and ideas involved are available as early as possible; the matter will be taken up in more detail later (Chapter 11).

For each species, and with a general distribution function $f(\mathbf{x}, \mathbf{v}, t)$, we define the *number density* and the *hydrodynamic* (*Eulerian*) *velocity* by

$$N(\mathbf{x}, t) = \int f(\mathbf{x}, \mathbf{v}, t)d^3v \tag{7.15}$$

and

$$\mathbf{u}(\mathbf{x}, t) = \frac{1}{N}\int \mathbf{v}f(\mathbf{x}, \mathbf{v}, t)d^3v. \tag{7.16}$$

Starting with Boltzmann's equation in the form (7.5):

$$\frac{\partial f}{\partial t} + \frac{\partial}{\partial x_j}(v_j f) + \frac{\partial}{\partial v_j}(a_j f) = 0, \tag{7.17}$$

and integrating over velocity space, we find

$$\frac{\partial N}{\partial t} + \frac{\partial}{\partial x_j}(Nu_j) = -\int \frac{\partial}{\partial v_j}(a_j f)d^3v = 0, \tag{7.18}$$

where in the last step we make use of the divergence theorem in velocity space and assume that $\mathbf{a}f \to 0$ as $\mathbf{v} \to \infty$, sufficiently rapidly that the integral over the "sphere at infinity" vanishes. Equation (7.18) is just the normal equation of continuity in physical space. It also ensures the continuity of charge for each species.

We define the *pressure tensor*,* p_{ij}, by

$$p_{ij}(\mathbf{x}, t) = \int m(v_i - u_i)(v_j - u_j) f(\mathbf{x}, \mathbf{v}, t) d^3v \tag{7.19}$$

$$= m \int v_i v_j f \, d^3v - mNu_i u_j. \qquad \text{(by (7.16))}$$

Now multiply (7.17) by v_i and integrate over velocity space. The first term is $\partial(Nu_i)/\partial t$. The second is

$$\frac{\partial}{\partial x_j} \int v_i v_j f \, d^3v = \frac{\partial}{\partial x_j}\left[\frac{1}{m} p_{ij} + Nu_i u_j\right]. \qquad \text{(by (7.19))}$$

The third is

$$\int v_i \frac{\partial}{\partial v_j}(a_j f) d^3v = -\int a_i f \, d^3v = -\frac{1}{m} F_i \quad \text{(say)},$$

where in the first step we carry out an integration by parts (again making use of conditions at infinity in velocity space) and the second step merely serves to define $\mathbf{F}$, the total body force per unit volume. Collecting these results and making use of (7.18) we have

$$\frac{\partial u_i}{\partial t} + u_j \frac{\partial u_i}{\partial x_j} = \frac{1}{Nm}\left(-\frac{\partial p_{ij}}{\partial x_j} + F_i\right). \tag{7.20}$$

This is the equation of momentum, and the left-hand side is the usual convective acceleration, Du_i/Dt. In the presence of electromagnetic and gravitational fields,

$$\mathbf{a} = \frac{e}{m}[\mathbf{E} + \mathbf{v} \times \mathbf{B}] + \mathbf{g},$$

where $\mathbf{g}$ is the acceleration due to gravity, and so

$$\mathbf{F} = m \int \mathbf{a} f \, d^3v = Ne(\mathbf{E} + \mathbf{u} \times \mathbf{B}) + Nm\mathbf{g} = \rho\mathbf{E} + \mathbf{j} \times \mathbf{B} + Nm\mathbf{g} \tag{7.21}$$

where ρ and $\mathbf{j}$ are as defined in (7.11).

Although we have deduced continuity and momentum equations for each species, we cannot claim to have produced a complete set of equations for magnetohydrodynamics, as p_{ij} is known only in terms of f. By multiplying Boltzmann's equation by two components of $\mathbf{v}$, one can obtain (see Problem 2) a somewhat complicated equation for $\partial p_{ij}/\partial t$, but only at the

* This is the *negative* of the *stress tensor* as usually defined in fluid mechanics or elasticity; it is, however, the quantity generally used in kinetic theory.

expense of introducing still more unknown quantities such as the *heat flow tensor*

$$Q_{ijk} = m\int (v_i - u_i)(v_j - u_j)(v_k - u_k) f \, d^3v. \tag{7.22}$$

The complete elimination of f in favor of a finite set of variables such as N, u_i, p_{ij}, etc., depending on space and time only, for which a closed set of equations are known, is not in general possible. Approximations which make it possible in particular cases will be explained later (Chapter 11); the considerable success achieved by applying some form of magnetohydrodynamics to problems concerning plasmas can be attributed to the validity of one or other of these approximations.

Finally, we define a "local temperature", $T(\mathbf{x}, t)$ by

$$3NKT = p_{jj} = m\int |\mathbf{v} - \mathbf{u}|^2 f \, d^3v. \tag{7.23}$$

The energy density of the random translational motion is thus $\frac{3}{2}NKT$, as usual.

The reader will easily verify that for the Maxwellian distribution with drift, the quantities N, $\mathbf{u}$, T introduced in (7.14) are consistent with those defined in the present section, and that the pressure tensor is isotropic, so that $p_{ij} = p\delta_{ij}$ where $p = NKT$ and δ_{ij} is the usual Kronecker delta.

7.5 THE SOLUTION OF BOLTZMANN'S EQUATION IN GIVEN FIELDS: JEANS'S THEOREM

The Vlasov equations, (7.8)–(7.11), form a closed set of non-linear integro-differential equations which with suitable initial and boundary conditions would in principle handle any problem of collision-free plasma dynamics. In practice the difficulties are formidable, and it is useful to consider first the more restricted problem of solving the Boltzmann equation for one species when the electromagnetic field (or other external field) is given, so postponing the question of making this field consistent with the sources produced by the plasma. Then we have to solve

$$\frac{\partial f}{\partial t} + v_i \frac{\partial f}{\partial x_i} + a_i \frac{\partial f}{\partial v_i} = 0, \tag{7.8}$$

with $\mathbf{a}(\mathbf{x}, \mathbf{v}, t)$ *given* since $\mathbf{E}(\mathbf{x}, t)$, $\mathbf{B}(\mathbf{x}, t)$ are given. This is a linear first-order homogeneous equation, and we can apply Lagrange's method (see R. Courant, 1962). We first solve the associated ordinary differential equations

$$\frac{dt}{1} = \frac{dx_1}{v_1} = \frac{dx_2}{v_2} = \frac{dx_3}{v_3} = \frac{dv_1}{a_1} = \frac{dv_2}{a_2} = \frac{dv_3}{a_3} \tag{7.24}$$

to construct the characteristic curves in the 7-dimensional space $(\mathbf{x}, \mathbf{v}, t)$. But these equations are $d\mathbf{x}/dt = \mathbf{v}$, $d\mathbf{v}/dt = \mathbf{a}$, i.e. the equations for the orbit in phase space of a particle subject to the given fields. Suppose the general solution is

$$\mathbf{x} = \mathbf{x}(\alpha_1, \ldots, \alpha_6, t), \qquad \mathbf{v} = \mathbf{v}(\alpha_1, \ldots, \alpha_6, t), \tag{7.25}$$

where $\alpha_1, \ldots, \alpha_6$ are six constants of integration. Then it must be possible to solve (7.25) in the form

$$\alpha_j = \alpha_j(\mathbf{x}, \mathbf{v}, t), \qquad (j = 1, \ldots, 6); \tag{7.26}$$

in other words it must be possible to discover the values of α_j which label the particular orbit passing through $\mathbf{x}$ with velocity $\mathbf{v}$ at time t. According to Lagrange's method the general solution of (7.8) is

$$f(\mathbf{x}, \mathbf{v}, t) = H(\alpha_1, \ldots, \alpha_6), \tag{7.27}$$

where H is an arbitrary function and the α_j's are to be eliminated by (7.26). (Actually H must of course be such that f is everywhere non-negative and satisfies the necessary conditions as $\mathbf{v} \to \infty$.) To verify that (7.27) is a solution, we write

$$\begin{aligned}\frac{\partial f}{\partial t} + v_j \frac{\partial f}{\partial x_j} + a_j \frac{\partial f}{\partial v_j} &= \sum_i \frac{\partial H}{\partial \alpha_i}\left[\frac{\partial \alpha_i}{\partial t} + v_j \frac{\partial \alpha_i}{\partial x_j} + a_j \frac{\partial \alpha_i}{\partial v_j}\right] \\ &= \sum_i \frac{\partial H}{\partial \alpha_i}\left[\frac{d\alpha_i}{dt}\right]_{\text{orbit}} = 0.\end{aligned}$$

Here $[d\alpha_i/dt]_{\text{orbit}}$ means the rate of change of α_i along the orbit given by (7.25); but since each α_i is constant on any orbit, this rate of change is zero. In the same notation, we can write (7.8) as

$$\left[\frac{df}{dt}\right]_{\text{orbit}} = 0. \tag{7.28}$$

To summarize, we may say that the orbits in phase space, expressed in terms of the time as a parameter, give the characteristic curves of Boltzmann's equation for given electromagnetic fields, and the solutions of the equation are obtained by writing f as an arbitrary function of the invariants of the orbits. These results are sometimes called "Jeans's Theorem".

In practice it may be quite difficult to obtain the general solution (7.25), but easy to find one or two invariants of the motion, such as the total energy in the case of an electrostatic field. In these circumstances, (7.27) often provides useful solutions even though H depends on fewer than six constants; but in such a case the most general solution of Boltzmann's equation has not been found. The next section contains a few examples of this kind. Another device worth noting is that of using approximate invariants to obtain solutions of Boltzmann's equation valid to whatever is the appropriate accuracy. In particular the first adiabatic invariant (see §4.3.2), and even the second and third invariants (§4.4.5), are available for this purpose.

7.6 EXAMPLES OF JEANS'S THEOREM

7.6.1 Field-free plasma

A trivial example of Jeans's theorem arises when we assume that there are no fields, so $\mathbf{E} = \mathbf{B} = 0$. The motion of a particle is then given by

$$\mathbf{x} = \mathbf{v}t + \mathbf{x}_0, \tag{7.29}$$

where $\mathbf{x}_0$ and $\mathbf{v}$ are constant vectors. This is the form taken by equations (7.25) in the present case, so we may take the six constants $\alpha_1, \ldots, \alpha_6$ to be simply the components of $\mathbf{x}_0$ and $\mathbf{v}$. Then solving as in (7.26), three of the α's are simply $\mathbf{v}$ itself, and the other three are given by $\mathbf{x}_0 = \mathbf{x} - \mathbf{v}t$. Any function

$$H(\mathbf{v}, \mathbf{x} - \mathbf{v}t) \tag{7.30}$$

gives a solution of Boltzmann's equation. When we apply this procedure to both the ions and electrons, and then ask how the total charge density and current density could be made zero to satisfy Maxwell's equations, we find of course that H in (7.30) may depend only on $\mathbf{v}$ and not on $\mathbf{x} - \mathbf{v}t$. The conditions become those of (7.13).

7.6.2 Uniform plasma in an external magnetic field

Suppose now we allow the magnetic field $\mathbf{B}$ to be non-zero, but source free and time independent. The orbits become more complicated than as shown in (7.29) and $\mathbf{v}$ is not now constant. But its magnitude, v, is of course constant, so we may take the Boltzmann function to be any $f(v)$, that is we may take any spatially uniform isotropic distribution no matter how complicated the magnetic field. Since f is isotropic, there is no current. The only remaining condition is that the ions and electrons have equal number density, to provide for charge neutrality.

7.6.3 Electrostatic fields; a model sheath

We look for solutions in which there is an electrostatic field:

$$\mathbf{E}(\mathbf{x}) = -\nabla\phi(\mathbf{x}). \tag{7.31}$$

(A source-free magnetic field may also be present, but as in §7.6.2 it does not affect the calculation.) The orbits have as an invariant the total energy

$$W = \tfrac{1}{2}mv^2 + e\phi, \tag{7.32}$$

where e is the charge. Boltzmann's equation is satisfied by taking f to be any non-negative function of W, but let us restrict attention to solutions representing thermal equilibrium at temperature T;

$$f(\mathbf{x}, \mathbf{v}) = \text{constant} \times \exp(-W/KT). \tag{7.33}$$

The distribution is then Maxwellian everywhere and necessarily positive. The number density is

$$N(\mathbf{x}) = N_0 \exp(-e\phi/KT), \tag{7.34}$$

where N_0 is constant. This is sometimes called the "scale height formula". (The density of an atmosphere under gravity obeys a similar law with $e\phi$ replaced by the potential energy mgz: the "scale height" is the characteristic length KT/mg.)

For ions and electrons with charges $\pm e$ we have

$$N_i = N_0 \exp(-e\phi/KT), \qquad N_e = N_0 \exp(e\phi/KT). \tag{7.35}$$

(N_0 may be taken the same for each by choice of the reference potential $\phi = 0$.)

A complete self-consistent solution will result if we can satisfy Poisson's equation

$$\nabla^2\phi = -\frac{e}{\varepsilon_0}(N_i - N_e) = \frac{2eN_0}{\varepsilon_0}\sinh(e\phi/KT),$$

i.e.

$$\nabla^2(e\phi/KT) = \lambda^2 \sinh(e\phi/KT), \tag{7.36}$$

where

$$\lambda^2 = 2N_0e^2/\varepsilon_0 KT.$$

We have a characteristic potential KT/e, and the characteristic length is $\lambda^{-1} = h/\sqrt{2}$ where h is the Debye length corresponding to density N_0.

General solution of (7.36) presents some difficulty, but in the one-dimensional case there exists an explicit solution. Writing $\psi = e\phi/KT$ and $\xi = \lambda x$, with no variations in the y- and z- directions, we have

$$\frac{d^2\psi}{d\xi^2} = \sinh\psi.$$

Integrating,

$$\frac{1}{2}\left(\frac{d\psi}{d\xi}\right)^2 = \cosh\psi + \text{constant}.$$

If the plasma extends to infinity in some direction, and is in equilibrium there, so that $N_i = N_e$ and $\psi = 0$, we must have $d\psi/d\xi = 0$ there also; hence the constant is -1 in this case, and so

$$\left(\frac{d\psi}{d\xi}\right)^2 = 2(\cosh\psi - 1) = 4\sinh^2 \tfrac{1}{2}\psi.$$

Choosing the undisturbed region to be at $x \to +\infty$, we have to take the negative sign for the square root:

$$\frac{d\psi}{d\xi} = -2 \sinh \tfrac{1}{2}\psi,$$

and integrating again,

$$\xi = -\log \tanh (\tfrac{1}{4}\psi),$$

the constant of integration being made zero by choice of the origin. Thus finally,

$$\psi = 4 \tanh^{-1} [\exp (-\xi)],$$

so

$$\phi = \frac{4KT}{e} \tanh^{-1} [\exp (-\lambda x)]. \tag{7.37}$$

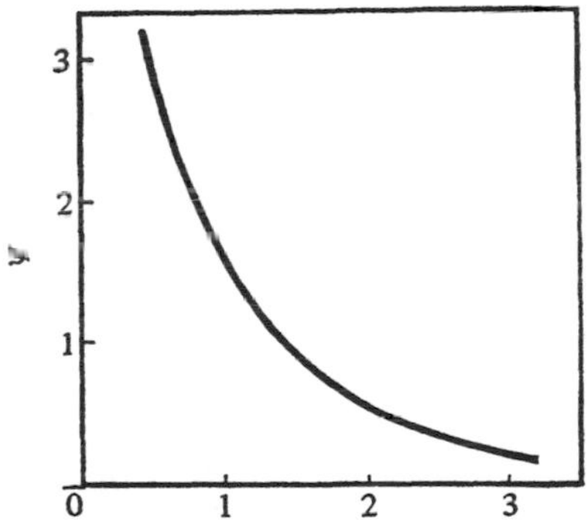

Fig. 7.1. Potential for a model sheath.

This potential is plotted, using the normalized variables, in Fig. 7.1. It has a singularity at the origin and does not exist for $x < 0$. For a physical solution, we may take any $x_1 > 0$ and regard the plasma as occupying the space $x > x_1$. At the plane $x = x_1$ there is a perfectly reflecting electrode raised to potential $\phi(x_1)$. Our calculation shows that the plasma is disturbed by the formation of a "sheath", of characteristic thickness λ^{-1}; beyond this the plasma is virtually undisturbed. In practice, a metallic electrode is not perfectly reflecting and the theory of formation of the sheath is more complicated, though the situation is qualitatively similar.

7.6.4 The Debye–Hückel theory

To find other solutions of (7.36), we resort to approximation by supposing that the disturbance is everywhere small, so that $\phi \ll KT/e$, and replace $\sinh (e\phi/KT)$ by $e\phi/KT$. Then we have the linear "Debye–Hückel equation"

$$\nabla^2\phi = \lambda^2\phi. \tag{7.38}$$

Solutions of this are readily obtained by the methods used for Laplace's equation and the wave equation. For example, a radially symmetric solution in polar coordinates is

$$\phi = \frac{q}{4\pi\varepsilon_0 r} \exp(-\lambda r), \tag{7.39}$$

where q is a constant. This represents the disturbance caused by a charge q embedded in the plasma at the origin, and again we notice that it is screened by the plasma, so that the potential falls off more rapidly than the usual Coulomb potential, with a characteristic distance of about one Debye length. Very close to the embedded charge, (7.39) becomes invalid on account of the linearization.

We have explained that in describing plasmas by means of the collision-free Boltzmann equation the particles are replaced by a continuous fluid in phase space, so that any effects arising from their discrete nature are lost. We can gain some insight into the approximation involved by fixing attention on one particle, called a "test particle", which retains its discrete nature, while all the other particles, "field particles", are treated by the collision-free equations. This is in general very complicated, but in the special case when the test particle is at rest the calculation is simply the one just carried out, with $q = e$, the elementary charge. If, for instance, a test electron is at rest at the origin, the distribution function for the field electrons is modified close to the origin as given by equations (7.33) and (7.39). Physically, the other electrons are repelled by the test electron, but this is effective only within a sphere of radius about λ^{-1}, owing to the shielding. (The reader will easily verify that in this case the linearization holds good for distances as small as $r = N^{-1/3}$, that is, up to the nearest field particle. This follows from the inequality $\alpha \ll 1$, where $\alpha = N_0^{1/3}e^2/(4\pi\varepsilon_0 KT)$, discussed in §1.2.)

Qualitatively similar results can be obtained for the case of a slowly moving particle, but the disturbance is no longer spherically symmetric. For a fast (suprathermal) particle the shielding disappears and is replaced by an "electrostatic wake" following the particle. These cases can only be treated by a much more sophisticated analysis.

It is sometimes said that any electron is surrounded by a cloud of positive charge, and any positive ion by negative charge, so as to screen the rest of the plasma. This is perhaps a little misleading as it is hard to imagine how it can be true for all the particles simultaneously. In fact, for a plasma in equilibrium, the distribution functions for the ions and electrons are uniform in physical space, as we know from earlier sections. The distribution function calculated in the present section is that for the field particles *given* the position and velocity of the test particle which we have singled out. We could call it a *conditional* distribution function by analogy with conditional probabilities. The formal theory of these various types of distribution function will be given in Chapter 12.

The particles substantially influenced by any one particle (assumed not suprathermal) are clearly those within its Debye sphere, with radius λ^{-1}. These are numerous (this again follows from $\alpha \ll 1$), and this is the fundamental reason why it is inadequate to regard the interactions between ions and electrons as a succession of binary collisions.

If we expand the exponential in (7.39) we find that close to a test particle with charge e

$$\phi \approx \frac{e}{4\pi\varepsilon_0}\left[\frac{1}{r} - \lambda + O(r)\right].$$

Hence the potential energy of the charge (omitting the infinite self-energy) is $-\lambda e^2/4\pi\varepsilon_0$. We can assume this is the characteristic potential energy for most of the particles, even if they are not at rest, and we see that it is negative whatever the sign of the charge. We find a typical potential energy $-\lambda N_0 e^2/4\pi\varepsilon_0$ per unit volume,* i.e. $-2^{1/2}N_0^{3/2}e^3/[4\pi\varepsilon_0^{3/2}(KT)^{1/2}]$, using the definition of λ. We can also write this $2^{3/2}\pi^{1/2}N_0KT\alpha^{3/2}$ where α is as previously. Since $\alpha \ll 1$ this potential energy is much smaller in magnitude than the kinetic energy $3N_0KT$; accordingly any corrections to the internal energy and the equation of state, representing departures from those of a perfect gas, are small (see Problem 4).

7.6.5 Momentum and angular momentum

Besides energy, momentum and angular momentum may sometimes be used as constants of the orbit. When there is a magnetic field, the generalized momenta must be used. Suppose the electromagnetic field is represented by scalar and vector potentials:

$$\mathbf{E} = -\nabla\phi - \dot{\mathbf{A}}, \quad \mathbf{B} = \text{curl}\,\mathbf{A}. \tag{7.40}$$

Constants of the motion are most readily obtained from the Lagrangian, equation (4.7)

$$\mathscr{L} = \tfrac{1}{2}mv^2 + e\mathbf{A}\cdot\mathbf{v} - e\phi. \tag{7.41}$$

---

* This step includes a factor $\frac{1}{2}$ arising from the fact that the mutual potential energy of a set of particles is counted twice if one simply sums the energies of the particles, but a further factor 2 as there are two species each with density N_0.

Forming Lagrange's equations in cartesian coordinates we find

$$\frac{d}{dt}\frac{\partial \mathscr{L}}{\partial v_x} = \frac{\partial \mathscr{L}}{\partial x},$$

i.e.

$$\frac{dp_x}{dt} = e\left(\mathbf{v}\cdot\frac{\partial \mathbf{A}}{\partial x} - \frac{\partial \phi}{\partial x}\right)$$

with two similar equations. Here $\mathbf{p} = m\mathbf{v} + e\mathbf{A}$ is the generalized momentum. Thus if $\mathbf{A}$ and ϕ are independent of x, p_x is invariant.

Similarly, writing down Lagrange's equations in cylindrical polar coordinates (r, θ, z) we find that if $\mathbf{A}$, ϕ are independent of θ, the generalized angular momentum $L = mr^2\dot{\theta} + erA_\theta$ is invariant.

There are several instances in which a component of $\mathbf{p}$, or L, have been used in Jeans's theorem, with interesting results. Some details of these appear in Problems 5–9. As in the model sheath given above, the possibility of success depends upon noticing rather fortuitous solutions of non-linear equations. The solutions include models of a self-contracting current sheet, cylindrical pinched plasmas carrying either axial or azimuthal currents, and a plasma–magnetic field interface. In each case the plasma is maintained in a non uniform state by its own magnetic field, there being no space charge or electric field. Although instructive, these models are of little practical importance as they actually represent highly unstable configurations.

7.7 A RELATIVISTIC FORMULATION OF BOLTZMANN'S EQUATION

The discussion of Boltzmann's equation so far has been non-relativistic in the sense that the Lorentz force per unit mass was identified, as in (7.6), with the Newtonian acceleration. It is in fact not difficult to allow for the dynamics of special relativity, and we now describe one way in which this can be done.

In place of the velocity $\mathbf{v}$ we introduce the "reduced" velocity

$$\mathbf{u} = \frac{\mathbf{v}/c}{\sqrt{(1 - v^2/c^2)}}, \tag{7.42}$$

which is the space part of the contravariant four-vector (2.157), namely

$$(\sqrt{(1 + u^2)}, \mathbf{u}). \tag{7.43}$$

The inverse relation to (7.42) is

$$\mathbf{v} = c\mathbf{u}/\sqrt{(1 + u^2)}. \tag{7.44}$$

We continue to denote the distribution function by f, but it is now regarded as a function of $\mathbf{x}$, $\mathbf{u}$ and t, and defined by the statement that

$$f(\mathbf{x}, \mathbf{u}, t)d^3x\, d^3u \tag{7.45}$$

is the number of particles in the range $\mathbf{x}$ to $\mathbf{x} + d\mathbf{x}$ with reduced velocities between $\mathbf{u}$ and $\mathbf{u} + d\mathbf{u}$. The counterpart of equation (7.5) is then

$$\frac{\partial f}{\partial t} + \frac{\partial}{\partial x_i}(v_i f) + \frac{\partial}{\partial u_i}(A_i f) = 0, \tag{7.46}$$

with summation over $i = 1, 2, 3$ assumed, where

$$A_i = du_i/dt; \tag{7.47}$$

and the relativistic equation of motion (2.187) shows that the counterpart of (7.6) is

$$\mathbf{A} = \frac{e}{mc}[\mathbf{E}(\mathbf{x}, t) + \mathbf{v} \times \mathbf{B}(\mathbf{x}, t)], \tag{7.48}$$

m being the rest mass. It is readily confirmed by substituting for $\mathbf{v}$ from (7.44) into (7.48) that $\partial A_i/\partial u_i = 0$, so the required relativistic equation may be written

$$\sqrt{(1 + u^2)}\frac{\partial f}{\partial t} + cu_i\frac{\partial f}{\partial x_i} + \sqrt{(1 + u^2)}A_i\frac{\partial f}{\partial u_i} = 0, \tag{7.49}$$

where for electromagnetic forces

$$\mathbf{A} = \frac{e}{mc}\left(\mathbf{E} + \frac{c}{\sqrt{(1 + u^2)}}\mathbf{u} \times \mathbf{B}\right). \tag{7.50}$$

It is instructive to confirm the Lorentz invariance of (7.49) explicitly.

First we show that f itself is an invariant. In a frame of reference moving with a particular group of particles the number of particles with coordinates and reduced velocity components within infinitesimal ranges is written

$$f_0\, d^3x_0\, d^3u_0.$$

The same number is also expressed by (7.45), and since (cf. (2.158))

$$d^3x_0 = \sqrt{(1 + u^2)}d^3x$$

it follows that

$$fd^3u = f_0\sqrt{(1 + u^2)}d^3u_0. \tag{7.51}$$

We need to relate the volume elements in u-space for two frames of reference, say S and S', which we do by computing the Jacobian of the transformation. If S' moves relative to S with uniform velocity V in the x_1-direction, the Lorentz transformation applied to the four-vector (7.43) gives (cf. (2.151))

$$u_1 = \frac{u_1' + (V/c)\sqrt{(1 + u'^2)}}{\sqrt{(1 - V^2/c^2)}}, \qquad u_2 = u_2',$$

$$u_3 = u_3', \qquad \sqrt{(1 + u^2)} = \frac{\sqrt{(1 + u'^2)} + (V/c)u_1'}{\sqrt{(1 - V^2/c^2)}}. \tag{7.52}$$

Hence

$$\frac{\partial u_1}{\partial u_1'} = \frac{1}{\sqrt{(1 - V^2/c^2)}}\left[1 + \frac{V}{c}\frac{u_1'}{\sqrt{(1 + u'^2)}}\right] = \sqrt{\left(\frac{1 + u^2}{1 + u'^2}\right)},$$

and since

$$\frac{\partial u_2}{\partial u_2'} = \frac{\partial u_3}{\partial u_3'} = 1, \qquad \frac{\partial u_j}{\partial u_i'} = 0, (i \neq j)$$

evidently

$$d^3u = \sqrt{\left(\frac{1 + u^2}{1 + u'^2}\right)}\, d^3u'.$$

Hence in (7.51) we have

$$d^3u = \sqrt{(1 + u^2)}d^3u_0,$$

which establishes that $f = f_0$.

Since f is an invariant

$$\left(\frac{\partial f}{\partial t}, c\frac{\partial f}{\partial x_i}\right) \tag{7.53}$$

is a covariant four-vector. The first two terms of (7.49) thus together form the scalar product of two four-vectors, (7.43) and (7.53), which is an invariant.

It remains to show that the last term of (7.49) is an invariant. To do this we note that, as in (2.186),

$$(\mathbf{A}\cdot\mathbf{u},\ \sqrt{(1 + u^2)}\mathbf{A})$$

is a contravariant four-vector, so that

$$\begin{aligned}\sqrt{(1 + u'^2)}A_1' &= \frac{\sqrt{(1 + u^2)}A_1 - (V/c)\mathbf{A}\cdot\mathbf{u}}{\sqrt{(1 - V^2/c^2)}},\\ \sqrt{(1 + u'^2)}A_i' &= \sqrt{(1 + u^2)}A_i \quad \text{for} \quad i = 2, 3.\end{aligned} \tag{7.54}$$

Also, from

$$\frac{\partial f}{\partial u_i'} = \frac{\partial f}{\partial u_j}\frac{\partial u_j}{\partial u_i'}$$

and the expressions for $\partial u_j/\partial u_i'$, it appears that

$$\begin{aligned}\frac{\partial f}{\partial u_1'} &= \sqrt{\left(\frac{1 + u^2}{1 + u'^2}\right)}\frac{\partial f}{\partial u_1},\\ \frac{\partial f}{\partial u_i'} &= \frac{V/c}{\sqrt{(1 - V^2/c^2)}}\frac{u_i'}{\sqrt{(1 + u'^2)}}\frac{\partial f}{\partial u_1} + \frac{\partial f}{\partial u_i} \text{ for } i = 2, 3.\end{aligned} \tag{7.55}$$

A little algebra, aided by the last equation of (7.52) with $u \leftrightarrow u'$, $V \to -V$, now confirms that the combination of (7.54) and (7.55) gives

$$\sqrt{(1+u'^2)}A'_i \frac{\partial f}{\partial u'_i} = \sqrt{(1+u^2)}A_i \frac{\partial f}{\partial u_i}.$$

PROBLEMS

1. One species of a gas is in the uniform Maxwellian distribution
$$f(\mathbf{x}, \mathbf{v}, t) = N\left[\frac{m}{2\pi KT}\right]^{3/2} \exp\left[-m(\mathbf{v}-\mathbf{u})^2/2KT\right].$$
Show that equations (7.15), (7.16), (7.19) and (7.23) give the correct density, drift velocity, pressure tensor and temperature, and that the heat-flow tensor, (7.22), vanishes.
2. Obtain the equation for $\partial p_{ij}/\partial t$ mentioned in §7.4. (See (11.19) for the answer.)
3. What are the one and two dimensional analogues of the approximate solution (7.39), and when would they arise?
4. According to §7.6.4, the internal energy density in a plasma in thermal equilibrium is
$$u(N, T) = 3NKT\left[1 - \frac{(8\pi)^{\frac{1}{2}}}{3}\alpha^{3/2}\right],$$
where α is small and is proportional to $N^{1/3}T^{-1}$. According to thermodynamics, the pressure $p(T, V)$ of a gas satisfies
$$\frac{\partial}{\partial T}\left(\frac{p}{T}\right) = \frac{1}{T^2}\frac{\partial U}{\partial V},$$
where U and V are the internal energy and volume of a *fixed mass* of the gas. Use these results to show that the corrected equation of state for the plasma is
$$p = 2NKT\left[1 - \frac{(2\pi)^{\frac{1}{2}}}{3}\alpha^{3/2}\right].$$
5. (See §7.6.5) A plasma is situated in the static magnetic field given by the vector potential $\mathbf{A} = (0, 0, A)$, where A is independent of z, and there is no electric field. Show that for each species
$$f \propto \exp\left\{-\frac{W - up_z}{KT}\right\}$$
is a solution of Boltzmann's equation, where W is the energy $\frac{1}{2}mv^2$ and $p_z = mv_z \pm eA$; interpret the constants u and T. Show that, on combining the ions and electrons, the plasma may be made everywhere neutral provided
$$\frac{u_i}{T_i} + \frac{u_e}{T_e} = 0.$$
Taking $T_i = T_e = T$ and $u_i = -u_e = u$, show that the particle density for each species may be written
$$N(x, y) = N_0 \exp(euA/KT),$$

where N_0 is constant, and that the solution is self-consistent provided

$$\nabla^2\psi + 2\lambda^2 \exp(\psi) = 0,$$

where $\psi = euA/KT$ and $\lambda^2 = \mu_0 e^2 u^2 N_0/KT$.

6. Show that, in plane geometry, with A depending only on x, the above equation is satisfied by

$$\psi = -2 \log \cosh \lambda x.$$

Calculate the corresponding magnetic field, plasma density and current. (This solution represents a plane current sheet.)

7. Show that, in cylindrical geometry, with A depending only on r, the equation obtained in Problem 5 is satisfied by

$$\psi = -2 \log(1 + \tfrac{1}{4}\lambda^2 r^2).$$

Calculate the corresponding magnetic field, plasma density and current. (This solution, given by W. H. Bennett in 1934, represents what is now usually called a "z-pinch", i.e. a self-confined axial current.)

8. Working in a way analogous to Problem 5, show that in cylindrical polar co-ordinates, with $A_\theta(r)$ the only non-vanishing component of **A**, Boltzmann's equation is satisfied if

$$f \propto \exp\left\{-\frac{W - \Omega L}{KT}\right\},$$

where $L = mrv_\theta \pm erA_\theta$ and Ω is a constant to be interpreted. Show that for neutrality one must take

$$\frac{T_e}{T_i} = -\frac{\Omega_e}{\Omega_i} = \gamma,$$

where γ is the mass ratio m_i/m_e, giving

$$N(r) = N_0 \exp\left\{\frac{e\Omega_i r A_\theta + \frac{1}{2}m_i r^2 \Omega_i^2}{KT_i}\right\}.$$

Show that for a self-consistent solution

$$\frac{d^2\psi}{d\xi^2} + 2\lambda^2 \exp[\psi + \delta\xi] = 0,$$

where

$$\psi = e\Omega_i r A_\theta/KT_i, \qquad \xi = r^2, \qquad \lambda^2 = -\mu_0 N_0 e^2 \Omega_i^2 (1 + \gamma)/2KT_i$$

and

$$\delta = m_i \Omega_i^2/2KT_i.$$

Show finally that this equation is satisfied by

$$\psi = -\delta\xi - 2 \log \cosh[\lambda(\xi - \xi_0)]$$

where ξ_0 is a constant. Calculate the corresponding magnetic field and plasma properties. (This gives a model of "θ-pinch". It should be noted that

$$[\nabla^2 \mathbf{A}]_\theta = \frac{d}{dr}\left[\frac{1}{r}\frac{d}{dr}(rA_\theta)\right]$$

for a vector field of the type involved here.)

9. Using the same notation and assumptions as in Problem 5, show that self-consistent solutions may be set up with A depending only on x, and

$$f \propto \exp\left\{-\frac{W + \lambda p_z^2/2m}{KT}\right\},$$

where the constant λ is different for the two species. In particular the boundary layer between a uniform magnetic field and a field-free hot plasma can be discussed in this way, though the plasma has to be slightly anisotropic. (See Lam (1967) for details.)

REFERENCES

Liouville's theorem and its role in statistical mechanics are discussed in many standard texts on the subject, for instance

J. E. MAYER and M. G. MAYER, *Statistical Mechanics*, Wiley (1940).

Boltzmann collected his researches on kinetic theory into a treatise which has recently been republished in an English translation:

L. BOLTZMANN, *Lectures On Gas Theory*, translated, and with an Introduction, by S. G. Brush, University of California (1964).

The basic properties of the Boltzmann equation are given in detail in the classic text

S. CHAPMAN and T. G. COWLING, *The Mathematical Theory of Non-Uniform Gases*, 2nd edn, Cambridge University Press (1952).

For Lagrange's method in partial differential equations:

R. COURANT, *Methods of Mathematical Physics*, vol. 2, p. 62, Interscience (1962).

For the relativistic Boltzmann equation, and a relativistic counterpart of Problem 7, see

O. BUNEMAN, in *Plasma Physics*, Chapter 7, ed. J. E. Drummond, McGraw-Hill (1961).

For details of the solution suggested in Problem 9, see

S. H. LAM, *Phys. Fluids*, **10**, 2454 (1967).

For the application of Jeans's theorem to stellar dynamics, see

S. CHANDRASEKHAR, *Principles of Stellar Dynamics*, p. 85. Dover Publications (1960).

For the Saha equation

J. A. FAY, *Molecular Thermodynamics*, Addison-Wesley (1965).

CHAPTER 8

WAVES IN IONIZED GASES: KINETIC THEORY

8.1 INTRODUCTION

In this chapter we resume our study, begun in Chapter 5, of waves of small amplitude propagating in an infinite plasma. Our starting point will be the Vlasov equation introduced in Chapter 7, and attention will be directed to the new effects arising from the introduction of a thermal spread in the particle velocities. We could, of course, proceed directly to the most general case, but as in our previous work we shall find it more instructive to begin with a simple problem (longitudinal plasma oscillations without magnetic field) containing many of the essential features of the subject. More complicated cases are then readily treated despite the rather involved analysis. But first we write down general equations.

The collision-free Boltzmann equation, in vector form, is

$$\frac{\partial f}{\partial t} + \mathbf{v}\cdot\frac{\partial f}{\partial \mathbf{x}} + \frac{e}{m}(\mathbf{E} + \mathbf{v} \times \mathbf{B})\cdot\frac{\partial f}{\partial \mathbf{v}} = 0 \tag{8.1}$$

(see (7.8) and (7.9); the charge is taken as $+e$). In the unperturbed state we shall have, for each species, $f = f_0$ throughout infinite physical space, where the functions f_0 depend only on $\mathbf{v}$, and the total charge and current densities are zero; $\mathbf{E} = 0$ and $\mathbf{B} = \mathbf{B}_0$ where $\mathbf{B}_0$ is a uniform magnetic field. Thus (8.1) requires

$$(\mathbf{v} \times \mathbf{B}_0)\cdot\frac{\partial f_0}{\partial \mathbf{v}} = 0. \tag{8.2}$$

If $\mathbf{B}_0 \neq 0$, f_0 must be symmetric about the magnetic field; for instance if $\mathbf{B}_0$ is along the z-axis, f_0 depends only on $v_x^2 + v_y^2$ and v_z. There is no condition on f_0 if $\mathbf{B}_0 = 0$ (except for the requirement that the total charge and current vanish).

During the perturbations we have

$$f = f_0 + f_1, \qquad \mathbf{B} = \mathbf{B}_0 + \mathbf{B}_1, \tag{8.3}$$

where f_1, $\mathbf{B}_1$ and the electric field $\mathbf{E}$ are small perturbations. The linearized equation obtained from (8.1) is

$$\frac{\partial f_1}{\partial t} + \mathbf{v}\cdot\frac{\partial f_1}{\partial \mathbf{x}} + \frac{e}{m}(\mathbf{v} \times \mathbf{B}_0)\cdot\frac{\partial f_1}{\partial \mathbf{v}} + \frac{e}{m}(\mathbf{E} + \mathbf{v} \times \mathbf{B}_1)\cdot\frac{\partial f_0}{\partial \mathbf{v}} = 0. \tag{8.4}$$

Our problem is to solve this equation (or, rather, such an equation for each species) combined with Maxwell's equations. We may take our perturbations to be harmonic in space, and so proportional to exp $(-i\mathbf{k}\cdot\mathbf{x})$. The wave number $\mathbf{k}$ will be regarded as real throughout; however most of the analysis holds for complex $\mathbf{k}$ provided the plasma is suitably restricted in space. The time dependence will also be harmonic, proportional to exp $(i\omega t)$, for much of the work. For some purposes we shall find it more convenient to use the Laplace transform with respect to time. The two representations are related essentially by the transformation $p \to i\omega$, where p is the variable introduced by the Laplace transform. So we have, effectively,

$$f_1, \mathbf{E}, \mathbf{B}_1, \propto e^{i(\omega t - \mathbf{k}\cdot\mathbf{x})}, \tag{8.5}$$

and we shall, as usual, omit the harmonic factor in most formulas.

In the present chapter we restrict attention to the case when the applied field $\mathbf{B}_0$ is zero, postponing to Chapter 9 the study of the effects of such a field.

8.2 THE RESPONSE OF A PLASMA TO LONGITUDINAL FIELDS

Throughout this and the following three sections we shall be concerned only with "longitudinal" fields and motion.* So $\mathbf{E}_1$ and $\mathbf{k}$ are parallel, say along the x-axis. There is no magnetic field, either constant or oscillating, so $\mathbf{B}_0 = \mathbf{B}_1 = 0$. Thus the y- and z-components of velocity for each particle remain constant, and the distribution of such velocities is irrelevant. Formally, we can integrate (8.1) with respect to v_y and v_z and deal only with the distribution of longitudinal velocity

$$f_{\text{long}}(x, v_x, t) = \iint f\, dv_y\, dv_z. \tag{8.6}$$

Our problem is purely one dimensional, and for the present we may omit the suffix "long", with similar interpretation for f_0 and f_1. Equation (8.4) thus becomes

$$\frac{\partial f_1}{\partial t} + v\frac{\partial f_1}{\partial x} = -\frac{e}{m}E\frac{df_0}{dv}. \tag{8.7}$$

---

* Though it is not very apparent at this stage, some approximation is involved in the procedure used here. For sufficiently asymmetric distributions the space charge produced may be convected perpendicular to $\mathbf{k}$ by the velocity components v_y and v_z, so causing transverse currents. These lead to transverse components of $\mathbf{B}_1$, which in turn require transverse components of $\mathbf{E}_1$ by the law of electromagnetic induction. The error is, however, very small (see §8.6).

This equation enables us to find the effect on the particles of the perturbing field E. Poisson's equation would tell us how the charge density produced by the particles gives rise to electric fields such as E, and combining the two equations (summing over species if need be) we could find a self-consistent solution representing free oscillations of the plasma; this would be just an extension of the theory of plasma oscillations already discussed in §5.2.2 for the cold plasma. Because of subtleties which will shortly appear, we shall find it prudent to concentrate first on (8.7) without regard for Poisson's equation; that is we regard E as a given forcing field, the right-hand side of (8.7) being therefore known.

Using (8.5) we have

$$i(\omega - kv)f_1 = -\frac{e}{m}E\frac{df_0}{dv},$$

with solution

$$f_1 = \frac{ie}{m}E\frac{df_0/dv}{\omega - kv}. \tag{8.8}$$

This, with its harmonic factor reintroduced, is certainly a particular integral of (8.7), but it is possible that a complementary function, that is, a solution of the field-free equation

$$\frac{\partial f_1}{\partial t} + v\frac{\partial f_1}{\partial x} = 0 \tag{8.9}$$

may have to be added. The solution of (8.9) is a very simple case of Jeans's theorem (see §7.6.1) and is

$$f_1(x, v, t) = H(vt - x, v), \tag{8.10}$$

where H is an arbitrary real function of its arguments. It represents nothing more than the convection of a disturbance by particles moving freely with constant velocity. So the general solution of (8.7) is

$$f_1(x, v, t) = \frac{ie}{m}E\frac{df_0/dv}{\omega - kv}e^{i(\omega t - kx)} + H(vt - x, v), \tag{8.11}$$

where H is determined by initial conditions. For electrodynamic purposes, we are not really interested in f_1 itself, but only in the charge and current densities

$$\rho = e\int f_1\, dv, \qquad j = e\int vf_1\, dv. \tag{8.12}$$

With a general choice of the function H, we should find that ρ and j were not themselves proportional to exp $[i(\omega t - kx)]$, and we could not complete our program of making the fields and sources consistent (except for the peculiar possibility that the non-harmonic part cancels on summing over species). In asking for the "response" of the species to a harmonic driving field, we have in mind only harmonic values of the quantities (8.12). One obvious possibility is to set $H = 0$ in (8.11), but as we shall see shortly it is not only one, and it is not always the correct one.

Before investigating further, it may be well to refer to a more familiar and closely analogous situation in which a similar difficulty arises. In Chapter 2 we calculated the electromagnetic field due to a given source, for instance a single moving particle. We noticed an ambiguity in the result, whereby the difference between any two possible solutions was simply a source-free field, in other words a complementary function. However, we selected a particular solution, the retarded potential, as the only correct one. The fundamental reason for this is still a matter for lively discussion, leading to cosmological and other very general questions; however we shall be content here to say that the choice should always be made with reference to the principle of causality. This asserts that for a stable system a "response," that is, the quantity to be calculated, must always *follow* the "source," that is the given or inhomogeneous term. The electromagnetic field produced by a moving charge appears after the motion, and similarly in our present problem the charge and current densities as given by (8.12) must follow the applied E. (The principle of special relativity implies the stronger statement that if the source and response are observed at points of finite spatial separation, the response appears a finite time later, namely the time of transit of light; this need not concern us at present however.)

In the linear case it is often convenient to handle all quantities by Fourier analysis, and it is not then quite so clear how to incorporate the principle of causality. Suppose we make H of (8.11) proportional to the magnitude of E and choose its form so that ρ and j are indeed proportional to $\exp[i(\omega t - kx)]$ Then ρ and j will be proportional to E; say

$$\rho = \eta(\omega, k)E \quad \text{and} \quad j = \sigma(\omega, k)E, \tag{8.13}$$

where η and σ may be complex. The principle of causality imposes on the functions η and σ certain analytic conditions; these merely ensure that however we superpose harmonic solutions to form a real disturbance, the response always follows the source. For example if the source is zero for all $t < 0$, so must be the response. To make the correct choice one regards η and σ as functions of complex ω and k. Following the discussion of §2.1.5, we first take k to be real but ω complex and in the lower half plane. Then the quantities proportional to $\exp(i\omega t)$ oscillate but also grow in time; we can imagine the disturbing field switched on in the infinite past, when the plasma was at rest, and growing to its present value, the problem being treated by spatial Fourier analysis if need be. For this case we must choose H to be zero, otherwise f_1 would not be identically zero as $t \to -\infty$. We then have a causal solution. The resulting values of η and σ will be analytic functions of

ω (otherwise we could not apply Fourier techniques) and there will be no poles when ω is in the lower half plane, as they would indicate instability. To find η and σ for any other values of ω and k we extend these results by analytic continuation.

This process gives the response coefficients η and σ consistent with the principle of causality. A mathematical expression of this, which could be used to test any proposed form for the coefficients, is that they satisfy the Kramers–Kronig relations;* see for instance Landau and Lifshitz (1958), §122.

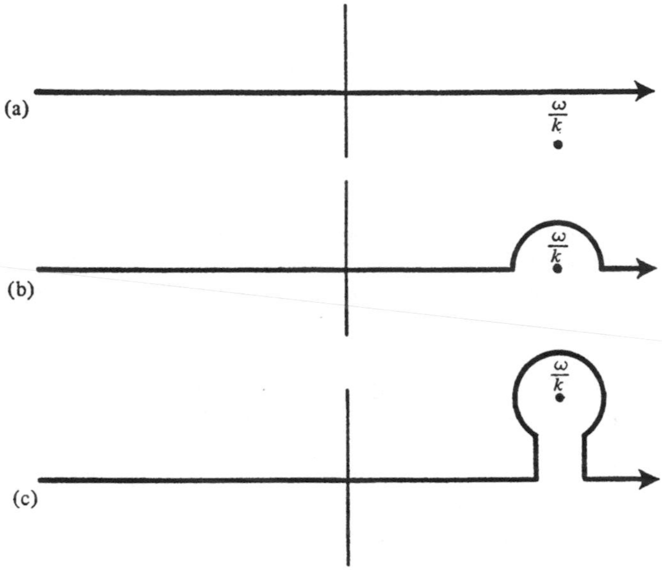

Fig. 8.1. The Landau contour.

To follow this procedure, we resume our calculations at equation (8.11). With $\text{Im}\,(\omega) < 0$ and k real we set $H = 0$. Combining this with (8.12) and (8.13) we find

$$\eta = \frac{ie^2}{m}\int_{-\infty}^{\infty} \frac{df_0/dv}{\omega - kv}\,dv,$$
$$\sigma = \frac{ie^2}{m}\int_{-\infty}^{\infty} \frac{vdf_0/dv}{\omega - kv}\,dv. \tag{8.14}$$

The second of these we naturally call the conductivity; the first is perhaps a little less familiar and has no established name. They are, of course, related by $\omega\eta = k\sigma$ on account of the conservation of the particles. In each of (8.14) the integrand has a singularity at $v = \omega/k$; however, this need not worry us

* Sometimes called Bode's Theorem in texts on circuit theory.

as it lies below the path of integration, as shown in Fig. 8.1(a). (We take $k > 0$ without loss of generality.) Now let the imaginary part of ω tend to zero. The situation becomes that of Fig. 8.1(b), and to preserve analyticity of η and σ the path of integration has to be indented upwards as shown. If we make Im $(\omega) > 0$ (representing a decaying field) the path becomes that shown in Fig. 8.1(c). We will refer to these contours collectively by the symbol C. The way in which the choice of the path of integration depends on Im (ω) is sometimes called the "Landau prescription." Provided df_0/dv can be regarded as an analytic function of complex v, this will automatically give us the analytic continuation required. (If not, our rule still stands, but it is more difficult to construct the required analytic continuation; we shall not consider such cases here.)

When ω is real, or in the upper half plane, our integrals follow the paths of Figs. 8.1(b) and (c), but we can write the results in the form

$$\begin{aligned} \int_C &= P\int_{-\infty}^{\infty} - \pi i \,(\text{residue at } v = \omega/k), \ (\omega \text{ real}) \\ \int_C &= \int_{-\infty}^{\infty} - 2\pi i \,(\text{residue at } v = \omega/k). \ (\text{Im}\,(\omega) > 0), \end{aligned} \tag{8.15}$$

In the first of these, the integral along the real axis is to be interpreted as the principal part, in view of the singularity at $v = \omega/k$. One may also write symbolically

$$\frac{1}{\omega - kv} = P\frac{1}{\omega - kv} + i\pi\delta(\omega - kv), \ (\omega \text{ real}). \tag{8.15a}$$

We can then reinterpret the contributions from the residues as due to a non-zero choice of H in (8.11), whilst the integrals along the real axis come from the particular integral (8.8). To see what this choice of H is, we remark that if H is to be proportional to exp $(-ikx)$, it must be

$$H(vt - x, v) = e^{ik(vt-x)}\, h(v), \tag{8.16}$$

where h is arbitrary. If ω/k is real, comparison with (8.15a) shows that the choice

$$h(v) = A\,\delta(\omega - kv)$$

is appropriate, where

$$A = -\frac{\pi E e}{m}\left[\frac{df_0}{dv}\right]_{\omega/k},$$

so we have a disturbance confined to the particles with velocity ω/k. When ω is in the upper half plane, the choice of $h(v)$ is not quite so obvious but will emerge from the theory later (§8.4); in that case h is not a δ-function, and H

is not as a whole proportional to exp $(i\omega t)$. Its temporal behavior is in fact like exp $(ikvt)$, so is harmonic but at a frequency depending on v. Nevertheless the contributions to (8.12) add up to make ρ and j proportional to exp $(i\omega t)$; these are therefore harmonic but damped. Incidentally the form of (8.16) shows explicitly why it is necessary to set $H \equiv 0$ to make the complementary function vanish as $t \to -\infty$.

One other representation of the response to longitudinal fields may be noted here as it is commonly used. Poisson's equation is

$$\varepsilon_0 \operatorname{div} \mathbf{E} = \rho_{\text{plasma}} + \rho_{\text{ext}},$$

where $\rho_{\text{plasma}} = \eta E$ is the quantity we have just been discussing and ρ_{ext} is the charge due to any other ("external") source. We may rewrite this

$$\operatorname{div} \mathbf{D} = \rho_{\text{ext}},$$

where $\mathbf{D} = \varepsilon_0 \varepsilon \mathbf{E}$ and (assuming all quantities spatially harmonic) the "dielectric constant" is the dimensionless quantity*

$$\varepsilon(\omega, k) = 1 + \eta/ik\varepsilon_0 \quad (= 1 + \sigma/i\omega\varepsilon_0). \tag{8.17}$$

(If necessary η is of course summed over species.) In this nomenclature one is regarding the charges of the plasma as giving rise to a polarization rather than a current; the analogy is then with a dielectric instead of a conducting medium.

The complications which we have discussed in this section are closely related to the singularity appearing in our expressions when $\omega = kv$. Physically this arises whenever there are particles whose velocity v is the same as the wave velocity. Returning for a moment to the three-dimensional formulation, the condition would be $\omega = \mathbf{k} \cdot \mathbf{v}$, which states that the particle velocity, resolved along the wave normal, is equal to the speed of the wave. We recognize this as the condition for Čerenkov radiation (§3.2). Particles and waves satisfying it have a strong resonant interaction with one another. Just as in the case of Čerenkov emission, we need to be very careful when we allow parameters to change continuously through the resonant condition; the mathematical ambiguity which enters has to be settled by considerations of causality. The result is an irreversible behavior even though the governing equations (and the physics they represent) are reversible. This will be explicit in the next section, where we find the dispersion relation for plasma oscillations.

---

* This quantity and its tensor generalization (see (8.50)) will be denoted by ε throughout the remainder of the book, though in Chapters 5 and 6 (see equation 5.47) we used κ.

8.3 PLASMA OSCILLATIONS AND LANDAU DAMPING

8.3.1 The dispersion relation

We now impose the condition that the field E, hitherto assumed given, arises solely from the space charge of the plasma itself. This will give us a self-consistent theory of a plasma oscillating freely under its own internal forces. For longitudinal motion in one dimension, we need only Poisson's equation

$$-\varepsilon_0 ikE = \rho. \tag{8.18}$$

Here ρ is the total space charge due to all the species present, but for simplicity we take it to be the charge of the electrons only. The ions, being much heavier, can with good approximation be supposed to be at rest and serve only to neutralize the static charge of the electrons. The corrections due to ion-motion will be considered in §8.8. Then as $\rho = \eta E$ by (8.13), we have $\eta = -\varepsilon_0 ik$, or in full

$$\frac{e^2}{\varepsilon_0 mk}\int_C \frac{df_0/dv}{\omega - kv}\,dv = -1. \tag{8.19}$$

Here C denotes a contour chosen according to the Landau prescription, i.e. the real axis if Im $(\omega) < 0$, but the indented contour if Im $(\omega) \geqslant 0$. Equation (8.19) may also be written $\varepsilon(\omega, k) = 0$, by virtue of (8.17).

Normally we shall think of k as fixed (and conventionally positive) and solve the dispersion relation (8.19) for ω. Once $f_0(v)$ is specified this is possible in principle, though often awkward in practice as (8.19) involves ω and k implicitly. The first question we might ask is whether there are solutions with ω in the lower half plane. These would represent growing oscillations, and the plasma would be inherently unstable, as the slightest perturbation would eventually grow to large amplitude (at which point our linearized theory would become inadequate). It is easy to see that there must be cases where this does happen. In the case of two cold interpenetrating plasma streams, we have already noted (§6.3.1) that plasma oscillations become unstable. In the present language, such a plasma would have f_0 composed of two δ-functions situated at the streaming velocities; by broadening the streams slightly we could make f_0 a continuous distribution which would still be unstable.

Later (§10.2) we shall use (8.19) to find a necessary and sufficient condition for a plasma with distribution f_0 to be stable against such instabilities; for the present we will be content with a sufficient condition which can be obtained by a very simple argument. The sufficient condition is that $f_0(v)$ should have not more than one maximum (remember that it is necessarily positive and tends to zero as $v \to \pm\infty$, so certainly has at least one maximum). To show this, let us work in a frame of reference in which the maximum is at $v = 0$, i.e. $f_0'(0) = 0$. Then $f_0'(v)$ has the same sign as $-v$. If

Im $(\omega) < 0$, our integration is simply along the real axis, and one finds that each contribution to the integrand of (8.19) lies in the shaded region of the complex plane as shown in Fig. 8.2. Such contributions cannot possibly add up to a negative real number; therefore there are no growing solutions in that case. (It is interesting to note that if the integral of (8.19) were to be calculated along the real axis when Im $(\omega) > 0$ also, a similar argument would show that there are no damped solutions either; hence ω would have to be real and the singularity fall on the path of integration. However, this impasse is avoided by the Landau prescription.)

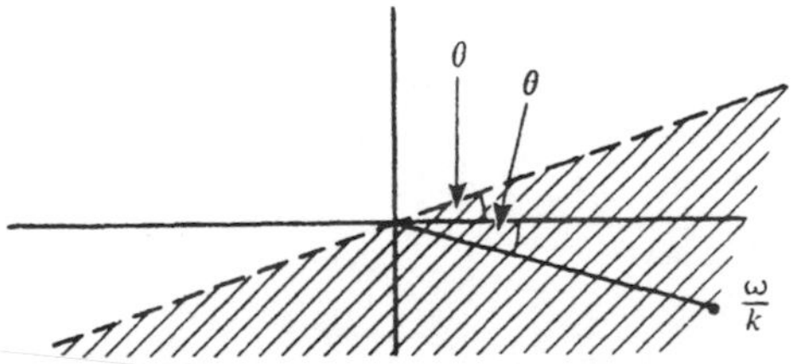

Figure 8.2

Our conclusion is that the distributions which we might call "single humped," of which one obvious example is the Maxwellian distribution, are certainly stable. For them, solutions must be sought with the contour C as in Figs. 8.1(b) or (c), and Im $(\omega) \geqslant 0$.

8.3.2 Long waves

A second general question we can ask about the dispersion relation (8.19) is this: in what circumstances does the cold plasma theory (§5.2.2) provide a good approximation? And in such cases, can we derive the next order correction, in which the thermal spread is regarded as small? Clearly we must restrict attention to cases in which $f_0(v)$ is stable, and for convenience we may work in the rest-frame of the electrons, so that

$$\int_{-\infty}^{\infty} v f_0(v)\, dv = 0. \tag{8.20}$$

We rewrite (8.19) as $D(\omega, k) = 0$, where

$$D(\omega, k) \equiv \int_C \frac{df_0/dv}{(\omega/k) - v}\, dv + \frac{\varepsilon_0 m k^2}{e^2}. \tag{8.21}$$

For a cold plasma, the dispersion relation is simply $\omega = \omega_p$ where ω_p is the plasma frequency. Suppose now we introduce the thermal spread, a typical thermal velocity being v_t. We may take it that for $|v| \gg v_t$, $f_0(v)$ is extremely small. With the thermal spread confined in this way, we expect it to be unimportant if the wave speed ω/k is much greater than v_t, so that nearly all the particles are practically at rest. As ω will then be approximately ω_p, we expect the condition $\omega_p/k \gg v_t$, i.e. $k \ll \omega_p/v_t$ to be the appropriate one. So the wavelength has to be sufficiently long, and the critical length v_t/ω_p is of course roughly the Debye length.

Let us assume that although $f_0(v)$ decreases rapidly and monotonically when $|v|$ exceeds v_t it does not actually reach zero until $v \to \pm\infty$. The same will then be true of df_0/dv. Then there can be no solutions with ω real; for in that case the principal part (see 8.15) gives a real contribution to (8.21) and the residue an imaginary one. So we assume that $\omega = \omega_r + i\gamma$ where $\omega_r - \omega_p$ and γ are small corrections to the real and imaginary parts of ω respectively; these are to be obtained by iteration. An *a priori* assumption about their relative magnitudes is required; in the most interesting case the relation $\gamma \ll |\omega_r - \omega_p|$ holds and is verified at the end. For the first correction we can therefore still ignore the residue and solve $D=0$ with the principal part only. Expanding the integrand, and integrating by parts

$$P\int_{-\infty}^{\infty} \frac{df_0/dv}{(\omega/k)-v}\,dv = \frac{k}{\omega}\int_{-\infty}^{\infty}\left[1+\frac{kv}{\omega}+\left(\frac{kv}{\omega}\right)^2+\dots\right]\frac{df_0}{dv}\,dv$$

$$= -\frac{k^2}{\omega^2}\int_{-\infty}^{\infty} f_0\left[1+2\frac{kv}{\omega}+3\left(\frac{kv}{\omega}\right)^2+\dots\right]dv.$$

Integrating term by term noting (8.20) and the definitions (see §7.4)

$$N = \int f_0\,dv, \qquad NKT = \int mv^2 f_0\,dv$$

we find

$$D(\omega, k) \approx \frac{\varepsilon_0 mk^2}{e^2} - \frac{Nk^2}{\omega^2}\left[1+\frac{3k^2KT}{m\omega^2}+\dots\right]. \tag{8.22}$$

(It should be noted that the temperature, T, introduced here is that for the longitudinal motions and may differ from the overall temperature if the distribution is anisotropic.) The expansion of $[(\omega/k) - v]^{-1}$ is valid only for $|v| < \omega/k$; but we are assuming that nearly all the particles lie in the range $|v| < v_t \ll \omega/k$, so that we have committed a small error only in respect of the few particles whose speeds exceed ω/k. If $f_0(v) \to 0$ as $|v| \to \infty$ faster than any power of v^{-1} (which it must if all the integrals arising in the term by term integration are to converge), it can be shown that (8.22) is an asymptotic expansion in powers of ω^{-1}. In the lowest approximation the equation $D(\omega, k) = 0$ reduces to $\omega^2 = Ne^2/\varepsilon_0 m = \omega_p^2$, as expected. In the next approximation we can put $\omega = \omega_p$ in the correcting term, so giving

$$\omega^2 = \omega_p^2 + 3KTk^2/m = \omega_p^2(1 + 3h^2k^2), \tag{8.23}$$

where h is the Debye length $(KT/m\omega_p^2)^{\frac{1}{2}}$. This process can be continued so long as we are satisfied about the neglect of the residue; at every stage ω is purely real.

Now suppose that we wish to include the residue at some stage in this procedure. We assume that the correction $i\gamma$ in ω is smaller than the real corrections calculated so far (otherwise the work would already be inconsistent). So ω_r is already known to sufficient accuracy. Evaluating $D(\omega, k)$ properly at $\omega = \omega_r$ gives

$$D(\omega_r, k) = PD(\omega_r, k) + i\pi\left[\frac{df_0}{dv}\right]_{v=\omega_r/k}.$$

So

$$D(\omega_r + i\gamma, k) = PD(\omega_r, k) + i\pi\left[\frac{df_0}{dv}\right]_{v=\omega_r/k} + i\gamma\left[\frac{\partial D}{\partial \omega}\right]_{\omega=\omega_r}. \tag{8.24}$$

Writing $D \equiv D_r + iD_i$, the last term becomes

$$\gamma\left[\frac{\partial D_r}{\partial \gamma} + i\frac{\partial D_i}{\partial \gamma}\right]_{\omega=\omega_r}.$$

The real part of this gives a further slight correction to ω_r, which we have agreed to neglect; the imaginary part must cancel the residue term in (8.24) if $D = 0$ is still to hold at the corrected frequency. By the Cauchy–Riemann equations, $\partial D_i/\partial\gamma = \partial D_r/\partial\omega_r$ as D is an analytic function. Hence

$$\gamma = -\pi\left[\frac{df_0}{dv}\right]_{v=\omega_r/k}\bigg/\frac{\partial D_r}{\partial \omega_r}. \tag{8.25}$$

But for real ω, D_r is just $P(D)$, so we can calculate $\partial D_r/\partial\omega_r$ as follows. Let $D_r = 0$ define $\omega_r(k)$; this is just the quantity for which we have calculated successive approximations such as (8.23). Differentiating this relation,

$$\frac{d\omega_r}{dk}\frac{\partial D_r}{\partial \omega_r} + \frac{\partial D_r}{\partial k} = 0.$$

But from the form of D, (8.21),

$$\frac{\partial D_r}{\partial k} = -\frac{\omega_r}{k}\frac{\partial D_r}{\partial \omega_r} + \frac{2\varepsilon_0 mk}{e^2}.$$

Thus

$$\left(\frac{d\omega_r}{dk} - \frac{\omega_r}{k}\right)\frac{\partial D_r}{\partial \omega_r} = -\frac{2\varepsilon_0 mk}{e^2}$$

and (8.25) becomes

$$\gamma = -\frac{\pi e^2}{2\varepsilon_0 mk}\left[\frac{df_0}{dv}\right]_{v=\omega_r/k}\left(\frac{\omega_r}{k} - \frac{d\omega_r}{dk}\right). \tag{8.26}$$

In the lowest approximation ω_r is simply ω_p and $d\omega_r/dk$ is zero; the result is then that of Landau. As γ is positive (df_0/dv being negative) our wave contains the factor $\exp(-\gamma t)$, and so is damped. This is called "Landau damping." Its physical nature is discussed in other ways below. For the present we may remark that the irreversible behavior which it exhibits springs not from the governing equations (which are reversible) but from our choice of complementary function discussed in the previous section. It is at this point that the time is treated asymmetrically.

When (8.23) is used for $\omega_r(k)$, the final factor of (8.26) becomes approximately $(\omega_p/k)(1 - \frac{3}{2}h^2k^2)$, and df_0/dv must be evaluated at a slightly corrected value of ω_r, namely $\omega_p(1 + \frac{3}{2}h^2k^2)$; some correction of Landau's formula is thus involved.

For the Maxwellian distribution, one finds

$$\gamma = \left(\frac{\pi}{8}\right)^{1/2} \frac{\omega_p}{h^3k^3} \exp\left(-\frac{1}{2h^2k^2} - \frac{3}{2}\right), \tag{8.26a}$$

the factor $\exp(-3/2)$ being the correction just mentioned. As $(KT/m)^{1/2}$ plays the part of v_t in our qualitative discussion, the condition $k \ll \omega_p/v_t$ becomes $hk \ll 1$. The first correction to ω_r (8.23) is in fact of order h^2k^2, and on continuing the procedure one would get a power series. On the other hand, (8.26a) cannot be expanded as a power series in hk; for the function $x^{-3}\exp(-1/x^2)$ tends to zero as $x \to 0$ and is continuous as a function of x, yet has no Taylor series about $x = 0$. For the Maxwellian distribution, hk being sufficiently small, the corrections to ω_r are more important than γ, as we anticipated. The same would be true of any distribution with the property that $f_0(v) \to 0$ as $|v| \to \infty$ faster than any power of v, i.e. a distribution with all moments finite. For such a plasma, any attempt to generate the corrections to $\omega(k)$ as a power series for small k (or, equivalently, small temperature) must overlook altogether the Landau damping.

More detailed results for the Maxwellian distributions will be given later (§8.7).

8.4 THE INITIAL VALUE PROBLEM

8.4.1 Formal solution

As an alternative approach to the theory of the two previous sections, we outline the original approach of Landau. In order to deal automatically with the difficulty of the choice of complementary function for the Boltzmann equation which we discussed in §8.2, Landau considered the full solution of an initial-value problem, using the Laplace transform. If $h(t)$ is any function of time, its Laplace transform is

$$\bar{h}(p) = \int_0^\infty e^{-pt}\, h(t)\, dt, \tag{8.27}$$

provided this exists. The inversion formula is

$$h(t) = \frac{1}{2\pi i}\int_S e^{pt}\,\bar{h}(p)\,dp. \tag{8.28}$$

Here S denotes the standard Laplace inversion contour in the p-plane. It goes from $-i\infty$ to $+i\infty$ along a path such that all singularities of $\bar{h}(p)$ lie to the left, and $\bar{h}(p)$ (defined originally for Re (p) sufficiently large and positive) is continued analytically, if necessary, to the region Re $(p) < 0$.

We have to solve the linearized Boltzmann equation (8.7), and Poisson's equation (8.18), simultaneously for f_1 and E. Let us assume spatial dependence proportional to $\exp(-ikx)$ but make no prior assumption about the dependence on time. Let $f_1(v, t)$ be some given function $g(v)$ at $t = 0$, and apply the Laplace transformation to our equations. We obtain

$$(p - ikv)\bar{f}_1 + \frac{e}{m}\bar{E}\frac{df_0}{dv} = g(v) \tag{8.29}$$

and

$$-\varepsilon_0 ik\bar{E} = e\int_{-\infty}^{\infty}\bar{f}_1\,dv. \tag{8.30}$$

Solving (8.29) for $\bar{f}_1$,

$$\bar{f}_1(v, p) = \frac{1}{p - ikv}\left\{g(v) - \frac{e}{m}\bar{E}\frac{df_0}{dv}\right\}, \tag{8.31}$$

and inserting this into (8.30) and solving for $\bar{E}$

$$\bar{E}(p) = \frac{ie}{\varepsilon_0 k}\,\frac{\displaystyle\int\frac{g(v)}{p - ikv}\,dv}{1 + \displaystyle\frac{ie^2}{\varepsilon_0 mk}\int\frac{df_0/dv}{p - ikv}\,dv} \tag{8.32}$$

Following the rule just mentioned, we start with Re $(p) > 0$, and the integrals of (8.32) are over the real range $(-\infty, \infty)$; the results are then to be continued analytically to Re $(p) \leqslant 0$. Identifying p with $i\omega$, one sees that the integrals will be governed by the "Landau prescription" which we found earlier, and described by Fig. 8.1. On this understanding, the denominator of (8.32) is simply $\varepsilon(\omega, k)$.

Applying the inversion formula (8.28), for $t \geqslant 0$, we displace the contour S leftwards, say into a position such as S' in Fig. 8.3, indenting the contour so as to capture any singularities of $\bar{E}(p)$. Then $E(t)$ will be a sum of terms proportional to $\exp(p_n t)$ where $p_1, p_2, \ldots$ are the singularities. These are just the zeros of the denominator of (8.32), i.e. they are given by $p = i\omega$ where ω satisfies the dispersion relation $\varepsilon(\omega, k) = 0$ we found before, (8.19). Thus the electric field appears as the superposition of one or more oscillations whose

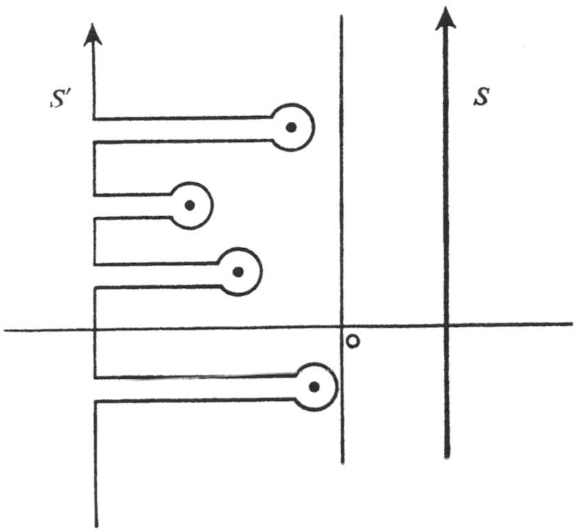

Fig. 8.3. The Laplace inversion contour.

complex frequency satisfies the dispersion relation, and this is just what we would expect. For a stable plasma they lie to the left of the imaginary axis, as shown; in the unstable case they occur on the right of it also.

We can apply the inversion formula to (8.31) also, and again deform the contour to the left. Fixing attention on any one value of v, and substituting $\bar{E}$ from (8.32), the singularities of $\bar{f}_1(v, p)$ will be those of $\bar{E}(p)$, already discussed, together with $p = ikv$. This lies on the imaginary axis, and gives a contribution proportional to exp $(ikvt)$; we recognize it as being just the term we discussed at equation (8.16) as the complementary function. As the present calculation is completely self-consistent, we have shown explicitly how a complementary function of the form (8.16) can give rise to an electric field proportional to exp $(i\omega t)$ where ω is complex.

8.4.2 Asymptotic behavior after a long time

One remarkable feature of these results is that if $f_0(v)$ is a stable distribution, the singularities of $\bar{E}$ represent decaying waves, whilst f_1 has additional singularities representing purely harmonic oscillations for each v. After a long time, therefore, the electric field strength decays (albeit very slowly indeed if the wavelength is long); but f_1 does not decay. If, therefore, we wait an extremely long time, the "plasma oscillation," in the sense we usually understand, will have disappeared, but there will still be a disturbance in the motion of the particles. This will consist simply of what we called "complementary functions," i.e. motions which can persist without an electric field, and so contrived that they produce no electric field. These field-free disturbances are, however, very peculiar, for $f_1(x, v, t)$ becomes increasingly pathological as a function of v, as is apparent from (8.16). The situation eventually becomes physically unreasonable, as f_1 acquires a finer structure in velocity space than can properly be included in any distribution function which actually represents discrete particles.

8.5 FURTHER DISCUSSION OF LANDAU DAMPING

8.5.1 The Van Kampen modes

Another approach to the initial-value problem is worth mentioning. As in any mechanical system, one can search for a complete set of harmonic solutions, and build up the solution corresponding to any initial situation by superposition. So once more we write down the linearized Boltzmann equation (8.7), and Poisson's equation, but this time treating them together, not decoupled as in §§8.2 and 8.3. As no dissipative effects are included in these governing equations, we expect the harmonic solutions to be all purely imaginary, or to occur in conjugate complex pairs. As before, we use Fourier analysis in space, and as we have an infinite plasma with no boundary conditions, every wave number, k, is possible. Fixing attention on a particular k, we observe that the corresponding contribution to f_1 is also a function of v in the range $-\infty < v < \infty$; so the space of functions required to form a complete set does not have an enumerable number of dimensions. This will result in a continuous range of normal frequencies ω, in contrast to the more usual case where discrete values of ω arise (a plasma made up of discrete streams, as considered in Chapter 6, would be a good example of this). The corresponding eigenfunctions $f_1(\omega, k, v)$ are also peculiar in that they involve improper functions (namely δ-functions) of v. As a result the mathematical discussion of the completeness and of the construction of the solution of an initial-value problem is rather more difficult than in the familiar discrete case. It has, however, been fully worked out by Van Kampen (1955) and by Case (1959).

Eliminating the electric field from (8.7) and (8.18) we have

$$i(\omega - kv)f_1 = \frac{e^2}{m\varepsilon_0 ik}\frac{df_0}{dv}\int_{-\infty}^{\infty} f_1(v')\,dv'$$

i.e.

$$(v - V)f_1(v) = -\eta(v)\int_{-\infty}^{\infty} f_1(v')\,dv', \tag{8.33}$$

where V is the wave speed, ω/k, and

$$\eta(v) = -\frac{e^2}{\varepsilon_0 mk^2}\frac{df_0}{dv} \tag{8.34}$$

(as defined by Case). We wish to find the eigenvalues, V, and the corresponding eigenfunctions $f_1(v)$ of the integral equation (8.33). But in fact any real V is an eigenvalue; for normalizing the eigenfunctions by

$$\int_{-\infty}^{\infty} f_1(v, V)\, dv = 1, \tag{8.35}$$

where the extra variable V serves to label them, (8.33) and (8.35) are satisfied if

$$f_1(v, V) = -P\frac{\eta(v)}{v - V} + \lambda(V)\, \delta(v - V), \tag{8.36}$$

where

$$\lambda(V) = 1 + P\int_{-\infty}^{\infty} \frac{\eta(v)\, dv}{v - V}. \tag{8.37}$$

Our eigenfunctions (8.36) are singular at $v = V$, but have been specified in such a way as to give quite definite results whenever macroscopic quantities are calculated by integration over v. We recognize (8.36) as an example of the general solution (8.11) with H as in (8.16) and h proportional to $\delta(v - \omega/k)$ as mentioned in the discussion at that point. The choice of λ given by (8.37) is simply the one required to make the solution consistent with Poisson's equation. The solutions given by (8.36) are known as "Van Kampen modes."

When the plasma is inherently unstable, there are also some discrete complex eigenvalues of (8.33), and the corresponding eigenfunctions have to be added to our collection to make up a complete set. These do not occur for distributions with a single maximum (as proved in §8.3.1) and we leave them aside for the present.

To solve the initial-value problem considered in the previous section we would need to reintroduce the time dependence into our eigenfunctions (8.36) and construct a superposition of them designed to give the correct initial conditions. This would take the form

$$f_1(v, t) = \int_{-\infty}^{\infty} f_1(v, V)\, e^{ikVt}\, C(V)\, dV, \tag{8.38}$$

where $C(V)$ is to be determined, $f_1(v, V)$ is as in (8.36) and we have omitted possible additional terms corresponding to discrete eigenvalues. To find $C(V)$ one considers the equation adjoint to (8.33); its eigenfunctions form a set orthogonal to those of (8.33). The calculation of the appropriate normalization constants together with a proof of the completeness theorem have been given by Case. This is rather lengthy, and it suffices here to say that when $C(V)$ is obtained in integral form and inserted into (8.38), the resulting double integral is, apart from notation, essentially the result obtained by combining (8.31), (8.32), and the inversion formula (8.28). The δ-function in the expression (8.36) for the normal mode $f_1(v, V)$ ensures that $f_1(v, t)$ as given by (8.38) includes a contribution proportional to $\exp(ikvt)$, which is just what we discovered in §8.4.1. Similarly from (8.38) we can construct the charge density by integrating over v; one finds in fact that

$$\int_{-\infty}^{\infty} f_1(v, t)\, dv = \int_{-\infty}^{\infty} C(v)\, e^{ikvt}\, dv, \tag{8.39}$$

some simplification resulting from (8.37). The electric field, $E(t)$ is just $ie/\varepsilon_0 k$ times this (see 8.18).

There is a technical difference between these two solutions to the initial value problem. In the normal mode approach our result appears as an ordinary Fourier transform, (8.38) with kV interpreted as ω, whereas in the Landau theory it appears in the form of Bromwich's integral for the inversion of a Laplace transform. To obtain exact correspondence, we should have used the two-sided Laplace transform instead of the standard one. This would have given the solution for all time, whereas the Bromwich integral only applies for $t \geqslant 0$, and is zero for $t < 0$.

This incidentally raises the interesting question of the nature of the plasma motion for $t < 0$ subject to the given conditions at $t = 0$. So far as f_1 is concerned the answer is that it contains contributions proportional to $\exp(ikvt)$; thus its behaviour as $t \to -\infty$ is just the same as that for $t \to \infty$, namely that it does not tend to zero, but is liable to become highly pathological. Further, E behaves in the limit $t \to -\infty$ just as it does for $t \to \infty$, for it tends to zero at a rate given by the Landau damping coefficient.

The interpretation of this is as follows. The plasma is capable of genuine undamped normal modes. These are given by (8.36) and (8.37) and include every combination of frequency and wavelength. However, they correspond to distributions which are singular in velocity space (at $v = \omega/k$), so by themselves are objectionable, especially in a linearized theory. These modes can be superposed to give solutions which are well behaved as functions of v near $t = 0$, but they too become pathological as $t \to \pm\infty$. When one looks at the corresponding density in physical space, N_1, and hence the accompanying electric field, it emerges that they are predominantly oscillatory for sufficiently large $|t|$, but tend to zero as $|t| \to \pm\infty$. For large wavelength the frequency of these oscillations is approximately the plasma frequency and the decay rate small. They are to be recognized as plasma oscillations, and are obtained by solving the dispersion relation (8.19) with the Landau prescription. But the motion so described is not a true normal mode of the system; it is merely the asymptotic state of the electric field which takes the same form for a wide variety of initial conditions.

When the plasma is unstable there are solutions of (8.19) with ω complex even for the straight line contour; these are true normal modes, corresponding to eigenvalues of (8.33); the preceding remarks then need modification, for the electric field does not tend to zero as $t \to \infty$. In such a case, however, the linearized theory is only meaningful in the early stages of development of the instability.

8.5.2 Physical discussion

To clarify the rather abstruse mathematical derivation, several attempts have been made to give a physical discussion of the Landau damping of plasma oscillations, and in particular to account for the occurrence of damping when no dissipation mechanism is available.

It seems to us that the key to the mystery lies in the fact, already referred to, that the perturbation in the distribution function, f_1, does not show damping, but persists for all time. It is f_1 which gives complete information (insofar as this is possible in the Boltzmann formulation) about the motion of particles. The electric field, E, and equivalently the density, N_1, give only greatly reduced information and do not determine their own future. If only N_1 is specified at $t = 0$, there are still infinitely many degrees of freedom in the choice of $f_1(v)$ at that time, and by selecting suitable Van Kampen modes we could make N_1 behave in any way we like; however, if we insist that $f_1(v)$ be a smooth function of v at $t = 0$, N_1 will, after a period of transient behavior, develop in the way predicted by Landau damping. We can illustrate this by considering the simpler case where the particles are uncharged and non-interacting. If we impose a wave like perturbation in density at $t = 0$

$$N_1(x, 0) \propto e^{ikx},$$

the corresponding $f_1(x, v, 0)$ could be made up in many ways, but if we assume that it is smoothly varying in velocity space, the solution is clearly of the type considered at (8.10) and (8.16), namely

$$f_1(x, v, t) = e^{ik(vt-x)}\, h(v). \tag{8.40}$$

This gives

$$N_1(x, t) = e^{-ikx} \int_{-\infty}^{\infty} e^{ikvt}\, h(v)\, dv. \tag{8.41}$$

Physically, this represents wave-like disturbances impressed upon each stream of particles and convected with velocity v; at time $t = 0$ the disturbances are in phase, but as time goes on the phases get "mixed" on account of the spread of velocities, and the amplitude of N_1 decreases through interference, eventually tending to zero as $t \to \infty$ provided $h(v)$ is continuous* (in fact it tends to zero the faster the more "smooth" $h(v)$ is). But by making $h(v)$ a delta-function, say $\delta(v - u)$, we could make $N_1(x, t)$ appear to be an undamped wave with any speed u; this is the analog of the Van

---

* Mathematically this follows from the Riemann–Lebesgue lemma; see page 200, or for more detail, Jeffreys and Jeffreys, *Methods of Mathematical Physics*, p. 431.

Kampen mode. When the particles are charged, the effect of the electric field is only to postpone the process of phase mixing, the more so the longer the wavelength. The particles can act collectively for a while, through the process of "bunching," just as in a cold plasma, but unless the distribution is unstable, phase mixing must eventually take over. This can only be avoided by imposing a singular perturbation initially, and the apparent irreversibility in N_1 arises from the assumption that $f_1(v)$ is not singular. This is a phenomenon characteristic of statistical mechanics generally. A sufficiently carefully "prepared" initial situation will not exhibit the usual irreversible behavior, but if the initial conditions are only partially specified and the missing information is presumed to be expressible in terms of a non-pathological function, irreversibility follows.

It might be asked whether it would be possible to "prepare" a distribution in the above sense, with initial perturbations $f_1(v)$ which are out of phase, so that there is virtually no electric field, but so contrived that they will subsequently come into phase. If so, the electric field would, for a time, build up in a way that could be described as "Landau growing", i.e. Landau damping in reverse. Eventually, of course, the electric field would decay again as the perturbations evolved still further and the phases mixed once more. At first sight it would seem unlikely that such a delicate operation could be carried out experimentally; yet this is just what happens in the phenomenon of the "plasma wave echo," recently reported observationally (Malmberg, Wharton, Gould and O'Neil, 1968). A full theoretical treatment has also been published (O'Neil and Gould, 1968). The essence of the phenomenon is the following. Suppose that at time $t = 0$ a plasma oscillation with spatial dependence $\exp(-ik_1x)$ is excited. As we have seen, there is after a long time a perturbation proportional to $f_1(v)\exp(-ik_1x + ik_1vt)$, though the electric field has decayed by Landau damping. Next suppose that a second wave with spatial dependence $\exp(+ik_2x)$ is imposed at time τ, where $\tau \gg \gamma_1^{-1}$ (γ_1 being the damping coefficient of the first wave), and $k_2 \neq k_1$, though k_1 and k_2 are of similar order. After a further long time, this wave will behave in a way similar to the first one, but it will also modulate the first to give a second order term proportional to

$$f_1(v)f_2(v)\exp[i(k_2 - k_1)x + ik_1vt - ik_2v(t - \tau)].$$

At time $t = \tau k_2/(k_2 - k_1)$ the coefficient of v vanishes, indicating that the perturbations have come back into phase; it is at about this time that the electric field reaches its maximum. The overall effect is that the two imposed waves are, in turn, started and die away, and a long time after the second one has disappeared (so far as E is concerned) a new wave (the "echo") mysteriously grows from nothing up to the time just calculated, then finally decays again. A variant of this can be constructed by reversing the roles of space and time; in this case one has continuous excitation at two real frequencies by probes separated by large distances, and the "echo" appears at a third point far removed from both of the probes. It is this version which has been the subject of experiment; we chose to describe the temporal echo since in the present chapter we have only treated the case of real k and complex ω.

Another physical approach is by way of energy. The total energy is of course conserved; formally this is most quickly shown by starting with an equation equivalent to Poisson's in the case of longitudinal fields, namely

$$\mathbf{j} + \varepsilon_0 \dot{\mathbf{E}} = 0,$$

which expresses the fact that the total current (real plus displacement) is zero. From this

$$\mathbf{E}\cdot\mathbf{j} + \frac{\partial}{\partial t}(\tfrac{1}{2}\varepsilon_0 E^2) = 0.$$

In our one dimensional case

$$\frac{\partial}{\partial t}(\tfrac{1}{2}\varepsilon_0 E^2) + e\int_{-\infty}^{\infty} Evf_1\, dv = 0 \tag{8.42}$$

to the second order. The mean rate of damping is given by averaging the second term over one period of the oscillation and over one wavelength. For long waves in a stable plasma (the only case in which this description is instructive) this is determined by the contribution of those particles whose unperturbed speed is close to that of the wave, ω_r/k, (ω_r being the real part of ω, as before), for the mean work done by the electric field is zero for all other particles. One divides the plasma into the "resonant" particles, for which $v \approx \omega_r/k$, and the "main plasma." To evaluate (8.42) for the resonant particles one inserts (8.8), but it is again necessary to argue that the Landau prescription be used. The residue (counted at half-value) then gives the damping coefficient, the principal part of the whole integral being simply the expression of the "main plasma." The mathematical steps are but a thinly disguised version of the iterative procedure which led to (8.26) and will not be repeated here. The physical interpretation of (8.26) is that particles with speed just less than ω_r/k are abstracting energy from the electric field of the wave, while those moving just a little faster are helping to support the wave. Provided df_0/dv is negative at this speed, there are more particles in the former category, and the wave is damped. When the reverse is true the wave grows and we would attribute this to a Čerenkov-like emission of plasma waves by the resonant particles. So Landau damping and the Čerenkov effect are closely related, basically through the condition $\omega = \mathbf{k}\cdot\mathbf{v}$ in three dimensional notation; they can be thought of as simply the inverses of each other. However, when ω/k has an appreciable imaginary part the relation is not so clear cut.

Another interpretation of the special role of the resonant particles follows from the remark that for a wave of finite amplitude, these are just the particles which are "trapped" in the potential troughs of the wave, whereas the main plasma particles pass right through the wave. Some writers were tempted to estimate the damping coefficient γ by considering the energy involved in the trapping process, and such arguments do appear to lead to a value of γ of about the right order. Nevertheless, they are fundamentally incorrect, for trapping is an essentially non-linear effect and ought not to be invoked to explain a phenomenon which appears in a linearized theory.

Trapping is, however, of interest in another connection. In principle, one can set up a wave of finite amplitude having any periodic (not necessarily harmonic) form in space and moving, without changing its form, at any speed. To see this (following Bernstein, Greene and Kruskal, 1957), one works in a frame of reference moving with the wave. The distribution of untrapped particles is determined by Jeans's Theorem, the constants of the motion in one dimension being total energy and the phase of the motion; the latter is discarded to make the resulting charge density stationary. The trapped particles do not sample all the wave field, so their distribution is not completely fixed by Jeans's theorem; it can always be adjusted to ensure that Poisson's equation is satisfied. The existence of these finite-amplitude undamped waves led some writers to doubt the validity of Landau damping. However, when one takes the limit in which the amplitude tends to zero and the particles are seen in their rest frame, the energy range for trapping tends to zero but the number of trapped particles remains finite. One obtains precisely the Van Kampen modes (8.36). The discussion of the initial-value problem is then as before—when the starting distribution is smooth with respect to velocity, the modes are superposed in such a way that the electric field decays. The existence of the *undamped* finite amplitude waves reinforces the above comment that trapping cannot account for Landau damping.

To summarize, we can say that if $f_1(x, v, t)$ is to be handled completely, there exists a doubly infinite family of normal modes corresponding to all real values of ω and k (together with some discrete complex modes in the unstable case). But if, as often happens, only the electric field E is of interest, the assumption that modes are to be superposed in a non-pathological way leads to special solutions which look very much like normal modes for E, and exhibit the irreversible property we have called Landau damping. These solutions can be obtained directly by calculating the response function for an externally applied field and taking the causal limit, as we did in §8.2. This is the method we shall use, without further comment, for the remainder of our work in this subject, in which we generalize the foregoing to include transverse fields and external magnetic fields. We could equally use the Laplace transform and regard the causal limit as following from the rule for inversion; many writers have presented the work in this way.

8.6 TRANSVERSE WAVES

For the cold plasma without applied magnetic field, we noted (§5.2) two kinds of "wave," the plasma oscillation with purely longitudinal electric field and the radio wave with purely transverse electromagnetic field. After dealing in great detail with the influence of thermal motions upon the former of these, we turn to a brief discussion of the latter.

For radio waves we need to calculate the current transverse to $\mathbf{k}$ to insert into Maxwell's curl equation; according to the preceding discussion it will be sufficient to work out the appropriate conductivity for each species, as a function of ω and $\mathbf{k}$, and apply the causal limit, i.e. start with ω in the lower half plane and proceed by analytic continuation. So we must return to the three dimensional formulation, and write down (8.4) omitting the term arising from $\mathbf{B}_0$. Adopting the harmonic variation in space and time, as in (8.5) we obtain

$$i(\omega - \mathbf{k}\cdot\mathbf{v})f_1 = -\frac{e}{m}(\mathbf{E} + \mathbf{v} \times \mathbf{B}_1)\cdot\frac{\partial f_0}{\partial \mathbf{v}}. \tag{8.43}$$

A new feature is the $\mathbf{v} \times \mathbf{B}_1$ term. In the important case where f_0 is isotropic it disappears, as $\partial f_0/\partial \mathbf{v}$ is parallel to $\mathbf{v}$; in other cases $\mathbf{B}_1$ has to be eliminated by using the equation of electromagnetic induction in the form

$$\omega\mathbf{B}_1 = \mathbf{k} \times \mathbf{E}. \tag{8.44}$$

In suffix notation, the r-component of $\mathbf{E} + \mathbf{v} \times \mathbf{B}_1$ becomes

$$\omega^{-1}[(\omega - \mathbf{k}\cdot\mathbf{v})\delta_{qr} + k_r v_q]E_q.$$

Solving (8.43) for f_1 and constructing the p-component of $\mathbf{j}$ as in (8.12) we find

$$j_p = \sigma_{pq}E_q, \tag{8.45}$$

where

$$\sigma_{pq} = \frac{ie^2}{m\omega}\int_C \frac{(\omega - \mathbf{k}\cdot\mathbf{v})\delta_{qr} + k_r v_q}{\omega - \mathbf{k}\cdot\mathbf{v}} v_p \frac{\partial f_0}{\partial v_r}\, d^3v \tag{8.46}$$

and C indicates that the integral over $\mathbf{v}$-space is to be carried out with the Landau prescription.

In the isotropic case, the numerator of the fraction may be replaced by $\omega\delta_{qr}$ on account of the remark made above, so that

$$\sigma_{pq} = \frac{ie^2}{m}\int_C \frac{1}{\omega - \mathbf{k}\cdot\mathbf{v}} v_p \frac{\partial f_0}{\partial v_q}\, d^3v. \tag{8.47}$$

Writing $f_0 = f_0(W)$, where $W = \frac{1}{2}m\mathbf{v}^2$ is the kinetic energy, this takes the form

$$\sigma_{pq} = ie^2\int_C \frac{v_p v_q}{\omega - kv_1}\frac{df_0}{dW}\, d^3v, \tag{8.47a}$$

where $\mathbf{k}$ has been directed along the x_1-axis, as before. Since f_0 is even in the components of $\mathbf{v}$, only the diagonal terms of σ_{pq} survive; σ_{11} is of course the quantity already dealt with (equation 8.14). For σ_{22} ($= \sigma_{33}$), (8.47) can also be written

$$\sigma_{22} = \sigma_{33} = -\frac{ie^2}{m}\int_C \frac{f_0}{\omega - kv_1}\, d^3v. \tag{8.47b}$$

When f_0 is not isotropic, perhaps the most convenient form of σ_{pq} is reached by noting that the contribution arising from δ_{qr} in (8.46) is

$$\frac{ie^2}{m\omega}\int_C v_p \frac{\partial f_0}{\partial v_q}\, d^3v,$$

which by an integration by parts becomes $(Ne^2/mi\omega)\delta_{pq}$, so is just the "cold" conductivity, for which analytic continuation presents no problem. Hence (8.46) is

$$\sigma_{pq} = \frac{Ne^2}{mi\omega}\,\delta_{pq} + \frac{ie^2k}{m\omega}\int_C \frac{v_p v_q}{\omega - kv_1}\frac{\partial f_0}{\partial v_1}\, d^3v. \tag{8.48}$$

More formally, equation (8.46) may be integrated by parts throughout, to yield

$$\sigma_{pq} = \frac{Ne^2}{mi\omega}\,\delta_{pq} + \frac{e^2}{mi\omega}\int_C\left\{\frac{k_p v_q + k_q v_p}{\omega - \mathbf{k}\cdot\mathbf{v}} + \frac{k^2 v_p v_q}{(\omega - \mathbf{k}\cdot\mathbf{v})^2}\right\} f_0\, d^3v. \tag{8.48a}$$

The conductivity for a discrete "stream" with $\mathbf{v} = \mathbf{U}$ may be recovered by setting $f_0(\mathbf{v}) = N\delta^3(\mathbf{v} - \mathbf{U})$ and trivially carrying out the integration; the result is the non-relativistic form of (6.24), but without special choice of axes.

To investigate wave propagation, we have Maxwell's equations (8.44) and

$$-i\mathbf{k} \times \mathbf{B}_1/\mu_0 = \mathbf{j} + i\omega\varepsilon_0\mathbf{E} = i\omega\mathbf{D}, \tag{8.49}$$

say, where $\mathbf{D}$ is an equivalent displacement vector, $\mathbf{D} = \varepsilon_0\mathbf{E} + \mathbf{j}/i\omega$. We then have

$$D_p = \varepsilon_0\varepsilon_{pq}E_q, \tag{8.50}$$

where the dielectric tensor is

$$\varepsilon_{pq} = \delta_{pq} + \frac{1}{i\omega\varepsilon_0}\,\sigma_{pq}. \tag{8.51}$$

Elimination of $\mathbf{B}_1$ between (8.44) and (8.49) yields

$$D_p = \varepsilon_0\mathscr{M}_{pq}E_q, \tag{8.52}$$

where

$$\mathscr{M}_{pq} = \frac{c^2}{\omega^2}(k^2\delta_{pq} - k_p k_q) \tag{8.53}$$

is a matrix representing Maxwell's equations. The equation of the normal modes for **E** (or "pseudo normal modes" as they should perhaps be called in view of the preceding discussion) is

$$(\varepsilon_{pq} - \mathscr{M}_{pq})E_q = 0, \tag{8.54}$$

so requiring

$$|\, \varepsilon_{pq} - \mathscr{M}_{pq}\,| = 0. \tag{8.55}$$

In the isotropic case (8.55) reduces to $\varepsilon_{11} = 0$ or $(\varepsilon_{22} - c^2k^2/\omega^2)^2 = 0$. The first of these is the equation for plasma oscillations, and the second describes radio waves, which may have any polarization. Explicitly, the second equation is

$$\frac{c^2k^2}{\omega^2} = 1 - \frac{e^2}{m\varepsilon_0\omega}\int_C \frac{f_0}{\omega - kv_1}\,d^3v. \tag{8.56}$$

In the limit of zero temperature, this reduces of course to $c^2k^2 = \omega^2 - \omega_p^2$. The thermal corrections are much less important here than in the case of plasma oscillations; for in the expansion analogous to (8.22) the factor kv/ω is extremely small for virtually all particles if ω/k is estimated by its value for the cold plasma. Indeed the phase speed ω/k was shown in §5.2 to be greater than c for a cold plasma, so that strictly speaking no particles can satisfy $v \approx \omega/k$. The true state of affairs can only be found from a relativistic treatment.

In the anisotropic case, (8.55) does not factorize in this neat way; however (assuming a non-relativistic temperature) the plasma oscillations are virtually decoupled from the radio waves. This is so because with **k** along the first axis, only the 22- and 33-components of $\mathscr{M}$ are non-zero, and they are very large. The small coupling is the error referred to early in §8.2; it disappears altogether if $\sigma_{12} = \sigma_{13} = 0$, and from (8.48a) it may be deduced that this requires

$$\int v_p f_0 \, dv_2\, dv_3 \equiv 0 \qquad (p = 2{,}3)$$

for all values of v_1. For the transverse waves, the argument just given still holds, so that there exist high-frequency solutions which to very good approximation are the same as for a cold plasma.

There remains the possibility that anisotropy may introduce completely fresh solutions which have no analog in the isotropic case. This does indeed occur, and leads to a class of instabilities which will be discussed in §10.3.

8.7 MAXWELLIAN DISTRIBUTIONS

The most important physical application of the work of this chapter is to the Maxwellian distribution. For plasma oscillations we only need the one dimensional version

$$f_0(v) = N\left[\frac{m}{2\pi KT}\right]^{1/2} \exp(-mv^2/2KT). \tag{8.57}$$

To handle the integrations arising from the solution of Boltzmann's equation we define a standard function

$$G(z) \equiv \pi^{-\frac{1}{2}} \int_C \frac{e^{-\zeta^2}\, d\zeta}{z-\zeta}. \tag{8.58}$$

Here ζ and z will correspond to v and ω/k respectively using the thermal speed $(2KT/m)^{1/2}$ as unit; C denotes the Landau contour, i.e. writing $z = x + iy$, C is the real axis if $y < 0$, and the real axis indented above $\zeta = z$ if $y \geqslant 0$. Similarly integrals with powers of ζ in the numerator (corresponding to powers of v, as for instance in (8.47)) are also needed; these are trivially related to G by using

$$\frac{\zeta}{z-\zeta} = \frac{z}{z-\zeta} - 1.$$

For instance

$$\pi^{-1/2} \int_C \frac{\zeta e^{-\zeta^2}\, d\zeta}{z-\zeta} \equiv zG(z) - 1. \tag{8.59}$$

By differentiating (8.58) and then integrating by parts,

$$G'(z) = -\pi^{-1/2} \int_C \frac{e^{-\zeta^2}\, d\zeta}{(z-\zeta)^2} = -2\pi^{-1/2} \int_C \frac{\zeta e^{-\zeta^2}\, d\zeta}{z-\zeta}. \tag{8.60}$$

Comparing the last two results we have the differential equation

$$\tfrac{1}{2}G' + zG - 1 = 0. \tag{8.61}$$

For real z, (8.58) takes the form noted in (8.15), and the principal part gives the real part of G while the residue term gives the imaginary part. When $z = 0$ the principal part vanishes by symmetry; consequently

$$G(0) = i\pi^{1/2}. \tag{8.62}$$

Equations (8.61) and (8.62) fix G uniquely. Solving in the elementary way

$$G(z) = e^{-z^2}\left[2 \int_0^z e^{p^2}\, dp + i\pi^{1/2}\right]. \tag{8.63}$$

This exhibits the real and imaginary parts when z is real; and as it is obviously analytic it also represents G as defined in (8.58) for the whole of the complex plane. (8.63) can be written still more briefly as

$$G(z) = 2\,e^{-z^2} \int_{-i\infty}^{z} e^{p^2}\,dp. \tag{8.64}$$

Thus G is simply an error function with argument iz. In its original form (8.58) it is essentially the Hilbert transform of the Gaussian function.

Expansion of (8.63) for small z readily yields the power series

$$G(z) = i\pi^{1/2}\,e^{-z^2} + 2z(1 - 2z^2/3 + 4z^4/15 + \ldots) \tag{8.65}$$

while expansion of (8.58) and term by term integration, as in §8.3.2, and addition of the residue as appropriate, yields an asymptotic expansion for large z

$$G(z) \approx i\pi^{1/2}\varepsilon\,e^{-z^2} + z^{-1}(1 + 1/2z^2 + 3/4z^4 + \ldots), \tag{8.66}$$

where ε is 0, 1 or 2 according as y is negative, zero or positive. (It should be noted that when $|\,y\,|$ is small but $|\,z\,|$ large, (8.66) does not give the best approximation to the imaginary part of G; this is just the situation encountered when calculating the Landau damping coefficient for long waves. See §8.3.2 for the correct method.) Corresponding expansions for $G'(z)$ follow easily.

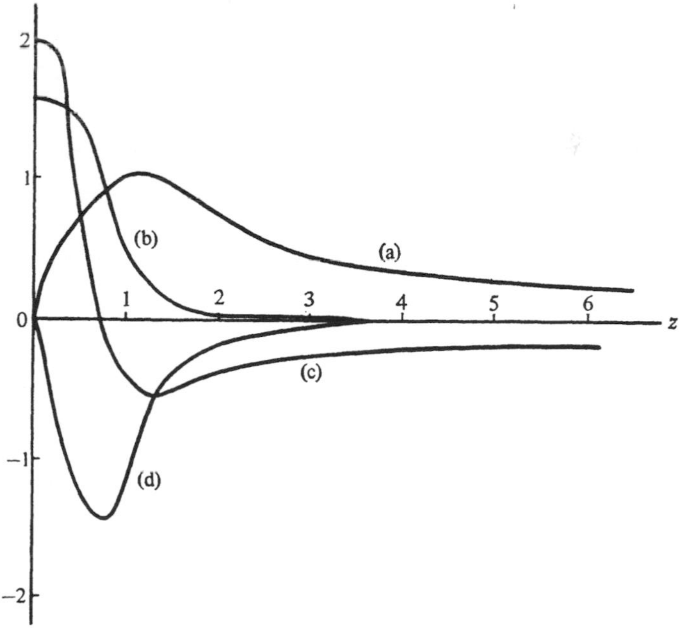

Fig. 8.4. G an G' as functions of x when $y = 0$. (a) Re(G); (b) Im(G); (c) Re(G'); (d) Im(G').

For fixed y, G has the following symmetry property in x: its real part is odd and its imaginary part even.

Tables of a function $Z(z)$ which is simply $G(-z)$ (the difference in sign arising from the opposite sign convention in the factor $\exp(i\omega t - ikx)$) have been published by Fried and Conte, and cover all complex values of z for which the early terms of (8.66) are not a good approximation. Figure 8.4 shows the real and imaginary parts of G and G' as functions of x for $y = 0$.

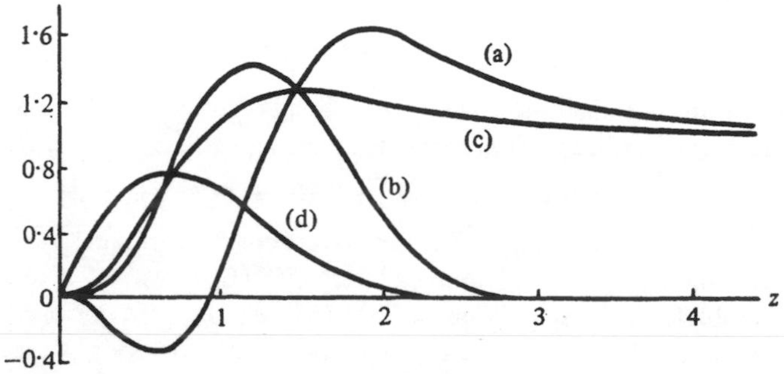

Fig. 8.5. The longitudinal and transverse conductivities for a Maxwellian plasma. (a) $\mathrm{Re}(\sigma_{11}/\sigma_c)$; (b) $\mathrm{Im}(\sigma_{11}/\sigma_c)$; (c) $\mathrm{Re}(\sigma_{22}/\sigma_c)$; (d) $\mathrm{Im}(\sigma_{22}/\sigma_{cc})$. [After J. P. Dougherty and D. T. Farley, *Proc. Roy. Soc.* **A259**, 96 (1960).]

The conductivity tensor (8.47) takes the following form for the Maxwellian plasma

$$\sigma_{11} = -z^2 G'(z)\sigma_c, \qquad \sigma_{22} = \sigma_{33} = zG(z)\sigma_c, \quad \text{other components zero,} \tag{8.67}$$

where σ_c is the cold conductivity, $Ne^2/mi\omega$, and z is the normalized phase speed

$$z = \frac{\omega}{k}\left(\frac{m}{2KT}\right)^{1/2}. \tag{8.68}$$

Figure 8.5 shows the real and imaginary parts of σ_{11}/σ_c and σ_{22}/σ_c, so giving a general idea of the importance of thermal effects. The quantity η, introduced in (8.14), is simply $k\sigma_{11}/\omega$, and can also be written

$$\eta = \frac{Ne^2 i}{2KTk} G'(z), \tag{8.69}$$

and the dielectric constant (see 8.17) is

$$\varepsilon(\omega, k) = 1 + \frac{1}{2h^2k^2} G'(z). \tag{8.70}$$

The dispersion relation for plasma oscillations, $\varepsilon(\omega, k) = 0$, thus becomes in our new notation

$$\left.\begin{aligned} \tfrac{1}{2}G'(z) + h^2k^2 &= 0 \\ \text{or equivalently} \qquad & \\ zG(z) &= 1 + h^2k^2. \end{aligned}\right\} \tag{8.71}$$

The solution for small hk has already been considered in §8.3.2. To see how it arises from Fig. 8.4, we observe that for large x, the real part of $G'$is small and negative, and the imaginary part much smaller. At the opposite extreme, we can search for solutions when $hk \gg 1$. Let us assume *a priori* that $y \gg 1$ while x is small but positive; (8.66) then leads to the approximate form

$$z\, e^{-z^2} = h^2k^2/2i\pi^{1/2},$$

or roughly

$$y\, e^{y^2 - 2ixy} = -h^2k^2/2\pi^{1/2}.$$

By making $2xy = (2n + 1)\pi$ where n is an integer, and

$$y\, e^{y^2} = h^2k^2/2\pi^{1/2},$$

the equation is approximately satisfied. This gives roughly

$$y \approx (2 \log hk)^{1/2} \quad \text{and} \quad x \approx (n + \tfrac{1}{2})\pi/y \tag{8.72}$$

and we have a whole family of solutions satisfying the initial hypothesis about x and y. They are, however, of little physical interest, as they are heavily damped in a time much shorter than one period.

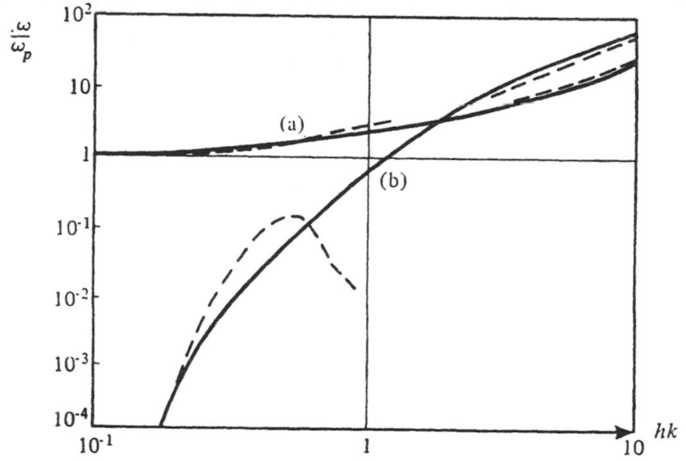

Fig. 8.6. The dispersion relation for a Maxwellian plasma. (a) $\text{Re}(\omega/\omega_p)$; (b) $\text{Im}(\omega/\omega_p)$. [After J. D. Jackson, *J. Nuclear Energy*, Part C, 1, 171 (1960).]

Numerically, one can readily solve (8.71) for all hk, and it is of interest to plot the real and imaginary parts of frequency as functions of the wave number for all real values of the latter. The results can be given in universal form by using the normalized variables hk and ω/ω_p ($= 2^{1/2}hkz$). This is done in Fig. 8.6, due to J. D. Jackson (1960). The dashed curves represent the approximate formulas (8.23), (8.26a) and (8.72) in the appropriate ranges; in the case of the last of these it is the branch $n = 1$ which leads to the undamped solutions when traced back to small values of hk; the branches $n = 2, 3, \ldots$ can also be traced but are always heavily damped and are not shown here. It will be noticed that the real and imaginary parts of ω are equal at $hk = 2.32$; for larger values (shorter wavelength) than this, plasma oscillations are virtually non-existent from the physical point of view.

8.8 THE EFFECT OF POSITIVE IONS. ION WAVES

So far we have included only the dynamics of electrons in the waves discussed in this chapter. It is straightforward to extend the work to the case of a uniform mixture of any number of species of charged particles. The conductivity tensor $\boldsymbol{\sigma}$ and the quantity η are simply the sum of those calculated for the separate species. One writes down the dispersion relations (8.55) (general waves) or (8.19) (longitudinal waves) as before, and the solutions again describe the asymptotic behavior of electromagnetic waves, though they do not give all the details of particle motions. In general the ions are found to participate very little in the types of wave we have discussed so far; this is, of course, due to their greater mass. The only hope for finding new effects of any real interest lies in the possibility that there might be new, undamped, waves at much lower frequency than the electron plasma frequency. This does indeed happen if the electron temperature is appreciably higher than the ion temperature.

This phenomenon can be adequately illustrated by including a single pecies of singly charged ions, with density N, charge e, mass m_i and temperature T_i. The corresponding quantities for the electrons are N, $-e$, m_e, T_e, and we assume both gases to have Maxwellian distributions. With other notation as before, (8.19) reads

$$\sum_a \eta_a + \varepsilon_0 ik = 0,$$

i.e., using (8.69)

$$\frac{1}{2}\sum_a \frac{G'(z_a)}{h_a^2k^2} + 1 = 0. \tag{8.73}$$

Here a labels the species, and z_a, η_a, h_a are as before, evaluated with the relevant parameters. Let us use z_e as the independent variable; then (8.73) gives

$$\tfrac{1}{2}G'(z_e) + \frac{T_e}{2T_i}G'(z_i) = -h_e^2k^2, \tag{8.74}$$

where $z_i = (m_iT_e/m_eT_i)^{\frac{1}{2}}z_e$, so usually $z_i \gg z_e$.

A glance at Fig. 8.4 will show that if $T_e \approx T_i$, the left-hand side of (8.74) is given essentially by the first term, except for very small z_e (such that $z_i \approx 1$), but even then the real part of $G'(z_i)$ is only about -0.5, whereas that of $G'(z_e)$ is about 2.0. Thus the left-hand side is positive and cannot equal $-h_e^2k^2$ at low frequency. But by increasing T_e/T_i the second term can be enhanced in this region to the extent that the real part of the left-hand side goes negative. By making z_i rather greater than 1 (say about 2.5) one can also arrange for the imaginary parts of both terms to be small, the first because z_e is small, the second because z_i is sufficiently large. There then appear virtually undamped waves traveling at speeds governed by the ions. The damping is in fact already quite small when $T_e \approx 4T_i$; hk is required to be small in these solutions. Since this is made possible by the effective cancelation of the two quantities η_i and η_e, these waves are characterized by very small space charge. Their speed is roughly $(KT_i/m_i)^{1/2}$, and they have long wavelength. So they can very properly be called "ion sound waves."

If $T_e \gg T_i$, the real part of the left-hand side of (8.74) goes substantially negative when $z_i \approx 1$ (this corresponds to an even smaller value of z_e than was the case in the previous paragraph); it remains negative until z_i is quite large. So larger values of $h_e^2k^2$ can be accommodated in the region where the imaginary part of the left-hand side is small, and the damping can be made extremely small by choosing z_i rather larger than unity, say $z_i \approx 3$ or more. Then we can regard z_e as small and z_i as large, so $G'(z_e) \approx 2$ and $G'(z_i) \approx -z_i^{-2}$. (8.74) then gives (cf. also (5.157))

$$\frac{\omega^2}{k^2} = \frac{KT_e}{m_i}\cdot\frac{1}{1 + h_e^2k^2}. \tag{8.75}$$

For long wave solutions the speed is therefore $(KT_e/m_i)^{1/2}$, that is the waves propagate at a sound speed fixed by the *electron* pressure and by the *ion* mass density. This implies $z_i \gg 1$, as assumed. But, as just noted, we may reduce z_i to about 3 without introducing damping, so that the right-hand side of (8.75) becomes about KT_i/m_i; so we can reduce the wavelength until $h_e^2k^2 \approx T_e/T_i \gg 1$, i.e. $h_i^2k^2 \approx 1$. For such waves (8.75) becomes approximately

$$\omega^2 = \omega_{pi}^2 = Ne^2/\varepsilon_0 m_i. \tag{8.76}$$

In this limit we have oscillations occurring at the ion plasma frequency.

The physical interpretation of this work is as follows. When $T_i \approx T_e$, the sound waves which we might have expected from the simple treatment involving pressure (§5.5.3) do not materialize. They are highly damped because many particles move with the speed of the wave. To put it another way, there is no way of giving the ions the necessary coherence so that when compressed the excess pressure is transmitted rapidly to all the ions. That can of course be achieved by collisions, which are not included in our work so far; that is the mechanism in ordinary gases. When $T_e \gg T_i$ the electrons are so mobile that they can immediately shield the ions on a scale greater than h_e, so maintaining approximate neutrality, just as in the Debye–Hückel theory. The electric field required to hold the electrons in place, against their own pressure, also acts on the ions, whose pressure is negligible. This provides the necessary restoring force, and explains the speed of propagation. But when the length scale is less than h_e, shielding breaks down. The electrons then act as a fixed background of negative charge and the ions perform plasma oscillations; so the picture is just like ordinary plasma oscillations but with the role of ions and electrons reversed—hence (8.76).

8.9 A RELATIVISTIC TREATMENT

8.9.1 The conductivity for longitudinal oscillations

We conclude this chapter by giving some indication of the way in which the relativistic form of the theory presented thus far can be developed from the relativistic treatment of the collisionless Boltzmann equation given in §7.7. We shall confine the discussion to longitudinal oscillations, as in §8.2, taking the electric field to be

$$\mathbf{E} = (E, 0, 0).$$

There is thus no magnetic field, and the Boltzmann equation (7.49) becomes

$$\sqrt{(1 + u^2)}\frac{\partial f}{\partial t} + cu_1\frac{\partial f}{\partial x} + \frac{eE}{mc}\sqrt{(1 + u^2)}\frac{\partial f}{\partial u_1} = 0. \tag{8.77}$$

With $f = f_0 + f_1$, and space–time factor $\exp[i(\omega t - kx)]$ for the perturbation quantities f_1 and E, the linearized form of (8.77) is

$$i[\omega\sqrt{(1 + u^2)} - cku_1]f_1 = -\frac{eE}{mc}\sqrt{(1 + u^2)}\frac{\partial f_0}{\partial u_1}. \tag{8.78}$$

The charge and current densities associated with the perturbation can now be written down from the formulas (8.12), namely

$$\rho = e\int f_1\, d^3u, \qquad j = ec\int\frac{u_1}{\sqrt{(1 + u^2)}}f_1\, d^3u. \tag{8.79}$$

The latter is, of course, in the x-direction; insertion of the solution of (8.78) for f_1 gives the longitudinal conductivity

$$\sigma = \frac{ie^2}{m} \int \frac{u_1\, \partial f_0/\partial u_1}{\omega\sqrt{(1+u^2)} - cku_1}\, d^3u. \tag{8.80}$$

This should be compared with (8.14) (it being immaterial whether we work with σ or η). Perhaps the most noticeable difference from (8.14) lies in the fact that it has not been possible to treat the problem as purely one dimensional by a device like (8.6), because universal integration with respect to u_2 and u_3 is frustrated by the presence of these variables in $\sqrt{(1+u^2)}$. One important case in which the triple integral of (8.80) is readily transformed to a single integral is when f_0 is isotropic in reduced velocity space; that is, is a function of u only. This assumption is adopted presently, with ultimate emphasis on an effectively Maxwellian distribution.

Another distinction between (8.80) and (8.14) is that in the former the integrand is not singular at some (real) value of **u** for *all* real values of ω and k; in fact it is clear that the denominator of the integrand cannot vanish, for real **u**, when $\omega > ck$. This feature is examined more closely after the reduction of σ to a single integral, which is now carried out.

When f_0 is a function of u only,

$$\frac{\partial f_0}{\partial u_1} = \frac{u_1}{u}\frac{df_0}{du}.$$

If, then, the change of integration variables

$$u_1 = u\cos\theta, \qquad u_2 = u\sin\theta\cos\phi, \qquad u_3 = u\sin\theta\sin\phi$$

is made, and the ϕ-integration effected, (8.80) becomes

$$\sigma = 2\pi i\frac{e^2}{m}\int_0^\infty\int_0^\pi \frac{\sin\theta\cos^2\theta\, d\theta}{\omega\sqrt{(1+u^2)} - cku\cos\theta}\, u^3\frac{df_0}{du}\, du. \tag{8.81}$$

Rather than evaluate the θ-integral it is neater to observe that

$$\frac{d}{du}\left\{\frac{u^3}{1+u^2}\int_0^\pi \frac{\sin\theta\cos^2\theta\, d\theta}{\omega\sqrt{(1+u^2)}-cku\cos\theta}\right\} = \frac{2\omega u^2}{(1+u^2)^{3/2}[\omega^2+(\omega^2-c^2k^2)u^2]},$$

and to introduce the function $F_0(u)$, satisfying

$$\frac{dF_0}{du} = (1+u^2)\frac{df_0}{du} \tag{8.82}$$

that tends to zero as $u \to \infty$. An integration by parts of (8.81) then leads to the required form

$$\sigma(\omega) = -\frac{4\pi ie^2}{m}\omega\int_0^\infty \frac{u^2F_0(u)\, du}{(1+u^2)^{3/2}[\omega^2+(\omega^2-c^2k^2)u^2]}. \tag{8.83}$$

Just as in the non-relativistic case, consideration of the initial-value problem shows that (8.83) must be used in the first instance to define $\sigma(\omega)$ for ω in the lower half plane (taking k real and positive). For ω in the upper half plane we extend the definition by analytic continuation. The quantity in square brackets in (8.83) takes the place of the familiar $\omega - kv$ of the non-relativistic case. This has the effect that if ω crosses the real axis from below with $\omega^2 > c^2k^2$, there is no need to indent the contour, as that expression does not vanish for any real u. But if $\omega^2 < c^2k^2$, indentation is necessary at the point $u = \omega/(c^2k^2 - \omega^2)^{1/2}$, and the contour for (8.83) is treated according to the Landau rule. Physically this behavior derives simply from the restriction that the particles have speed less than c according to the theory of special relativity; as we shall see in the next section this leads to the existence of a class of waves for which there is no Landau damping.

Mathematically, the change in behavior at $\omega = \pm ck$ leads to a branch point in $\sigma(\omega)$ at each of those two points. The structure of $\sigma(\omega)$ near these branch points depends in turn on the form of $F_0(u)$ as $u \to \infty$. If we now write $p = i\omega$ and consider the inversion of the Laplace transform for large values of t, as in §8.4, the contour, Fig. 8.3, needs modification to take account of the branch points at $p = \pm ick$. It will in fact be indented to stay to the right of these points, in addition to any singularities already shown in the figure. The expression for the electric field thus consists of a contribution due to the branch cut integrals, together with residues at poles which, as before, are located by solving the dispersion relation. In some cases the branch cut contribution decays more slowly than the wave-like contribution, but this depends in detail on $F_0(u)$ and on k; this point is investigated by Aaron and Currie (1966). In the following section we confine the discussion to the solution of the dispersion relation.

The treatment of the initial-value problem by setting up the relativistic analog of the Van Kampen modes, following the procedure of §8.5, has been carried out by Lerche (1967).

8.9.2 The dispersion relation

The relation $j = \sigma E$ is taken in conjunction with the Maxwell equation

$$i\omega\varepsilon_0 E + j = 0$$

expressing that the total current, displacement plus convection, is zero, and, of course, with charge conservation, equivalent to Poisson's equation (8.18). The resulting dispersion relation is $i\omega\varepsilon_0 + \sigma = 0$; that is, using (8.83),

$$\frac{4\pi e^2}{\varepsilon_0 m}\int_0^\infty \frac{u^2 F_0}{(1+u^2)^{3/2}[\omega^2 + (\omega^2 - c^2k^2)u^2]}\,du = 1. \tag{8.84}$$

We first show that if the path of integration is kept strictly along the positive real axis, then, for given real k, a solution of (8.84) for ω must be real and such that $\omega^2 > c^2k^2$. It is easy to do this by introducing the refractive index $\mu = ck/\omega$, and proving that if k is real then so is μ: for if $\mu^2 = a + ib$, (8.84) can be written

$$\frac{4\pi e^2}{\varepsilon_0 m}(a+ib)\int_0^\infty \frac{[1+(1-a)u^2+ibu^2]u^2F_0}{(1+u^2)^{3/2}\{[1+(1-a)u^2]^2+b^2u^4\}}\,du = c^2k^2\,; \quad (8.85)$$

clearly a necessary condition for the imaginary part of the left-hand side of (8.85) to vanish is $b = 0$; moreover it does vanish as $b \to 0$ if $a < 1$, but not if $a > 1$; and furthermore, when $a < 1$ and $b = 0$ the left-hand side is positive if and only if $0 < a < 1$.

Moreover, such solutions with ω and k real, and $\omega^2 > c^2k^2$, certainly exist; for if any value of a between 0 and 1 is chosen, (8.85), with $b = 0$, gives the corresponding value of k. This is in contrast to the non-relativistic treatment, where there were no solutions with both ω and k real. The reason for the difference has already been noted. In the relativistic treatment the fact that the particles necessarily travel with speed less than c is built into the analysis; a wave whose phase speed exceeds c therefore eludes the special interaction that arises when particle and wave velocities are the same or nearly so.

Next we obtain the asymptotic form of the dispersion relation corresponding to the treatment of §8.3.2, and valid under the condition there discussed, $\omega/k \gg \sqrt{(KT/m)}$, and the further condition $KT \ll mc^2$. The procedure is simply to develop the function multiplying F_0 in the integrand of (8.84) as a power series in u^2, and integrate term by term. The first two terms of the expansion give

$$4\pi\int_0^\infty [1+(\mu^2-5/2)u^2]u^2F_0\,du = \frac{\varepsilon_0 m}{e^2}\,\omega^2,$$

which from (8.82) is, to the same order of approximation,

$$4\pi\int_0^\infty [1+(\mu^2-5/6)u^2]u^2f_0\,du = \frac{\varepsilon_0 m}{e^2}\,\omega^2. \quad (8.86)$$

But

$$4\pi\int_0^\infty u^2f_0\,du = N,$$

and we can define the temperature T by

$$4\pi\int_0^\infty u^4f_0\,du = \frac{3KT}{mc^2}.$$

Then (8.86) gives

$$\frac{\omega^2}{\omega_p^2} = 1 + \frac{3KT}{m}\left(\frac{k^2}{\omega^2} - \frac{5}{6c^2}\right), \quad (8.87)$$

which may be compared with the vanishing of (8.22). Since (8.87) is only valid when the temperature term is a small correction, ω on the right-hand side can be replaced by ω_p, and then

$$\omega^2 = \omega_p^2\left[1 - \frac{5KT}{2mc^2} + 3h^2k^2\right], \tag{8.88}$$

to be compared with (8.23).

The asymptotic result takes no account of possible Landau damping. With k real there are, as we have seen, undamped solutions if and only if $\omega^2 > c^2k^2$, which is satisfied by (8.88) when $c^2k^2 < \omega_p^2(1 + KT/2mc^2)$ approximately. On the other hand it is certainly possible to have $|\omega|^2 < c^2k^2$ with (8.88) remaining substantially valid; in this case the solutions for ω have a small positive imaginary part and the path of integration in (8.84) must be distorted to pass above the pole at $u = \omega/\sqrt{(c^2k^2 - \omega^2)}$. Rather than give the details of the damping correction to the asymptotic analysis, which follow precisely the lines of §8.3.2, we conclude this section by considering briefly the relativistic counterpart of the treatment of Maxwellian distributions in §8.7.

For mathematical convenience take

$$f_0 = N(\lambda/\pi)^{3/2}\, e^{-\lambda u^2},$$

where

$$\lambda = mc^2/2KT. \tag{8.89}$$

Then from (8.82)

$$F_0 = N(1 + 1/\lambda + u^2)f_0,$$

so that (8.84) is

$$4\pi\frac{(\lambda/\pi)^{3/2}}{1-\mu^2}\int_0^\infty \frac{u^2(1 + 1/\lambda + u^2)}{(1+u^2)^{3/2}[(1-\mu^2)^{-1} + u^2]}\, e^{-\lambda u^2}\, du = \frac{\omega^2}{\omega_p^2} \tag{8.90}$$

with $\mu = ck/\omega$ as before, and we have in mind in the first instance the case where μ is real and $\mu^2 > 1$.

Since the integrand in (8.90) is negligible when u^2 is much greater than $1/\lambda$, it is legitimate to regard u^2 in the non-exponential part of the integrand as of order $1/\lambda$. If we now assume $\lambda \gg 1$, the non-exponential part of the integrand may therefore be replaced, to order $1/\lambda$, by

$$\frac{u^2(1 + 1/\lambda - \frac{1}{2}u^2)}{(1-\mu^2)^{-1} + u^2}; \tag{8.91}$$

and this is without prejudice to subsequent investigation of the case when $|\mu|^2$ is comparable with or greater than λ, that is $k^2 \gtrsim m|\omega|^2/(2KT)$. Since (8.91) can in turn be written

$$-\tfrac{1}{2}u^2 + \left[\frac{1}{2(1-\mu^2)} + 1 + 1/\lambda\right]\left[1 - \frac{(1-\mu^2)^{-1}}{(1-\mu^2)^{-1} + u^2}\right], \tag{8.92}$$

the integral in (8.90) can be expressed in terms of the error integral

$$E(w) = 2e^{w^2}\int_w^\infty e^{-x^2}\,dx = \frac{2}{\sqrt{\pi}}w\int_0^\infty \frac{e^{-x^2}}{w^2+x^2}\,dx. \tag{8.93}$$

The equality of the two forms in (8.93) holds if the real part of w is positive; it is clearly related to (8.63), and can be established in much the same way.

The substitution of (8.92) into (8.90), the evaluation of the simple integrals, and the use of (8.93), leads to the dispersion relation

$$2w^2\left(1+\frac{1}{\lambda}+\frac{w^2}{2\lambda}\right)[1-wE(w)]-\frac{w^2}{2\lambda}=\frac{\omega^2}{\omega_p^2}, \tag{8.94}$$

where

$$w=\sqrt{\left(\frac{\lambda}{1-\mu^2}\right)}. \tag{8.95}$$

Since $E(w)$ is an analytic function of w for all values of w in the complex plane, the form (8.94) is of universal application. In particular, when ω has a real part less than ck (taken positive), and a small positive imaginary part, the appropriate branch for w, in agreement with the prescription previously discussed, is indicated by writing

$$w = iz',$$

where

$$z'=\sqrt{\left(\frac{\lambda}{\mu^2-1}\right)}=\frac{z}{\sqrt{(1-z^2/\lambda)}}, \tag{8.96}$$

with

$$z=\sqrt{\left(\frac{m}{2KT}\right)}\frac{\omega}{k}$$

as in (8.68). Reference to (8.64) shows that

$$E(iz)=-iG(z),$$

so (8.94) is

$$-\left(1+\frac{1}{\lambda}-\frac{z'^2}{2\lambda}\right)[1-z'G(z')]+\frac{1}{2\lambda}=\left(1-\frac{z^2}{\lambda}\right)h^2k^2. \tag{8.97}$$

Evidently, for any given value of z, (8.71) is recovered by letting $\lambda\to\infty$.

It might perhaps be thought that, even though $\lambda \gg 1$, (8.97) would differ appreciably from (8.71) when z^2 was comparable with λ, since then z' would differ appreciably from z. But this is not so. Both z and z' would be large, and from the asymptotic form of G it can indeed be verified that (8.97) and (8.71) yield (8.88) and (8.23) respectively.

PROBLEMS

1. Verify equation (8.26a).
2. Construct details of the argument immediately following equation (8.42). [See J. Dawson (1961).]
3. Derive equation (8.48a).
4. Derive equations (8.65) and (8.66).
5. Show that for a Maxwellian distribution of electrons, the dielectric constant for longitudinal fields $\varepsilon(\omega, \mathbf{k})$ tends to the limit

$$1 + \frac{1}{h^2k^2}$$

when $\omega \to 0$. Hence solve, by spatial Fourier analysis, the problem of the field due to a point charge q embedded in a uniform plasma with a uniform stationary background of ions. Show also that if the dynamics of the ions (supposed at the same temperature as the electrons) is included in the calculation, the results are the same provided h is replaced by $h/\sqrt{2}$ (in other words the Debye length is calculated on the basis of the *total* number density). Your results should agree with those of §7.6.4.
[Note that

$$\left[\int \frac{e^{-i\mathbf{k}\cdot\mathbf{r}}}{\lambda^2 + k^2}\, d^3k = 2\pi^2\frac{e^{-\lambda r}}{r}.\right]$$

REFERENCES

The key paper which initiated our understanding of this subject was

L. D. LANDAU, *J. Phys. U.S.S.R.* **10,** 25 (1946).

Other papers cited in the text are

I. B. BERNSTEIN, J. M. GREENE and M. D. KRUSKAL, *Phys. Rev.* **108,** 546 (1957).

K. M. CASE, *Ann. Phys. (N.Y.)* **7,** 349 (1959).

J. DAWSON, *Phys. Fluids* **4,** 869 (1961).

J. D. JACKSON, *J. Nuclear Energy Part C,* **1,** 171 (1960).

J. H. MALMBERG, C. B. WHARTON, R. W. GOULD and T. M. O'NEIL, *Phys. Fluids* **11,** 1147 (1968).

T. M. O'NEIL and R. W. GOULD, *Phys. Fluids* **11,** 134 (1968).

N. G. VAN KAMPEN, *Physica* **21,** 949 (1955).

For the relativistic theory, see for instance

P. C. CLEMMOW and A. J. WILLSON, *Proc. Roy. Soc.* **A237,** 117 (1956).

R. AARON and D. CURRIE, *Phys. Fluids* **9,** 1423 (1966).

I. Lerche, *J. Math. Phys.* **8**, 1838 (1967).

Books mentioned in passing:

L. D. Landau and E. M. Lifshitz, *Statistical Physics*, Pergamon (1958).

H. Jeffreys and B. S. Jeffreys, *Methods of Mathematical Physics*, Cambridge University Press (1956).

For tables of the "G-function":

B. D. Fried and S. D. Conte, *The Plasma Dispersion Function*, Academic Press (1961).

CHAPTER 9

WAVES IN IONIZED GASES: KINETIC THEORY (CONTINUED)

The work of the previous chapter can be extended in a number of directions, the most important being the inclusion of an externally applied magnetic field. We shall devote most of this chapter to that problem, and the remainder to the approximate treatment of the effect of collisions.

Throughout this work we shall be interested only in the electromagnetic effects of the waves rather than the details of the distribution function for particles. We shall therefore simply find what we called the "response" (i.e. charge and current density) of the particles for harmonic fields, and apply the procedure developed in §8.2. The response is first calculated for fields switched on in the past and continued analytically to provide what we called the "causal limit." We then use Maxwell's equations to make the fields consistent with the space-charge and current. But we shall not construct the analogs of the Van Kampen modes or discuss the asymptotic form of the distribution function.

9.1 SOLUTION OF BOLTZMANN'S EQUATION IN AN EXTERNAL MAGNETIC FIELD

9.1.1 Integration along unperturbed orbits

As in Chapter 8, we imagine a uniform plasma in an equilibrium distribution f_0, this time with an externally applied uniform magnetic field $\mathbf{B}_0$. Perturbations f_1, $\mathbf{E}$, $\mathbf{B}_1$ are then considered, as in §8.1, so that equations (8.1) to (8.4) hold; then $f_0(\mathbf{v})$ satisfies

$$(\mathbf{v} \times \mathbf{B}_0) \cdot \frac{\partial f_0}{\partial \mathbf{v}} = 0, \tag{9.1}$$

and $f_1(\mathbf{x}, \mathbf{v}, t)$ is determined by

$$\frac{\partial f_1}{\partial t} + \mathbf{v} \cdot \frac{\partial f_1}{\partial \mathbf{x}} + \frac{e}{m}(\mathbf{v} \times \mathbf{B}_0) \cdot \frac{\partial f_1}{\partial \mathbf{v}} = -\frac{e}{m}[\mathbf{E} + \mathbf{v} \times \mathbf{B}_1] \cdot \frac{\partial f_0}{\partial \mathbf{v}}. \tag{9.2}$$

Equation (9.1) requires

$$f_0 = f_0(v_\perp, v_\parallel) \tag{9.3}$$

where $v_\perp = (v_x^2 + v_y^2)^{\frac{1}{2}}$, $v_\parallel = v_z$ are the components of velocity perpendicular and parallel to $\mathbf{B}_0$ respectively, taking the z-axis along $\mathbf{B}_0$. Equation (9.2) can be written, by analogy with (7.28),

$$\left[\frac{df_1}{dt}\right]_u = h(\mathbf{x}, \mathbf{v}, t), \tag{9.4}$$

where the operator $[d/dt]_u$ is defined as the rate of change following an unperturbed orbit in phase space. So the left-hand side is the rate of change of f_1 as "seen" by a particle which at time t is at the phase point $(\mathbf{x}, \mathbf{v})$ under observation, but which is moving through phase space in the way fixed by the external field $\mathbf{B}_0$. The function $h(\mathbf{x}, \mathbf{v}, t)$ is simply the right-hand side of (9.2). Adopting the approach of §8.2 we take $\mathbf{E}$ and $\mathbf{B}_1$ as given perturbing fields for which the response is to be calculated; subsequently the fields are to be made consistent with the charge and current densities, after summing over species where necessary. So at this stage $h(\mathbf{x}, \mathbf{v}, t)$ is a known perturbing function.

Formally, the solution of (9.4) is

$$f_1(\mathbf{x}', \mathbf{v}', t') = \int_{-\infty}^{t'} h(\bar{\mathbf{x}}, \bar{\mathbf{v}}, t)\, dt. \tag{9.5}$$

Here $\bar{\mathbf{x}}(t)$, $\bar{\mathbf{v}}(t)$ is the unperturbed orbit which will arrive at $(\mathbf{x}', \mathbf{v}')$ when t reaches t'. We have made the lower limit $t = -\infty$ under the assumption that the plasma is undisturbed in the infinite past and the perturbing fields grow from zero; as we have seen this ensures the correct causal properties.

Equivalently, (9.5) may be written in terms of a Green's function, G,

$$f_1(\mathbf{x}', \mathbf{v}', t') = \int_{\mathbf{x}}\int_{\mathbf{v}}\int_t G(\mathbf{x}, \mathbf{v}, t, \mathbf{x}', \mathbf{v}', t')\, h(\mathbf{x}, \mathbf{v}, t)\, d^3x\, d^3v\, dt, \tag{9.6}$$

where

$$G(\mathbf{x}, \mathbf{v}, t, \mathbf{x}', \mathbf{v}', t') = \delta^3(\mathbf{x} - \bar{\mathbf{x}})\delta^3(\mathbf{v} - \bar{\mathbf{v}})\varepsilon(t' - t), \tag{9.7}$$

and $\varepsilon(\tau)$ is 1 if $\tau > 0$, and 0 if $\tau < 0$. G is simply the δ-function transferring the disturbance $h(\mathbf{x}, \mathbf{v}, t)$ to all subsequent points on the same unperturbed orbit; mathematically, the unperturbed orbits are the characteristic curves for the inhomogeneous equation (9.2).

It may be noted that if the unperturbed state consists of f_0 with external fields $\mathbf{E}_0$ and $\mathbf{B}_0$, all of which may be non-uniform and time dependent, our work so far will still hold good for small perturbations about this state, provided that (9.1) is replaced by the appropriate zeroth order equation, and the "unperturbed orbits" are correctly interpreted. In what follows, however, only uniform stationary states will be considered.

9.1.2 Harmonic perturbations

In order to deal with plane waves, we make all the perturbation quantities such as f_1 harmonic in space and time, and represent them by a complex amplitude with the factor $\exp(i\omega t - i\mathbf{k}\cdot\mathbf{x})$ assumed, as usual. Let us introduce a new notation

$$[\mathbf{X}(t), \mathbf{V}(t)] = \text{the orbit passing through } \mathbf{x} = 0 \text{ with velocity } \mathbf{v} \text{ at } t = 0. \quad (9.8)$$

This orbit is, of course, just a uniform spiral, and as our situation is independent of space and time any other orbit can be written in terms of $\mathbf{X}$ and $\mathbf{V}$.

To give specific form to (9.7), the δ-function merely has to state that $(\mathbf{x}, \mathbf{v})$ and $(\mathbf{x}', \mathbf{v}')$ be on the same unperturbed orbit at the times concerned. It is slightly more convenient to use the earlier point as the label for the orbit (no new normalizing constant is necessary since the Jacobian of the transformation between points on an orbit is unity by Liouville's theorem). The required orbit is, at time t',

$$\mathbf{x} + \mathbf{X}(t' - t), \qquad \mathbf{V}(t' - t), \tag{9.9}$$

where $\mathbf{X}$ and $\mathbf{V}$ are evaluated with $\mathbf{v}$ as the initial velocity. So (9.7) reads in the new notation

$$G(\mathbf{x}, \mathbf{v}, t, \mathbf{x}', \mathbf{v}', t') = \delta^3[\mathbf{x}' - \mathbf{x} - \mathbf{X}(t' - t)]\,\delta^3[\mathbf{v}' - \mathbf{V}(t' - t)]\,\varepsilon(t' - t). \tag{9.10}$$

We are really interested not in f_1 itself but in the charge and current densities

$$\rho = e\int f_1(\mathbf{v}')\,d^3v', \qquad \mathbf{j} = e\int \mathbf{v}' f_1(\mathbf{v}')\,d^3v'. \tag{9.11}$$

More concisely, let us calculate the integral

$$J(\mathbf{a}) = e\int f_1(\mathbf{v}')\exp(i\mathbf{a}\cdot\mathbf{v}')\,d^3v'. \tag{9.12}$$

This and its first derivative with respect to the components of $\mathbf{a}$, evaluated at $\mathbf{a} = 0$, will give the required information.

We now combine (9.6), (9.10) and (9.12). Retaining for a moment the harmonic factors

$$J\exp(i\omega t' - i\mathbf{k}\cdot\mathbf{x}') = e\int_{\mathbf{x}}\int_{\mathbf{v}}\int_{\mathbf{v}'}\int_t h(\mathbf{v})\exp(i\omega t - i\mathbf{k}\cdot\mathbf{x} + i\mathbf{a}\cdot\mathbf{v}')\, G(\mathbf{x}, \mathbf{v}, t, \mathbf{x}', \mathbf{v}', t')\,d^3x\,d^3v\,d^3v'\,dt.$$

Using (9.10) to carry out trivial integrations over $\mathbf{x}$ and $\mathbf{v}'$, and writing $t' - t = t''$, say, but subsequently dropping the double-prime,

$$J = e\int_{\mathbf{v}}\int_{t=0}^{\infty} \exp[-i\omega t + i\mathbf{k}\cdot\mathbf{X}(t) + i\mathbf{a}\cdot\mathbf{V}(t)]\,h(\mathbf{v})\,d^3v\,dt. \tag{9.13}$$

We recall that

$$h(\mathbf{v}) = -\frac{e}{m}[\mathbf{E} + \mathbf{v} \times \mathbf{B}_1]\cdot\frac{\partial f_0}{\partial \mathbf{v}}, \tag{9.14}$$

where $\mathbf{E}$ and $\mathbf{B}_1$ are now the complex representatives of harmonic fields. It is usually inconvenient to work with both $\mathbf{E}$ and $\mathbf{B}_1$, and one can either use scalar and vector potentials or simply eliminate $\mathbf{B}_1$ by means of the equation of electromagnetic induction, which yields

$$\mathbf{B}_1 = \mathbf{k} \times \mathbf{E}/\omega. \tag{9.15}$$

As in §8.6 we adopt this latter course, and use

$$h(\mathbf{v}) = -\frac{e}{m}\left[\delta_{rj} + \frac{1}{\omega}\varepsilon_{rsl}\varepsilon_{lnj}v_s k_n\right]\frac{\partial f_0}{\partial v_r}E_j. \tag{9.16}$$

It is worth noting that the additional terms arising from $\mathbf{B}_1$ in (9.14) vanish in several common cases: (a) if f_0 is isotropic, as $\partial f_0/\partial \mathbf{v}$ is then parallel to $\mathbf{v}$, (b) in the "electrostatic approximation" in which the velocity of light is regarded as infinite, so that Maxwell's equations may be replaced by Poisson's equation, leading to purely electrostatic fields and $\mathbf{B}_1 \approx 0$, (c) the "cold" plasma, already treated in Chapter 5.

Substituting (9.16) into the integral (9.13), we find that J, and therefore ρ and $\mathbf{j}$, are linear in $\mathbf{E}$, and the coefficients relating them are fourfold integrals involving ω, $\mathbf{k}$, $f_0(\mathbf{v})$, and the unperturbed orbits as fixed by $\mathbf{B}_0$. It is usually difficult to complete all these four integrations analytically, though one can often do two or three of them at the cost of rather involved analysis. The integrand takes several forms depending on the components of $\mathbf{E}$ and $\mathbf{j}$ to be related, and the most expeditious order in which to attempt the integrations may depend on f_0 and on the type of problem in hand. Consequently there is a rather extensive literature on calculations of this type. As the working is of no intrinsic interest, it would be inappropriate to attempt an exhaustive treatment here. The work of the next three sections is therefore intended to cover some of the most commonly quoted results and to illustrate the type of procedure necessary.

9.2 ELECTROSTATIC WAVES IN A MAXWELLIAN PLASMA

9.2.1 Gordeyev's equation

Let us assume the electrostatic approximation ($c \to \infty$), so that the waves should be analogous to the plasma oscillations treated in Chapter 8; we shall comment on the validity of this procedure later.* Then we have

* At the end of §9.3.5.

$\mathbf{E} = -\nabla\phi = i\mathbf{k}\phi$, where ϕ is the potential, and $\mathbf{B}_1 = 0$; we need only calculate ρ, so we can put $\mathbf{a} = 0$ in (9.13), leading to

$$\rho = -\frac{ie^2\phi}{m}\int_{\mathbf{v}}\int_{t=0}^{\infty} \exp\{-i\omega t + i\mathbf{k}\cdot\mathbf{X}(t)\}\mathbf{k}\cdot\frac{\partial f_0}{\partial \mathbf{v}}\, d^3v\, dt. \tag{9.17}$$

As the components of $\mathbf{X}(t)$ are linear in those of $\mathbf{v}$, we can write

$$\mathbf{k}\cdot\mathbf{X}(t) \equiv \mathbf{v}\cdot\mathbf{p}(t), \tag{9.18}$$

where $\mathbf{p}(t)$ is the same for all particles and is easily calculated. An integration by parts with respect to $\mathbf{v}$ then transforms (9.17) to

$$\rho = -\frac{e^2\phi}{m}\int_{\mathbf{v}}\int_{t=0}^{\infty} \mathbf{k}\cdot\mathbf{p}(t)\exp\{-i\omega t + i\mathbf{p}\cdot\mathbf{v}\}f_0(\mathbf{v})\, d^3v\, dt. \tag{9.19}$$

Let us now make f_0 a Maxwellian distribution. At little extra trouble we can take the temperatures along and across the magnetic field to be unequal, with (9.1) still satisfied, so that

$$f_0 = N\left[\frac{m}{2\pi KT_\parallel}\right]^{\frac{1}{2}}\left[\frac{m}{2\pi KT_\perp}\right]\exp\left[-\frac{mv_\parallel^2}{2KT_\parallel} - \frac{mv_\perp^2}{2KT_\perp}\right], \tag{9.20}$$

where $T_\parallel$ and $T_\perp$ are the respective temperatures. Then the integration with respect to $\mathbf{v}$ is simply the Fourier transform of a normal (or Gaussian) distribution. Carrying it out yields

$$\rho = -\frac{Ne^2\phi}{m}\int_{t=0}^{\infty}\mathbf{k}\cdot\mathbf{p}(t)\exp\left\{-i\omega t - \frac{KT_\parallel p_\parallel^2}{2m} - \frac{KT_\perp p_\perp^2}{2m}\right\}dt, \tag{9.21}$$

where $p_\parallel$ and $p_\perp$ are the corresponding components of $\mathbf{p}$, in obvious notation

It is easily verified that the orbit $\mathbf{X}(t)$ is actually

$$\mathbf{X}(t) = \left(\frac{v_x}{\Omega}\sin\Omega t + \frac{v_y}{\Omega}(1-\cos\Omega t), -\frac{v_x}{\Omega}(1-\cos\Omega t) + \frac{v_y}{\Omega}\sin\Omega t, v_z t\right), \tag{9.22}$$

where Ω is the gyro-frequency eB_0/m. We can without loss of generality write $\mathbf{k} = (k_\perp, 0, k_\parallel)$. From the definition (9.18) we then have

$$\mathbf{p}(t) = \left(\frac{k_\perp}{\Omega}\sin\Omega t, \frac{k_\perp}{\Omega}(1-\cos\Omega t), k_\parallel t\right). \tag{9.23}$$

Thus

$$\rho = -\frac{Ne^2\phi}{m}\int_0^{\infty}\left[\frac{k_\perp^2\sin\Omega t}{\Omega} + k_\parallel^2 t\right]\exp\{-i\omega t - g(t)\}\, dt, \tag{9.24}$$

where

$$g(t) \equiv \frac{KT_\parallel k^2 t^2}{2m} + \frac{KT_\perp k_\perp^2(1-\cos\Omega t)}{m\Omega^2}. \tag{9.25}$$

When $T_\perp = T_\parallel$, the first factor in the integrand of (9.24) is proportional to dg/dt, so that one can integrate by parts to write the result more concisely. When $T_\perp \neq T_\parallel$ this can still be done provided one includes a correcting term. The result is

$$\rho = -\frac{Ne^2\phi}{KT_\perp}\left\{1 - i\left[\omega + \frac{K(T_\parallel - T_\perp)k_\parallel^2}{m}\frac{\partial}{\partial\omega}\right]\int_0^\infty \exp\{-i\omega t - g(t)\}\,dt\right\}. \tag{9.26}$$

The dispersion relation is obtained by summing over the species and combining with Poisson's equation

$$\varepsilon_0 k^2\phi = \sum_{\text{species}} \rho, \tag{9.27}$$

where $k^2 = k_\parallel^2 + k_\perp^2$. In the case of a single species (e.g. electrons with a stationary background of positive ions) the result is

$$1 + h_\perp^2 k^2 = i\left[\omega + \frac{K(T_\parallel - T_\perp)k_\parallel^2}{m}\frac{\partial}{\partial\omega}\right]\int_0^\infty \exp[-i\omega t - g(t)]\,dt, \tag{9.28}$$

where $h_\perp = (\varepsilon_0 KT_\perp/Ne^2)^{1/2}$ is the Debye length relative to the transverse temperature. This equation (for $T_\perp = T_\parallel$) was first obtained by Gordeyev, in 1952. To explore its consequences, the "Gordeyev integral"

$$I = \int_0^\infty \exp[-i\omega t - g(t)]\,dt \tag{9.29}$$

must be calculated as a function of ω, $k_\parallel$ and $k_\perp$, possibly as complex variables. Unfortunately there is no concise analytic expression for it. For general values of the parameters it seems best to proceed from this point (or from the form appearing in (9.24)) by numerical methods. Some cases which are amenable to further analytical work leading to interesting results are discussed in the next two subsections.

9.2.2 Expansion in series: the Bernstein modes

A simple case is that of waves traveling along $\mathbf{B}_0$, so $k_\perp = 0$; we expect to recover our previous results since the perturbations consist entirely of additional motions parallel to the magnetic field. This follows at once by making the change of variable $p = z - \frac{1}{2}i\tau$ in equation (8.64), giving

$$G(z) \equiv i\int_0^\infty \exp(-iz\tau - \tfrac{1}{4}\tau^2)\,d\tau. \tag{9.30}$$

Comparing this with the required integral, identifying $z = (\omega/k_\parallel)(m/2TK_\parallel)^{1/2}$ and $\omega t = z\tau$, (9.28) becomes

$$1 + h_\perp^2 k^2 = \left[z + \frac{T_\parallel - T_\perp}{2T_\parallel}\frac{\partial}{\partial z}\right]G(z). \tag{9.31}$$

This reduces at once to (8.71) when $T_\parallel = T_\perp$; if $T_\parallel \neq T_\perp$ it should of course be

$$1 + h_\parallel^2 k^2 = zG(z), \tag{9.32}$$

where $h_\parallel$ is the Debye length relative to $T_\parallel$, and this is easily shown by using (8.61), or more directly by slightly different manipulation of (9.24).

For waves traveling across the magnetic field, $k_\parallel = 0$, (9.29) diverges for real ω, but we recall that ω should first be in the lower half plane (so securing convergence) and the results continued analytically. Writing

$$\lambda = \frac{KT_\perp k_\perp^2}{m\Omega^2}, \tag{9.33}$$

we make use of the expansion

$$\exp(\lambda \cos \Omega t) = \sum_{n=-\infty}^{\infty} I_n(\lambda) \exp(in\Omega t), \tag{9.34}$$

where I_n is the Bessel function of imaginary argument $\big($i.e. $I_n(\lambda) = i^{-n}J_n(i\lambda)\big)$, and (9.29) becomes

$$I = \int_0^\infty \sum_{-\infty}^{\infty} I_n(\lambda) e^{-\lambda - i\omega t + in\Omega t}\, dt = -ie^{-\lambda} \sum_{-\infty}^{\infty} \frac{I_n(\lambda)}{\omega - n\Omega}$$

and the dispersion relation is

$$1 + h_\perp^2 k^2 = \omega e^{-\lambda} \sum_{-\infty}^{\infty} \frac{I_n(\lambda)}{\omega - n\Omega}. \tag{9.35}$$

Making further use of (9.34) with $t = 0$, and noting $h_\perp^2 k^2 = \lambda\Omega^2/\omega_p^2$ we find

$$\frac{\omega_p^2}{\Omega\lambda} e^{-\lambda} \sum_{-\infty}^{\infty} \frac{nI_n(\lambda)}{\omega - n\Omega} = 1. \tag{9.36}$$

This equation was extensively investigated by Bernstein (1958), who showed that it has solutions with both ω and k real. These are often called the "Bernstein modes."

As $I_{\pm 1}(\lambda) \approx \frac{1}{2}\lambda$, while $I_{\pm n} = O(\lambda^n)$ for small λ, we can arrange that only the terms $n = \pm 1$ contribute significantly to (9.36); this leads to

$$\omega^2 = \omega_p^2 + \Omega^2. \tag{9.37}$$

There are therefore solutions with $\lambda \to 0$ (that is, long waves, as $k \to 0$, with frequency given by (9.37), and known as the hybrid frequency. But we could also take $\omega \approx n\Omega$ for $n \geqslant 2$, and satisfy (9.35) by arranging that only the nth term contributes, which it will if $\omega - n\Omega = O(\lambda^{n-1})$. So there are long waves at each harmonic of the gyro-frequency, though not at the fundamental. There are also short wave solutions ($\lambda \to \infty$) with $\omega \approx n\Omega$ for all n (except zero), which are obtained by noting that $e^{-\lambda} I_n(\lambda) = O(\lambda^{-\frac{1}{2}})$ for large λ.

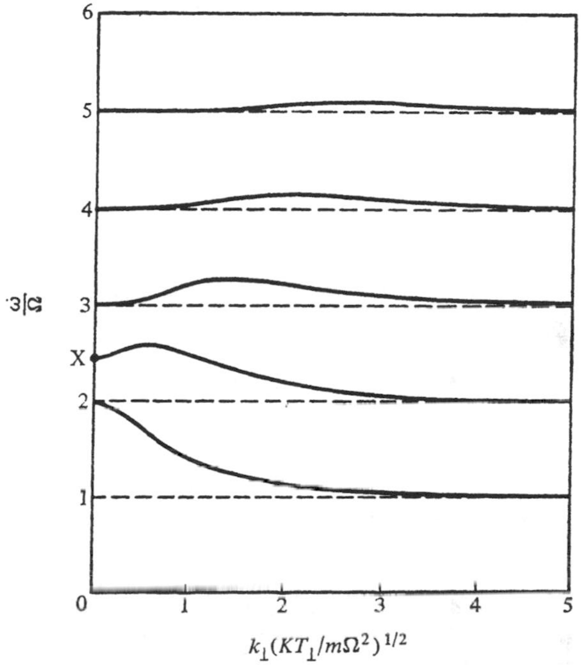

Fig. 9.1 Dispersion relation for the propagation of plasma oscillations normal to a magnetic field when $\omega_p = 5^{1/2}\Omega$. [After F. W. Crawford, 1965.]

The way in which these solutions are connected through intermediate values of k is made clear by numerical work. Figure 9.1 shows some results due to F. W. Crawford (1965). The branch $\omega \approx n\Omega$ at small k, when followed through increasing k, becomes the branch $\omega \approx n\Omega$ at large k, provided $n\Omega \geqslant (\omega_p^2 + \Omega^2)^{1/2}$; it has one maximum in frequency, which for large n is only slightly larger than $n\Omega$. The branch which starts at the hybrid frequency at small k becomes the branch $\omega \approx n\Omega$ at large k, where $n\Omega$ is the harmonic next below the hybrid frequency. The remaining branches start at $n\Omega$ for small k and become $(n - 1)\Omega$ at large k, where $n \geqslant 2$. It will be noted that below each harmonic of the gyro-frequency that is above the hybrid frequency there is a range of ω in which no real propagation is possible. These "gaps" were first noticed by Gross (1951). This description of the behavior of the modes holds generally, though our figure deals only with the case $\omega_p = 5^{1/2}\Omega$ by way of illustration. Plots for other values of ω_p/Ω are given in Crawford's paper.

When propagation is to take place in a general direction, we can use (9.34) and (9.30) to expand (9.29) in the form

$$I = \int_0^\infty \exp\left[-i\omega t - \lambda + \lambda\cos\Omega t - KT_\parallel k_\parallel^2 t^2/2m\right] dt$$

$$= -\frac{i}{k_\parallel}\left(\frac{m}{2KT_\parallel}\right)^{1/2} e^{-\lambda}\sum_{-\infty}^{\infty} I_n(\lambda)G(z_n), \tag{9.38}$$

where

$$z_n = \frac{\omega - n\Omega}{k_\parallel}\left(\frac{m}{2KT_\parallel}\right)^{1/2}. \tag{9.39}$$

Considering for simplicity the case $T_\parallel = T_\perp$, the dispersion relation becomes

$$1 + h^2k^2 = z_0\, e^{-\lambda} \sum_{n=-\infty}^{\infty} I_n(\lambda)G(z_n). \tag{9.40}$$

There are now no longer solutions in which both ω and k are real, as G is complex for real z. It will emerge later, during our discussion of instabilities, that for the Maxwellian distribution the waves are all damped when k is real; this is just the generalization of Landau damping, described earlier. Its physical origin is the same as that occurring in the case of no magnetic field, namely a resonant interaction between particles and waves. We shall explore this in more detail shortly; for the present we comment that the disappearance of Landau damping when the wave travels exactly orthogonally to the magnetic field can be traced to the fact that no particles can keep in step with the wave in that direction, as the motion in the xy-plane is circular. When $\mathbf{k}$ has a component along the field the free motion of the particles in that direction does enable resonance to occur; however if $k_\parallel$ is small the phase varies very slowly along the line of force, and a particle has to have a large v_z to keep in step with the wave. If the required velocity exceeds the typical thermal speed, the damping is slight. The rough criterion for this is $(KT/m)^{1/2} \ll \omega/k_\parallel$, or if θ is the angle between $\mathbf{k}$ and $\mathbf{B}_0$,

$$\cos\theta \ll \frac{\omega}{k}\left(\frac{m}{KT}\right)^{1/2}. \tag{9.41}$$

In the interesting case where ω is comparable with Ω or a low harmonic of it, these conditions are equivalent to $\cos\theta \ll 1/R_L k$ where R_L is the characteristic Larmor radius $(KT/m\Omega^2)^{1/2}$. These same conclusions also follow from a qualitative discussion of I.

When $T_\parallel \neq T_\perp$ and the direction of propagation is general, a new kind of instability may occur. This will be treated later (§10.4.2).

9.2.3 Waves involving positive ions

It is a straightforward extension of this work to include the motion of onc or more species of positive ions. Consider for example just one such species, and assume that both the electrons and ions have isotropic distributions with the same temperature T. In general the ions introduce only a small correction to the dispersion relation (just as in the case of no magnetic field), but new effects occur in the case of waves traveling almost, but not exactly, perpendicular to the magnetic field, the frequency being real and of the order of the ion gyro-frequency. To see this, suppose ω, k and θ are chosen so that $\cos\theta$ is small, and

$$\left(\frac{KT}{m_i}\right)^{1/2} \ll \frac{\omega}{k_\parallel} \ll \left(\frac{KT}{m_e}\right)^{1/2} \ll \frac{\Omega_e}{k_\parallel}\,;\ \frac{\omega}{\Omega_i} = O(1).$$

These are consistent requirements since $m_e \ll m_i$. By (9.39) z_n is large for the ions, but is still larger (by a factor $(m_i/m_e)^{1/2}$) for the electrons, except that z_{0e} is small. Using the expansions (8.65) and (8.66) for the G-function, as appropriate, we find that (9.38) reads

$$I \approx \frac{1}{k_\parallel}\left(\frac{\pi m_e}{2KT}\right)^{1/2} e^{-\lambda_e}\, I_0(\lambda_e)$$

for the electrons, and

$$I \approx -i\, e^{-\lambda_i} \sum_{n=-\infty}^{\infty} \frac{I_n(\lambda_i)}{\omega - n\Omega_i}$$

for the ions. The dispersion relation obtained from (9.27) is then

$$2 + h^2k^2 = \omega\, e^{-\lambda_i} \sum_{n=-\infty}^{\infty} \frac{I_n(\lambda_i)}{\omega - n\Omega_i} + i\pi^{1/2} z_{0e}\, e^{-\lambda_e} I_0(\lambda_e). \tag{9.42}$$

As z_{0e} is small and the function $e^{-\lambda} I_0(\lambda)$ is bounded, the last term can be omitted, and one has an equation similar to (9.35) with approximately real solutions close to each harmonic of the ion gyro-frequency.* There is, however, no analog here of the oscillation at the " hybrid " frequency which we noted in connection with the Bernstein modes.

The physical explanation of this is as follows. We have chosen a frequency well below Ω_e so that no resonance effects associated with the electrons can occur. Then we chose the direction of propagation so that (9.41) was satisfied by the ions, but the reverse for the electrons. So the ions have only unimportant motions along the field lines, while the electrons are very mobile in that direction. The situation is then rather similar to that of §8.8; the

* It is usually verified at the end that the frequency is not so close to a harmonic of Ω_i as to vitiate the assumption that each z_{ni} is large.

electrons can easily shield the ions by slight adjustment parallel to the field lines, yet the ions are moving too slowly to give significant Landau damping. The motion of the ions in the plane perpendicular to the field lines is periodic, so gives no Landau damping; any perturbations in ion density are freely convected by the ions, but recur after time $2\pi/\Omega_i$, so giving rise to harmonics of the gyro-frequency. If $\cos\theta$ is made still smaller, the electrons cannot move fast enough to shield the ions, and large electrostatic forces are set up; if $\cos\theta$ is made larger the waves are Landau damped by the ions.

Ion waves are possible for a much wider range of θ if $T_e \gg T_i$, the theory being closely analogous to that of §8.8. Still more generally, a great variety of waves and instabilities occurs if all four of $T_{i\parallel}$, $T_{e\parallel}$, $T_{i\perp}$, $T_{e\perp}$ can be arbitrarily chosen.

9.3 THE CONDUCTIVITY TENSOR FOR A MAXWELLIAN PLASMA

9.3.1 Formal expression

If the electrostatic approximation is not applicable, as for example in the treatment of Alfvén or radio waves, we must calculate the oscillating current $\mathbf{j}$ for use in Maxwell's curl equations instead of using ρ and Poisson's equation as in the last section. To do this we resume the calculation left unfinished in §9.1.2. We will again take f_0 to be the Maxwellian distribution (9.20); the possibility that $T_\parallel \neq T_\perp$ rather complicates the analysis but is included for later reference.

Equations (9.11)–(9.16) are summarized by*

$$j_i = \left[\frac{\partial J}{i\partial a_i}\right]_{\mathbf{a}=0}, \tag{9.43}$$

where

$$J = -\frac{e^2}{m} E_j \int_{\mathbf{v}} \int_0^\infty \left(\delta_{rj} + \frac{1}{\omega}\, \varepsilon_{rsl}\varepsilon_{lnj} v_s k_n\right) \frac{\partial f_0}{\partial v_r} \times \exp\left[-i\omega t + \mathbf{k} i\cdot \mathbf{X}(t) + i\mathbf{a}\cdot\mathbf{V}(t)\right] d^3v\, dt. \tag{9.44}$$

To carry out the integrations over $\mathbf{v}$ we note that $\mathbf{X}$ and $\mathbf{V}$ are linear in $\mathbf{v}$ so define $\mathbf{p}(t)$ by

$$\mathbf{p}\cdot\mathbf{v} \equiv \mathbf{k}\cdot\mathbf{X} + \mathbf{a}\cdot\mathbf{V} \tag{9.45}$$

as a generalization of (9.18). Also, for a Maxwellian distribution we have

$$\frac{\partial f_0}{\partial v_r} = -\frac{m v_r}{K T^{(r)}}\, f_0, \tag{9.46}$$

* The use of i and j as suffixes need cause no confusion with their use elsewhere as algebraic quantities.

where $T^{(r)}$ is $T_\perp$ if $r = 1$ or 2 and $T_\parallel$ if $r = 3$; suffixes bracketed do not count in applying the summation convention. Then the integrations over $\mathbf{v}$ take the form

$$\int v_r e^{i\mathbf{p}\cdot\mathbf{v}} f_0 \, d^3v \quad \text{and} \quad \int v_r v_s e^{i\mathbf{p}\cdot\mathbf{v}} f_0 \, d^3v.$$

We can handle these by considering the integral

$$\int \exp\,[i(\mathbf{p}+\mathbf{b})\cdot\mathbf{v}] f_0 \, d^3v. \tag{9.47}$$

This and its derivatives with respect to $\mathbf{b}$, evaluated at $\mathbf{b} = 0$, give the required results. When f_0 is the Maxwellian distribution, (9.47) is

$$N \exp\left[-\frac{KT^{(q)}}{2m}(p_q + b_q)(p_q + b_q)\right]. \tag{9.48}$$

Combining these formulas, we find a generalized Ohm's Law

$$\mathbf{j} = \boldsymbol{\sigma}\cdot\mathbf{E} \tag{9.49}$$

where the conductivity tensor is given by

$$\sigma_{ij} = -\frac{Ne^2}{KT^{(r)}}\frac{\partial^2}{\partial a_i \partial b_r}\left(\delta_{rj} + \frac{1}{i\omega}\varepsilon_{rsl}\varepsilon_{lnj}k_n\frac{\partial}{\partial b_s}\right)I \tag{9.50}$$

evaluated at $\mathbf{a} = \mathbf{b} = 0$, and

$$I = \int_0^\infty \exp\left[-i\omega t - \frac{KT^{(q)}}{2m}(p_q + b_q)(p_q + b_q)\right] dt \tag{9.51}$$

(note that summations over r, s, l, n and q are implied). I is the generalization of (9.29), to which it reduces when $\mathbf{a} = \mathbf{b} = 0$, and our problem is now converted to that of evaluating I to sufficient accuracy in $\mathbf{a}$ and $\mathbf{b}$ when they are small.

The second term in the bracket of (9.50) is the part due to the $\mathbf{B}_1$ term in Boltzmann's equation. Since it contains the factor

$$\frac{1}{T^{(r)}}\varepsilon_{rsl}\frac{\partial^2}{\partial b_r \partial b_s}$$

it vanishes, as expected, if all the $T^{(r)}$'s are the same. It follows that the second derivative with respect to $\mathbf{b}$ need be calculated only for indices r and s such that $T^{(r)} \neq T^{(s)}$; in other words when one component is parallel to the unperturbed magnetic field and the other perpendicular to it.

9.3.2 Explicit formulas; polarized coordinates

We can write our conductivity tensor (9.50) explicitly as a single integral by taking the differentiations indicated by (9.50) inside the integration of (9.51), carrying them out and setting $\mathbf{a} = \mathbf{b} = 0$; the result takes the form

$$\sigma_{ij} = \int_0^\infty F_{ij}(t) \exp\left[-i\omega t - g(t)\right] dt, \tag{9.52}$$

where $g(t)$ is as before, (9.25), and F_{ij} consists of a variety of elementary functions. Constructing F_{ij} in Cartesian coordinates, using (9.45) together with the formulas for $\mathbf{X}(t)$ and $\mathbf{V}(t) = \dot{\mathbf{X}}(t)$ is rather tedious owing to the asymmetric form of (9.22).

A device, due in its original form to O. Buneman (1961), for preserving tensor notation and writing the results in compact form, is the use of "polarized" coordinates. We define the new coordinates by

$$x^1 = (x + iy)/2^{1/2}, \qquad x^0 = z, \qquad x^{-1} = (x - iy)/2^{1/2}, \tag{9.53}$$

where the old ones are denoted by (x, y, z). Similar formulas hold for components of velocity, acceleration, etc. The numbering of the new coordinates is particularly convenient as it turns out that the suffixes can also be used as algebraic quantities. We use Greek suffixes for vector and tensor components in the new system. The transformation (9.53) is not orthogonal, although it is unitary. We therefore distinguish between contravariant vectors, such as $\mathbf{x}$ in the above transformation, and covariant vectors, by the usual convention of writing the suffixes as superscript or subscript respectively. It is easily verified that the scalar product of two contravariant vectors $\mathbf{A}$ and $\mathbf{B}$ is $\sum_\lambda A^\lambda B^{-\lambda}$ in these coordinates, and this must be identified with $\sum_\lambda A^\lambda B_\lambda$. Clearly the metric tensor is $\delta_{\lambda,-\mu}$, and raising or lowering of the suffixes is achieved simply by changing the sign. (Note that this is not in general the same as taking the complex conjugate; it is so for a real vector but not for a vector which is the complex representative of a harmonic quantity.)

To convert all the previous formulas to the new notation one must decide for each suffix whether it is covariant or contravariant. This can be done merely by recalling that any pair of contracted suffixes must be one of each type and that in the case of derivatives, differentiation with respect to a contravariant component yields a covariant result and vice versa; for instance (9.49)–(9.51) would be written

$$i^\mu = \sigma^\mu_\nu E^\nu, \tag{9.54}$$

where*

$$\sigma^\mu_\nu = -\frac{Ne^2}{KT^{(\alpha)}} \frac{\partial^2}{\partial a_\mu\, \partial b^\alpha} \left\{ \delta^\alpha_\nu + \frac{1}{i\omega} \varepsilon^{\alpha\beta\gamma} \varepsilon_{\gamma\lambda\nu} k^\lambda \frac{\partial}{\partial b^\beta} \right\} I, \tag{9.55}$$

* The use of the alternating tensor in these coordinates requires some care; however, there is no ambiguity where two such tensors are contracted, as the product may be regarded as synonymous with the usual expansion; here for instance it is $\delta^\alpha_\lambda \delta^\beta_\nu - \delta^\alpha_\nu \delta^\beta_\lambda$.

and

$$I = \int_0^\infty \exp\left\{-i\omega t - \frac{KT^{(\tau)}}{2m}(p^\tau + b^\tau)(p_\tau + b_\tau)\right\} dt. \tag{9.56}$$

Here $T^{(\pm 1)} = T_\perp$ and $T^{(0)} = T_\parallel$.

To find the unperturbed orbits we solve

$$\dot{\mathbf{V}} = \frac{e}{m}\mathbf{V} \times \mathbf{B}_0, \qquad \dot{\mathbf{X}} = \mathbf{V}$$

where $\mathbf{V} = \mathbf{v}$, $\mathbf{X} = 0$ at $t = 0$. In the new coordinates these equations are

$$\frac{dV^\lambda}{dt} = -\, i\Omega\lambda V^\lambda, \qquad \frac{dX^\lambda}{dt} = V^\lambda, \tag{9.57}$$

with solution

$$V^\lambda = v^\lambda\, e^{-i\Omega\lambda t}, \qquad X^\lambda = \frac{v^\lambda}{i\Omega\lambda}(1 - e^{-i\Omega\lambda t}). \tag{9.58}$$

Here it will be noted that for $\lambda = 0$, X^λ apparently becomes indeterminate, but the correct result is given by taking the limit $\lambda \to 0$. Thus $X^0 = v^0 t$. (In applying the summation convention the appearance of a suffix as an algebraic quantity is disregarded.)

The defining relation (9.45) for $\mathbf{p}$ is

$$p_\lambda v^\lambda = k_\lambda X^\lambda + a_\lambda V^\lambda,$$

and so

$$p_\lambda = \frac{k_\lambda}{i\Omega\lambda}(1 - e^{-i\Omega\lambda t}) + a_\lambda\, e^{-i\Omega\lambda t}. \tag{9.59}$$

To find p^λ we of course raise the suffix, noting that this involves a change in sign of λ where it appears as an algebraic quantity.

It is a straightforward matter to carry out the differentiations with respect to $\mathbf{a}$ then $\mathbf{b}$, and set $\mathbf{a} = \mathbf{b} = 0$ to obtain an expression of the type (9.52), which we should now write

$$\sigma^\mu_\nu = \int_0^\infty F^\mu_\nu(t) \exp\left[-i\omega t - g(t)\right] dt, \tag{9.60}$$

and F^μ_ν has the fairly compact form

$$F^\mu_\nu = \frac{Ne^2}{m} e^{-i\Omega\mu t}\left\{\delta^\mu_\nu - \frac{KT^{(\mu)}}{m}p^\mu p_\nu + \frac{iK(T_\perp - T_\parallel)}{m\omega}\left[(\mu^2 - \lambda^2)k^\lambda p_\lambda \delta^\mu_\nu \right.\right.$$
$$\left.\left. + (\nu^2 - \mu^2)p_\nu k^\mu + \frac{KT^{(\mu)}}{m}(\lambda^2 - \nu^2)p^\mu p_\nu p_\lambda k^\lambda\right]\right\} \tag{9.61}$$

with **p** as in (9.59) but the term in **a** now omitted; we observe that p_0 is simply $k_{\|}t$. The rather complicated part in square brackets in (9.61) arises from the $\mathbf{B}_1$ term, and is removed if $T_{\perp} = T_{\|}$, as already noted. For general purposes one would evaluate the integrals (9.60) numerically, having written out the various F^{μ}_{ν}'s explicitly; for instance when $T_{\perp} = T_{\|}$,

$$F^0_0 = (Ne^2/m)[1 - KTk^2_{\|}t^2/m].$$

To investigate wave propagation in such a medium, the procedure is as in §8.6. We have a dielectric tensor (cf. (8.51))

$$\varepsilon^{\mu}_{\nu} = \delta^{\mu}_{\nu} + \frac{1}{i\omega\varepsilon_0} \sum_{\text{species}} \sigma^{\mu}_{\nu} \tag{9.62}$$

and Maxwell's equations are represented by the matrix

$$\mathscr{M}^{\mu}_{\nu} = \frac{c^2}{\omega^2}(k^2\delta^{\mu}_{\nu} - k^{\mu}k_{\nu}), \tag{9.63}$$

the normal modes being given by

$$(\varepsilon^{\mu}_{\nu} - \mathscr{M}^{\mu}_{\nu})E^{\nu} = 0 \tag{9.64}$$

leading to the dispersion relation

$$| \varepsilon^{\mu}_{\nu} - \mathscr{M}^{\mu}_{\nu} | = 0. \tag{9.65}$$

It would, however, be a formidable task to deal with a representative selection of cases, owing to the large number of parameters involved, and the fact that each element of the determinant (9.65) involves an integral of the type (9.60). These general formulas have not therefore been much used, but some special cases are noted in the next three subsections.

9.3.3 The low temperature approximation

Setting all temperatures equal to zero in (9.61) and noting that g then vanishes, (9.60) becomes, in the causal limit,

$$\sigma^{\mu}_{\nu} = \frac{Ne^2}{mi(\omega + \mu\Omega)}\delta^{\mu}_{\nu}, \tag{9.66}$$

so that

$$\varepsilon^{\mu}_{\nu} = \left[1 - \sum_{\text{species}} \frac{\omega_p^2}{\omega(\omega + \mu\Omega)}\right]\delta^{\mu}_{\nu}. \tag{9.67}$$

This is just the equivalent of (5.49) expressed in polarized coordinates, so we recover all our previous results for the cold plasma. In this formulation, $\boldsymbol{\epsilon}$ appears in diagonal form, and this is particularly convenient for waves traveling along the magnetic field. For waves traveling at an angle θ to the magnetic field we may write $\mathbf{k} = k(\sin\theta, 0, \cos\theta)$ in the Cartesian coordinates, so that $k^{\pm 1} = k_{\pm 1} = k \sin\theta/2^{1/2}$; manipulation starting from (9.64) yields the dispersion relation in the form (Problem 5)

$$\frac{\varepsilon_0^0 \cos^2\theta}{\varepsilon_0^0 - k^2c^2/\omega^2} + \tfrac{1}{2}\sin^2\theta\left(\frac{\varepsilon_1^1}{\varepsilon_1^1 - k^2c^2/\omega^2} + \frac{\varepsilon_{-1}^{-1}}{\varepsilon_{-1}^{-1} - k^2c^2/\omega^2}\right) = 0, \quad (9.68)$$

which is a form of the Appleton–Hartree equation (see Problem 9).

Next we can introduce the temperatures as a small correction, in which we work to the first order. As defined in §9.2.1, $g(t)$ can be written $KT^{(\lambda)}p^\lambda p_\lambda/2m$, so $\exp(-g) \approx 1 - g$, and when (9.60) and (9.61) are approximated in this way, all the required integrals involve only exponential functions, bearing in mind that $\mathbf{p}$ is given by (9.59) with $\mathbf{a} = 0$. Carrying out the integrations in the causal limit and tidying the algebra leads to

$$\sigma_\nu^\mu = \frac{Ne^2}{im(\omega + \mu\Omega)}\left\{\delta_\nu^\mu + \frac{KT^{(\lambda)}k^\lambda k_\lambda \delta_\nu^\mu}{m(\omega + \mu\Omega)(\omega + \mu\Omega + \lambda\Omega)} + \frac{KT^{(\mu)}k^\mu k_\nu}{m(\omega + \nu\Omega)}\left(\frac{1}{\omega} + \frac{1}{\omega + \mu\Omega + \nu\Omega}\right) + \frac{K(T_\perp - T_\parallel)}{m\omega}\left[\frac{(\mu^2 - \lambda^2)k^\lambda k_\lambda \delta_\nu^\mu}{\omega + \mu\Omega + \lambda\Omega} + \frac{(\nu^2 - \mu^2)k^\mu k_\nu}{\omega + \mu\Omega + \nu\Omega}\right]\right\}. \quad (9.69)$$

This correction to $\boldsymbol{\sigma}$ for low temperatures turns out to be similar to the addition of pressure effects to the cold theory, as we did in §5.5. Here, however, the pressure tensor is in general anisotropic and obeys the double adiabatic equations rather than the ordinary law $pN^{-5/3} =$ constant. We shall derive these relations later (§11.3.4) (see also Buneman, 1961).

To discuss wave propagation with the corrected $\boldsymbol{\sigma}$, it is necessary to solve the dispersion relation in the cold approximation and use the results to evaluate the temperature-dependent part of (9.69), then solve again in iteration. A simple example would be plasma oscillations with $\mathbf{k}$ parallel to $\mathbf{B}_0$, where one recovers (8.23). The general case requires much tedious calculation, but one other interesting example is that of Alfvén waves traveling along the magnetic field. Taking $k_0 = k$, $k_1 = k_{-1} = 0$, (9.69) shows that $\boldsymbol{\sigma}$ is still diagonal in its corrected form, and

$$\sigma_1^1 = \frac{Ne^2}{im(\omega + \Omega)}\left\{1 + \frac{KT_\parallel k^2}{m(\omega + \Omega)^2} + \frac{K(T_\perp - T_\parallel)k^2}{m\omega(\omega + \Omega)}\right\}. \quad (9.70)$$

(For σ_{-1}^{-1}, we reverse the sign of Ω, and for electrons we reverse the sign of Ω throughout as the charge is negative but Ω is conventionally defined to be positive.) We are interested in the case where $\omega \ll \Omega$ for all species. In the cold approximation, assuming a single species of ions, we have

$$\sum \sigma_1^1 = \frac{Ne^2}{im_i(\omega + \Omega_i)} + \frac{Ne^2}{im_e(\omega - \Omega_e)} = \frac{Ne^2\omega}{i(\omega + \Omega_i)(\omega - \Omega_e)}\left(\frac{1}{m_i} + \frac{1}{m_e}\right)$$

(using $m_i\Omega_i = m_e\Omega_e$). At low frequency this becomes $iNe^2\omega(m_i + m_e)/\Omega_i\Omega_e m_i m_e$. It can also be written $i\omega/\mu_0 V_A^2$ where $V_A = B_0/[\mu_0 N(m_i + m_e)]^{1/2}$ is the Alfvén speed. Equations (9.62)–(9.64) then show that circularly polarized waves exist, satisfying

$$\frac{c^2k^2}{\omega^2} = 1 + \frac{c^2}{V_A^2}. \tag{9.71}$$

These are Alfvén waves, already discussed in §5.4.2. The leading correction to σ_1^1 arising from (9.70) when $\omega \ll \Omega$ is $Ne^2K(T_\perp - T_\parallel)k^2/im^2\Omega^2\omega$, i.e. $(p_\perp - p_\parallel)k^2/iB^2\omega$, where $p_\perp$ and $p_\parallel$ are the pressures orthogonal to and parallel to the field, and summing over the species gives the same result with $p_\perp$ and $p_\parallel$ summed. Thus (9.71) is to be corrected by

$$\frac{c^2k^2}{\omega^2} = 1 + \frac{c^2}{V_A^2} - \frac{c^2k^2\mu_0(p_\perp - p_\parallel)}{\omega^2 B^2}. \tag{9.72}$$

Omitting the term unity under the usual assumption that V_A and ω/k are much less than c, we have

$$\omega^2 = k^2\left[\frac{B^2/\mu_0 + p_\perp - p_\parallel}{\rho_m}\right], \tag{9.73}$$

where ρ_m is the mass density. Here it would appear at first sight that $p_\perp - p_\parallel$ must be a small correction in (9.73) if the foregoing process is to be justified, but actually all that is required is that the temperature-dependent terms in (9.70) be small, and this can be achieved by making k sufficiently small, i.e. the wavelength long. (The reason why the "corrections" appear to be of the same order as the original terms in (9.73) is that for $\omega \ll \Omega$ the large terms in σ_1^1 approximately cancel on summing over the species.) Accordingly, it is possible for $p_\perp$ and $p_\parallel$ to be comparable with the magnetic pressure B_0^2/μ_0, and if

$$p_\parallel > p_\perp + B_0^2/\mu_0 \tag{9.74}$$

(9.73) has solutions with ω purely imaginary and k real, thus showing possible growth in time. This is known as the "firehose" instability; we shall meet it again in the discussion of hydromagnetic equations. It is of interest to note that the instability can be traced back to the dynamical effect of the perturbation magnetic field, $\mathbf{B}_1$.

The expansion procedure used here can be continued to higher powers of the temperature, but at no stage do any dissipative effects like Landau damping make their appearance. This is just what we found in the case of no magnetic field (§8.3.2); the Landau damping, even when it is small, is an effect not representable as a power series expansion.

9.3.4 Propagation along the magnetic field

The case of propagation parallel to the magnetic field leads to quite tractable results even without the approximation developed in the last section. For if $k_\perp = 0$, F_ν^μ of (9.61) is diagonal, with

$$F_0^0 = \frac{Ne^2}{m}\left[1 - \frac{KT_\parallel}{m} k^2 t^2\right],$$

$$F_1^1 = \frac{Ne^2}{m}\left[1 + \frac{iK(T_\perp - T_\parallel)k^2 t}{m\omega}\right] \exp(-i\Omega t),$$

while (9.25) reads

$$g(t) = \frac{KT_\parallel k^2 t^2}{2m}.$$

Using (9.30) and other relations for the G-function developed earlier, we find

$$\sigma_0^0 = \frac{i\omega Ne^2}{KT_\parallel k^2}[1 - z_0 G(z_0)], \tag{9.75}$$

and

$$\sigma_1^1 = -\frac{iNe^2}{mk}\left(\frac{m}{2KT_\parallel}\right)^{1/2}\left[G(z_{-1}) + \left(\frac{T_\perp}{T_\parallel} - 1\right)\frac{z_{-1}G(z_{-1}) - 1}{z_0}\right], \tag{9.76}$$

where z_n are defined as before (9.39). There is a similar expression for σ_{-1}^{-1}, which has z_1 in place of z_{-1}. Equations (9.64) again separate into three normal modes with just one of E^0, $E^{\pm 1}$ non-zero respectively. The longitudinal mode is just the plasma oscillation, and the other two are circularly polarized electromagnetic waves satisfying the dispersion relation

$$\frac{c^2k^2}{\omega^2} = 1 + \frac{1}{i\omega\varepsilon_0}\sum_{\text{species}} \sigma_\mu^\mu \qquad (\mu = \pm 1). \tag{9.77}$$

For high frequency radio waves, involving effectively only the electrons, the phase speed of the waves is usually so high compared with the thermal speed that the variables z_0, z_1 and z_{-1} are very large; applying the appropriate expansion of the G-function leads to the cold theory as a very good approximation and to (9.70) as a first correction; the same holds for Alfvén waves of long wavelength. In other cases, Landau damping can be important, and one must proceed numerically from (9.76); for instance we could investigate the competition between Landau damping and the tendency for instability when (9.74) holds.

The damping is especially severe if $z_{\pm 1}$ is of order unity. An example is the whistler mode close to $\omega = \Omega$, where the cold theory predicts waves whose speed is much less than c, so that even in a plasma of modest temperature, like the ionosphere, $(\omega - \Omega)/k$ and $(KT_{\parallel}/m)^{1/2}$ may be comparable. Physically the damping is due to particles whose speed along the field is such that the wave frequency, as Doppler shifted, appears to be the gyro-frequency. The electric field vector of the wave rotates in the same sense as the gyration of the particle, so leading to resonant interaction, as will become clearer in §9.4. When the temperatures are unequal, (9.77) reveals unstable radio waves in some cases, in addition to the "firehose" instability already described.

9.3.5 Expansion in series: cyclotron resonances

As a final way of looking at the conductivity tensor for a Maxwellian plasma, we follow the procedure, already introduced in §9.2.2 (electrostatic approximation), in which the resonant effects at the harmonics of the gyro-frequency are exhibited in a series expansion. The integral I, equation (9.51), evaluated for small $\mathbf{a}$ and $\mathbf{b}$, contains all the required information, $\mathbf{p}$ being defined by (9.45) with $\mathbf{a}$ not yet set equal to zero. So we need an explicit formula for

$$-KT^{(q)}(p_q + b_q)(p_q + b_q)/2m. \qquad (9.78)$$

The contribution from the component parallel to the field is

$$-KT_{\parallel}(k_{\parallel}t + a_z + b_z)^2/2m, \qquad (9.79)$$

where x, y, z will be used to denote ordinary Cartesian coordinates. Writing $\mathbf{k} = (k_{\perp}, 0, k_{\parallel})$ as before and using (9.22) and $\mathbf{V} = \dot{\mathbf{X}}$, the remainder of (9.78) turns out to be

$$-KT_{\perp}(\mathbf{p}_{\perp} + \mathbf{b}_{\perp})^2/2m = -\lambda + \alpha + (\lambda + \beta)\cos\Omega t + \gamma \sin \Omega t. \qquad (9.80)$$

Here λ is as previously, (9.33), and α, β, γ are small quantities

$$\begin{aligned}
\alpha &= -\frac{KT_{\perp}}{m}\left[\tfrac{1}{2}(\mathbf{a}_{\perp}^2 + \mathbf{b}_{\perp}^2) - \frac{k_{\perp}}{\Omega}(a_y - b_y)\right] \\
\beta &= -\frac{KT_{\perp}}{m}\left[\mathbf{a}_{\perp}\cdot\mathbf{b}_{\perp} + \frac{k_{\perp}}{\Omega}(a_y - b_y)\right] \qquad (9.81) \\
\gamma &= -\frac{KT_{\perp}}{m}\left[a_x b_y - a_y b_x + \frac{k_{\perp}}{\Omega}(a_x + b_x)\right].
\end{aligned}$$

To take the same step as at (9.38), we need the expansion, analogous to (9.34), of an expression of the type $\exp(\lambda \cos \Omega t + \gamma \sin \Omega t)$. This is rather cumbersome, but fortunately we only need it for small γ. From (9.34) itself one can obtain

$$e^{(\lambda\cos\Omega t + \gamma \sin \Omega t)} = \sum_{\infty=-n}^{\infty}\left[I_n - \frac{in\gamma}{\lambda}I_n + \tfrac{1}{2}\gamma^2(I_n - I_n'')\right]e^{in\Omega t} + O(\gamma^3) \qquad (9.82)$$

(the Bessel functions all having argument λ). There is therefore an expansion of the form

$$e^{-KT_\perp(\mathbf{p}_\perp+\mathbf{b}_\perp)^2/2m} \equiv e^{-\lambda} \sum_{-\infty}^{\infty} U_n \, e^{in\Omega t} \tag{9.83}$$

and some rather tedious algebra shows that, to the first order in each of $\mathbf{a}$ and $\mathbf{b}$ (but including products of one component from each of these vectors)

$$\begin{aligned} U_n = I_n &+ \frac{iKT_\perp k_\perp}{m\Omega}\left[\frac{nI_n}{\lambda}(a_x + b_x) + i(I_n' - I_n)(a_y - b_y)\right] \\ &+ \frac{KT_\perp}{m}\Big[-\frac{n^2}{\lambda}I_n a_x b_x + in(I_n' - I_n)(a_x b_y - a_y b_x) \\ &\qquad + \left(2\lambda I_n' - 2\lambda I_n - \frac{n^2}{\lambda}I_n\right)a_y b_y\Big]. \end{aligned} \tag{9.84}$$

Working to the same approximation in a_z and b_z our integral (9.51) becomes

$$\begin{aligned} I \approx \left(1 - \frac{KT_\parallel}{m} a_z b_z\right) \exp(-\lambda) \sum_{-\infty}^{\infty} U_n \int_0^\infty \exp\Big[-i(\omega - n\Omega)t \\ - \frac{KT_\parallel}{m} k_\parallel (a_z + b_z)t - \frac{KT_\parallel}{2m} k_\parallel^2 t^2\Big]\, dt. \end{aligned} \tag{9.85}$$

The integral in this last formula is of our standard type (9.30) and is in fact

$$-i(m/2KT_\parallel k_\parallel^2)^{1/2}\, G[z_n - i(KT_\parallel/2m)^{1/2}(a_z + b_z)], \tag{9.86}$$

where z_n is as before (9.39).

These formulas give I to sufficient accuracy to enable us to evaluate (9.50); for we only need first derivatives with respect to $\mathbf{a}$, and the only second derivatives with respect to $\mathbf{b}$ required are those involving one component along the magnetic field and one across it, as remarked just after (9.51). Carrying this out, (9.51) yields

$$\boldsymbol{\sigma} = \frac{Ne^2}{mik_\parallel}\left(\frac{m}{2KT_\parallel}\right)^{1/2} e^{-\lambda} \sum_{n=-\infty}^{\infty} \mathbf{A}_n + \frac{Ne^2}{mi\omega}\left(1 - \frac{T_\perp}{T_\parallel}\right)e^{-\lambda} \sum_{n=-\infty}^{\infty} G'(z_n)\, \mathbf{B}_n, \tag{9.87}$$

where

$$\mathbf{A}_n = \begin{bmatrix} \dfrac{n^2 I_n G(z_n)}{\lambda} & -in(I_n' - I_n)G(z_n) & -\left(\dfrac{T_\perp}{T_\parallel}\right)^{1/2} \dfrac{nI_n G'(z_n)}{(2\lambda)^{1/2}} \\ in(I_n' - I_n)G(z_n) & \left(\dfrac{n^2 I_n}{\lambda} + 2\lambda I_n - 2\lambda I_n'\right)G(z_n) & -i\left(\dfrac{T_\perp}{T_\parallel}\right)^{1/2} \dfrac{\lambda^{1/2}(I_n' - I_n)G'(z_n)}{2^{1/2}} \\ -\left(\dfrac{T_\parallel}{T_\perp}\right)^{1/2} \dfrac{nI_n G'(z_n)}{(2\lambda)^{1/2}} & i\left(\dfrac{T_\parallel}{T_\perp}\right)^{1/2} \dfrac{\lambda^{1/2}(I_n' - I_n)G'(z_n)}{2^{1/2}} & I_n G(z_n) \end{bmatrix}$$

$$\mathbf{B}_n = \begin{bmatrix} \dfrac{n^2 I_n}{2\lambda} & -\dfrac{in(I_n' - I_n)}{2} & -\dfrac{1}{2\lambda}\dfrac{k_\perp}{k_\parallel} n^2 I_n \\ \dfrac{in(I_n' - I_n)}{2} & \dfrac{1}{2}\left(\dfrac{n^2 I_n}{\lambda} + 2\lambda I_n - 2\lambda I_n'\right) & \dfrac{-in(I_n' - I_n)k_\perp}{2k_\parallel} \\ \left(\dfrac{T_\parallel}{T_\perp}\right)^{1/2} \dfrac{nI_n z_n}{(2\lambda)^{1/2}} & -i\left(\dfrac{T_\parallel}{T_\perp}\right)^{1/2} \dfrac{\lambda^{1/2}(I_n' - I_n) z_n}{2^{1/2}} & -\left(\dfrac{T_\parallel}{T_\perp}\right)^{1/2} \dfrac{k_\perp}{k_\parallel} \dfrac{nI_n z_n}{(2\lambda)^{1/2}} \end{bmatrix}$$

It will be noted that the second series disappears if $T_\perp = T_\parallel$; it can be traced back to the $\mathbf{v} \times \mathbf{B}_1$ term in Boltzmann's equation.

These results are too cumbersome to enable us to explore, in a book of reasonable length, all the possible modes of propagation of electromagnetic waves in plasmas with all values of the fundamental ratios $\Omega : \omega_p$ and $T_\perp : T_\parallel$. Equation (9.87) is given in full only for reference. The most important cases in which it simplifies to something manageable have already been treated more directly: they are the "cold" approximation (Chapter 5), and the "warm" correction (§9.3.3), the electrostatic approximation (§9.2), and waves along the magnetic field (§9.3.4). (In this last case, we have $\lambda \to 0$, and the properties of $I_n(\lambda)$ and $I_n'(\lambda)$ ensure that only the terms $n = 0, \pm 1$ survive in the summation; there are, therefore, no resonances at multiples of the gyro-frequency, but only at the fundamental. This is what we found earlier, cf. equation (9.76). For waves in any other direction all harmonics give resonant effects.) One remaining tractable case is that of propagation perpendicular to the magnetic field, in which we can enquire into the effect upon the hybrid and Bernstein modes (§9.2.2) of the finite value of c; or alternatively the effect upon the magneto-ionic waves (equations 5.69 and 5.70) of a finite temperature. So we let $k_\parallel \to 0$, making use of (8.66) for the G-function. This leads to

$$\boldsymbol{\sigma} = \frac{Ne^2}{mi} e^{-\lambda} \sum_{-\infty}^{\infty} \frac{1}{\omega - n\Omega} \begin{bmatrix} \dfrac{n^2 I_n}{\lambda} & -in(I_n' - I_n) & 0 \\ in(I_n' - I_n) & \dfrac{n^2}{\lambda} I_n + 2\lambda I_n - 2\lambda I_n' & 0 \\ 0 & 0 & \left[1 + \dfrac{n\Omega}{\omega}\left(\dfrac{T_\parallel}{T_\perp} - 1\right)\right] I_n \end{bmatrix} \tag{9.88}$$

Considering only a single species, say electrons, and carrying out the procedure described by equations (8.51)–(8.55) leads to two possibilities, the "ordinary" wave, which satisfies

$$\frac{c^2 k^2}{\omega^2} = 1 - \frac{\omega_p^2}{\omega} e^{-\lambda} \sum_{-\infty}^{\infty} \frac{1 + \dfrac{n\Omega}{\omega}\left(\dfrac{T_\parallel}{T_\perp} - 1\right)}{\omega - n\Omega} I_n \tag{9.89}$$

and has its electric field purely in the z-direction, and the "extraordinary" wave, satisfying

$$\left\{1 - \frac{\omega_p^2}{\omega}\frac{e^{-\lambda}}{\lambda}\sum_{-\infty}^{\infty}\frac{n^2 I_n}{\omega - n\Omega}\right\}\left\{1 - \frac{c^2k^2}{\omega^2} - \frac{\omega_p^2 e^{-\lambda}}{\omega}\sum_{-\infty}^{\infty}\frac{n^2 I_n/\lambda + 2\lambda I_n - 2\lambda I_n'}{\omega - n\Omega}\right\}$$

$$= \left\{\frac{\omega_p^2 e^{-\lambda}}{\omega}\sum_{-\infty}^{\infty}\frac{n(I_n' - I_n)}{\omega - n\Omega}\right\}^2 \tag{9.90}$$

with its electric field in the xy-plane.

So far as the ordinary wave is concerned, the familiar Appleton–Hartree solution arises, of course, from letting $\lambda \to 0$ in (9.89), whereupon only the $n = 0$ term survives. However, λ is very small whenever

$$k_\perp \ll \Omega(m/KT_\perp)^{1/2}, \tag{9.91}$$

i.e. the wavelength is greater than the thermal Larmor radius. So in practice the Appleton–Hartree value $k = (\omega^2 - \omega_p^2)^{1/2}/c$ is nearly always a very good approximation to one of the solutions of (9.89). Also, when the reverse of (9.91) holds we have essentially free space propagation, $k = \omega/c$, where $\omega \gg \Omega(mc^2/KT_\perp)^{1/2}$, a result which would almost certainly agree again with the Appleton–Hartree formula since this latter frequency is very likely to be well above the plasma frequency. The only chance for new solutions is to let $\omega \approx n\Omega$ for $n = 1, 2, \ldots$, in which case (9.89) may be satisfied by suitable adjustment of the small quantity $\omega - n\Omega$. This has to be extremely small indeed except when $\lambda \approx 1$, that is when the wavelength is comparable with the Larmor radius. The situation is similar to that of the Bernstein modes (§9.2.2); there is a branch of the dispersion relation close to each line $\omega = n\Omega$. This time the solutions represent radio waves, and $n = 1$ is included.

As for the extraordinary wave, the "cold" results again follow from $\lambda \to 0$, as is readily verified (this time the $n = \pm 1$ terms survive); the "hot electrostatic" results follow from $c \to \infty$, which simply requires the vanishing of the first bracket on the left of (9.90), so reverting to (9.36) (using $\Sigma n I_n = 0$). The circumstances in which these are valid approximations follow from reconsidering (9.90): at wavelengths much greater than the Larmor radius the cold approximation is good except when the frequency is very close to a harmonic of the gyro-frequency, while if $\omega \ll ck$ the electrostatic approximation is good.

Assuming that $KT_\perp \ll mc^2$ (as is inherent in our non-relativistic treatment), our diagram (Fig. 9.1) needs modification only in an extremely narrow region, $k \ll \omega/c$, close to the vertical axis. The exact form of this modification depends in a rather complicated way on ω_p/Ω. The solution at what we called the "hybrid" frequency (9.37) appears in the limit $k \to 0$ in the hot electrostatic approximation, yet as $k \to \infty$ in the cold electrodynamic approximation (Fig. 5.5). It still appears in the exact theory, since what is really required is

$$(\omega_p^2 + \Omega^2)^{1/2}/c \ll k \ll \Omega(m/KT_\perp)^{1/2}$$

and such a range of k exists in the non-relativistic case, assuming ω_p/Ω is not unreasonably large.

We can also identify the extraordinary wave as the one involved in synchrotron radiation. We noted in §3.3 that for individual high-energy particles such radiation is emitted mainly in the plane of gyration, and at harmonics of the gyro-frequency; it is easily checked that the polarization is the correct one. To discuss possible collective effects (e.g. the appropriate optical depth) in the synchrotron radiation due to a very hot gas a relativistic version of (9.90) would be needed. Such considerations are important both in controlled fusion work, where synchrotron radiation may cause excessive loss of energy, and in radio astronomy. Even when $KT_\perp \ll mc^2$ the behavior of the dispersion relation very near to the harmonics of the gyro-frequency needs a relativistic discussion.

Finally, these results give some insight into the regions of validity of the cold and electrostatic approximations. Unfortunately one cannot easily predict *a priori* whether either or both can be applied to a given problem. Roughly speaking the cold treatment is justified if in the final results the phase speed of the waves is well in excess of the thermal speeds of most of the particles, except that for frequencies sufficiently close to harmonics of the gyro-frequency the approximation breaks down. The electrostatic approximation is usually justified if the phase speed turns out to be well below the speed of light, but the basis of the approximation is that one takes account only of the interaction through space charge, and not magnetic effects. An exception is therefore necessary for waves traveling along the unperturbed field, for which the longitudinal and transverse fields are completely decoupled; there are then pure plasma oscillations with arbitrarily high phase speed and pure radio waves (in the whistler mode) with very low phase speed. Another exception is necessary for cases where the space charge effects cancel on summing over the species, so leaving "slow" waves which are nevertheless electromagnetic—for instance low-frequency Alfvén waves in arbitrary directions. For further details of the waves governed by (9.90), or its generalization to include positive ions, or relativistic effects, the reader is referred to the papers of Shkarofsky (1966, 1968) and Fredricks (1968).

The reader may have noted that the limits $T \to 0$ and $c \to \infty$ both involve mathematical non-uniformity. The same is true of $\theta \to 0$ or $\theta \to \frac{1}{2}\pi$, and even more so for $\mathbf{B}_0 \to 0$. If one applies two or more of these limits (insofar as they are compatible) the results may depend on the order in which they are carried out. Reintroducing the effects which are ignored in these limits involves singular perturbations. (Yet another example, not treated here, would be the consequences of a slight spatial inhomogeneity.) This fact underlies many of the puzzling features of plasma physics. The most general treatment of waves is clearly impossibly cumbersome, yet if one considers the various limits in turn (applying them at the start of each calculation) it is hard to fit the results into a unified picture. This difficulty is of course well known in other fields, for instance in fluid dynamics, where viscosity and compressibility both involve singular perturbations.

9.4 GENERAL DISTRIBUTIONS

9.4.1 Formal calculation

We have so far concentrated on Maxwellian distributions because of their natural importance, and the fact that for many purposes they are sufficiently illustrative. We now turn to a brief treatment of the most general unperturbed distributions permissible when there is a magnetic field, namely those of the type $f_0(v_\perp, v_\parallel)$. Formulas for the conductivity tensor will be given, together with its reduction in the electrostatic approximation, and we shall gain more insight into the interaction between waves and individual particles; this was obscured in the previous work as we integrated over velocity space at a rather early stage.

We return to the fundamental equations (9.11)–(9.15). The integral J (a scalar quantity) has to be calculated to the first order in the small parameter $\mathbf{a}$ in order to give the charge and current densities ρ and $\mathbf{j}$, in the form $J = \rho + i\mathbf{a}\cdot\mathbf{j}$. As before, it will be convenient to set $\mathbf{k} = (k_\perp, 0, k_\parallel)$, and owing to the axial symmetry with respect to the magnetic field, to write $\mathbf{v} = (v_\perp \cos\phi, v_\perp \sin\phi, v_\parallel)$. The unperturbed orbit, $\mathbf{X}(t)$, has already been given, (9.22), and leads to

$$\mathbf{k}\cdot\mathbf{X}(t) = \frac{k_\perp v_\perp}{\Omega}[\sin(\Omega t - \phi) + \sin\phi] + k_\parallel v_\parallel t. \tag{9.92}$$

Further, we can write $\mathbf{a}\cdot\mathbf{V}(t)$ in a concise form using the polarized coordinates and (9.58):

$$\mathbf{a}\cdot\mathbf{V} = a_\mu V^\mu = a_\mu v^\mu\, e^{-i\mu\Omega t} = a_\mu \bar{v}^\mu\, e^{i\mu(\phi-\Omega t)}.$$

Here $\bar{v}^\mu$ means $v_\parallel$ if $\mu = 0$, and $v_\perp/2^{1/2}$ if $\mu = \pm 1$. So we can write

$$e^{i\mathbf{a}\cdot\mathbf{V}} \approx 1 + ia_\mu \bar{v}^\mu\, e^{i\mu(\phi-\Omega t)}. \tag{9.93}$$

Lastly, we need the quantity $h(\mathbf{v})$ given by (9.14) and (9.15). Using $\mathbf{B}_1 = \omega^{-1}\mathbf{k}\times\mathbf{E}$ and noting that

$$(\mathbf{v}\times\mathbf{B}_1)\cdot\frac{\partial f_0}{\partial \mathbf{v}} = \mathbf{B}_1\cdot\left(\frac{\partial f_0}{\partial \mathbf{v}}\times\mathbf{v}\right) = \left[v_\perp\frac{\partial f_0}{\partial v_\parallel} - v_\parallel\frac{\partial f_0}{\partial v_\perp}\right]\mathbf{B}_1\cdot(-\sin\phi, \cos\phi, 0),$$

one readily obtains

$$(\mathbf{E} + \mathbf{v} \times \mathbf{B}_1)\cdot\frac{\partial f_0}{\partial \mathbf{v}} = \sum_{1,0,-1} C^\nu e^{-i\nu\phi}, \tag{9.94}$$

where

$$C^0 = \frac{\partial f_0}{\partial v_\parallel} E^0$$

and for $\nu = \pm 1$,

$$2^{1/2}C^\nu = \frac{\partial f_0}{\partial v_\perp}E^\nu + \omega^{-1}\left[v_\perp \frac{\partial f_0}{\partial v_\parallel} - v_\parallel \frac{\partial f_0}{\partial v_\perp}\right](k_\parallel E^\nu - 2^{-1/2}k_\perp E^0). \tag{9.95}$$

Thus to the required approximation our integral J becomes

$$J \approx -\frac{e^2}{m}\int_{v_\parallel=-\infty}^{\infty}\int_{v_\perp=0}^{\infty}\int_{\phi=0}^{2\pi}\int_{t=0}^{\infty} \exp\left\{-i(\omega - k_\parallel v_\parallel)t + \frac{ik_\perp v_\perp}{\Omega}[\sin(\Omega t - \phi) + \sin\phi]\right\}$$

$$\times \sum_{\mu,\nu}\{1 + ia_\mu \bar{v}^\mu \exp[i\mu(\phi - \Omega t)]\} C^\nu \exp(-i\nu\phi)\, v_\perp\, dv_\perp\, dv\, d\phi\, dt. \tag{9.96}$$

In contrast to the previous development, we take the integral over t first, then that over ϕ. The other two integrations follow when f_0 is specified. So consider

$$\int_{\phi=0}^{2\pi}\int_{t=0}^{\infty} \exp\left\{-i(\omega - k_\parallel v_\parallel + \mu\Omega)t + i(\mu - \nu)\phi + i\zeta[\sin(\Omega t - \phi) + \sin\phi]\right\} d\phi\, dt, \tag{9.97}$$

where $\zeta = k_\perp v_\perp/\Omega$. The t-integration is elementary if one uses the identity (cf. 9.34)

$$e^{i\zeta \sin\theta} \equiv \sum_{n=-\infty}^{\infty} e^{in\theta} J_n(\zeta). \tag{9.98}$$

When this is carried out (in the causal limit), the integration over ϕ which remains is of the type

$$\int_0^{2\pi} \exp(i\zeta \sin\phi - im\phi)\, d\phi = 2\pi J_m(\zeta), \tag{9.99}$$

i.e. merely the inverse Fourier relation to (9.98). On completing this, (9.97) becomes

$$-2\pi i \sum_{n=-\infty}^{\infty} \frac{J_n(\zeta) J_{n+\nu-\mu}(\zeta)}{\omega - k_\parallel v_\parallel - (n - \mu)\Omega}.$$

Renaming n, this can also be written

$$-2\pi i \sum_{n=-\infty}^{\infty} \frac{J_{n+\mu}(\zeta)J_{n+\nu}(\zeta)}{\omega - k_\parallel v_\parallel - n\Omega}. \tag{9.100}$$

Thus

$$J = \frac{2\pi i e^2}{m}\int_{v_\parallel=-\infty}^{\infty}\int_{v_\perp=0}^{\infty} \sum_{\mu,\nu} (\delta_0^\mu + i a_\mu \bar{v}^\mu)C^\nu \times \sum_{n=-\infty}^{\infty} \frac{J_{n+\mu}(\zeta)J_{n+\nu}(\zeta)}{\omega - k_\parallel v_\parallel - n\Omega} v_\perp \, dv_\perp \, dv_\parallel, \tag{9.101}$$

the summations over μ and ν extending through 1, 0 and -1. These summations can be carried out by noting (9.95) and the recurrence formulas

$$J_{n+1} + J_{n-1} = \frac{2n}{\zeta} J_n, \qquad J_{n-1} - J_{n+1} = 2J_n'.$$

One finds (returning now to Cartesian components)

$$\sum_\mu (\delta_0^\mu + i a_\mu \bar{v}^\mu) J_{n+\mu} = \left(1 + \frac{i n v_\perp a_x}{\zeta} + i v_\parallel a_z\right) J_n - a_y v_\perp J_n' \tag{9.102a}$$

$$\sum_\nu C^\nu J_{n+\nu} = \left\{\frac{\partial f_0}{\partial v_\parallel} E_z + \frac{n}{\zeta}\left[E_x \frac{\partial f_0}{\partial v_\perp} + \frac{1}{\omega}\left(v_\perp \frac{\partial f_0}{\partial v_\parallel} - v_\parallel \frac{\partial f_0}{\partial v_\perp}\right)(k_\parallel E_x - k_\perp E_z)\right]\right\} J_n$$
$$- i\left\{\frac{\partial f_0}{\partial v_\perp} + \frac{k_\parallel}{\omega}\left(v_\perp \frac{\partial f_0}{\partial v_\parallel} - v_\parallel \frac{\partial f_0}{\partial v_\perp}\right)\right\} E_y J_n'. \tag{9.102b}$$

These are to be inserted into (9.101); the nth term then involves the Bessel function of solely the nth order. To find the charge density we simply set $\mathbf{a} = 0$, and as one generally needs this only in the electrostatic approximation, we have $\mathbf{E} = i\phi(k_\perp, 0, k_\parallel)$, where ϕ is the electrostatic potential. This leads to

$$\frac{\rho}{\phi} = -\frac{2\pi e^2}{m}\int_{v_\parallel=-\infty}^{\infty}\int_{v_\perp=0}^{\infty} \sum_{-\infty}^{\infty} \frac{J_n^2(\zeta)\left[\dfrac{\partial f_0}{\partial v_\parallel} k_\parallel + \dfrac{n}{\zeta}\dfrac{\partial f_0}{\partial v_\perp} k_\perp\right]}{\omega - k_\parallel v_\parallel - n\Omega} v_\perp \, dv_\perp \, dv_\parallel \tag{9.103}$$

and the dispersion relation for plasma oscillations is obtained by combining this with

$$\varepsilon_0 k^2 \phi = \sum_{\text{species}} \rho.$$

When $k_\perp = 0$, so $\zeta = 0$, all Bessel functions save J_0 vanish, and the results for no magnetic field are easily recovered. When $k_\parallel = 0$, we find

$$\frac{\rho}{\phi} = -\frac{2\pi e^2 \Omega}{m}\int_0^\infty \sum_{-\infty}^{\infty} \frac{n J_n^2(\zeta)(dF/dv_\perp)\, dv_\perp}{\omega - n\Omega} \tag{9.104}$$

where $$F(v_\perp) = \int_{-\infty}^{\infty} f_0(v_\perp, v_\parallel)\, dv_\parallel$$

and we notice the resonant effects which occur for $\omega = n\Omega$.

In the same way (9.101) also gives the usual equation $\mathbf{j} = \boldsymbol{\sigma}\cdot\mathbf{E}$, where σ_{pq} is obtained from the coefficients of ia_p and E_q in (9.101), after using equations (9.102) again; thus

$$\boldsymbol{\sigma} = \frac{2\pi e^2 i}{m} \int_{v_\perp=0}^{\infty} \int_{v_\parallel=-\infty}^{\infty} \sum_{-\infty}^{\infty} \frac{v_\perp\, dv_\perp\, dv_\parallel}{\omega - k_\parallel v_\parallel - n\Omega}$$

$$\times \begin{bmatrix} \frac{v_\perp}{\zeta^2}\left(P + \frac{k_\parallel}{\omega}Q\right)n^2 J_n^2 & -\frac{iv_\perp}{\zeta}\left(P + \frac{k_\parallel}{\omega}Q\right)nJ_nJ_n' & \frac{v_\perp}{\zeta}\left(R - \frac{k_\perp n}{\omega\zeta}Q\right)nJ_n^2 \\ \frac{iv_\perp}{\zeta}\left(P + \frac{k_\parallel}{\omega}Q\right)nJ_nJ_n' & v_\perp\left(P + \frac{k_\parallel}{\omega}Q\right)J_n'^2 & iv_\perp\left(R - \frac{k_\perp n}{\omega\zeta}Q\right)J_nJ_n' \\ \frac{v_\parallel}{\zeta}\left(P + \frac{k_\parallel}{\omega}Q\right)nJ_n^2 & -iv_\parallel\left(P + \frac{k_\parallel}{\omega}Q\right)J_nJ_n' & v_\parallel\left(R - \frac{k_\perp n}{\omega\zeta}Q\right)J_n^2 \end{bmatrix} \quad (9.105)$$

where

$$P = \frac{\partial f_0}{\partial v_\perp}, \qquad Q = v_\perp \frac{\partial f_0}{\partial v_\parallel} - v_\parallel \frac{\partial f_0}{\partial v_\perp}, \qquad R = \frac{\partial f_0}{\partial v_\parallel}, \quad (9.106)$$

the Bessel functions having argument ζ. By a trivial manipulation of the summation, it can be shown that the first two members of the last column of the matrix in (9.105) can be replaced respectively by the complex conjugates of the first two members in the bottom row, so leaving the matrix in Hermitian form. This is what would be expected from microscopic reversibility. Nevertheless $\boldsymbol{\sigma}$ itself is not anti-Hermitian, owing to the Landau prescription, and as we know this leads to macroscopic irreversibility.

Waves traveling along the magnetic field again involve some simplification, for they include the simple plasma oscillations, together with circularly polarized radio waves. To see this it is best to return to (9.101), as $\boldsymbol{\sigma}$ is diagonal in the polarized co-ordinates. The summation over n reduces to a single term, as J_0 is the only non-vanishing Bessel function. Noting again (9.95) one finds (for $\mu = \pm 1$)

$$\sigma_\mu^\mu = \frac{\pi e^2 i}{m} \int_{v_\parallel=-\infty}^{\infty} \int_{v_\perp=0}^{\infty} \frac{\dfrac{\partial f_0}{\partial v_\perp} + \dfrac{k_\parallel}{\omega}\left(v_\perp \dfrac{\partial f_0}{\partial v_\parallel} - v_\parallel \dfrac{\partial f_0}{\partial v_\perp}\right)}{\omega - k_\parallel v_\parallel + \mu\Omega}\, v_\perp^2\, dv_\perp\, dv_\parallel. \quad (9.107)$$

There is also considerable simplification of (9.105) when $k_\parallel = 0$, whereby the $v_\parallel$ integration can easily be carried out; this leads to the generalization of (9.88).

All the sums arising in (9.105) can be related, by elementary identities, to a single type, which can itself be summed explicitly. To be specific, there is an identity*

$$\sum_{n=-\infty}^{\infty} \frac{J_n^2(\zeta)}{\xi - n} \equiv \pi \operatorname{cosec} \pi\xi \, J_\xi(\zeta)\, J_{-\xi}(\zeta) \tag{9.108}$$

in which we would set $\xi = (\omega - v_\| k_\|)/\Omega$. By this means, $\boldsymbol{\sigma}$ can be expressed in a form free of summations; this does not, however, seem to have been found very convenient in practice, as the integrals over $v_\perp$ and $v_\|$ then involve Bessel functions where both the order and argument are continuous variables.

9.4.2. Discussion

As with the Maxwellian case, the general formula for $\boldsymbol{\sigma}$ is rather formidable and has not been used in full generality. Calculations have been carried out for electrostatic waves traveling perpendicular to the magnetic field (equation 9.104), mainly to search for instabilities, and (9.107) has been used in connection with generation of noise by energetic particles in natural or laboratory plasmas. These topics will be mentioned in Chapter 10.

An important aspect of this calculation is that it reveals which particles are involved in strong interaction with the wave in cases where $\mathrm{Im}(\omega) \approx 0$. In the absence of a magnetic field the relevant particles are those for which $\omega = \mathbf{k}\cdot\mathbf{v}$, and their role was discussed in connection with Landau damping in the previous chapter. In the presence of a magnetic field the resonant particles are those for which

$$\omega = k_\| v_\| + n\Omega \tag{9.109}$$

where n is an integer; they bring about the singularities in the integrands of (9.103) and (9.105). When $k_\perp = 0$, only $n = 0, \pm 1$ arise, while if $k_\| = 0$ either there are no resonant particles ($\omega \neq n\Omega$) or all particles are resonant ($\omega = n\Omega$) leading to cyclotron resonances. The physical interpretation of (9.109) is as follows. For particles satisfying it, the wave frequency, as Doppler shifted by the steady motion along the magnetic field, is an exact multiple of the gyro-frequency. The particle's motion relative to its guiding center introduces a periodic modulation in the field which it samples. When **k** is in a general direction this modulation is not purely harmonic, but can be built up from harmonics, resonance occurring if one of these coincides with the apparent wave frequency. When **k** is parallel to the magnetic field, the wave field sampled by the particle is either unaffected by the gyration (plasma oscillation) or modulated harmonically (radio wave). Hence only $n = 0, \pm 1$ arise in that case.

---

* Obtained by integrating the function $\pi \operatorname{cosec} \pi z\, J_z(\zeta) J_{-z}(\zeta)\cdot(\xi - z)^{-1}$ around a large circle in the z-plane.

This description was lost in our previous calculation for Maxwellian distributions, where it was more expeditious to integrate over velocity space first, leaving the algebraic complexity concentrated on the t-integration. Our earlier results obtained in that way can of course be recovered from those of the present section, but this involves further laborious application of Bessel function identities. The reader who wishes to try will find a little assistance in Problem 8.

If it is suspected *a priori* that Im (ω) is small, the exciting or damping effect of the particles just referred to can be estimated by the same procedure as in §8.3.2. The real part of ω is determined by the bulk of the particles, which are non-resonant, and can be described to sufficient accuracy by the "cold" or "warm" (§9.3.3) dispersion relation. The small imaginary part is determined by the resonant particles, and can be found by an iteration using the Cauchy–Riemann equations. This will involve the residue of integrals such as (9.105), at the resonant velocity.

9.5 EFFECT OF COLLISIONS

9.5.1 Model equations

Another extension of our work on waves in a hot uniform plasma is to include collisions. As discussed in §7.2, this involves the use of an equation of the form

$$\frac{\partial f}{\partial t} + \mathbf{v}\cdot\frac{\partial f}{\partial \mathbf{x}} + \frac{e}{m}(\mathbf{E} + \mathbf{v} \times \mathbf{B})\cdot\frac{\partial f}{\partial \mathbf{v}} = \left(\frac{\partial f}{\partial t}\right)_c \tag{9.110}$$

as the starting point. Here $(\partial f/\partial t)_c$ represents the interactions between rather close particles, for which the smoothing-out procedure involved in using the self-consistent fields **E** and **B** is inaccurate. We have already mentioned that the actual construction of $(\partial f/\partial t)_c$ is a matter of great difficulty, and in any event different forms are required for the various types of collision, such as electron–electron, electron–neutral molecule, etc. Most expressions for $(\partial f/\partial t)_c$ are quite formidable, involving integral functionals of f itself, as for example in the original Boltzmann equation for neutral gases, where one integrates over all the possible binary collisions which the particle under observation might suffer.

Once $(\partial f/\partial t)_c$ is known, the procedure is as before. We take $f = f_0 + f_1$, etc., where f_0 satisfies the zero-order equation; this will in fact dictate that f_0 be a uniform Maxwellian distribution if there is no zero-order electric field. Then we consider harmonic perturbations of f, **E** and **B**, and solve in the causal limit, in the familiar way. Unfortunately this is prohibitively complicated for any properly derived form of $(\partial f/\partial t)_c$.

However, it is only in rather unusual circumstances that the results of such detailed calculations are required. In studying radio wave propagation, both for laboratory or natural plasmas, the thermal speed is usually unimportant since the wave speed is so high, and the magneto-ionic theory, with collisions included, as in §5.6, is adequate; this amounts to neglecting $\mathbf{v}\cdot\partial f/\partial\mathbf{x}$. For low-frequency, large-scale phenomena, collisions may be dominant so that a hydrodynamic treatment applies (see Chapter 11), and the left-hand side of (9.110) is a small perturbation. For rarefied plasmas, the right-hand side is negligible, leading to the work we have just carried out, with the typical features of Landau damping and the cyclotron resonances. Cases where all terms of (9.110) are of comparable order do not often arise, and when they do results of great accuracy are not required. All that is needed is to interpolate roughly between these extreme cases to see how the features of one give way to those of another as the collision frequency is changed. This can be achieved by using for $(\partial f/\partial t)_c$ a simpler formula arranged to have a similar mathematical form to the real one and of about the same magnitude. We shall call these "model" equations. Two examples of this procedure have been worked out in some detail; they cover different physical situations and are not to be regarded as rival models.

9.5.2 Close collisions. The BGK model

Bhatnagar, Gross and Krook (1954) (see also Gross and Krook, 1956) and many later writers have made use of the model equation

$$\left(\frac{\partial f}{\partial t}\right)_c = -\nu(f - f_{\max}). \tag{9.111}$$

Here $f_{\max}$ is a suitable Maxwellian distribution and ν is an empirical "collision frequency," which can depend on velocity if desired, but is usually left constant (as we shall do). This is often called the "BGK" equation. The idea behind it is that $-\nu f$ represents the rate of loss of particles, due to collisions, from a small element of phase space, while $+\nu f_{\max}$ represents the corresponding rate of gain of particles as the end product of collisions. The approximation involved in (9.111), especially if ν is regarded as constant, is that of disregarding the detailed statistics and dynamics of the collisions, and the fact that the velocity after a collision is correlated with that before. We specify $f_{\max}$ by fixing the density, the mean velocity and the temperature; they will depend in some way on f itself evaluated at the same point of physical space, and the same time. Thus (9.111) represents a purely local effect whereby particles are transferred abruptly across velocity space only, and at a rate ν. It therefore simulates the effect of close binary collisions in which there is a substantial change of velocity. The collisions could be imagined to be a Poisson process, occurring with probability $\nu\,dt$ in the time interval $t, t + dt$, and (9.111) clearly tends to establish a Maxwellian distribution in a time of the order a few times ν^{-1}. Since the collisions are binary, the equation is not really suitable for those between charged particles, though it could be used as a rough guide in that case. It is suitable for collisions between charged and neutral particles.

Where several species are involved, we extend (9.111) as follows

$$\left(\frac{\partial f_a}{\partial t}\right)_c = -\sum_b \nu_{ab}(f_a - f_{ab}). \tag{9.112}$$

This gives the rate of change of the distribution function for "a"-particles as a sum of the effect of collisions with "b"-particles. ν_{ab} is an appropriate collision frequency and f_{ab} is a Maxwellian distribution for the a-molecules, the parameters being determined by the local values of f_a and f_b. Clearly ν_{ab} will be proportional to N_b, the local number density of b-particles, so that $\nu_{ab} = C_{ab}N_b$, say. Then $C_{ab} = C_{ba}$ since the total number of a–b collisions per unit volume is the same whichever way it is calculated; apart from this C_{ab} can be chosen at will. f_{ab} is known once its density, drift velocity and temperature, say N_{ab}, $\mathbf{u}_{ab}$ and T_{ab} are specified. Now since the a–b collisions do not change the number of a-particles (we assume the collisions are elastic, so there are no chemical reactions, ionization, etc.), we must have

$$\int (f_a - f_{ab})\, d^3v = 0,$$

so that in fact $N_{ab} = N_a$ for each b. Further, the momentum gained by species a must equal that lost by species b. This leads to

$$m_a \nu_{ab} N_a(\mathbf{u}_a - \mathbf{u}_{ab}) + m_b \nu_{ba} N_b(\mathbf{u}_b - \mathbf{u}_{ba}) = 0,$$

i.e.

$$m_a(\mathbf{u}_a - \mathbf{u}_{ab}) + m_b(\mathbf{u}_b - \mathbf{u}_{ba}) = 0. \tag{9.113}$$

This imposes a constraint on $\mathbf{u}_{ab}$ and $\mathbf{u}_{ba}$ but does not completely determine them except that $\mathbf{u}_{aa} = \mathbf{u}_a$ in the case of collisions among like particles. Similar considerations of energy transfer yield a relation between T_{ab} and T_{ba}, namely

$$m_a(\mathbf{u}_a^2 - \mathbf{u}_{ab}^2) + 3K(T_a - T_{ab}) = m_b(\mathbf{u}_b^2 - \mathbf{u}_{ba}^2) + 3K(T_b - T_{ba}), \tag{9.114}$$

so that $T_{aa} = T_a$. So far as collisions between unlike particles are concerned, the freedom of choice still available after taking account of (9.113) and (9.114) may be deployed according to any other knowledge (theoretical or experimental) about the mechanics of collisions.

Our concern here is to use the BGK equation for wave propagation in a uniform plasma. We therefore need to solve it for harmonic perturbations with respect to a uniform state, in the now familiar way. In principle one must carry this out for every species (charged or neutral) present, and the equations will all be coupled through the $(\partial f/\partial t)_c$ terms. As an illustration we will consider the simplified case in which (9.110) is solved only for the electrons, the positive ions forming a fixed background, and the only collisions to be included are those made by the electrons with some other species, for instance stationary neutral molecules. For simplicity, let us take f_{max} of (9.111) to have the same density as f, zero mean velocity, and temperature equal to the ambient temperature. Writing $f = f_0 + f_1$ where f_0 is the unperturbed Maxwellian distribution, we have

$$f_{max}(\mathbf{x}, \mathbf{v}, t) = N(\mathbf{x}, t) f_0(\mathbf{v})/N_0, \tag{9.115}$$

where $N = N_0 + N_1$ in obvious notation. This has the effect that the electrons lose momentum at a rate $Nm\nu\mathbf{u}$ per unit volume. This is just what we assumed in the cold magneto-ionic theory (§5.6), and our present work yields the same results in the limit $T \to 0$. After linearization

$$\left(\frac{\partial f}{\partial t}\right)_c = -\nu[f_1(\mathbf{x}, \mathbf{v}, t) - N_1(\mathbf{x}, t) f_0(\mathbf{v})/N_0]. \tag{9.116}$$

The first term in the brackets is very easy to incorporate into our earlier work as it only involves the formal modification of replacing the angular frequency ω by $\omega - i\nu$; if this were all we had to do any solution $\omega = \omega_1$ of the collision-free dispersion equation would be replaced by a solution $\omega = \omega_1 + i\nu$, so leading to an additional damping factor $\exp(-\nu t)$. Some writers have indeed used this simple representation of collisions, though this is not really satisfactory since without its last term (9.116) does not conserve particles locally; the tendency for the electrons to relax towards a uniform Maxwellian distribution is represented in an exaggerated way since the equation requires particles to disappear where there is an excess and reappear where there is a deficiency in physical space as well as in velocity space. We ought to solve the full equation, which can be written

$$i(\omega - i\nu - \mathbf{k}\cdot\mathbf{v})f_1 + \frac{e}{m}(\mathbf{v} \times \mathbf{B}_0)\cdot\frac{\partial f_1}{\partial \mathbf{v}} = -\frac{e}{m}\mathbf{E}\cdot\frac{\partial f_0}{\partial \mathbf{v}} + \frac{N_1}{N_0}f_0. \tag{9.117}$$

Here there is no term from the perturbation magnetic field, $\mathbf{B}_1$, since we have been obliged to take f_0 isotropic. Let us write the left-hand side as $\mathscr{D}f_1$, where $\mathscr{D}$ is a differential operator in velocity space (or simply an algebraic factor if $\mathbf{B}_0 = 0$), and we have already learnt in great detail how to invert it. So formally

$$f_1(\mathbf{v}) = \mathscr{D}^{-1}\left[-\frac{e}{m}E_a\frac{\partial f_0}{\partial v_a} + \nu\frac{N_1}{N_0}f_0\right]. \tag{9.118}$$

Here $\mathscr{D}^{-1}$ operates on f_0 and $\partial f_0/\partial v_a$ but commutes with $\mathbf{E}$ or N_1 since these are constants as far as velocity space is concerned. Nevertheless (9.118) is not by itself an explicit solution of (9.117) as N_1 is unknown. To eliminate N_1, we integrate over velocity space

$$N_1 = \int f_1(\mathbf{v})\, d^3v = -\frac{e}{m}E_q \int \mathscr{D}^{-1} \frac{\partial f_0}{\partial v_q}\, d^3v + \frac{\nu N_1}{N_0} \int \mathscr{D}^{-1} f_0\, d^3v.$$

Solving for N_1

$$N_1 = \frac{-(e/m)E_q \int \mathscr{D}^{-1}(\partial f_0/\partial v_q)\, d^3v}{1 - (\nu/N_0) \int \mathscr{D}^{-1} f_0\, d^3v}. \tag{9.119}$$

This gives at once the space charge $N_1 e$. For f_1 itself we substitute N_1 back into (9.118), so leading to the current, $\mathbf{j}$, as usual. The conductivity tensor is easily found to be

$$\sigma_{pq} = \frac{e^2}{KT}\left[\int v_p \mathscr{D}^{-1}(v_q f_0)\, d^3v + \frac{\nu\left\{\int \mathscr{D}^{-1}(v_q f_0)\, d^3v\right\}\left\{\int v_p \mathscr{D}^{-1} f_0\, d^3v\right\}}{N_0 - \nu \int \mathscr{D}^{-1} f_0\, d^3v}\right] \tag{9.120}$$

(noting (9.46)).

When there is no magnetic field $\mathscr{D}$ is simply $i(\omega - i\nu - \mathbf{k}\cdot\mathbf{v})$ and the results are easily expressible in terms of the G-function, (9.30). Taking $\mathbf{k} = (k, 0, 0)$, $\boldsymbol{\sigma}$ is diagonal, and in the case of σ_{22} $(= \sigma_{33})$ the second term of (9.120) vanishes by symmetry, so that for radio waves we have just the first term. In this case the prescription of replacing ω of the collision-free theory by $\omega - i\nu$ is correct after all; this is not surprising as the extra term involving N_1 in (9.116) disappears for transverse waves. For σ_{11} the situation is slightly more complicated; however, for plasma oscillations it is more natural to construct the space charge from (9.119). It is

$$\rho = \frac{N_0 e^2 E}{KTk} \frac{zG(z) - 1}{i - \dfrac{\nu}{k}\left(\dfrac{m}{2KT}\right)^{1/2} G(z)}, \tag{9.121}$$

where

$$z = \frac{\omega - i\nu}{k}\left(\frac{m}{2KT}\right)^{1/2}$$

and E is the appropriate component (compare with the results of §8.7).

When there is a magnetic field the procedure is similar; the integrals involving $\mathscr{D}^{-1}$ are now of the Gordeyev type. For waves which are primarily transverse it again amounts to adding $-i\nu$ to ω, while for plasma oscillations in the electrostatic approximation there is a result analogous to (9.121) with G replaced by iI, I being the integral (9.29), again with ω modified as before.

These formulas do indeed show how damping (in addition to Landau damping) is introduced by collisions with a stationary background. Plasma oscillations are heavily damped if ν is comparable with ω_p; for waves traveling orthogonally to the magnetic field the cyclotron resonance effect is lost if ν is comparable with Ω. Numerical work showing the onset of damping has been carried out by several authors, but presentation of the details is beyond the scope of this book.

In more general cases where the Maxwellian distribution to be used in $(\partial f/\partial t)_c$ is to be given a drift velocity, or a temperature different from the ambient, (9.116) has additional terms, but these are easily handled in the same way. Having solved in velocity space as at (9.118), one has to construct N_1, $\mathbf{u}$ and T_1 for each species by integration over f_1 (T_1 being the perturbation in temperature), and finally eliminate the quantities we called $\mathbf{u}_{ab}$ and T_{ab} using (9.113) and (9.114) and the additional relations needed to specify the system, as mentioned earlier.

9.5.3 Distant collisions. A model Fokker–Planck equation

When the particles of a gas interact according to an inverse square law, collisions are not predominantly binary, that is to say $(\partial f/\partial t)_c$ for a "test" particle at $(\mathbf{x}, \mathbf{v}, t)$ does not derive mainly from the possibility that other particles approach very closely and abruptly deflect it. The cumulative effect of more distant, but more numerous, particles is more important. Of course the influence of particles considerably removed from the test particle is already included in the self-consistent field, and $(\partial f/\partial t)_c$ is supposed to account only for the error in neglecting the discreteness of the particles. In fact the major contributions to $(\partial f/\partial t)_c$ are due to electrons and ions within the Debye sphere of the test particle, but beyond the interparticle spacing, $N^{-1/3}$. We shall discuss this quantitatively, at least in simple cases, in Chapter 12. Here it suffices to say that when a test particle is subject to simultaneous "grazing" collisions in this way its progress in velocity space becomes a random walk, superposed on the ordered motion due to the macroscopic field. This yields an expression of the Fokker–Planck type

$$\left(\frac{\partial f}{\partial t}\right)_c = \frac{\partial}{\partial v_i}\left[-A_i f + \frac{1}{2}\frac{\partial}{\partial v_j}(B_{ij} f)\right]. \qquad (9.122)$$

Here A_i and B_{ij} are the "friction" and "diffusion" coefficients respectively; besides being functions of the $(\mathbf{x}, \mathbf{v}, t)$ at which $(\partial f/\partial t)_c$ is to be calculated they are complicated functionals of f (at the same $\mathbf{x}$ and t) itself. Continuing in the same spirit as the BGK equation, we look for simple coefficients A_i and B_{ij} to serve as a model for approximate calculations.

Such coefficients can indeed be found, and the problem of wave propagation reconsidered. Here we give only a brief sketch of the work involved, which is lengthy and laborious. For simplicity, consider only a single species, so that the collisions are among like particles. Then to ensure that their number, momentum and energy are conserved, we need

$$\int (1, \mathbf{v}, v^2)\left(\frac{\partial f}{\partial t}\right)_c d^3v = 0 \tag{9.123}$$

respectively. The first of these is automatic, whatever the form of A_i and B_{ij}. The other two are satisfied if

$$\begin{aligned} A_i &= -\nu(v_i - u_i) \\ B_{ij} &= (2\nu KT/m)\delta_{ij} \end{aligned} \tag{9.124}$$

where $\mathbf{u}$, T are the local drift velocity and temperature, and ν is a constant, which can be called the "collision frequency." (9.124) also ensures that $(\partial f/\partial t)_c$ vanishes identically for a Maxwellian distribution. The kinetic equation is then

$$\frac{\partial f}{\partial t} + \mathbf{v}\cdot\frac{\partial f}{\partial \mathbf{x}} + \frac{e}{m}(\mathbf{E} + \mathbf{v} \times \mathbf{B})\cdot\frac{\partial f}{\partial \mathbf{v}} = \nu\frac{\partial}{\partial \mathbf{v}}\cdot\left[(\mathbf{v} - \mathbf{u})f + \frac{KT}{m}\frac{\partial f}{\partial \mathbf{v}}\right]. \tag{9.125}$$

When this equation is linearized, the solution in velocity space for harmonic perturbations is rather more difficult than that for the collision-free equation, as the new equation is second order in the velocity. A procedure analogous to that of §9.1 is possible, however. The Green's function which takes the place of (9.10) is not simply a δ-function, but is rather a Gaussian function in both physical space and velocity space, as one might expect would follow from the introduction of a diffusion term. This leads eventually to a generalization of the Gordeyev integral, (9.29); in which the function $g(t)$ is replaced by

$$\frac{KT_0}{m}\left\{\frac{k_\parallel^2}{\nu^2}(\nu t - 1 + e^{-\nu t}) + \frac{(1 + \cos\chi)k^2}{2\Omega^2}[\cos\chi + \nu t - e^{-\nu t}\cos(\Omega t - \chi)]\right\}, \tag{9.126}$$

where $\chi = 2\tan^{-1}(\nu/\Omega)$.

Comparing with the collision-free expression, (9.25), it is not hard to see that the modifications due to the collisions introduce damping. In particular, it is of interest that one can gain some insight into the destruction of the cyclotron resonance for waves propagating normal to the field, i.e. $k_\parallel = 0$. The resonance occurs when (9.126) is to good approximation periodic as a function of t, and it is the terms νt and $e^{-\nu t}$ which tend to prevent this. The latter is effective if $\nu \gtrsim \Omega$, while the importance of the former depends on the quantity $\nu KT_0 k_\perp^2/m\Omega^3$. In all, the resonance survives if

$$\nu \ll \Omega \text{ or } \Omega^3 m/KT_0 k_\perp^2 \tag{9.127}$$

whichever is the less.

The physical interpretation here is the following. The resonance owes its existence to the periodic motion of the particles in the plane perpendicular to the magnetic field. When $\Omega^2 m/KTk^2 \gg 1$, that is the wavelength is greater than the Larmor radius, (9.127) simply reads $\nu \ll \Omega$. Just as for binary collisions, the question is whether a particle has a good chance of completing a Larmor rotation without suffering appreciable deflection. In the opposite case, the second of (9.127) applies. The reason why a more stringent condition than $\nu \ll \Omega$ is demanded is this. In the course of one Larmor rotation the particle is subjected to numerous slight deflections, causing an error in its velocity. This error is proportional to $(\nu t)^{1/2}$ (t being the time), as is characteristic of a diffusion process. A corresponding error in its position is accumulated, proportional to $\nu^{1/2}t^{3/2}$. For a typical thermal particle this deflection is about $\nu^{1/2}\Omega^{-3/2}(KT_0/m)^{1/2}$ after one cycle. If the wave is to maintain coherence, this deflection must be much less than the wavelength, so leading to $\nu \ll \Omega^3 m/KT_0 k^2$, even though the error in velocity will be small if $\nu \ll \Omega$. Here then is an example where "grazing" collisions have a quite different effect from an equivalent number of large-angle collisions.

This loss of the cyclotron resonance, through ion–ion collisions, can be serious for the ion waves discussed in §9.2.3, and has been of some experimental interest. In this connection we have published elsewhere (Dougherty, 1964) the details of the calculation outlined in the present subsection.

PROBLEMS

1. Derive equation (9.38).
2. Derive equation (9.61).
3. Give explicit form to the coefficients F_ν^μ of equation (9.61) in the case $T_\perp = T$, carrying out any limiting processes required if μ or ν vanish.
4. Derive the conductivity tensor (9.66) for a cold plasma by starting from first principles, neglecting thermal effects throughout, and using the $(1, 0, -1)$ coordinates.
5. Using the explicit form of (9.63), show that if ϵ is diagonal in $(1, 0, -1)$ coordinates, the three scalar components of (9.64) can be summed, after multiplying by suitable quantities, so that only $\mathbf{k}\cdot\mathbf{E}$ is involved. Hence obtain the dispersion relation (9.68).
6. Supply the details omitted in the course of §9.4.1.
7. Another method for solving (9.2), after introducing the harmonic factors in space and time, is to use cylindrical polar coordinates in velocity space. The only derivative of f_1 occurring is that with respect to the polar angle, so that the solution for f_1, in the form of an integral, is straightforward. The integration involved is essentially the same as that over t in the Green's function method. Investigate this, and show that the construction of the charge and current densities leads to fourfold integrals similar to (9.96). (See Bernstein, 1958.)

8. In a development which proceeds direct to general distributions, so leading to (9.103) and (9.105), the recovery of our results for the Maxwellian case must be achieved by carrying out the integrations over $v_{\parallel}$ and $v_{\perp}$. It should be possible to obtain (9.40) and (9.87) in this way. The reader may like to try, for example, the former of these! (Identities such as

$$\int_0^\infty x \exp(-x^2/2\lambda)[J_n(x)]^2\, dx \equiv \lambda \exp(-\lambda) I_n(\lambda)$$

are needed to expedite the integrations over $v_{\perp}$.)

9. Show that for a cold plasma in which only the electrons are mobile, (9.67) may be written

$$\varepsilon_\nu^\lambda = \left[1 - \frac{X}{1 - \lambda Y}\right] \delta_\nu^\lambda$$

where $X = \omega_p^2/\omega^2$, $Y = \Omega/\omega$, as in Chapter 5 (note Y changes sign for the negative charge). Now write $\mu = ck/\omega$ for the refractive index and show that (9.68) may be solved to yield (5.81); the substitution

$$\mu^2 = 1 - \frac{X}{1 + \zeta},$$

where ζ is a new variable, helps to keep the analysis concise.

REFERENCES

I. B. Bernstein, *Phys. Rev.*, **109**, 10 (1958).

P. L. Bhatnagar, E. P. Gross and M. Krook, *Phys. Rev.*, **94**, 511 (1954).

O. Buneman, *Phys. Fluids*, **4**, 669 (1961).

F. W. Crawford, *Radio Science*, **69D**, 789 (1965).

J. P. Dougherty, *Phys. Fluids*, **7**, 1788 (1964).

R. W. Fredricks, *J. Plasma Phys.*, **2**, 365 (1968).

G. V. Gordeyev, *J. Exp. Theor. Phys.* (*U.S.S.R.*), **6**, 660 (1952).

E. P. Gross, *Phys. Rev.*, **82**, 232 (1951).

E. P. Gross and M. Krook, *Phys. Rev.*, **102**, 593 (1956).

I. P. Shkarofsky, *Phys. Fluids*, **9**, 561 and 570 (1966).

I. P. Shkarofsky, *J. Geophys. Res.*, **73**, 4859 (1968).

CHAPTER 10

MICROINSTABILITIES

10.1 INTRODUCTION

In any dynamical system, when a state of equilibrium or steady motion has been discovered, an important question is whether the state will be stable or unstable. In the unstable case, a further important question is that of the rate of growth of the unstable modes, especially the most rapidly growing. In the case of plasmas the investigation of possible instabilities is of great practical interest, but unfortunately very involved. In the attempts to confine plasmas in the laboratory so that controlled nuclear fusion may take place, it is essential to devise a configuration which is either stable or, if unstable, is such that the fastest growth rate allows time for the nuclear reactions to occur before the plasma is much disrupted. Instabilities may also be responsible for such phenomena as anomalous diffusion, whereby plasmas can move across magnetic fields at a rate faster than one would expect from straightforward theory. In the astrophysical and space context there are many observations which seem to call for explanation in terms of instabilities —the emission of radio waves by galaxies, stars or the magnetosphere; sunspots; the generation of high energy particles, and several others.

One large class of instabilities to which plasmas are susceptible is the "hydromagnetic." These depend only on the more gross features of the plasma configuration, and may be investigated with the usual hydromagnetic equations, without a detailed consideration of the distribution in velocity space. The geometry in physical space is the essential feature of these instabilities. Examples are the flute instability, which can occur when there is a curved interface between a strong magnetic field and a plasma, and the various instabilities ("kinks" and "sausages") to which a self-pinched discharge is liable. This class of instabilities will not be considered here, as applications of the hydromagnetic equations are outside the general plan of this book. The subject is treated extensively elsewhere, for example in the book by S. Chandrasekhar (1961).

A second class of instabilities, to which we devote the present chapter, comprises those which arise essentially from considerations in velocity space. These are usually called "microinstabilities." In the simplest examples, the equilibrium situation is uniform in space and time, and the stability is usually investigated by considering (in a linearized way) all the possible plane waves. One therefore writes down the dispersion relation, as developed in the previous two chapters, and searches for solutions with **k** real and $\text{Im}(\omega) < 0$. The existence of such a solution or solutions indicates instability, since the slightest inhomogeneity (due, for example, to thermal fluctuations) at the initial instant will grow exponentially, assuming the relevant mode is excited. For plasmas with finite spatial extent, it may also be of interest to consider solutions with **k** complex, as discussed in §6.1; however, we shall not do so here. A still further question concerns the classification of the instabilities as convective or nonconvective, as defined in §6.1; again this would take us too far afield.

Three well-known examples of microinstabilities have already been described. The two-stream instability (§6.3.1) arises when two streams of "cold" electrons interpenetrate, though as we shall see shortly it makes only a minor difference if the streams have a thermal spread, provided it is not so great that they practically merge into one. Physically, the process is simply that of plasma oscillations in which the restoring force due to the bunching of the electrons occurs at an inappropriate phase on account of the streaming. The instability for radio waves propagating at right angles to the streams in such an arrangement was also derived (§6.3.2); this can be explained as a tendency for the streams to separate into a number of self-contracting "pinches." The firehose instability (§9.3.4) concerns transverse waves at very low frequency traveling parallel to an imposed magnetic field when the pressure is sufficiently anisotropic (equation 9.74). A physical picture for this is given in §11.6.3. The firehose instability could also be classed as "hydromagnetic" inasmuch as it can be derived from hydromagnetic equations provided these take account of the peculiar stress tensor (such equations will be given in the next chapter), but it differs from the other hydromagnetic instabilities in that it occurs in a uniform infinite medium. The same remarks apply to the "mirror" instability (§11.6.3).

There are also microinstabilities which depend on a non-uniform equilibrium state (e.g. an inhomogeneous magnetic field), yet cannot be derived hydromagnetically. These require the solution of Boltzmann's equation with spatial gradients in the unperturbed state, so involving work at a stage more complex than we have attempted here. Though progress has been made in this area, it is as yet rather sketchy, and we shall not enter into details in this book. The term "universal" has been applied to these instabilities. Their growth rates are often smaller than those typical of hydromagnetic or pure velocity-space instabilities, yet they are important in the sense that they are likely to be present as a residual effect when steps have been taken to exclude the others.

10.2 LONGITUDINAL OSCILLATIONS IN A FIELD-FREE PLASMA. THE PENROSE STABILITY CRITERION

By far the most complete discussion is available for those microinstabilities which are simply longitudinal plasma oscillations in the absence of a magnetic field. The governing equations for these waves have been given in Chapter 8. It should be recalled (see §8.6) that, except in cases where f_0 has sufficient symmetry, purely longitudinal and purely transverse waves do not exist; however, there are usually plasma oscillations which are almost completely longitudinal and may be described by Poisson's equation; this is just the electrostatic approximation, and will be assumed in this section. We may therefore take our waves along the x-axis and use a one-dimensional distribution function, as defined by (8.6). The dispersion relation (8.19) may be written

$$Z(V) = k^2, \tag{10.1}$$

where

$$Z(V) \equiv \sum_{\text{species}} \frac{e^2}{\varepsilon_0 m} \int_C \frac{df_0/dv}{v - V}\, dv. \tag{10.2}$$

Here $V = \omega/k$, and we have introduced a summation over species. With k real, and conventionally positive, C is determined by the usual Landau rule, so is simply the real axis if $\mathrm{Im}(V) < 0$ but has to be indented upwards otherwise. For an instability, (10.1) must be satisfied with $\mathrm{Im}(V) < 0$, and this

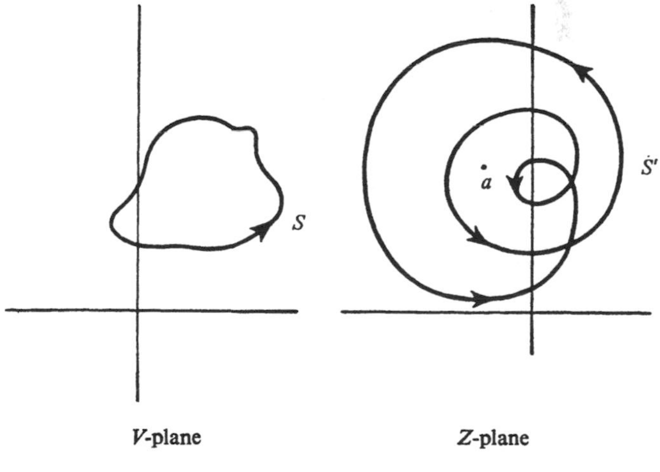

Fig. 10.1. Mapping of the V-plane onto the Z-plane.

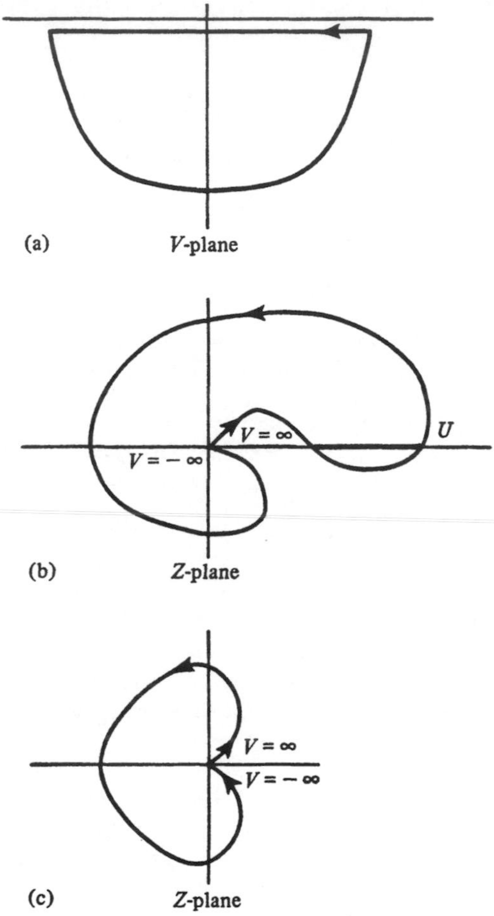

Fig. 10.2. The Nyquist diagram.

will be possible if $Z(V)$ can be real and positive with V in the lower half plane. We can find whether this is so by means of the well-known technique for locating zeros and poles of a function in the complex plane, often called the Nyquist diagram method. We wish to find the number of times a function $Z(V)$ attains a given value, say a, for V within a given domain D bounded by a simple curve S. We can do this by counting the number of zeros of the function $Z(V) - a$. A simple corollary of Cauchy's theorem* gives the number of zeros minus the number of poles; however, in our case we shall find

* See for instance E. T. Copson, *Theory of Functions of a Complex Variable* (Oxford University Press), p. 118.

that poles of $Z(V)$ do not arise. Accordingly the number of zeros is $(2\pi)^{-1}$ times the total increase in the argument of $Z(V) - a$ when V moves once, anticlockwise, around the boundary of D. This can be exhibited graphically as in Fig. 10.1. In the V-plane, V is taken once around S in the sense shown, and the corresponding values of $Z(V)$ plotted to give a curve S' in the Z-plane. The number of roots within D of the equation $Z(V) = a$ is simply the number of times the point a is encircled (anticlockwise) by S'; it would be two in the case shown.

To find unstable roots of (10.1) we take D to be the lower half plane* bounded by a large semicircle, as in Fig. 10.2(a). For physically permissible distribution functions f_0, $Z(V) \to 0$ as $V \to \infty$, so that the semicircle is all mapped onto the origin, and S' takes some form such as Figs. 10.2(b) or (c). All we have to decide is whether any part of the positive real axis is enclosed by S', as would be the case in Fig. 10.2(b) but not in Fig. 10.2(c). This can be determined by actually plotting S'. For instance, when we have electrons in a Maxwellian distribution and infinitely massive ions, this only involves the familiar G-function (compare with (8.71) and Fig. 8.4), and results in Fig. 10.2(c), so indicating stability as we would expect.

O. Penrose (1960) showed that in this problem the plotting of the Nyquist diagram, S', can be avoided, and a necessary and sufficient condition for instability established, as follows. We consider a single species and divide by the positive real factor $e^2/\varepsilon_0 m$. For real V, (10.2) becomes

$$\frac{\varepsilon_0 m}{e^2} Z(V) = P \int \frac{f_0'(v)\,dv}{v - V} - i\pi f_0'(V), \tag{10.3}$$

where P denotes the principal part. This exhibits $Z(V)$ in its real and imaginary parts. Now for instability, S' must cross the positive real axis in the Z-plane at a point (or points) such as U in Fig. 10.2(b) where S' travels upwards when traced in the sense fixed as above. It is easily verified that then and only then can S' encircle part of the positive real axis in an anticlockwise sense. There must therefore be a real V such that $f_0'(V) = 0$, and $-f_0'(v)$ changes from negative to positive as v decreases through the value V. In other words

$$f_0(v) \text{ has a minimum at some } v = V. \tag{10.4}$$

* That $Z(V)$ has no poles within D follows from the definition (10.2) and the assumption that each $f_0(v)$ is analytic for real v. Though not relevant for our immediate purpose, it is of interest to note that $Z(V)$ *could* have poles in the *upper* half plane. For $f_0(v)$, as a function of complex v, may have poles off the real axis, and when V is in the upper half plane, and approaches a pole of f_0, the Landau rule indicates that poles of the integrand of (10.2) approach C from either side, and in the limit meet on C. This then leads to a pole of $Z(V)$. The working of Problem 1 shows this happening in practice.

Further, the real part of (10.3) must be positive at this point, i.e.

$$P \int_{-\infty}^{\infty} \frac{f_0'(v)\, dv}{v - V} > 0. \tag{10.5}$$

Here the instruction "P" is really superfluous as $f_0'(V) = 0$. A somewhat more revealing form of (10.5) follows from integration by parts. This gives

$$\left[\frac{f_0(v) - A}{v - V}\right]_{V+\varepsilon}^{V-\varepsilon} + P \int_{-\infty}^{\infty} \frac{f_0(v) - A}{(v - V)^2}\, dv > 0,$$

where A is a constant of integration, and the limit $\varepsilon \to 0$ will give the principal part (contributions from the limits $v \to \pm\infty$ cannot arise). Choosing $A = f_0(V)$ removes the first term and makes the symbol P again unnecessary, so that (10.5) becomes

$$\int_{-\infty}^{\infty} \frac{f_0(v) - f_0(V)}{(v - V)^2}\, dv > 0. \tag{10.6}$$

Equations (10.4) and (10.6) comprise Penrose's necessary and sufficient condition for instability. The requirement that f_0 should have a minimum implies that it has two or more maxima, and this means that the plasma contains at least two identifiable "streams." Such of course was the case for the two-stream instability (§6.3.1), and our new work shows that some broadening of

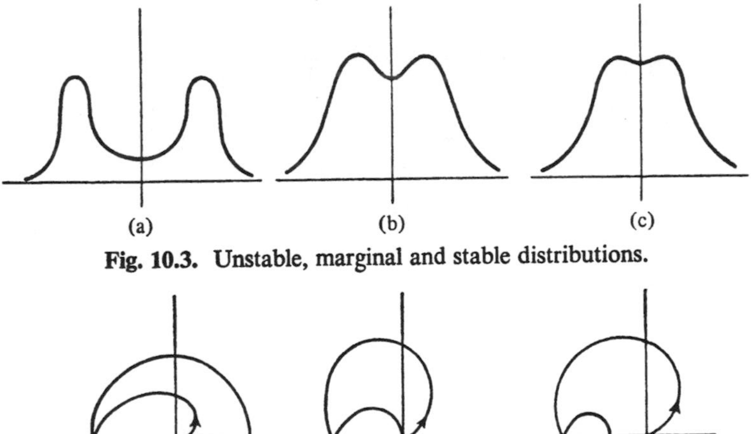

Fig. 10.3. Unstable, marginal and stable distributions.

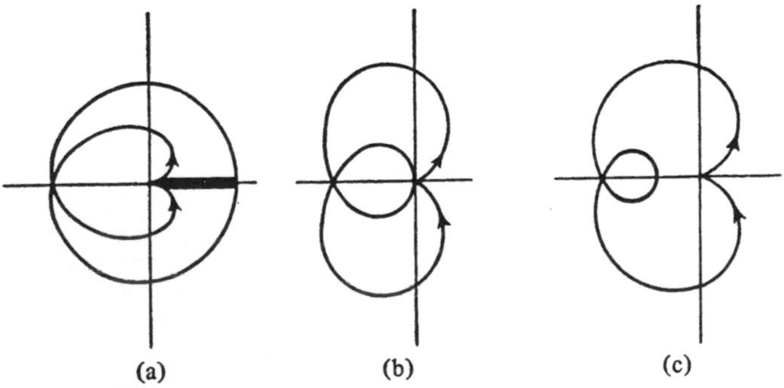

Fig. 10.4. Nyquist diagrams corresponding to Fig. 10.3.

the streams is permissible. However, if too much broadening is allowed, the instability disappears because (10.6) is not satisfied. This second part of the criterion may be interpreted by saying that the minimum in f_0 must be sufficiently pronounced, either in its depth or the width between the maxima, or some combination of these features. Figure 10.3 illustrates three distributions which are unstable, marginal, and stable respectively; the corresponding Nyquist diagrams are shown in Fig. 10.4. Here the two streams of electrons have the same density and temperature, and we choose three possible relative velocities. The critical relative velocity is roughly the same as the mean thermal speed for either stream, for instance it is about $0.9(KT/m_e)^{1/2}$ for Maxwellian distributions. For streams with different densities and temperatures there is a great variety of results. It should also be noted that, given a three-dimensional distribution, the Penrose criterion will in general have to be applied afresh for each direction of wave normal.

For cases where the growth rate is small, the physical explanation of the instability is closely related to that of Landau damping. Writing $\omega = \omega_r + i\gamma$, the real phase speed ω_r/k is easily seen to be close to the value V where f_0 has its minimum, but slightly displaced so that f_0' is positive at that speed. The derivation leading to (8.26) applies, and we have the growth rate

$$-\gamma = \frac{\pi e^2}{2\varepsilon_0 mk}\left[\frac{df_0}{dv}\right]_{v=\omega_r/k}\left(\frac{\omega_r}{k} - \frac{d\omega_r}{dk}\right) \tag{10.7}$$

(though $\omega_r(k)$ is not now given by (8.23)). The wave gains its energy from the kinetic energy of the particles with speed just greater than ω_r/k, but loses energy to those just slower. This is just the reverse of the situation discussed in §8.5.2. When the growth rate is large, most of the particles are actively concerned in the instability and one cannot single out a group of "resonant" particles in this way. However, it still appears that the wave speed is of the same order as the speed at which f_0 has its minimum; this incidentally verifies that the electrostatic approximation is valid except in relativistic cases. The frequency and growth rate tend to be roughly ω_p in strongly unstable cases, as we saw in Chapter 6; it follows that the wavelengths involved are of the order of the Debye length.

When we have to deal with several species, the results are exactly the same provided f_0 in (10.4) and (10.6) is interpreted as

$$\sum_{\text{species}} \frac{e^2 f_0}{m}$$

(as is clear from (10.2)). One interesting possibility is therefore that of a current-carrying plasma in which ions and electrons, each with a thermal distribution and equal number density, have a relative drift. (Such a situation does not, strictly speaking, satisfy the zero-order equations in that a d.c.

current produces a self-magnetic field, leading to the pinch effect and non-uniformity; however, the length scale involved in that process, c/ω_p, is so much larger than the scale of our instabilities that the objection is not serious.) There will be a critical drift speed, V_c, above which the plasma would be violently perturbed by the instabilities, and there is therefore a maximum current density NeV_c which can be carried by a plasma. V_c is easily calculated by the procedure just given. E. A. Jackson (1960) has done this for ions and electrons in Maxwellian distributions, allowing unequal temperatures T_i and T_e. The critical speed V_c, expressed in units of $(2KT_e/m_e)^{1/2}$ has an interesting dependence on T_e/T_i, shown in Fig. 10.5 (the ions being protons). At equal temperatures V_c is approximately the electron thermal speed, but when $T_e \gg T_i$ it falls to a much lower value, reaching eventually the ion thermal speed. This might have been guessed from the results given in §8.8. We noted there that for stationary ions Landau damping is greatly reduced if $T_e \gg T_i$, so resulting in "ion waves" with frequency typically ω_{pi}. If we think of the present situation as a competition between Landau damping and the tendency for growth due to counter-streaming, we would deduce that when $T_e \gg T_i$ it requires a smaller relative velocity to bring about the instability. Equation (10.6) enables these results to be obtained quantitatively, and for Maxwellian distributions the integral is expressible in terms of the G-function.

As with all instabilities, the linear theory fails to answer the question of

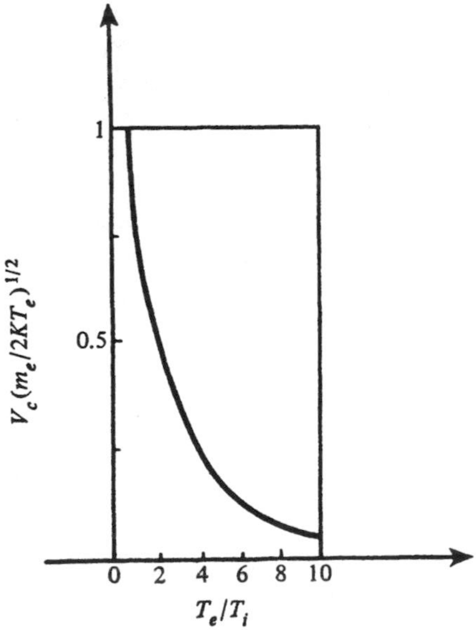

Fig. 10.5. Critical drift speed in a current-carrying plasma. (After E. A. Jackson, *Phys. Fluids*, **3**, 786 (1960).)

how the system will behave eventually if we bring about (or attempt to bring about) the unstable situation. What happens if a plasma initially contains two streams, or if we fire a beam of electrons into a stationary ionized gas? The corresponding non-linear analysis is formidable, but one way to make progress is by *numerical experiments*. The actual orbits of a large number of ions and electrons, moving in one dimension and regarded as plane sheets of charge, are calculated numerically. O. Buneman (1959) has demonstrated in this way that when the two-stream instability (for electrons flowing through ions) is followed beyond the linear stage the initial ordered kinetic energy is converted, within a few plasma periods, into the energy of apparently random motion. The final distribution function closely resembles a Maxwellian one, at an increased temperature, so accounting for the additional energy. It is often conjectured that, for microinstabilities generally, the eventual effect of the growth of the instability is to lower the growth rate and make the system evolve towards a stable state. Thus two streams of electrons would be heated, with consequent reduction in the relative velocity, until (10.6) fails to be satisfied. It is therefore doubtful whether a highly unstable situation could be brought about in the first place. In the example just mentioned, any beam of electrons fired at a plasma in such a way that if it penetrated far an unstable situation would result, would in practice be stopped within a short distance. Some writers have suggested that a mechanism of this type might be relevant to the structure of shock waves in field-free and collision-free plasmas.

Another approach to this problem is that of *quasi-linear theory*. This is applicable only in the marginal or slightly unstable case. Linearized theory is clearly the first step in an expansion in powers of the amplitude of the perturbation, and the idea is to continue the expansion to the second step. This leads to the consideration of wave–wave interaction. In particular, some combinations of Fourier components result in a secular change in the ambient distribution f_0 (which of course was regarded as constant in the linear approximation). If this secular change is such as to lead f_0 towards a stable distribution, one can gain some insight into the process just described. A survey of this work is given by Kadomtsev (1965).

10.3 UNSTABLE TRANSVERSE WAVES IN A FIELD-FREE PLASMA

Besides the familiar electrostatic two-stream instability, a plasma without magnetic field may also be susceptible to unstable waves or oscillations which are partly or wholly transverse. This has already been illustrated in §6.3.2. One of the earliest cases to be noted (by E. S. Weibel, in 1959) is that of two streams of electrons, with density $\frac{1}{2}N$ and unperturbed velocity $(0, \pm v, 0)$, with stationary ions. For a wave with $\mathbf{k} = (k, 0, 0)$ and $\mathbf{E} = (0, E, 0)$, with associated magnetic field $\mathbf{B}_1 = \mathbf{k} \times \mathbf{E}/\omega$, equation (6.73) yields

$$\frac{c^2k^2}{\omega^2} = 1 - \frac{\omega_p^2}{\omega^2}\left[1 + \frac{k^2v^2}{\omega^2}\right] \tag{10.8}$$

in the non-relativistic approximation ($v \ll c$). We plot k^2 as a function of ω^2 in Fig. 10.6. This differs from Fig. 6.8 on account of the neglect of relativistic effects; however, the difference is very slight for $\omega > \omega_p$, when the usual result for a cold plasma is recovered, and it is also slight in the region $-v^2\omega_p^2/c^2 < \omega^2 < 0$, where we find the instability to be discussed in this section. Here we have purely imaginary ω, indicating a non-propagating disturbance whose growth rate is, at the most, $(v/c)\omega_p$, so is much smaller than that typical of the electrostatic instabilities. A physical interpretation was given in §6.3.2.

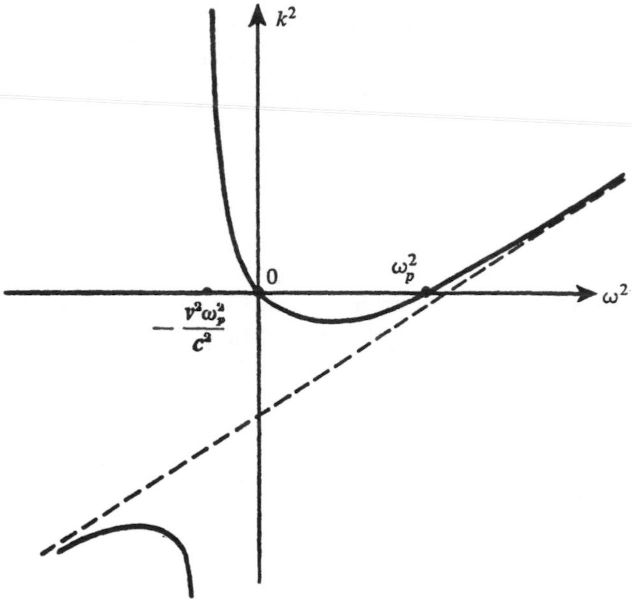

Fig. 10.6. Dispersion relation for Weibel's instability.

We would expect that a distribution of velocities in the x-direction would tend to weaken or destroy the instability, in the manner of Landau damping; referring again to §6.3.2, it would impede the formation of "pinched" regions. So it is desirable to consider general distributions $f_0(\mathbf{v})$ (which are not subject to any restriction in the field-free case), and write down the complete dispersion relation. We may then exclude the longitudinal plasma oscillations and the "fast" radio waves, and search for "slow" waves which are primarily transverse. We ask what class of f_0's are liable to this new type of instability. We have already given the necessary formulation in §8.6; see in particular (8.48). Under the assumptions $|\omega| \ll \omega_p$ and $|V| \ll c$, where $V = \omega/k$, and with the non-relativistic approximation, it is readily shown that the matrix $\varepsilon_{pq} - \mathcal{M}_{pq}$ is proportional to

$$\begin{pmatrix} \omega_p^2 & 0 & 0 \\ 0 & \omega_p^2 + c^2k^2 & 0 \\ 0 & 0 & \omega_p^2 + c^2k^2 \end{pmatrix} + \frac{e^2}{\varepsilon_0 m} \int \frac{v_p v_q}{v_1 - V} \frac{\partial f_0}{\partial v_1} \, d^3v. \qquad (10.9)$$

(If more than one species is to be included, all terms except c^2k^2 are summed.) The dispersion relation calls for the vanishing of the determinant of this matrix, and the new waves, if any, are those with k of order ω_p/c. (Where one or both of p, q is unity the integral can be slightly simplified.)

As in the previous section, the question is whether we can choose k real and positive in such a way as to make $\mathrm{Im}(V) < 0$, but it does not seem possible to derive a necessary and sufficient condition for this by the rather elegant method which led to the Penrose criterion. By arguments somewhat too lengthy to be given here, F. D. Kahn (1962) has obtained a sufficient but not necessary condition for the existence of an instability. Kahn restricts attention to distributions with "central symmetry," i.e. those for which $f_0(\mathbf{v}) = f_0(-\mathbf{v})$, and which are stable with respect to electrostatic waves. Then he defines

$$F_n(\theta, \phi) = \int_0^\infty v^n f_0(\mathbf{v}) \, dv \qquad (n = 0, 1, \ldots), \qquad (10.10)$$

where (v, θ, ϕ) are spherical polar coordinates for $\mathbf{v}$. Electromagnetic instability certainly occurs unless F_1 and F_2 are independent of the direction (θ, ϕ). But it may occur in other cases also. Now the isotropy of F_1 and F_2 is not so stringent a condition as complete isotropy for f_0, since that would make all the integrals F_n $(n = 0, 1, \ldots)$ independent of direction; but in practice an anisotropic distribution would almost certainly have F_1 and/or F_2 dependent on direction, and so be unstable. Kahn also conjectures that distributions without central symmetry in any frame of reference are unstable.

Since electromagnetic instabilities are so much more slowly growing than the electrostatic, they are only of interest if the latter are absent. From this point of view our first example was somewhat ill chosen. However, Kahn's result shows that it is easy to give examples which are stable electrostatically (as tested by the Penrose criterion), and we will content ourselves by doing this. Let us take electrons in a Maxwellian distribution with three unequal temperatures, the principal axes being along the coordinate axes,

$$f_0(\mathbf{v}) = N_0 \prod_{j=1}^{3} \left(\frac{m}{2\pi KT_j}\right)^{1/2} \exp\left(-mv_j^2/2KT_j\right). \qquad (10.11)$$

The integral in (10.9) vanishes for $p \neq q$, so the matrix is diagonal, and its determinant factorizes. The first of the three factors is meaningless since its vanishing should describe plasma oscillations, but we have already made approximations incompatible with that. The other two factors give the dispersion relations for plane electromagnetic waves traveling along the x_1-axis with their electric fields along the x_2- and x_3-axes respectively. For instance if $p = q = 2$, the integral becomes $-\frac{1}{2}(T_2/T_1)\omega_p^2 G'(z)$, where $z = V(m/2KT_1)^{1/2}$ and the dispersion relation is

$$\frac{c^2k^2}{\omega_p^2} = \frac{T_2}{2T_1}G'(z) - 1. \tag{10.12}$$

So we need only plot a Nyquist diagram for the right-hand side of this equation. Comparing with the dispersion relation for plasma oscillations in the Maxwellian case (8.71),

$$h^2k^2 = -\tfrac{1}{2}G'(z),$$

for which the Nyquist diagram is Fig. 10.2(c), the new diagram can be sketched at once as in Fig. 10.7, noting that $G'(0) = 2$. Provided $T_2 > T_1$, unstable waves exist, with

$$k^2 \leqslant \left[\frac{T_2}{T_1} - 1\right]\frac{\omega_p^2}{c^2},$$

as given by the part of the positive real axis enclosed by Fig. 10.7. If $T_1 \to 0$, all wave numbers are unstable, giving a situation very similar to that of our first example. A velocity distribution in the x_1-direction opposes the instability, as one would expect, and if $T_1 \geqslant T_2$ it is completely damped—but then it starts to appear in the waves along Ox_2, in agreement with Kahn's criterion. On solving (10.12) for k^2 in the range just quoted, one finds that z is in fact purely imaginary, so, as in our earlier example, the wave grows but does not propagate. (This is not, however, the case generally.) The growth

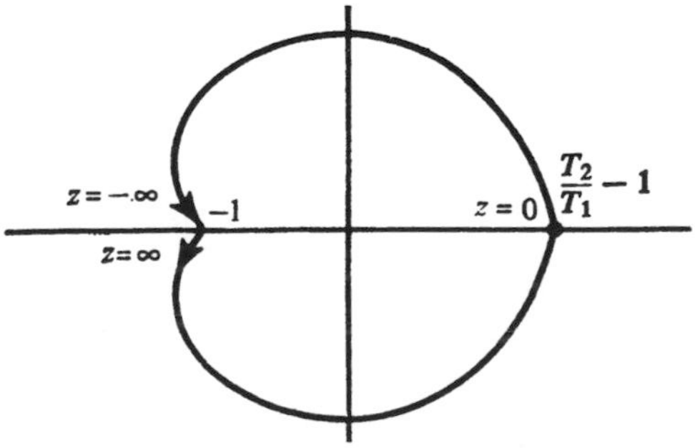

Fig. 10.7. Nyquist diagram for an example of unstable transverse waves.

rate is readily found by use of Fried and Conte's tables, and is of order $(KT_2/mc^2)^{1/2}\omega_p$. When the temperatures are only slightly unequal, only the very long waves are unstable and the maximum growth rate can be found by using (8.66) as z is small; it is found to be

$$\left(\frac{32KT_2}{27\pi mc^2}\right)^{1/2}(1 - T_1/T_2)^{3/2}\omega_p. \tag{10.13}$$

In more recent work, Kahn (1964) has dealt with waves whose propagation vector is in a general direction, in an ellipsoidal f_0 such as this, and has shown that the most rapidly growing wave travels along the principal axis corresponding to the smallest temperature. This is what one would expect intuitively.

Further investigations of these instabilities are reported by Kalman, Montes and Quemada (1968).

10.4 ELECTROSTATIC INSTABILITIES IN THE PRESENCE OF A MAGNETIC FIELD

10.4.1 Physical introduction

We naturally turn next to microinstabilities in plasmas with an imposed magnetic field. We have already developed in Chapter 9 the necessary dispersion relations both for Maxwellian and for general distributions. In a sense there is less freedom of choice here than in the case of no magnetic field, for f_0 is now restricted to be of the form $f_0(v_\perp, v_\parallel)$ only, and the only new parameter is the strength of the field; nevertheless there is a greater variety in the types of instability and the physical processes underlying them, and there does not seem to be any simple criterion similar to that of Penrose, or even a partial treatment comparable with Kahn's theory of slow transverse waves in the field-free case. The subject is still incompletely investigated and difficult to survey.

As before, there are waves which are essentially longitudinal and can be treated in the electrostatic approximation, and other waves which are essentially transverse. As emphasized by M. N. Rosenbluth (1964), the former are usually more important, and grow more rapidly, especially in so-called "low-β" plasmas. Here β is the ratio of particle to magnetic pressure, $2\mu_0 NKT/B^2$. Small β simply means that it is rather difficult for the plasma to perturb the imposed field, so that $\mathbf{B}_1$, and hence curl $\mathbf{E}$, remain small. This is the important case for thermonuclear work since the plasma is supposed to be contained by the magnetic field. It also applies in the magnetosphere. The transverse waves include fast radio waves, which for non-relativistic plasmas are well described in the cold approximation, but also include the whistler and Alfvén modes (for which we have already noted the firehose instability), and possibly others, like the Weibel–Kahn waves, whose existence could not be suspected from the cold theory.

In the electrostatic approximation, to which we devote this section, the starting point is equation (9.27), with (9.26) for Maxwellian distributions or (9.103) for general distributions; in the former case the possibility $T_\perp \neq T_\parallel$ is already included and we can also allow for a drift $\mathbf{V}$ along the magnetic field if we replace ω by $\omega - \mathbf{k}\cdot\mathbf{V}$ everywhere. We choose a distribution with some destabilizing feature, such as counter-streaming or anisotropy, and the calculation accounts quantitatively for the competition between this and Landau damping.

In practice, the most interesting instabilities are those involving both the ions and electrons, with frequencies comparable with the ion gyro-frequency, Ω_i. The existence of these is easy to understand. We mentioned in §8.8 that ion waves, which would normally be heavily damped, occur very readily if $T_e \gg T_i$, and in §10.2 we noted that the two-stream instability could therefore be excited much more easily in that case. Now let us recall that in §9.2.3 we showed that even with $T_i = T_e$ there are some almost undamped waves near $\omega = \Omega_i$ and its harmonics, the propagation being nearly (but not exactly) orthogonal to the field and the wavelength comparable with the ion Larmor radius. Clearly it should be fairly easy to make these unstable. But this is not the only possibility—there are also instabilities characteristic of the electrons only. In the following subsections we illustrate the above remarks by means of examples; these are far from exhaustive.

10.4.2 Anisotropic distributions; the Harris instabilities

The fact that anisotropy of the distribution function may lead to electrostatic instabilities in the presence of a magnetic field was pointed out by E. G. Harris (1959, 1961). One of the simplest examples he gave was that of electrons all traveling at the same speed, V, perpendicular to the magnetic field, the ions being infinitely massive. We can represent this by

$$f_0(\mathbf{v}) = N_0(2\pi v_\perp)^{-1}\,\delta(v_\parallel)\,\delta(v_\perp - V). \tag{10.14}$$

When this is inserted into (9.103) and the integrations carried out (integrate by parts to convert δ' into δ), and using $\varepsilon_0 k^2\phi = \rho$, we find the dispersion relation

$$\frac{\omega_p^2}{\Omega^2}\sum_{-\infty}^{\infty}\left\{\frac{(k_\parallel/k)^2 J_n^2(b)}{\left(\dfrac{\omega}{\Omega} - n\right)^2} + \frac{\left(\dfrac{k_\perp}{k}\right)^2 \dfrac{n}{b}\dfrac{d}{db}[J_n^2(b)]}{\dfrac{\omega}{\Omega} - n}\right\} = 1, \tag{10.15}$$

where $b = k_\perp V/\Omega$. It is not easy to see at a glance whether there are solutions with ω in the lower half plane (indicating growth), but Harris used the Nyquist diagram to show that when $k_\parallel = 0$ and $b > 1.84$ (the first maximum of $J_1(b)$), there are indeed growing waves whose frequency and growth rate are comparable with Ω, provided ω_p is sufficiently large; he showed also that more rapid growth can be obtained by taking $k_\parallel \neq 0$. A sufficient condition on ω_p is $\omega_p > \Omega$. When ion motion is included, the left-hand side of (10.15) is summed over species, and if $\omega_{pe} > \Omega_i$ (which is nearly always so), there are instabilities with $\omega \approx \Omega_i$.

Anisotropic Maxwellian distributions have been extensively studied in this connection. The charge density for a species is given by (9.26), namely

$$\frac{\rho}{\varepsilon_0 k^2 \phi} = -\frac{1}{h_\perp^2 k^2}\left\{1 - i\left[\omega + \frac{K(T_\parallel - T_\perp)k_\parallel^2}{m}\frac{\partial}{\partial\omega}\right]I\right\}, \tag{10.16}$$

where I is the Gordeyev integral (9.29). There is no instability for $k_\parallel = 0$ since the term proportional to $T_\parallel - T_\perp$ disappears in that limit and we are left with our former results, the Bernstein modes. It is therefore necessary to take both $k_\parallel$ and $k_\perp$ non-zero, and it is the term just referred to which introduces instability. To show this in detail involves rather heavy calculation, but we will sketch the procedure in the case of the ion instabilities mentioned in our opening discussion (§10.4.1). We make the same approximations as in §9.2.3, so that (in the notation used there) $z_{ne} \gg 1$ except that $z_{0e} \ll 1$, and we can assume $\lambda_e \ll 1$, as the interesting wave numbers are those for which $\lambda_i \approx 1$. Using $G(z) \approx i\pi^{1/2} + 2z$ for small z, we find eventually

$$\frac{\rho_e}{\varepsilon_0 k^2 \phi} \approx -\frac{1}{h_{\parallel e}^2 k^2}\left[1 - \frac{i\omega T_{\parallel e}}{k_\parallel T_{\perp e}}\left(\frac{m\pi}{2KT_\parallel}\right)^{1/2}\right]. \tag{10.17}$$

The second term in the bracket gives the Landau damping due to electron motion parallel to the field, and can be assumed small provided $T_{\parallel e}$ and $T_{\perp e}$ are not grossly dissimilar. Hence

$$\frac{\rho_e}{\varepsilon_0 k^2 \phi} \approx -\frac{1}{h_{\parallel e}^2 k^2}$$

to good approximation, and our results are independent of $T_{\perp e}$ over a wide range of its values. For the ions one needs the full expression (9.38) for I since it is anticipated that λ_i and some of the z_{ni}'s are of order unity. However, we can suppose for simplicity that $T_{\parallel i}$ is very small, for this will give the best chance of instability as ion motion parallel to the field is liable to increase the Landau damping. This leads to

$$I = -i\, e^{-\lambda_i} \sum_{-\infty}^{\infty} \frac{I_n(\lambda_i)}{\omega - n\Omega_i}. \tag{10.18}$$

When this is inserted into (10.16) and the results applied to Poisson's equation, we find

$$\sum_{-\infty}^{\infty} e^{-\lambda_i} I_n(\lambda_i)\left[\frac{\varpi}{\varpi - n} + \frac{\eta}{(\varpi - n)^2}\right] = 1 + \frac{T_{\perp i}}{T_{\parallel e}} + h_{\perp i}^2 k^2, \tag{10.19}$$

where $\varpi = \omega/\Omega_i$ and $\eta = KT_{\perp i}k_{\parallel}^2/m_i\Omega_i^2$. Now λ_i and η are merely $k_{\perp}^2$ and $k_{\parallel}^2$ on a normalized scale, and the question is whether we can choose them so that the normalized frequency ϖ is in the lower half plane, all of these quantities being of order unity. It turns out that assuming the Debye length to be small (i.e. the plasma density is sufficient), instability is indeed possible, with $\frac{1}{2} < \varpi < 1$ (and often at higher values of ϖ also). This can be seen roughly by retaining only the terms $n = 0$ and 1 in (10.19).

A. V. Timofeev (1961) has reported an investigation in which all of $T_{\perp i}$, $T_{\parallel i}$ and $T_{\parallel e}$ are allowed to take general values. A region of instability is plotted, which to good approximation is represented by

$$\frac{2}{T_{\perp i}} + \frac{5}{T_{\parallel e}} < \frac{1}{T_{\parallel i}}. \tag{10.20}$$

As expected, low $T_{\parallel i}$ and high $T_{\parallel e}$ help to promote the instability. However high $T_{\parallel e}$ is, $2T_{\parallel i} < T_{\perp i}$ is required for an instability. A sufficient condition for stability is therefore $T_{\perp i} < 2T_{\parallel i}$, a result frequently quoted. More recently, G. K. Soper and E. G. Harris (1965) have shown that if $T_{\parallel e} < 5T_{\parallel i}$, so that (10.20) would indicate certain stability, there is actually a fresh possibility for instability, but only for $T_{\parallel i} < T_{\perp i}/(1+\gamma)$, where $\gamma^2 = m_iT_{\parallel e}/m_eT_{\parallel i}$ so normally $\gamma \gg 1$. Timofeev's diagram, as amended by Soper and Harris, is shown in Fig. 10.8.

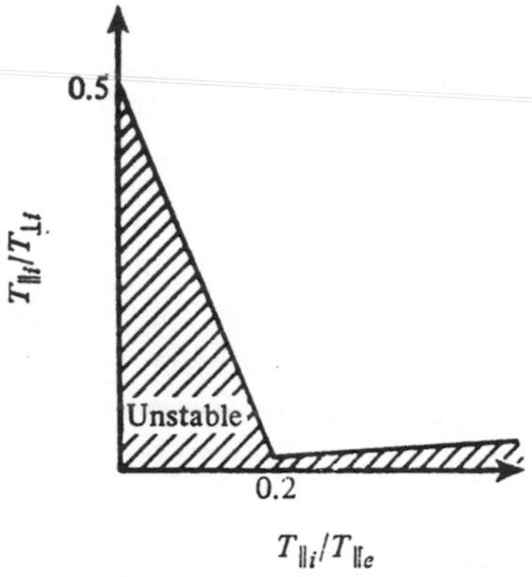

Fig. 10.8. The region of instability for Maxwellian plasmas in a magnetic field. (After G. K. Soper and E. G. Harris, *Phys. Fluids*, **8**, 984 (1965).)

Another anisotropic distribution to have attracted attention (Rosenbluth and Post, 1965) is the "loss-cone" type. Here f_0 is zero within the cone $v_{\perp} < |v_{\parallel}| \tan \alpha$ for some fixed α, but isotropic for $\mathbf{v}$ outside the cone. Distributions of this form occur when a plasma is confined between magnetic mirrors, as is readily understood from the conservation of the first adiabatic invariant in a strong field (see Chapter 4). The "loss cone" constitutes the particles which have so small a pitch angle that they can escape through the magnetic mirrors, i.e. they are not reflected even by the strongest field present. The angle α varies with position in such a way that $\sin^2 \alpha/B$ is constant, but for many purposes the discussion based on a uniform state is adequate. It is not difficult to see that with this f_0 the distribution function of say v_x (integrated over v_y and v_z) is double humped, having a central minimum due to the "lost" particles. It is therefore natural to expect that this feature would indeed have a destabilizing effect on the ion waves discussed in §9.2.3, and this is just what happens.

10.4.3 Two-stream instability in the presence of a magnetic field

Here we again consider the situation of §9.2.3 and adopt the same approximations, but the electrons have a mean velocity V along the magnetic field. All temperatures are supposed equal. Replacing ω by $\omega - k_{\parallel}V$ in (10.17), the electrons are described by

$$\frac{\rho_e}{\varepsilon_0 k^2 \phi} \approx -\frac{1}{h^2k^2}\left\{1 - i\left(\frac{\omega}{k_{\parallel}} - V\right)\left(\frac{m_e\pi}{2KT}\right)^{1/2}\right\}, \tag{10.21}$$

while the ions are again handled by (10.18), under the approximation $(\omega/k_{\parallel}) \gg (KT/m_i)^{1/2}$, this time due to $k_{\parallel}$ being small (rather than T, i.e. $T_{\perp i}$, small as in the last section). Landau damping due to the ions is therefore avoided. This leads to the dispersion relation

$$\omega\, e^{-\lambda_i} \sum_{-\infty}^{\infty} \frac{I_n(\lambda_i)}{\omega - n\Omega_i} + i\left(\frac{\omega}{k_{\parallel}} - V\right)\left(\frac{m_e\pi}{2KT}\right)^{1/2} = 2 + h^2k^2 \tag{10.22}$$

as a simple extension of (9.42). The imaginary term here normally describes the Landau damping due to the electrons, and changes sign when $V > \omega/k_{\parallel}$. It is easily verified that there are then growing solutions, with $\mathrm{Re}(\omega)$ a little above each harmonic $n\Omega_i$. To ensure the validity of the approximation used to treat the ions, we require that $\omega/k_{\parallel}$ substantially exceed $(KT/m_i)^{1/2}$. It is not, however, necessary for V to exceed $(KT/m_e)^{1/2}$ (as would be the case for waves traveling along the field); in fact it is already assumed that $(\omega/k_{\parallel}) \ll (KT/m_e)^{1/2}$.

As anticipated in §10.4.1, we conclude that for waves traveling at a large angle to the magnetic field the two-stream instability can easily be excited by a drift of electrons along the field.

10.5 UNSTABLE TRANSVERSE WAVES IN A MAGNETIZED PLASMA

When we do not avail ourselves of the electrostatic approximation, waves in a hot plasma with magnetic field have to be investigated by using the full conductivity tensor derived in Chapter 9—a daunting task! However, some cases (in addition to those already considered) are amenable to analysis, and lead to still further instabilities. At the high frequency end ($\omega > \Omega_e$, ω_{pe}) the cold treatment is adequate and any meaningful improvement on this should be carried out relativistically. At the other extreme, one has the "hydromagnetic" limit $\omega \ll \Omega_i$, with wavelengths long. Our results of §9.3.3 (low temperature approximation) contain all that is required here; in particular one would expand (9.69) under the approximation $\omega \ll \Omega_i \ll \Omega_e$. We have already dealt, in §9.3.3, with one microinstability arising in this limit, the well-known "firehose" instability for waves with **k** parallel to $\mathbf{B}_0$ provided $p_\parallel - p_\perp > B_0^2/\mu_0$. By taking **k** in other directions a further instability known as the "mirror" instability can be obtained. The details of this are deferred to Chapter 11, where we investigate the hydromagnetic treatment of plasmas generally (see §11.6.3). In the present chapter we are concerned with phenomena explicable only in terms of Boltzmann's equation.

To simplify the discussion of waves for the whole spectrum of frequencies, we restrict attention to propagation along the magnetic field. We noted that when $k_\perp = 0$ the conductivity tensor assumes diagonal form in the polarized coordinates, and we have two circularly polarized transverse waves satisfying the dispersion relation (9.77),

$$\frac{c^2k^2}{\omega^2} = 1 + \frac{1}{i\omega\varepsilon_0} \sum_{\text{species}} \sigma_\mu^\mu \quad (\mu = \pm 1), \tag{10.23}$$

where σ_μ^μ (no summation) is given by (9.107):

$$\sigma_\mu^\mu = \frac{\pi e^2 i}{m} \int_{v_\parallel = -\infty}^{\infty} \int_{v_\perp = 0}^{\infty} \frac{\dfrac{\partial f_0}{\partial v_\perp} + \dfrac{k}{\omega}\left(v_\perp \dfrac{\partial f_0}{\partial v_\parallel} - v_\parallel \dfrac{\partial f_0}{\partial v_\perp}\right)}{\omega - kv_\parallel + \mu\Omega} v_\perp^2 \, dv_\perp \, dv_\parallel. \tag{10.24}$$

For Maxwellian distributions the integrations may be carried out in terms of the G-function, leading to (9.76). For particles with negative charge, the sign of Ω is reversed. The stability is discussed by means of the Nyquist diagram as developed in §10.2. We write (10.23) in the form

$$k^2 = \frac{\omega}{i\varepsilon_0(c^2 - V^2)} \sum_{\text{species}} \sigma_\mu^\mu, \tag{10.25}$$

where $V = \omega/k$ is the complex velocity. Now when f_0 and $\mathbf{B}_0$ are given, the expression $\omega\sigma$ depends on ω and k only through V together with the real quantities

$$U_j = \Omega_j/k = | e_j B/m_j k |, \tag{10.26}$$

where j labels the species. So (10.25) becomes

$$k^2 = Z(V, U_j). \tag{10.27}$$

The most convenient procedure is to fix the U_j's—this only involves one choice since their ratio is known—and then plot the usual Nyquist diagram for $Z(V)$ obtained by letting V describe the locus of Fig. 10.2(a). If any portion of the positive real axis is encircled, possible values of k^2 for instability are indicated, but as the U_j's are already chosen, the instability is only established for the corresponding values of B given by (10.26). One then repeats the process for other values of the U_j's, and in this way all values of k and B are eventually tried.

As an illustration (due, independently, to R. N. Sudan (1963) and to P. D. Noerdlinger (1963)), consider the anisotropic Maxwell distribution for electrons, the ions being infinitely massive. The instabilities will be closely related to those discussed in §10.2, equation (10.12), the difference being that here we have the magnetic field as an additional parameter, but we are obliged to take the "temperature" isotropic in the plane perpendicular to the field, and to consider circularly polarized waves. Equation (9.76) gives

$$Z^{(\mu)}(V, U) = -\frac{\omega_p^2}{c^2 - V^2}\left[z_0 G(z_\mu) + \left(\frac{T_\perp}{T_\parallel} - 1\right)(z_\mu G(z_\mu) - 1)\right], \quad (10.28)$$

where

$$z_\mu = (V - \mu U)(m/2KT_\parallel)^{1/2}, \qquad (\mu = 0, \pm 1) \quad (10.29)$$

and $Z^{(1)}$, $Z^{(-1)}$ refer to the two polarizations (the change of sign of Ω has been incorporated). The form of Nyquist diagram obtained from (10.28) when $T_\perp > T_\parallel$ and $U \ll c$ is shown in Fig. 10.9. With our sign convention, V is to be traced from $+\infty$ to $-\infty$ just below the real axis (Fig. 10.2a) and this results in the locus $OAA'PBB'O$ in the Z-plane, the portions AA', BB' being completed by large semicircles. These extra branches at infinity, not present in Fig. 10.7, are due to the factor $(c^2 - V^2)^{-1}$, but are of no consequence so far as instabilities are concerned. The effect of the magnetic field is to distort the portion encircling the origin, but it still encloses a part of the real axis, OP. To find P we set $\text{Im}(Z) = 0$ and use (8.63). This leads to the vanishing of the coefficient of $G(z_\mu)$ in (10.28), namely

$$V = \mu(1 - T_\parallel/T_\perp)U. \quad (10.30)$$

Figure 10.9 is correct only if this V satisfies $|V| < c$, and we assume this to be so; the crossing points in the neighborhood of Q arise from the factor $(c^2 - V^2)^{-1}$ and depend critically on how the contour is indented near $V = \pm c$; they are of no significance. The actual value of $Z^{(\mu)}$ at this point

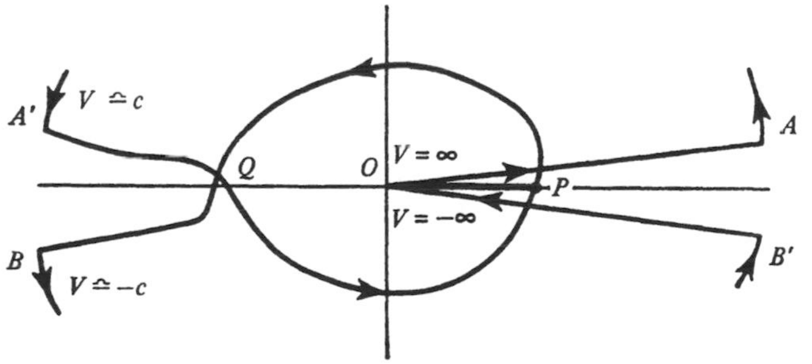

Fig. 10.9. Nyquist diagrams for transverse waves traveling along a magnetic field.

P is, accordingly

$$\frac{\omega_p^2}{c^2 - V^2}\left(\frac{T_\perp}{T_\parallel} - 1\right).$$

Substituting for V from (10.30) and remembering $U = \Omega/k$ and $\mu^2 = 1$, we can solve (10.27) for k^2 to find a critical value k_c^2 given by

$$c^2k_c^2 = \omega_p^2\left(\frac{T_\perp}{T_\parallel} - 1\right) + \left(1 - \frac{T_\parallel}{T_\perp}\right)^2\Omega^2. \tag{10.31}$$

Wave numbers $k < k_c$ will be associated with growing waves provided $T_\perp > T_\parallel$. It is readily checked that when $k = k_c$, with $U = \Omega/k_c$ accordingly, V as given by (10.30) is indeed smaller than c. For smaller values of k (well within the loop of Fig. 10.9) there may be trouble in this respect, and a relativistic treatment should be used. Multiplying (10.30) by k, we note also that for the waves on the threshold of instability,

$$\omega = \pm(1 - T_\parallel/T_\perp)\Omega. \tag{10.32}$$

Discussion of the growth rates is rather more complicated. We have to keep in mind that the Nyquist diagram was plotted for fixed U. When we consider solutions of the dispersion relation for real values of k inside the loop (i.e. on OP) we are examining various unstable waves but at the same time changing the magnetic field strength B_0 (other plasma parameters being held constant). To find how the growth rate varies with k at fixed Ω would need more extensive calculations, with results depending in a rather complicated way on the ratios $\omega_p : \Omega$ and $p_\perp : p_\parallel : B_0^2/\mu_0$. For moderate isotropy and moderate values of ω_p/Ω, the growth rates tend to be of the order $0.1\ (KT/mc^2)^{1/2}\omega_p$ (compare (10.13)).

A similar procedure applies when ions with finite mass are included. For frequencies around or below Ω_i new results, principally the firehose instability, appear. Even if there is insufficient anisotropy for the firehose type, there are instabilities due essentially to the inverse Landau damping of ions satisfying $\omega - kv_z \approx \Omega_i$ (see (10.24)), but the growth rate in the case of slight anisotropy is rather small, and given by M. N. Rosenbluth (1964) as

$$\Omega_i \exp\left[-\frac{1}{\beta}\left(\frac{T_\perp}{T_\parallel} - 1\right)^2\right], \tag{10.33}$$

where as before $\beta = 2NKT/\mu_0B_0^2$. As discussed in §10.4.1, electromagnetic instabilities are usually unimportant in low-β plasmas, as (10.33) is very small and the firehose and mirror instabilities do not exist at all. The Harris instabilities are more "dangerous" so far as attempts to confine plasmas are concerned. But the unstable waves discussed in this section are of interest in connection with the production of radio noise by plasmas, notably in the earth's magnetosphere.

10.6 SOME OTHER MICROINSTABILITIES

Our work so far by no means exhausts all the possibilities for instability latent in the general dispersion relations which we set up in Chapters 8 and 9; nor have we, for the most part, given the precise conditions, wave numbers, frequencies and growth rates for all those mentioned. A really comprehensive treatment would be not only very long but also fairly turgid. Here, as befits a textbook, we have tried rather to sketch some of the more commonly quoted results in the field, explain the physical processes involved and illustrate the type of analysis used. We ought, however, to mention briefly some other types of microinstabilities.

Firstly collisions can (paradoxically) bring about instability. In the absence of any imposed electric or gravitational field, the only strict equilibrium solution available is the Maxwellian, with temperatures the same for all species, and that situation is stable. But now let us add an external electric field, perpendicular to a magnetic field, both fields being uniform. Without collisions, both species would drift at the same velocity, $\mathbf{E} \times \mathbf{B}/B^2$ (see §4.2.1), but when there are collisions, for example with neutral molecules in a partially ionized gas, the drifts are modified so that there is a relative velocity between the two species. The most effective situation is when the gyro-frequency exceeds the effective collision frequency for the electrons but the reverse is true for the ions. The relative velocity (i.e. current flow) is then approximately orthogonal to the magnetic field. Electrostatic waves traveling in this same direction are then subject to the two-stream instability and at the same time are able to enjoy the reduction in ion Landau damping by virtue of propagation across the field; a situation impossible in the collision-free case. The collisions provide a new source of damping, but even so, quite a small drift, of the order $(KT/m_i)^{1/2}$ can excite ion cyclotron oscillations. D. T. Farley (1963) investigated this instability as an explanation of radar scattering by small-scale structure in the equatorial ionosphere. This phenomenon provides a striking confirmation of the theory. At the geomagnetic equator an electric current (the electrojet) builds up gradually during the morning, at a height where the collision frequencies have the property mentioned above. The scattering is observed to start when the current reaches the critical strength. A. Simon (1963) discussed a similar instability, but involving non-uniformity in space also. The BGK model (§9.5.2) proves to be very useful in this type of work.

Considerable attention has been paid to the "universal" instabilities. A typical case is a plasma with slightly non-uniform magnetic field, say $(0, 0, B)$, where $B = B_0(1 + \varepsilon x)$. An equilibrium solution can be set up by means of Jeans's theorem. As we know from the guiding-center theory (§4.3.1) the two species will drift in the y-direction, but with different speed. One therefore looks for solutions proportional to $\exp [i(\omega t - ky)]$, and it emerges that there are indeed growing waves, with the real and imaginary parts of ω of the order kv_D where v_D is the drift velocity. Several variants of this idea have been considered but they are beyond the scope of this book; their treatment involves a return to the basic principles of §9.1.1 starting with the construction of the orbits in a non-uniform field and integration along them.

At the other extreme, relativistic particles are able to induce instability in high-frequency electromagnetic waves. For example the relativistic version of (10.24), governing waves traveling along the magnetic field, has been considered by several writers. I. Lerche (1965) has estimated growth rates for these waves in several cases. The treatment of waves traveling across the field is also important in connection with collective effects in synchrotron radiation; this has been considered by W. E. Drummond and M. N. Rosenbluth (1960), and by B. A. Trubnikov (1961).

10.7 STABILITY OF ISOTROPIC MONOTONIC DISTRIBUTIONS

We have seen in this chapter how a great variety of microinstabilities can occur in collision-free plasmas, and one might very reasonably begin to suspect that the only stable distribution is the uniform Maxwellian one, i.e. the state of thermal equilibrium. But this is not so. We give below a proof that any isotropic uniform distribution in which f_0 is a decreasing function of the energy, $\frac{1}{2}mv^2$, is stable in a uniform (possibly a zero) magnetic field. As a matter of fact we have not yet even shown that the Maxwellian distribution is stable, but that is a special case of the theorem, and was proved in this way by W. A. Newcomb.* The extension to a general decreasing function of the energy is due to M. N. Rosenbluth (1964). A further variant of this has been given by G. Rowlands (1966).

We first prove a preliminary result. Let $d\tau$ be any element of phase space which we identify by means of the particles in it and so follow during any motion of the gas. By conservation of particles $f\,d\tau$ is constant, but the collision-free Boltzmann equation states that f and $d\tau$ are separately invariant following the motion of the particles. It follows that, if $G(f)$ is any function of f, the quantity

$$S = \int G(f)\, d\tau \tag{10.34}$$

(the integral ranges over all phase space) is invariant; for evaluating it at two instants simply involves the same summation but with the elements $d\tau$ relabeled. For example, if $G(f) \equiv f \log f$, S is proportional to the entropy, while if $G(f) \equiv f$, S becomes the total number of particles.

* Cited in an appendix to the paper of Bernstein (1958).

The total kinetic energy is of course

$$T = \int Wf\,d\tau, \tag{10.35}$$

where $W = \frac{1}{2}mv^2$. Suppose, with $G(f)$ chosen, we wished to find that $f(\mathbf{x}, \mathbf{v})$ which minimized T subject to given values of S and of the total number of particles. By Lagrange's undetermined multipliers we would find

$$W + \alpha + \beta\frac{dG}{df} = 0, \tag{10.36}$$

where α and β are constants. Since this has to hold for all $(\mathbf{x}, \mathbf{v})$, f must be a function of W only, as we would expect, and this can be obtained explicitly once $G(f)$ is specified.

To prove the theorem itself we now suppose instead that an unperturbed $f_0(\mathbf{x}, \mathbf{v})$ of the form $f_0(W)$ is proposed; we can find the particular $G(f)$ for which f_0 is the distribution satisfying (10.36), by writing

$$dG = -\frac{W + \alpha}{\beta}\,df_0.$$

This can be integrated unambiguously provided the inverse function $W(f_0)$ is single valued. This is so if the given $f_0(W)$ is monotonic, and therefore necessarily decreasing since f_0 must be positive but tends to zero as $W \to \infty$.

Once this is done, S as defined by (10.34), using the G just calculated, is a constant of the motion, and is so contrived that the minimum value of the kinetic energy is precisely the energy of the specified f_0, if f is subject to the constraints that the number of particles and the value of S are those of f_0. Moreover, the electrostatic energy already has its minimum value (zero), and the magnetic energy is also minimum subject to the condition that it should have the correct flux at infinity (write $\mathbf{B} = \mathbf{B}_0 + \text{curl}\,\mathbf{A}$, where $\mathbf{A} \to 0$ at infinity). So, given the initial state $f = f_0(W)$, $\mathbf{E} = 0$, $\mathbf{B} = \mathbf{B}_0$, each form of energy is already minimum subject to known invariants of the subsequent motion. But since the *total* energy is also conserved, this can only mean that exponentially growing waves cannot arise from small perturbations, for no form of energy can grow at the expense of another.

In the particular case where f_0 is the Maxwell distribution, the relevant $G(f)$ is just $f \log f$.

Variational procedures like this give a very concise approach to stability criteria, as we know from other branches of mechanics. It would be very useful if the argument could be extended to deal with anisotropic or even non-uniform plasmas, so avoiding the laborious analysis involved in solving the dispersion relation. Unfortunately this has not so far been achieved, owing to the difficulty of constructing suitable constants of the motion.

PROBLEMS

1. Consider a plasma composed of several streams (which may be of various species) each stream having a velocity distribution of the type

$$f_s(v) = \frac{N_s c_s}{\pi} \frac{1}{(v - u_s)^2 + c_s^2}.$$

(Thus N_s is the number density, u_s the drift velocity and c_s a measure of the thermal broadening; such a distribution is a fair approximation to the Maxwellian.)

As noted in the text the Penrose criterion should be applied to the function

$$f_0 = \sum_s \frac{e_s^2 f_s}{m_s}.$$

Show that f_0 does indeed have a minimum, occurring at $v = 0$, provided

$$\sum_s \frac{\omega_s^2 u_s c_s}{(u_s^2 + c_s^2)^2} = 0 \qquad \text{(a)}$$

and

$$\sum_s \frac{\omega_s^2 c_s (3u_s^2 - c_s^2)}{(u_s^2 + c_s^2)^3} > 0, \qquad \text{(b)}$$

where ω_s is the plasma frequency of the sth stream.

Show further that, assuming these conditions are met, the second part of the Penrose criterion, (10.6), requires that

$$\sum_s \frac{\omega_s^2 (u_s^2 - c_s^2)}{(u_s^2 + c_s^2)^2} > 0 \qquad \text{(c)}$$

for instability to occur.

(The integral is expeditiously carried out by means of a semicircular contour in the upper half plane; note that there is no need to regard $v = 0$ as a singularity owing to the term $f_0(V)$ in (10.6).)

2. Use these results to show that for two equal streams of electrons with drift $\pm u$ and equal values of c, with static ions, f_0 has the required minimum if $u > c/3^{1/2}$, but that instability occurs only if $u > c$.

3. Show that if there are just two streams, the marginal condition (i.e. equation (a) and equation (c) with equality) is achieved if

$$u_1 = \omega_1 c_2/\omega_2, \qquad u_2 = -\omega_2 c_1/\omega_1$$

assuming (b) is satisfied.

We may apply these results to a plasma consisting of singly charged ions (species 1) and electrons (species 2) in equal density. Writing $\varepsilon^2 = m_e/m_i$, and interpreting $(\varepsilon c_e/c_i)^2$ as T_e/T_i, show that the critical drift velocity is

$$[\varepsilon + (T_i/T_e)^{1/2}]c_e$$

and that (b) is satisfied provided $(T_i/T_e) < (3 + \varepsilon^3)/(1 + 3\varepsilon^3) \approx 3$.

Note that the first of these results is very close to that of Fig. 10.5.

4. Consider a plasma composed of several streams, as in Problem 1, but with Maxwellian distributions. By retracing the steps of §10.2 as far as equation (10.5) and expressing the required functions in terms of the real and imaginary parts of the function $G(z)$ introduced in §8.7, give conditions for instability which could be applied numerically given the data of Fig. 8.4. Use your method to find the relative drift velocity for the situation of Problem 2, with Maxwellian distributions.
5. Obtain equation (10.13).
6. Show that the theorem of §10.7 can be extended, by minor modification of the proof, to include relativistic dynamics.

REFERENCES

I. B. BERNSTEIN, *Phys. Rev.*, **109**, 10 (1958).

O. BUNEMAN, *Phys. Rev.*, **115**, 503 (1959).

S. CHANDRASEKHAR, *Hydrodynamic and Hydromagnetic Stability*, Oxford University Press (1961).

W. E. DRUMMOND and M. N. ROSENBLUTH, *Phys Fluids*, **3**, 45, and 491 (1960).

D. T. FARLEY, *J. Geophys. Res.*, **68**, 6083 (1963).

E. G. HARRIS, *Phys. Rev. Lett.*, **2**, 34 (1959).

E. G. HARRIS, *J. Nuclear Energy Part C*, **2**, 138 (1961).

E. A. JACKSON, *Phys. Fluids*, **3**, 786 (1960).

B. B. KADOMTSEV, *Plasma Turbulence*, Academic Press (1965).

F. D. KAHN, *J. Fluid Mech.*, **14**, 321 (1962).

F. D. KAHN, *J. Fluid Mech.*, **19**, 210 (1964).

G. KALMAN, C. MONTES and D. QUEMADA, *Phys. Fluids*, **11**, 1797 (1968).

I. LERCHE, *Proc. Roy. Soc.*, **A283**, 203 (1965).

P. D. NOERDLINGER, *Ann. Phys.* (*N.Y.*), **22**, 12 (1963).

O. PENROSE, *Phys. Fluids*, **3**, 258 (1960).

M. N. ROSENBLUTH in *Advanced Plasma Theory* (Ed. Rosenbluth), p. 137, Academic Press (1964).

M. N. ROSENBLUTH and R. F. POST, *Phys. Fluids*, **8**, 547 (1965).

G. ROWLANDS, *Phys. Fluids*, **9**, 2528 (1966).

A. SIMON, *Phys. Fluids*, **6**, 382 (1963).

G. K. SOPER and E. G. HARRIS, *Phys. Fluids*, **8**, 984 (1965)

R. N. SUDAN, *Phys. Fluids*, **6**, 57 (1963).

A. V. TIMOFEEV, *J. Exp. Theor. Phys.* (translations), **12**, 281 (1961).

B. A. TRUBNIKOV, *Phys. Fluids*, **4**, 195 (1961).

E. S. WEIBEL, *Phys. Rev. Lett.*, **2**, 83 (1959).

CHAPTER 11

DERIVATION OF MAGNETOHYDRODYNAMICS

11.1 INTRODUCTION

For many purposes the treatment of gases by means of Boltzmann's equation is far too cumbersome and unnecessarily detailed. In aerodynamics, most practical problems can be very accurately treated by means of the ordinary equations of momentum, thermodynamic state, heat conduction, and so on, the only common exception being the very difficult question of the structure of shock waves. For plasmas there is a rather richer variety of phenomena which involve velocity space, for example Landau damping, cyclotron resonances, and many of the microinstabilities. But despite this, there are many other situations where plasmas do behave in an essentially hydrodynamic way. It ought, therefore, to be possible to obtain the equations of magnetohydrodynamics (MHD) from the Boltzmann equation by means of suitable approximations.

In the hydrodynamic description we replace the Boltzmann function $f(\mathbf{x}, \mathbf{v}, t)$ by the number density $N(\mathbf{x}, t)$, mean velocity $\mathbf{u}(\mathbf{x}, t)$, and so on, as defined in §7.4. The success of the approximation depends on whether the information so discarded is irrelevant, not merely in the sense that it is not of interest, but also in the sense that the future development of hydrodynamic variables is virtually unaffected by it. If this is so the hydrodynamic equations will form a closed set. We attempted to find such a set in §7.4, and though we made some progress, we concluded that unless some approximation were made we would not succeed.

The physical content of the required approximation is that there should be sufficient "localization" of the particles in physical space. For if we hope to obtain differential equations in space and time, the rate of change of hydrodynamic quantities at a point must depend on their present values at that point or in its neighborhood. This is not possible if particles are able to move substantial distances in a short time, as the evolution of local properties is then influenced by the arrival from various directions of fresh particles from quite separate parts of the flow field. The only distant influence which can be included in the hydrodynamic scheme is a self-consistent field, which appears as a body force.

In ordinary gases this localization is provided by collisions. Even a fast particle has very little chance of traveling more than two or three mean free paths without being deflected, so that if the flow field is slowly varying in both space and time (the mean free path L_c and the collision time τ_c being the relevant orders of magnitude), the particles are effectively localized. The equations of an ideal gas, introducing only first derivatives with respect to space, are often quite adequate, and the correction which allows for the incompleteness of the localization, i.e. the finite mean free path, simply introduces viscosity and heat conduction, so leading to second derivatives in space. The formal derivation of the ideal equations is described in §11.2. The general method of obtaining the first correction is explained in §11.5.1; here the detailed calculations require the specific form of $(\partial f/\partial t)_c$. For neutral gases this is the expression describing binary collisions; we shall not pursue this here, but reference may be made to the treatise of Chapman and Cowling. The workings of the method may, however, be illustrated by a simple model, given in §11.5.2.

In plasmas the same method is applicable, in principle. In the lowest approximation the result is the same as for neutral gases, and we have ideal gas behavior. However, at this stage, all species have the same drift velocity, so that assuming electrical neutrality is maintained (which as we shall see is usually justifiable in the hydrodynamic context), the current, **j**, vanishes. It is, therefore, important to consider the next approximation, and whilst the procedure for doing so is similar to that for neutral gases, the work tends to be even more complicated. Firstly one must find the appropriate form for $(\partial f/\partial t)_c$. This is the subject of Chapter 12, where we find it is an expression of the Fokker–Planck type with somewhat involved coefficients, and some uncertainty is already introduced by approximations. Next one must apply the Chapman–Enskog expansion, and it is of course necessary to deal with at least two species. A further source of complication is that if there is an appreciable magnetic field, anisotropy is introduced, so for example the coefficient of viscosity becomes a tensor. In earlier work (for example the final chapter of Chapman and Cowling) the binary collision form was used for $(\partial f/\partial t)_c$, notwithstanding that this is inappropriate for plasmas, and that the result is divergent unless integrals are "cut off" by imposing a maximum impact parameter. This actually yields fairly accurate results if the Debye length is used for the cut off. More recent work, in particular that of Robinson and Bernstein (1962) has been based on the Fokker–Planck expression. A sketch of this subject is given in §11.5.3–4, and the results outlined in §11.5.5, a full treatment being too lengthy for the present work.

For many purposes, notably astronomical work, spatial and temporal gradients are indeed so small that the fluid picture is adequate, even in the lowest approximation, despite the fact that the mean free path, L_c, (meaning the rough order of magnitude given in §1.2) is quite large. Also it is often possible to regard the conductivity as infinite. (This may seem surprising as the conductivity is a transport coefficient which should vanish in the lowest approximation; this paradox will be explained in §11.6.) Many phenomena can therefore be treated by a quite simple set of magnetohydrodynamic equations.

There are other situations where the scale of variation is smaller than the mean free path. For example in laboratory tubes L_c may well be longer than the apparatus, and in interplanetary space L_c is comparable with the astronomical unit (sun–earth distance), so the flow of plasma around the earth's neighborhood seems hardly to satisfy the condition for fluid behavior. Nevertheless many of these problems too may be successfully treated by MHD. This is because a new source of localization is present when there is a strong magnetic field: a small Larmor radius prevents directed motion across the field lines, and if this applies to both ions and electrons there is a tendency for fluid-like behavior. As there is no localization along the field lines, hydrodynamic equations will only apply if special circumstances ensure that the plasma properties are extremely slowly varying in that direction. When this is so a new type of MHD emerges. It differs from the usual formulation in having unequal pressures along and across the magnetic field, and these satisfy new adiabatic equations. This theory will be derived in §11.3. As might be expected, these results are closely related to the guiding-center and adiabatic-invariant theory used in Chapter 4 to describe the motion of individual particles. The connection will be demonstrated in §11.4.

The remainder of the chapter will deal briefly with some applications of MHD. An extensive treatment of the solutions of the equations of MHD would be out of place here, being itself the subject of several existing texts.

11.2 THE ADIABATIC APPROXIMATION FOR COLLISION-DOMINATED GASES

11.2.1 Simple gas

Let us take first the case of a gas composed of a single monatomic species. We start with Boltzmann's equation for $f(\mathbf{x}, \mathbf{v}, t)$

$$\frac{\partial f}{\partial t} + v_k \frac{\partial f}{\partial x_k} + a_k \frac{\partial f}{\partial v_k} = \left(\frac{\partial f}{\partial t}\right)_c, \tag{11.1}$$

where $\mathbf{a}$ is the acceleration due to external forces, e.g. gravity, assumed independent of $\mathbf{v}$. The number density, mean velocity, pressure tensor and heat flux tensor were defined in §7.4; we repeat them here for reference.

$$N(\mathbf{x}, t) = \int f \, d^3v \tag{11.2}$$

$$u_i(\mathbf{x}, t) = N^{-1} \int v_i f \, d^3v \tag{11.3}$$

$$p_{ij}(\mathbf{x}, t) = m\int c_i c_j f\, d^3v \tag{11.4}$$

$$Q_{ijk}(\mathbf{x}, t) = m\int c_i c_j c_k f\, d^3v, \tag{11.5}$$

where $\mathbf{c} = \mathbf{v} - \mathbf{u}$. The local temperature is

$$T(\mathbf{x}, t) = p_{jj}/3NK. \tag{11.6}$$

The heat flux vector, whose significance will be clear shortly, is defined as

$$q_i(\mathbf{x}, t) = \tfrac{1}{2}Q_{ijj}. \tag{11.7}$$

Three important results, the "moment equations" follow if we multiply (11.1) by 1, mc_i and $\frac{1}{2}mc^2$ in turn, and integrate over velocity space. The expressions obtained on the left are identified by means of (11.2)–(11.7). On the right we have zero since for a single species the number density, momentum and energy are conserved locally during collisions (compare with (9.123)). The first two equations so obtained are identical with (7.18) and (7.20) since the inclusion of collisions has no effect. Thus

$$\frac{\partial N}{\partial t} + \frac{\partial}{\partial x_k}(Nu_k) = 0 \tag{11.8}$$

$$Nm\frac{Du_i}{Dt} + \frac{\partial p_{ik}}{\partial x_k} - Nma_i = 0 \tag{11.9}$$

where

$$\frac{D}{Dt} \equiv \frac{\partial}{\partial t} + u_k\frac{\partial}{\partial x_k}.$$

The third equation, not given in §7.4 but readily obtained, is

$$\frac{D}{Dt}(\tfrac{3}{2}NKT) + \frac{\partial q_k}{\partial x_k} + \frac{\partial u_j}{\partial x_k}p_{jk} + \tfrac{3}{2}NKT\frac{\partial u_k}{\partial x_k} = 0. \tag{11.10}$$

These are equations of continuity, momentum and internal energy respectively. As discussed in §7.4, they are not in themselves a complete set of equations as there are too many unknowns. Now suppose that in the Boltzmann equation, (11.1), $(\partial f/\partial t)_c$ is the dominant term, in the sense that if f departs appreciably from the condition $(\partial f/\partial t)_c = 0$ then $(\partial f/\partial t)_c$ would become very large compared with the second and third terms on the left, and f would change very rapidly towards a solution of $(\partial f/\partial t)_c = 0$. Then in the lowest approximation (11.1) is simply $(\partial f/\partial t)_c = 0$. The most general solution, whatever may be the type of interaction involved in the collisions, must be the Maxwellian

$$f = f^{(0)}(\mathbf{v}, N, \mathbf{u}, T) \equiv N\left(\frac{m}{2\pi KT}\right)^{3/2} \exp\left[-m(\mathbf{v} - \mathbf{u})^2/2KT\right], \tag{11.11}$$

where N, $\mathbf{u}$ and T may be any (slowly varying) functions of $\mathbf{x}$ and t. Equation (11.11) is of course consistent with (11.2), (11.3) and (11.6); and equations (11.4) and (11.7) show further that

$$p_{ij}^{(0)} = p\delta_{ij}, \quad q_i^{(0)} = 0, \quad \text{where} \quad p = NKT, \tag{11.12}$$

the superscript (0) denoting that p_{ij}, q_i have been calculated in the approximation $f \simeq f^{(0)}$.

When $p_{ij}^{(0)}$, $q_i^{(0)}$ are substituted in (11.9) and (11.10) as appropriate, we find

$$Nm\frac{Du_i}{Dt} + \frac{\partial p}{\partial x_i} - Nma_i = 0 \tag{11.13}$$

and

$$\frac{D}{Dt}(\tfrac{3}{2}NKT) + \tfrac{5}{2}NKT\frac{\partial u_j}{\partial x_j} = 0.$$

Combining this latter equation with (11.8) yields

$$\frac{D}{Dt}(pN^{-5/3}) = 0. \tag{11.14}$$

Equations (11.8), (11.13) and (11.14) are the usual equations for a perfect gas with neglect of viscosity and heat conduction. They form a complete set for the variables N, $\mathbf{u}$, p, subject to appropriate initial and boundary values.

The manner in which these results form the lowest stage of an approximation procedure will become clearer in §11.5.

11.2.2 Gas mixture

For a gas composed of a mixture of species labeled $r, s, \ldots$, a similar development is possible. $(\partial f/\partial t)_c$ consists of a sum of terms due to all possible kinds of collisions, and each term separately is classified as "large." The equation of continuity (11.8), still holds for each species (though we must affix a label r onto N and $\mathbf{u}$), since collisions neither create nor destroy particles. The equation of momentum (11.9), for a single species now includes terms representing the transfer by collisions of momentum between that species and all the others; however, if we sum these equations to get an expression for the rate of change of *total* momentum, these additional terms must clearly cancel, by Newton's Third Law, whatever may be the detailed mechanism of the collisions. An identical remark applies to the energy equation (11.10). When we make the assumption that $(\partial f/\partial t)_c$ is the dominant term for each species, we are led in the lowest approximation to the solution (11.11), where $\mathbf{u}$ and T are the same for every species (though the N_r may be different). In other words, we again have local thermal equilibrium. It follows that equations (11.12) hold for each species, and we may conveniently form a total pressure tensor by summing p_{ij} over the species, as each p_{ij} is calculated with respect to the same mean velocity $\mathbf{u}$. Equations (11.9) and (11.10) may also be easily summed since $\mathbf{u}$, T, and the operator D/Dt are the same for each species, leading to corresponding forms for (11.13) and (11.14) for the gas as a whole.

A gas mixture thus behaves exactly like a simple gas to this approximation. The composition of each packet of gas remains constant, as $\mathbf{u}$ is the same in each equation of continuity (11.8). The usual equations of mass conservation, momentum, and adiabatic state for a perfect gas are applicable to such a mixture. However, if, as we expect, almost perfect electrical neutrality is maintained, there will be no electromagnetic force at all since there is no space charge or current; the dynamics is then completely decoupled from the electromagnetic field—a situation of no interest so far as plasma physics is concerned. This conclusion is of course changed by consideration of the next approximation.

11.3 THE ADIABATIC APPROXIMATION FOR A COLLISIONLESS PLASMA IN A STRONG MAGNETIC FIELD

11.3.1 Solution in a strong magnetic field

As foreshadowed in the introduction, we turn now to a different situation, namely a plasma with a strong magnetic field, and such that the Larmor radii for both species are much smaller than the typical scale of variation. The required localization is thus provided by the field rather than collisions; the latter will in fact be ignored. Our objective in this section is to obtain hydrodynamic equations in the lowest approximation in which the field is regarded as dominant, proceeding in a way analogous to §11.2.

Consider first one species; we rewrite (11.1)

$$\frac{\partial f}{\partial t} + v_k \frac{\partial f}{\partial x_k} = -a_k \frac{\partial f}{\partial v_k} \tag{11.15}$$

with $\mathbf{a} = (e/m)(\mathbf{E} + \mathbf{v} \times \mathbf{B})$; the right-hand side is to be regarded as one order larger than the left-hand side and $\mathbf{E}$, $\mathbf{B}$ are specified. The moments are defined as before, equations (11.2)–(11.7), and the first two moment equations (11.8) and (11.9) are

$$\frac{DN}{Dt} + N\frac{\partial u_j}{\partial x_j} = 0 \tag{11.16}$$

$$Nm\frac{Du_i}{Dt} + \frac{\partial p_{ij}}{\partial x_j} = NeE_i', \tag{11.17}$$

where

$$\mathbf{E}' = \mathbf{E} + \mathbf{u} \times \mathbf{B} \tag{11.18}$$

(the momentum equation involves slight amendment to the derivation, as $\mathbf{a}$ is now velocity dependent). $\mathbf{E}'$ is of course the electric field in the frame moving with the fluid velocity. For the remaining moment equation we need something more general than (11.10); it is obtained by multiplying (11.15) by $c_i c_j$ and integrating with respect to velocity. After some tedious algebra, this yields

$$\frac{Dp_{ij}}{Dt} + \frac{\partial Q_{ijk}}{\partial x_k} + \frac{\partial u_i}{\partial x_k} p_{jk} + \frac{\partial u_j}{\partial x_k} p_{ik} + \frac{\partial u_k}{\partial x_k} p_{ij} = \frac{eB_k}{m}[\varepsilon_{ink} p_{jn} + \varepsilon_{jnk} p_{in}], \tag{11.19}$$

where $\boldsymbol{\varepsilon}$ is the usual alternating tensor. This reverts to (11.10) on contracting i and j. In (11.17) and (11.19) the terms on the right are just those arising from the right-hand side of (11.15).

In the lowest approximation, $f = f^{(0)}$ must satisfy

$$[\mathbf{E} + \mathbf{v} \times \mathbf{B}]\cdot\frac{\partial f^{(0)}}{\partial \mathbf{v}} = 0, \tag{11.20}$$

and from (11.17) we must have $\mathbf{E}' \simeq 0$, i.e.

$$\mathbf{E} + \mathbf{u}^{(0)} \times \mathbf{B} \simeq 0. \tag{11.21}$$

This is only possible if $\mathbf{E}\cdot\mathbf{B} \simeq 0$; this is an integrating condition, without which the process cannot start. The approximation sign is used to indicate that an error of one order smaller could be involved, for example $\mathbf{E}$ could have a small component parallel to $\mathbf{B}$. The reason for adding the superscript to $\mathbf{u}$ will be mentioned shortly. From (11.20) and (11.21) we have

$$(\mathbf{c} \times \mathbf{B})\cdot\frac{\partial f^{(0)}}{\partial \mathbf{v}} \simeq 0.$$

The situation is the same as in §9.1.1, so we have

$$f^{(0)} = f^{(0)}(c_\parallel, c_\perp, \mathbf{x}, t) \tag{11.22}$$

where $c_\parallel, c_\perp$ are the components of $\mathbf{c}$ along and perpendicular to $\mathbf{B}$. The right-hand side of (11.19) then vanishes to this approximation, as it should. All we have done so far is to ensure that $f^{(0)}$ is not varying rapidly (at the gyro-frequency); yet we already have a significant result, for (11.21) applies to all species, and we shall show below that this implies that the field lines are frozen into the plasma. We may also write

$$\mathbf{u}^{(0)} = \mathbf{u}_E + \mathbf{u}_\parallel \tag{11.23}$$

where, as in §4.4.2,

$$\mathbf{u}_E = \mathbf{E} \times \mathbf{B}/B^2. \tag{11.24}$$

This is the *electric drift* velocity, the same for all species. No condition is yet imposed on $\mathbf{u}_{\|}$, the drift parallel to $\mathbf{B}$.

Our solution (11.22) imposes special forms on $p_{ij}^{(0)}$ and $Q_{ijk}^{(0)}$. Clearly

$$\mathbf{p}^{(0)} = \begin{pmatrix} p_\perp & 0 & 0 \\ 0 & p_\perp & 0 \\ 0 & 0 & p_\| \end{pmatrix}$$

provided $\mathbf{B}$ is along the third axis, and there are simple integral expressions for the pressures $p_\perp$ and $p_\|$. We can write this in tensor form as

$$p_{ij}^{(0)} = p_\| \hat{B}_i \hat{B}_j + p_\perp(\delta_{ij} - \hat{B}_i \hat{B}_j), \tag{11.25}$$

where $\hat{\mathbf{B}}$ is the unit vector $\mathbf{B}/|B|$. Also

$$Q_{ij\,k}^{(0)} = q_\| \hat{B}_i \hat{B}_j \hat{B}_k + q_\perp(\delta_{ij}\hat{B}_k + \delta_{jk}\hat{B}_i + \delta_{ki}\hat{B}_j), \tag{11.26}$$

where $q_\|, q_\perp$ are further integrals over $f^{(0)}$. We may note that the heat flow vector $\frac{1}{2}Q_{jjk}$ is $\frac{1}{2}(q_\| + 5q_\perp)\hat{B}_k$, so there is no thermal conduction across the field lines to this order.

Following the development of §11.2, we are to substitute the zero-order quantities into the moment equations. This will give hydrodynamic equations whose role is that of integrating conditions for the next stage of the approximation. The equation of continuity is simply (11.16) applied to $\mathbf{u}^{(0)}$. The other equations are considered in turn in the next subsections.

11.3.2 The momentum equation; Parker's modified hydromagnetic equation

The momentum equation is simply (11.17), with $u_i^{(0)}$ and $p_{ij}^{(0)}$ used on the left. It should not, however, be supposed that the right-hand side vanishes, for although $\mathbf{E} \simeq -\mathbf{u}^{(0)} \times \mathbf{B}$, we must, by hypothesis, evaluate the terms on the right more accurately as they contain the "large" factors $\mathbf{E}$ and $\mathbf{B}$. Consequently we have

$$Nm\frac{Du_i^{(0)}}{Dt} + \frac{\partial p_{ij}^{(0)}}{\partial x_j} = [\rho\mathbf{E} + \mathbf{j} \times \mathbf{B}]_i \tag{11.27}$$

and the condition (11.21) merely ensures that the right-hand side is not one order too large. Here $\rho = Ne$ and $\mathbf{j} = Ne\mathbf{u}$ are the charge and current for one species. So $\mathbf{j}$ is not, as yet, known to sufficient accuracy for (11.27) to be useful. However, this difficulty can be obviated by summing over the species. The right-hand side is then formally the same, ρ and $\mathbf{j}$ being now the summed quantities. The left-hand side is awkward to handle unless $\mathbf{u}^{(0)}$ is the same for each species, for if not $\mathbf{p}^{(0)}$ is calculated relative to a different mean velocity in each case, and the summing of the acceleration term will also be inconvenient. Now we showed that to lowest order all species have the same drift, $\mathbf{u}_E$, across the field. The only assumption required is therefore that $\mathbf{u}_\|$ be the same for all species in the first approximation. In the simple case where there is only one kind of ion, this is equivalent to saying that we will not allow substantial currents to flow along the field lines. However $\mathbf{u}$ is not *exactly* the same for each species, and this is why we introduced the symbol $\mathbf{u}^{(0)}$ for the approximate common value. Granting this, we have

$$\rho_m \frac{Du_i^{(0)}}{Dt} + \frac{\partial p_{ij}^{(0)}}{\partial x_j} = [\rho \mathbf{E} + \mathbf{j} \times \mathbf{B}]_i, \tag{11.28}$$

where ρ_m is the total mass density, $\mathbf{p}^{(0)}$ the total pressure tensor. The quantities ρ and $\mathbf{j}$ may now be eliminated by means of Maxwell's equations

$$\rho = \varepsilon_0 \operatorname{div} \mathbf{E}, \qquad \mathbf{j} = \mu_0^{-1} \operatorname{curl} \mathbf{B} - \varepsilon_0 \frac{\partial \mathbf{E}}{\partial t} \tag{11.29}$$

so that the terms on the right of (11.28) may be written solely in terms of $\mathbf{E}$ and $\mathbf{B}$. At this point it is usual in magnetohydrodynamics to make the approximation that the contributions from $\rho\mathbf{E}$ and the displacement current are negligible. It is readily shown that this is so for slowly varying phenomena with $u \ll c$; the point really is that space charge effects are important only at much higher frequency (the plasma frequency), and can be ignored in a hydrodynamic treatment. This is, however, a quite separate approximation from that of the strong magnetic field on which our iteration procedure is based; the two do not necessarily stand or fall together. It remains only to evaluate the divergence of $\mathbf{p}^{(0)}$ explicitly, using (11.25) and noting $\operatorname{div} \mathbf{B} = 0$. Omitting the superscript (0), we have finally

$$\rho_m \frac{D\mathbf{u}}{Dt} = -\nabla(p_\perp + B^2/2\mu_0) + (\mathbf{B}\cdot\nabla)\left[\left(\frac{1}{\mu_0} + \frac{p_\perp - p_\parallel}{B^2}\right)\mathbf{B}\right]. \tag{11.30}$$

This differs from the usual hydromagnetic equation for a perfectly conducting inviscid fluid only through the term $(p_\perp - p_\parallel)/B^2$. It was derived (in a quite different way) by E. N. Parker in 1957, though it is implicit in earlier work.

11.3.3 Freezing-in of field lines

It will be useful to insert here a comment on equation (11.21)

$$\mathbf{E} + \mathbf{u} \times \mathbf{B} = 0. \tag{11.31}$$

The usual aim in magnetohydrodynamics is to eliminate $\mathbf{E}$, $\mathbf{j}$ and ρ so that $\mathbf{B}$ is the only variable additional to those of ordinary hydrodynamics. We have achieved this in (11.30). It only remains to eliminate $\mathbf{E}$ from (11.31) by means of the law of electromagnetic induction,

$$\operatorname{curl} \mathbf{E} = -\frac{\partial \mathbf{B}}{\partial t}. \tag{11.32}$$

Thus

$$\frac{\partial \mathbf{B}}{\partial t} = \operatorname{curl}(\mathbf{u} \times \mathbf{B}). \tag{11.33}$$

Expanding the right-hand side and omitting the term in div **B**,

$$\frac{D\mathbf{B}}{Dt} = (\mathbf{B}\cdot\nabla)\mathbf{u} - \mathbf{B}\,\mathrm{div}\,\mathbf{u}. \tag{11.34}$$

Two well-known results follow from this equation. Let P and Q be any two neighboring points, and write $\overrightarrow{PQ} = d\mathbf{r}$. If P and Q are convected with the fluid velocity $\mathbf{u}(\mathbf{x}, t)$, the rate of change of the vector $d\mathbf{r}$ due to non-uniformity of $\mathbf{u}$ is $(d\mathbf{r}\cdot\nabla)\mathbf{u}$. Thus

$$\frac{D}{Dt}(dr_i) = dr_j\frac{\partial u_i}{\partial x_j}. \tag{11.35}$$

Again, if $d\mathbf{S}$ is a surface element composed of points which are convected with the fluid

$$\frac{D}{Dt}(dS_i) = dS_i\,\mathrm{div}\,\mathbf{u} - \frac{\partial u_j}{\partial x_i}\,dS_j. \tag{11.36}$$

(This second result is most easily obtained by noting that if dV is the volume $d\mathbf{r}\cdot d\mathbf{S}$, the required formula must be such that $D(dV)/Dt = dV\,\mathrm{div}\,\mathbf{u}$.) Combining (11.34) with (11.35) shows that

$$\frac{D}{Dt}(d\mathbf{r}\times\mathbf{B}) = -d\mathbf{r}\times\mathbf{B}\,\mathrm{div}\,\mathbf{u}, \tag{11.37}$$

which vanishes identically if $d\mathbf{r}$ is parallel to **B**. Consequently if P and Q were initially on the same line of force, they will remain so. So whilst we are not obliged to make any identification of lines of force at different instants, we can very conveniently regard the lines as convected with the fluid. This is the "freezing-in" theorem. Similarly, by combining (11.34) with (11.36), we find the second result

$$\frac{D}{Dt}(\mathbf{B}\cdot\mathbf{dS}) = 0. \tag{11.38}$$

The flux of **B** through $d\mathbf{S}$ is thus constant, and by integration the same holds for surfaces of finite size if convected with the fluid. (Alternatively, apply Stokes' theorem to (11.33)). These results are useful for visualizing the way in which the magnetic field varies as a result of the flow, without the need to consider the electric current. This has been exploited by Alfvén, and by many others.

For later reference, the convective rate of change of the *magnitude* of **B** is also of interest. The scalar product of $\mathbf{B}/B^2$ with (11.34) yields

$$\frac{1}{B}\frac{DB}{Dt} = \hat{B}_i\hat{B}_k\frac{\partial u_i}{\partial x_k} - \mathrm{div}\,\mathbf{u}. \tag{11.39}$$

11.3.4 The double-adiabatic equations of Chew, Goldberger and Low

We still lack equations for the rates of change of $p_\parallel$ and $p_\perp$. This information can be expected to come from (11.19) (used to lowest order) if we are to follow the procedure of §11.2. The right-hand side of the equation involves $\mathbf{p}$ evaluated more accurately than $\mathbf{p}^{(0)}$, since it is multiplied by a "large" factor (essentially the gyro-frequency). We can obviate this by contracting the equation with δ_{ij} and $\hat{B}_i\hat{B}_j$ in turn as the right-hand side then vanishes identically, and this will give us two independent scalar equations; this is just what is needed. The contribution from $\partial Q^{(0)}_{ijk}/\partial x_k$ creates a new difficulty compared with the collision-dominated case; for using (11.26) we find that the contributions to the contracted equations will vanish only if $q_\parallel/B$ and $q_\perp/B$ are constant throughout space. This is another reflection of the fact that the strong magnetic field localizes the particles only in the transverse direction; they are free to move large distances along the field. To make progress we assume that the distribution function possesses sufficient symmetry, and varies so gradually along the field, that the effect of these terms (which are akin to heat conduction) is negligible.

Carrying out these contractions, with some simplification arising from $\hat{\mathbf{B}}$ being a unit vector, leads to

$$\frac{D}{Dt}(p_\parallel + 2p_\perp) + 2(p_\parallel - p_\perp)\hat{B}_i\hat{B}_k\frac{\partial u_i}{\partial x_k} + (p_\parallel + 4p_\perp)\,\mathrm{div}\,\mathbf{u} = 0, \quad (11.40)$$

and

$$\frac{Dp_\parallel}{Dt} + 2p_\parallel\hat{B}_i\hat{B}_k\frac{\partial u_i}{\partial x_k} + p_\parallel\,\mathrm{div}\,\mathbf{u} = 0. \quad (11.41)$$

From these,

$$\frac{Dp_\perp}{Dt} - p_\perp\hat{B}_i\hat{B}_k\frac{\partial u_i}{\partial x_k} + 2p_\perp\,\mathrm{div}\,\mathbf{u} = 0. \quad (11.42)$$

Equations (11.41) and (11.42) enable $p_\parallel$ and $p_\perp$ to be calculated. To write them in a more succinct form, we use (11.39) together with the equation of continuity

$$\frac{1}{\rho_m}\frac{D\rho_m}{Dt} + \mathrm{div}\,\mathbf{u} = 0 \quad (11.43)$$

(which follows from (11.16), remembering that to lowest order $\mathbf{u}$ is supposed to be the same for all species). We can use these to eliminate div $\mathbf{u}$ and $B_iB_k\,\partial u_i/\partial x_k$, giving eventually

$$\frac{D}{Dt}\left(\frac{p_\parallel B^2}{\rho_m^3}\right) = 0 \quad (11.44)$$

$$\frac{D}{Dt}\left(\frac{p_\perp}{\rho_m B}\right) = 0. \quad (11.45)$$

These so-called "double adiabatic equations" are due to G. F. Chew, M. L. Goldberger and F. E. Low (1956), and take the place of the familiar

$$\frac{D}{Dt}(p\rho_m^{-\gamma}) = 0 \tag{11.46}$$

of gas dynamics, γ being the ratio of specific heats (compare with (11.14)). The same equations hold if ρ_m is replaced by N, the density of any one species.

A more intuitive approach to these results can be obtained considering a small cylinder, of length L and cross-section S, with axis parallel to the field lines. Let δ denote the change occurring in time δt if the cylinder is convected with the fluid. Then by (11.35) and (11.36)

$$\frac{\delta L}{L} = \hat{B}_i\hat{B}_k\frac{\partial u_i}{\partial x_k}\,\delta t \quad \text{and} \quad \frac{\delta S}{S} = \left(\operatorname{div}\mathbf{u} - \hat{B}_i\hat{B}_k\frac{\partial u_i}{\partial x_k}\right)\delta t. \tag{11.47}$$

Now the internal energy density due to motion parallel to the field is $\frac{1}{2}p_\parallel$, and equating this to work done by pressure forces across an end section, assuming the process adiabatic, we find

$$\delta(\tfrac{1}{2}p_\parallel LS) = -p_\parallel S\,\delta L.$$

Similarly the internal energy density due to transverse motion is $p_\perp$, and its change is given by the work on the curved surfaces, so

$$\delta(p_\perp LS) = -p_\perp L\,\delta S.$$

When these are combined with (11.47), our former equations (11.41) and (11.42) follow at once.

In view of the assumption which we were obliged to make to remove the heat conduction term, the most natural application of this work is to the purely two-dimensional case with straight field lines and all motion perpendicular to the field. This can be pictured as the motion of flux tubes identified by the particles they contain. This leads to $B_iB_k\,\partial u_i/\partial x_k = 0$, and ρ_m/B is constant for each tube. The adiabatic equations reduce to

$$\frac{D}{Dt}\left(\frac{p_\parallel}{\rho_m}\right) = \frac{D}{Dt}\left(\frac{p_\perp}{\rho_m^2}\right) = 0. \tag{11.48}$$

Comparing with (11.46), γ takes the value 1 parallel to the field and 2 across it. The momentum equation (11.30) reduces to

$$\rho_m\frac{D\mathbf{u}}{Dt} = -\nabla(p_\perp + B^2/2\mu_0), \tag{11.49}$$

and as $p_\perp \propto B^2$ the total pressure (kinetic plus magnetic) may be assigned the value $\gamma = 2$.

On the other hand, if the only variations are parallel to the field (e.g. for sound waves traveling along the field lines), B stays constant and we have

$$\frac{D}{Dt}\left(\frac{p_{\parallel}}{\rho_m^3}\right) = \frac{D}{Dt}\left(\frac{p_{\perp}}{\rho_m}\right) = 0, \tag{11.50}$$

so that γ has the values 3 along the field and 1 across it. This case must be treated with more caution in view of our additional approximations.

Summarizing the whole of this section, we have produced a complete scheme of magnetohydrodynamics for a plasma, consisting of the equations of mass conservation (11.43), freezing in of field lines (11.34), momentum (11.30) and adiabatic state for $p_{\parallel}$ and $p_{\perp}$ (11.44 and 11.45). These hold for low-frequency phenomena in a strong magnetic field, subject to the assumption (whose justification will depend on the boundary conditions) that there are not heavy electric currents, or a considerable heat flux, along the field lines. It has been tacitly assumed that the equation of mass conservation implies a continuity equation for each species. This is certainly true if there is only one species of positive ion and we accept the approximation that the space charge is very small. If, however, there is more than one species of ion, the equations just listed provide a complete set only if relative diffusion between ions of different type may be assumed unimportant.

In certain circumstances it is possible to derive results which are analogs of Bernoulli's theorem of ordinary fluid mechanics—see Problems 4 and 5.

11.4 CONNECTION WITH PARTICLE ORBIT THEORY

The reader may well have suspected that the development of adiabatic hydromagnetic equations in a strong magnetic field must be related to the theory of particle orbits in such a field, as given in Chapter 4. The two are indeed closely connected, as we will now digress to show.

The motion of a single particle was divided into the motion of its "guiding center" together with the motion relative to the guiding center. The latter part is the usual Larmor rotation at angular velocity Ω, and it will be convenient to call the velocity $c_{\perp}$ (this is simply the modulus of $\mathbf{v} - \mathbf{u}$ as defined in Chapter 4). The circling motion is completely specified, once the path of the guiding center is known, by the constancy of the magnetic moment

$$M = mc_{\perp}^2/2B. \tag{11.51}$$

The path of the guiding center is determined by equations (4.88) and (4.89), which we repeat with minor changes of notation:

$$\begin{aligned}\mathbf{u}_{\perp} = \mathbf{u}_E + \frac{m}{eB}\hat{\mathbf{B}} \times \Bigg\{&\frac{M}{m}\nabla B + \frac{v_{\parallel}^2}{B^2}(\mathbf{B}\cdot\nabla)\mathbf{B} \\ &+ \frac{v_{\parallel}}{B}\left[\frac{\partial \mathbf{B}}{\partial t} + (\mathbf{B}\cdot\nabla)\mathbf{u}_E + (\mathbf{u}_E\cdot\nabla)\mathbf{B}\right] + (\mathbf{u}_E\cdot\nabla)\mathbf{u}_E + \frac{\partial \mathbf{u}_E}{\partial t}\Bigg\}\end{aligned} \tag{11.52}$$

$$\frac{dv_\parallel}{dt} = -\frac{M}{m}(\hat{\mathbf{B}}\cdot\nabla)B + \frac{e}{m}\hat{\mathbf{B}}\cdot\mathbf{E} + \frac{1}{B}\mathbf{u}_E\cdot\left\{\frac{\partial \mathbf{B}}{\partial t} + [(v_\parallel\hat{\mathbf{B}} + \mathbf{u}_E)\cdot\nabla]\mathbf{B}\right\}. \quad (11.53)$$

Here $\mathbf{u}_E$ is the electric drift, as before; $\mathbf{u}_\perp$ is the drift perpendicular to the magnetic field, and $v_\parallel$ is the particle's speed along the field. (Note that the second term on the right of (11.53) is permitted only if $\hat{\mathbf{B}}\cdot\mathbf{E}$ is small, as it is multiplied by e/m; this is in accord with our assertion that $\mathbf{B}\cdot\mathbf{E} \simeq 0$ to lowest order.) The idea we wish to pursue is that if a plasma consists of a large number of particles all of which have very small Larmor radii in the field to be considered, we can think of them as a fluid of fictitious particles at the corresponding guiding centers, endowed with an additional property, the magnetic moment. When we examine a small volume element $d\tau$, its hydromagnetic properties will be obtained by summing over the (fictitious) particles within. As the first term of (11.52) is one order larger than the rest, and is the same for all particles, the hydrodynamic velocity must be, in the lowest approximation, $\mathbf{u}_E$ together with an unknown part $\bar{v}_\parallel\hat{\mathbf{B}}$, where $\bar{v}_\parallel$ is the mean of $v_\parallel$ for the particles in $d\tau$. This agrees with (11.21) and (11.23). As in §11.3.2 we can take it that the charge density cancels out when we sum over the species, but we would like to calculate the current density. The part perpendicular to the field, $\mathbf{j}_\perp$, can be found by summing (11.52) over the particles in $d\tau$ to account for the drift of guiding centers, and adding to that the current "due" to the dipole moments M, representing the motion relative to the guiding center. Let us do this for one species. We shall need, noting (11.51):

$$\begin{aligned} p_\perp\, d\tau &= \tfrac{1}{2}m\sum_{d\tau} c_\perp^2 = B\sum_{d\tau} M \\ p_\parallel\, d\tau &= m\sum_{d\tau}(v_\parallel - \bar{v}_\parallel)^2. \end{aligned} \quad (11.54)$$

(The factor $\frac{1}{2}$ is involved as $c_\perp$ represents two orthogonal components; the summations range over all the particles in $d\tau$.) We also introduce a magnetization, $\mathbf{I}$,

$$\mathbf{I}\, d\tau = -\hat{\mathbf{B}}\sum_{d\tau} M,$$

i.e.

$$\mathbf{I} = -(p_\perp/B^2)\mathbf{B}, \quad (11.55)$$

the negative sign indicating that the sense of gyration is such as to make the plasma appear to be diamagnetic. The current due to the gyration is just curl $\mathbf{I}$, so we have

$$\mathbf{j}_\perp\, d\tau = \sum_{d\tau} e\mathbf{u}_\perp + (\text{curl}\,\mathbf{I})_\perp\, d\tau,$$

which after some vector manipulation leads to

$$\mathbf{j}_\perp = Ne\mathbf{u}_E + \frac{Nm}{B}\hat{\mathbf{B}} \times \left\{\frac{p_\perp}{NmB}\nabla B + \frac{\overline{(v_\parallel^2)}}{B^2}(\mathbf{B}\cdot\nabla)\mathbf{B} + \frac{\bar{v}_\parallel}{B}\left[\frac{\partial \mathbf{B}}{\partial t} + (\mathbf{B}\cdot\nabla)\mathbf{u}_E + (\mathbf{u}_E\cdot\nabla)\mathbf{B}\right] + (\mathbf{u}_E\cdot\nabla)\mathbf{u}_E + \frac{\partial \mathbf{u}_E}{\partial t}\right\} + \hat{\mathbf{B}} \times \left\{\nabla\left(\frac{p_\perp}{B}\right) - \frac{p_\perp}{B^3}(\mathbf{B}\cdot\nabla)\mathbf{B}\right\}. \tag{11.56}$$

The right-hand side of (11.56) has 10 distinct terms; numbering them in order in which they appear they are

(1) The *electric drift* current, which is formally large but will cancel out on summing over species. A corresponding effect in the presence of a gravitational field would not, however, cancel.

(2) The *gradient* current, being the drift of charged particles due to the shear in the magnetic field lines.

(3) The *curvature* current, due to centrifugal reaction against curved lines of force.

(4)–(7) Various terms due to rates of change of **B** and $\mathbf{u}_E$; no definite names seem to have been attached to these. They are absent in many simple situations, and are omitted in some accounts of the subject.

(8) The term in $\partial\mathbf{u}_E/\partial t$ could also be grouped with (4)–(7), but is sometimes called the *polarization* current, and has the following interpretation. Suppose **B** is large and essentially constant but there is a small fluctuating **E** orthogonal to **B**. Then $\mathbf{u}_E = -\hat{\mathbf{B}}\times \mathbf{E}/B$, so $\partial\mathbf{u}_E/\partial t = -(\hat{\mathbf{B}}/\mathbf{B}) \times \dot{\mathbf{E}}$ and the polarization current is $(Nm/B^2)\dot{\mathbf{E}}$. We could regard this current as the rate of charge of a polarization $\mathbf{P} = (Nm/B^2)\mathbf{E}$, so the medium behaves as a dielectric, with dielectric constant $1 + Nm/\varepsilon_0B^2$, i.e. $1 + Nm\mu_0c^2/B^2$. This is precisely the description for a medium carrying Alfvén waves (see §5.4.2), after allowing for summation over species.

(9)–(10) The *magnetization* current, resulting from curl **I**.

Equation (11.56) is the mathematical expression of what is often a useful physical approach to the low-frequency properties of plasmas. This is especially so for low-β plasmas, i.e. those for which the particle pressures are much less than the magnetic pressure. In that case the magnetic field is simply imposed from outside, to a first approximation, and the particles drift in a known field to produce the current; the magnetic effect of the current then follows from Maxwell's equations. However, our purpose here is to point out that (11.56) is really identical (as regards the component at right angles to **B**) with the momentum equation (11.27). To show this we note that $\mathbf{u} = \mathbf{u}_E + \bar{v}_\parallel\hat{\mathbf{B}}$ to lowest order, so an expression for $D\mathbf{u}/Dt$ is readily constructed. The term $\partial p_{ij}/\partial x_j$ is expanded just as in (11.30). The vector product of (11.27) with $\mathbf{B}$ then leads to (11.56). (The conscientious reader who wishes to fill in the derivation will need to note that

$$p_{\|} = Nm\overline{(v_{\|} - \bar{v}_{\|})^2} = Nm[(\overline{v_{\|}^2}) - (\bar{v}_{\|})^2],$$

and that $\hat{\mathbf{B}} \times$ (a derivative of $\mathbf{B}$) may be written $\mathbf{B} \times$ (the same derivative of $\hat{\mathbf{B}}$).) It can also be shown that summation over (11.53) leads to a result equivalent to the component of (11.27) along the field. When summing over species, the same reservations are necessary as in §11.3.2; granting this the particle orbit theory thus leads eventually to (11.30)—indeed it was by this means that E. N. Parker originally derived that equation.

The double adiabatic equations can also be approached in this way. For instance (11.45), for one species, states that $p_{\perp}/NB$ is constant following a "packet" of fluid. But $p_{\perp}/NB$ is just $\bar{M}$, the mean magnetic moment for the packet, and M is constant for each particle. In the two-dimensional case, with straight field lines, this establishes a complete connection between the two approaches, but in other cases it does not because D/Dt is not the same as d/dt following particles. One cannot avoid introducing the distribution function for the parallel velocities, and the adiabatic equation for $p_{\perp}$ will only follow if a suitable assumption, equivalent to the vanishing of $q_{\perp}$ is made. The general equation for $p_{\|}$ is also rather difficult to obtain from the orbit equations, though the first part of (11.50) can be explained as follows. Imagine a "magnetic bottle" consisting of a long straight flux tube bounded by magnetic mirrors. Suppose the length, L, is very slowly varied by moving the mirrors. For most of the length of the bottle the field stays constant. By the second adiabatic invariant (§4.4.5) $\oint v_{\|}\, ds$ is constant, so that $v_{\|} \propto L^{-1}$, and $\frac{1}{2}mv_{\|}^2 \propto L^{-2}$; $v_{\perp}$ remains constant by the first adiabatic invariant and the Larmor radius is unchanged. So $N \propto L^{-1}$, and $p_{\|} \propto L^{-3}$, and we have $p_{\|}\rho_m^{-3} = \text{constant}$.

The reader may have noticed a bizarre interchange in the familiar physical interpretation of the hydromagnetic equations. For $\mathbf{E} + \mathbf{u} \times \mathbf{B} = 0$ is a vestigial Ohm's law for a perfectly conducting fluid, yet it determines the principal drift velocity $\mathbf{u}_{\perp}$ rather than $\mathbf{j}$. On the other hand if we calculate carefully the current $\mathbf{j}$ by summing over all the numerous drifts, the result is the transverse component of the momentum equation! There is a similar oddity about boundary conditions. Consider the boundary between a plasma and vacuum, with magnetic field parallel to the surface. The mechanical boundary condition is that $p_{\perp} + B^2/2\mu_0$ be continuous. But using the particle orbit picture the relevant consideration is that the magnetization $\mathbf{I}$ produces an equivalent surface current $\mathbf{I} \times \mathbf{n}$ (carried by the particles circulating just on the surface). Applying the usual boundary conditions (continuity of tangential H) and using (11.55), we find $\mu_0^{-1}B + p_{\perp}/B$ must be constant. If the change in B is much less than B itself (as it must be in this treatment) this gives $p_\perp + B^2/2\mu_0$ continuous.

When trying to gain physical insight into a situation, the choice between the fluid picture and the orbits picture depends on temperament and on the problem in hand. For practical calculation, the fluid equations are generally more convenient, though the orbit theory is in any case useful for tracing a particular batch of particles, e.g. those with high energy. Within the approximations and assumptions specified here, the two methods are entirely equivalent when considering the plasma as a whole.

11.5 THE CHAPMAN–ENSKOG EXPANSION

So far in this chapter we have dealt only with the adiabatic approximation to fluid equations. In this section, we will show how the adiabatic equations form the first step in a scheme of successive approximation, and give some account of the calculations required in taking the second step. The work of the first two subsections, while not in itself relevant to plasmas, will be helpful in exhibiting the main principles of the method in a simpler context.

11.5.1 The expansion procedure for a simple gas

We resume the discussion, begun in §11.2.1, of a simple collision-dominated gas governed by (11.1). In that earlier section, the gas was described by $f = f^0$, defined as the Maxwellian distribution corresponding to the local values of N, $\mathbf{u}$ and T; this was equivalent to ignoring the left-hand side of (11.1). Yet the hydrodynamic equations can actually be traced to the left-hand side since they originated as moment equations, for which the effect of $(\partial f/\partial t)_c$, though formally large, vanished by virtue of the conservation laws. This peculiar situation is perhaps best understood by regarding the hydrodynamic equations as integrability conditions under which the next stage of approximation can be carried out. Thus if we write* $f = f^0 + f^1$ and iterate to obtain an equation for the correction f^1, this will only be soluble if the moment equations are satisfied in the approximation $f = f^0$. If, then, we solve for f^1, we can use it to obtain improved moment equations which, in turn, would form integrability conditions for a further correction, say f^2.

To show how this works in detail, it is convenient to write

$$f = f^0 + \varepsilon f^1 + \varepsilon^2 f^2 + \cdots \tag{11.57}$$

Here f^0 is the Maxwellian distribution which at a time t has the correct values of N, $\mathbf{u}$ and T, and $f^1, f^2, \ldots$ are successive corrections in our scheme; ε will eventually be unity but its inclusion helps us to handle the expansion consistently, higher powers of ε denoting smaller terms.

---

* From this point onwards we shall for the most part discontinue placing parentheses around the superscripts 0, 1, 2, . . . , which refer to the order of approximation.

As the right-hand side of (11.1) is to be regarded as one order larger than the left, we write the equation

$$\frac{\partial f}{\partial t} + v_k \frac{\partial f}{\partial x_k} + a_k \frac{\partial f}{\partial v_k} = \frac{1}{\varepsilon}\left(\frac{\partial f}{\partial t}\right)_c . \tag{11.58}$$

Before inserting (11.57) into (11.58) and equating coefficients of ε, we need to give further thought to the meaning of the $\partial/\partial t$ term. At a given instant f^0 is fixed by assigning N, $\mathbf{u}$ and T throughout space, but the time derivatives of those quantities are not determined in the lowest approximation; as we saw in §11.2 the next approximation is involved, via equations (11.8), (11.9) and (11.10). Now of these, only (11.8) is complete, in the sense that $\partial N/\partial t$ can be expressed in terms of the present values of N, $\mathbf{u}$ and T. The other two are not yet adequate as p_{ik} and q_k are not yet known, except to the lowest order. Our hypothesis is, however, that the collisions will at all times keep the gas in a state close to the Maxwellian, so that only N, $\mathbf{u}$ and T are needed to specify its state. Such solutions of the Boltzmann equation are called "normal." There will therefore be successive improvements to the expressions for $\partial \mathbf{u}/\partial t$ and $\partial T/\partial t$ as the iteration proceeds and p_{ik} and q_k are determined to greater accuracy. We can express this by writing

$$\frac{\partial}{\partial t} = \frac{\partial^{(0)}}{\partial t} + \varepsilon \frac{\partial^{(1)}}{\partial t} + \varepsilon^2 \frac{\partial^{(2)}}{\partial t} + \cdots \tag{11.59}$$

where the symbols $\partial^{(r)}/\partial t$ are not themselves operators but denote successive improvements to the time derivative of the quantity to which they are applied. We are now ready to substitute (11.57) and (11.59) into (11.58) and equate coefficients of ε, giving

$$0 = \left(\frac{\partial f^0}{\partial t}\right)_c \tag{11.60a}$$

$$\frac{\partial^{(0)} f^0}{\partial t} + v_k \frac{\partial f^0}{\partial x_k} + a_k \frac{\partial f^0}{\partial v_k} = C^{(1)}(f^0, f^1) \tag{11.60b}$$

$$\frac{\partial^{(0)} f^1}{\partial t} + \frac{\partial^{(1)} f^0}{\partial t} + v_k \frac{\partial f^1}{\partial x_k} + a_k \frac{\partial f^1}{\partial v_k} = C^{(2)}(f^0, f^1, f^2), \tag{11.60c}$$

where $C^{(1)}$, $C^{(2)}$, . . . are appropriate functionals which are known when the form of $(\partial f/\partial t)_c$ is known explicitly (usually $(\partial f/\partial t)_c$ is quadratic in f, and so therefore are the expressions for $C^{(1)}$, etc.). From now on we may set $\varepsilon = 1$. Now (11.60a) is automatically satisfied as f^0 is Maxwellian. Accordingly, f^0 is fixed by the given values of N, $\mathbf{u}$ and T. The moment equations, applied to the lowest order, then give $\partial^{(0)}N/\partial t$, $\partial^{(0)}\mathbf{u}/\partial t$ and $\partial^{(0)}T/\partial t$; this is what we did in §11.2. We therefore know $\partial^{(0)}f^0/\partial t$ and can attempt to solve (11.60b) for

f^1 at the time under consideration. Since f^1 must not contribute to N, $\mathbf{u}$ and T, it is also subject to

$$\int f^1\, d^3v = 0, \qquad \int \mathbf{c} f^1\, d^3v = 0, \qquad \int c^2 f^1\, d^3v = 0. \qquad (11.61)$$

It can be shown that (11.60b) and (11.61) possess a unique solution. We then return to the moment equations and evaluate p_{ik} and q_k more accurately, using $f = f^0 + f^1$. This gives improved expressions for $\partial\mathbf{u}/\partial t$ and $\partial T/\partial t$, the new terms being recognized as those describing viscosity and heat conduction. We thereby have the correction $\partial^{(1)} f^0/\partial t$. By solving (11.60b) for f^1 in this way at two successive instants, we find $\partial^{(0)} f^1/\partial t$. We are then ready to solve (11.60c) for f^2 subject to conditions similar to (11.61); and so the process continues.

In practice, one stage of improvement on our results of §11.2 will suffice, so only (11.60b) need be tackled. The direct procedure would be to solve for f^1 and so construct the corrections p^1_{ik}, q^1_k to the pressure tensor and heat flow vector. Usually, this is not analytically possible, and f^1 is expanded in a series of functions chosen in such a way that only a few early terms of the series are needed to give p^1_{ik} and q^1_k to reasonable accuracy.

11.5.2 An illustration

The method outlined above was devised independently by D. Enskog and S. Chapman, and applied by them, and by others, to the case where $(\partial f/\partial t)_c$ represents binary collisions with various laws of interaction. This is indeed appropriate for neutral monatomic gases, but as explained in our introduction we need not go into the details of such calculations here, and reference may be made to the textbook of Chapman and Cowling. However, it may be helpful to exhibit some of the workings of the method by means of a simple, though fictitious, example. Consider a gas governed by the BGK equation (see §9.5.2),

$$\left(\frac{\partial f}{\partial t}\right)_c = -\nu(f - f^0),$$

where ν is a collision frequency, itself determined in some way by N and T, while f^0, as before, is the Maxwellian distribution with the same N, $\mathbf{u}$ and T as f. The right-hand side of (11.58) becomes

$$-\nu(f^1 + \varepsilon f^2 + \cdots),$$

so that (11.60b) reads

$$\frac{\partial^{(0)} f^0}{\partial t} + v_k \frac{\partial f^0}{\partial x_k} + a_k \frac{\partial f^0}{\partial v_k} = -\nu f^1. \qquad (11.62)$$

The major difficulty of the method, that of solving for f^1, does not arise here as (11.62) involves f^1 in the simplest possible way. The left-hand side is readily evaluated as f^0 is the Maxwellian, (11.11). The time and space derivatives of N, $\mathbf{u}$ and T are involved, but the time derivatives are to be eliminated by using the lowest-order equations (11.8), (11.13) and (11.14). This yields

$$-\nu f^1 = f^0\left\{\frac{m}{KT}(c_k c_l - \tfrac{1}{3}c^2\delta_{kl})\frac{\partial u_l}{\partial x_k} + c_k\left(\frac{mc^2}{2KT} - \frac{5}{2}\right)\frac{1}{T}\frac{\partial T}{\partial x_k}\right\}. \quad (11.63)$$

As ν is large on the appropriate scale, $f^1 \ll f^0$ is verified; conditions (11.61) are also readily checked. The required expressions

$$p_{ij}^1 = m\int c_i c_j f^1\, d^3v, \qquad q_i^1 = \tfrac{1}{2}m\int c_i c^2 f^1\, d^3v \quad (11.64)$$

are easily evaluated, and adding in the zero-order pressure (11.12) we have finally

$$\begin{aligned} p_{ij} &= p\delta_{ij} - \mu\left\{\frac{\partial u_i}{\partial x_j} + \frac{\partial u_j}{\partial x_i} - \tfrac{2}{3}\delta_{ij}\,\mathrm{div}\,\mathbf{u}\right\} \\ q_i &= -\lambda\frac{\partial T}{\partial x_i}, \end{aligned} \quad (11.65)$$

where

$$\mu = NKT/\nu, \qquad \lambda = 5NK^2T/2m\nu \quad (11.66)$$

are the coefficients of viscosity and heat conduction respectively. As the specific heat at constant volume is $C_v = 3K/2m$ per unit mass, the dimensionless ratio* $\mathrm{f} = \lambda/\mu C_v$ takes the value $\frac{5}{3}$. This is actually rather lower than the true value for monatomic gases (around 2.5) but this is of no real consequence as our purpose here is merely to illustrate the method; our working would be unaffected up to (11.63) if we made ν a function of $\mathbf{c}$, and by choice of the function we could easily adjust f. The results (11.65), when substituted into (11.9) and (11.10), lead to the corrected versions of the equations of momentum and internal energy, which are of the standard form for a viscous thermally conducting gas.

When a proper expression for $(\partial f/\partial t)_c$ is used, the equation for f^1 has the same right-hand side as (11.63), but at the left is a complicated (though linear and homogeneous) operator acting on f^1. It is the approximate solution of this equation that is laborious.

11.5.3 The expansion procedure for a plasma: first stage

We are now ready to consider the same procedure as applied to a multi-component plasma, with allowance for the effect of a magnetic field. As

* The name "Prandtl Number" has been given to the quantity 1/f.

before, the labels 0, 1, . . ., written in the upper position, denote orders in the expansion scheme, while tensor suffixes are always in the lower position. Further suffixes r and s will generally be used to label the species; these will be written either in the upper or lower position, the choice being purely for typographical convenience. The definitions (11.2)–(11.7) of moments have to be generalized as follows. The number density of the rth species is

$$N_r = \int f_r \, d^3v, \tag{11.67}$$

and for the plasma as a whole we define

$$N = \sum_r N_r, \qquad \rho = \sum_r N_r e_r, \qquad \rho_m = \sum_r N_r m_r, \tag{11.68}$$

the total number density, charge density and mass density respectively. The mean velocity of the rth species is

$$\mathbf{u}_r = (N_r)^{-1} \int \mathbf{v} f_r \, d^3v \tag{11.69}$$

and that of the whole gas is defined as

$$\mathbf{u} = (\rho_m)^{-1} \sum_r N_r m_r \mathbf{u}_r, \tag{11.70}$$

while the total electric current is of course

$$\mathbf{j} = \sum_r N_r e_r \mathbf{u}_r. \tag{11.71}$$

In dealing with a gas mixture it is convenient to define $\mathbf{c}$ as the velocity relative to $\mathbf{u}$ rather than to the particular $\mathbf{u}_r$, i.e. $\mathbf{c} = \mathbf{v} - \mathbf{u}$. We then write

$$p_{ij}^r = m_r \int c_i c_j f_r \, d^3v \tag{11.72}$$

$$q_k^r = \tfrac{1}{2} m_r \int c^2 c_k f_r \, d^3v, \tag{11.73}$$

with corresponding totals

$$p_{ij} = \sum_r p_{ij}^r, \qquad q_k = \sum_r q_k^r. \tag{11.74}$$

Finally, the kinetic energy density W and temperature T are defined, relative to the local mean velocity $\mathbf{u}$, by

$$W = \tfrac{1}{2} p_{ii} = \tfrac{3}{2} NKT. \tag{11.75}$$

Boltzmann's equation for species r is

$$\frac{\partial f_r}{\partial t} + v_k \frac{\partial f_r}{\partial x_k} + \left\{ \mathbf{g}_r + \frac{e_r}{m_r}(\mathbf{E} + \mathbf{v} \times \mathbf{B}) \right\}_k \frac{\partial f_r}{\partial v_k} = \sum_s \left(\frac{\partial f_r}{\partial t} \right)_{cs}. \tag{11.76}$$

Here $\mathbf{g}_r$ is the acceleration due to any non-electromagnetic external force (e.g. gravity), and $(\partial f_r/\partial t)_{cs}$ is the rate of change of f_r due to collisions with particles of species s. One must decide on the ordering of the various terms in this equation, by analogy with (11.58). We shall assume here that every separate collision term, and the Lorentz force term, are all large for each species. Using again the artificial parameter ε, eventually set equal to unity, we write

$$\frac{\partial f_r}{\partial t} + v_k \frac{\partial f_r}{\partial x_k} + a_k^r \frac{\partial f_r}{\partial v_k} = \frac{1}{\varepsilon}\left\{\sum_s \left(\frac{\partial f_r}{\partial t}\right)_{cs} - \frac{e_r}{m_r}(\mathbf{c} \times \mathbf{B})_k \frac{\partial f_r}{\partial v_k}\right\}. \quad (11.77)$$

Here

$$\mathbf{a}^r = \mathbf{g}^r + \frac{e_r}{m_r}\mathbf{E}', \quad \text{where} \quad \mathbf{E}' \equiv \mathbf{E} + \mathbf{u} \times \mathbf{B} \quad (11.78)$$

is the macroscopic acceleration after deducting the $\mathbf{c} \times \mathbf{B}$ term (a slight change compared with our previous notation), and is velocity-independent. Clearly there are many variants of (11.77) in which some but not all the terms are transferred from the right to the left; but it would be tedious to deal with every possible ordering that could arise. Our own choice is consistent only if the quantity $\mathbf{a}$ really is correctly classified as $O(1)$ rather than $O(\varepsilon^{-1})$. Usually $\mathbf{g}^r$ is relatively unimportant, so one can conclude that although $\mathbf{B}$ itself is large, $\mathbf{E}$ and $\mathbf{u}$ are so arranged that $\mathbf{E} + \mathbf{u} \times \mathbf{B}$ is small, and we may write

$$\mathbf{E}' \approx 0 \quad (11.79)$$

just as at (11.21). The field lines are therefore "frozen" into the fluid to this order, since the argument of §11.3.3 followed solely from this last equation together with Maxwell's equations.

We are now ready to form moment equations by integration of (11.77) over velocity space. Integrating the equation as it stands, we find as usual

$$\frac{\partial N_r}{\partial t} + \frac{\partial}{\partial x_k}(N_r u_k^r) = 0 \quad (11.80)$$

after noting that

$$\int \left(\frac{\partial f_r}{\partial t}\right)_{cs} d^3v = 0$$

for all (r, s), since collisions conserve particles. This combined with (11.70) also implies an equation of total mass conservation, namely (11.43). Again, multiplying (11.77) by $m_r c_i$, and summing over r as well as integrating over $\mathbf{v}$, we find

$$\rho_m\left(\frac{\partial u_i}{\partial t} + u_k \frac{\partial u_i}{\partial x_k}\right) - \sum_r N_r m_r a_i^r + \frac{\partial p_{ik}}{\partial x_k} = \frac{1}{\varepsilon}\sum_r N_r e_r[(\mathbf{u}_r - \mathbf{u}) \times \mathbf{B}]_i. \quad (11.81)$$

Here we have used the fact that collisions between all pairs of species conserve momentum, that is, that

$$m_r \int v_i \left(\frac{\partial f_r}{\partial t}\right)_{cs} d^3v$$

is antisymmetric in (r, s). One should also bear in mind that the transformation from $\mathbf{v}$ to $\mathbf{c}$ involves $\mathbf{u}$, which varies with $\mathbf{x}$ and t. As we shall see, each $\mathbf{u}_r - \mathbf{u}$ will be small as a result of our expansion procedure, so that the right-hand side is not abnormally large. When, finally, ε is placed equal to unity, (11.81) reads

$$\rho_m \frac{Du_i}{Dt} - \sum_r N_r m_r g_i^r - \rho E_i - (\mathbf{j} \times \mathbf{B})_i + \frac{\partial p_{ik}}{\partial x_k} = 0, \qquad (11.82)$$

which is of course the expected momentum equation. Similarly, multiplying (11.77) by $\frac{1}{2}m_r \mathbf{c}^2$, integrating over $\mathbf{v}$ and summing over species (using the fact that

$$\tfrac{1}{2}m_r \int v^2 \left(\frac{\partial f_r}{\partial t}\right)_{cs} d^3v$$

is antisymmetric in (r, s) by conservation of energy in collisions), there results

$$\frac{\partial W}{\partial t} + u_k \frac{\partial W}{\partial x_k} + W \frac{\partial u_k}{\partial x_k} + p_{ik} \frac{\partial u_i}{\partial x_k} + \frac{\partial q_k}{\partial x_k} - \sum_r N_r e_r (\mathbf{u}_r - \mathbf{u}) \cdot (\mathbf{E} + \mathbf{u} \times \mathbf{B}) = 0, \qquad (11.83)$$

the equation of internal energy. Here there are no terms in ε^{-1}. An equation of total energy can be obtained by combining (11.82) and (11.83)—see Problem 7.

The moment equations given so far are all that we can obtain knowing only that the collision operators satisfy the conservation conditions which we have used. The equations give the rates of change of N_r, $\mathbf{u}$ and W (or, equivalently, T), which are just the variables normally used in treating a gas-mixture hydrodynamically. However, just as before, numerous other unknown quantities ($\mathbf{u}^r$, p_{ik}, q_k) enter, and a closed system is achieved only by the expansion procedure. Results at the lowest order follow solely from the fact that the collision operators tend to set up equilibrium distribution functions (the H-theorem), while the first correction requires calculations involving the detailed structure of $(\partial f_r/\partial t)_{cs}$. So we again write

$$f_r = f_r^0 + \varepsilon f_r^1 + \varepsilon^2 f_r^2 + \dots. \qquad (11.84)$$

To lowest order, (11.77) requires

$$\sum_s \left(\frac{\partial f_r^0}{\partial t}\right)_{cs} - \frac{e_r}{m_r}(\mathbf{c} \times \mathbf{B})_k \frac{\partial f_r^0}{\partial v_k} = 0. \qquad (11.85)$$

In the non-magnetic case one argues that the most general solution of (11.85) is a set of Maxwellian distributions with arbitrary densities N_r and common drift $\mathbf{u}$ and temperature T. For in any other state an equation

$$\frac{\partial f_r}{\partial t} = \sum_s \left(\frac{\partial f_r}{\partial t}\right)_{cs}, \qquad r = 1, 2, \ldots$$

leads to the H-theorem

$$\frac{dH}{dt} \leqslant 0, \quad \text{where} \quad H = \sum_s \int f_s \log f_s \, d^3v, \tag{11.86}$$

with equality occurring only when the Maxwellian state has been achieved. The addition of the magnetic term in (11.85) in no way alters this argument since that term vanishes too in the Maxwellian state (compare with (11.22)), and moreover the magnetic term has no effect on the H-theorem. So each f_r^0 is Maxwellian, as in (11.11), and $\mathbf{u}_r^0 = \mathbf{u}$, $\mathbf{j}^0 = \rho\mathbf{u}$, $p_{ij}^0 = p\delta_{ij}$, where $p = NKT$, while $q_k^0 = 0$. Since the parameters specifying f_r^0 are at our disposal, it is convenient to start our expansion by letting f_r^0 already imply the correct N_r for each r, and the correct $\mathbf{u}$ and T; however, the individual $\mathbf{u}_r$, and all higher moments, will show corrections at higher orders.

We may now consider the moment equations. Any terms of $O(\varepsilon^{-1})$ automatically vanish if our procedure has been correctly started, and this we noticed in respect of (11.81). The terms of $O(1)$ give the lowest approximation to the rates of change of N_r, $\mathbf{u}$ and T, and act as integrating conditions for the next stage in the solution for f_r. The equation of continuity (11.80) becomes

$$\frac{\partial N_r}{\partial t} + \operatorname{div}(N_r\mathbf{u}) = 0. \tag{11.87}$$

The composition of each fluid packet is thus invariant; that is, there is no diffusion. The equation of momentum (11.81), becomes

$$\rho_m \frac{D\mathbf{u}}{Dt} - \rho_m\mathbf{g} - \rho\mathbf{E}' - \mathbf{j}^1 \times \mathbf{B} + \nabla p = 0, \tag{11.88}$$

where $\mathbf{E}'$ is $O(1)$ by (11.79), and the total charge density ρ is usually small anyway in hydrodynamic phenomena. The term $\mathbf{j}^1 \times \mathbf{B}$ stems from the right-hand side of (11.81). The electric current is expanded as $\mathbf{j} = \mathbf{j}^0 + \varepsilon\mathbf{j}^1 + \cdots$, so that

$$\mathbf{j}^1 = \sum_r N_r e_r \mathbf{u}_r^1, \tag{11.89}$$

where $\mathbf{u}_r = \mathbf{u} + \mathbf{u}_r^1 + \cdots$. The diffusion between species, while unimportant in (11.87), is significant in the $\mathbf{j} \times \mathbf{B}$ term simply because $\mathbf{B}$ is large. However, this does not prevent us from setting up a closed set of hydrodynamic equations provided $\mathbf{j}$ is understood to mean $\mathbf{j}^0 + \mathbf{j}^1$ (which actually reduces to $\mathbf{j}^1$ when space charge is negligible). In this respect the situation is the same as in §11.3.2. The equation of internal energy (11.83) reduces to

$$\frac{D}{Dt}(\tfrac{3}{2}NKT) + \tfrac{5}{2}NKT\,\mathrm{div}\,\mathbf{u} = 0, \tag{11.90}$$

as $q_k = 0$ and the final term has been omitted since $\mathbf{u}_r - \mathbf{u}$ is $O(\varepsilon)$. Equation (11.90) is of course just the usual adiabatic law.

These results together with the freezing-in equation (11.33) constitute the governing equations for N_r, $\mathbf{u}$, T and $\mathbf{B}$ (using curl $\mathbf{B} = \mu_0\mathbf{j}$ to eliminate $\mathbf{j}$). They are just the equations of a perfect and *perfectly conducting* gas. We leave it to the reader (Problem 8) to sketch the modifications to be made to the present work if the magnetic field is not, after all, to be classified as large; he should find that the $\mathbf{j}^1 \times \mathbf{B}$ term disappears from (11.88). In that case we have a *perfectly insulating* gas; the only current would be a $\mathbf{j}^0$ due to the convection of space charge permanently embedded in the gas, and as before this could be expected to vanish. And, with (11.79) no longer applying, we should then have the electromagnetic field completely decoupled from the fluid variables. The extreme contrast between these two cases appears paradoxical, but is of course mitigated when the next correction in our expansion is considered; see also §11.6.

11.5.4 The expansion procedure for a plasma: second stage

On proceeding to the next order we insert the expansion (11.84) into (11.77) and equate terms of $O(1)$, using the lowest order moment equations to eliminate $\partial f_r^0/\partial t$. This leads to the following equation for f_r^1:

$$\left\{\frac{m_r}{KT}(c_ic_k - \tfrac{1}{3}c^2\delta_{ik})\frac{\partial u_i}{\partial x_k} + \left(\frac{m_rc^2}{2KT} - \frac{5}{2}\right)c_k\frac{\partial}{\partial x_k}\log T + \frac{N}{N_r}d_k^r c_k \right.$$
$$\left. + \frac{m_rc_k}{KT\rho_m}(\mathbf{j}^1 \times \mathbf{B})_k\right\}f_r^0 = C^r(N_s, \mathbf{u}, T, f_s^1) - \frac{e_r}{m_r}(\mathbf{c} \times \mathbf{B})_k\frac{\partial f_r^1}{\partial c_k}. \tag{11.91}$$

Here, following Chapman and Cowling, the vectors $\mathbf{d}^r$ are defined by

$$\mathbf{d}^r = \nabla\left(\frac{N_r}{N}\right) + \left(\frac{N_r}{N} - \frac{N_rm_r}{\rho_m}\right)\nabla\log(NKT)$$
$$+ \frac{N_rm_r}{NKT}\left[\frac{1}{\rho_m}\sum_s\left(N_sm_s\mathbf{a}_s\right) - \mathbf{a}_r\right], \tag{11.92}$$

and we note the identity

$$\sum_r \mathbf{d}^r = 0. \tag{11.93}$$

The functional C^r is defined as $(\partial f_r/\partial t)_c$ evaluated to the first order in ε, where $f_s = f_s^0 + \varepsilon f_s^1$ and $\{f_s^0\}$ is the set of Maxwellian distributions with parameters $\{N_s, \mathbf{u}, T\}$. It is, therefore, linear in $\{f_s^1\}$. The current $\mathbf{j}^1$ is also a linear functional of $\{f_s^1\}$. The set (11.91) constitute linear integro-differential equations to be solved in **c**-space for $\{f_r^1\}$, subject to the additional conditions

$$\int f_r^1\, d^3v = 0, \qquad (r = 1, 2, \ldots)$$
$$\sum_r m_r \int \mathbf{c} f_r^1\, d^3v = 0 \tag{11.94}$$
$$\sum_r m_r \int c^2 f_r^1\, d^3v = 0$$

which state that $\{f_r^1\}$ do not contribute to N_r, $\mathbf{u}$ and T.

Our equation (11.91) is the analog of (11.63) of our "simple illustration" (§11.5.2). Besides the more general collision term (as yet unspecified), we now have the following new features. The vectors $\mathbf{d}^r$, which clearly disappear when the number of species reverts to one, represent new effects occurring if there is a concentration gradient, or if the masses of the species are unequal, or if there are external forces leading to unequal acceleration of the species. These are just the circumstances which one would expect to give rise to relative diffusion. The magnetic field, besides providing the obvious term on the right-hand side, introduces the $\mathbf{j}^1 \times \mathbf{B}$ term on the left. So far as velocity space is concerned this is of identical form to the $\mathbf{d}$ term, so does not increase the difficulty of solving the equation, though it adds to the algebraic complexity.

Suppose for the moment that the solution of these equations has been achieved. Then we know $f_r = f_r^0 + \varepsilon f_r^1$, and can compute the corrections εu_r^1 $(r = 1, 2, \ldots)$, εp_{ij}^1, εq_k^1 and $\varepsilon \mathbf{j}^1$ by applying the definitions (11.69)–(11.74). We can then write down the moment equations (11.80), (11.81) and (11.83) correct to order ε, time derivatives being again interpreted as at (11.59). In principle, these would act as integrability conditions for the next correction (which, however, we do not even propose to describe!). On setting $\varepsilon = 1$, these equations are

$$\frac{\partial N_r}{\partial t} + \operatorname{div}(N_r \mathbf{u}_r) = 0, \text{ with } \mathbf{u}_r = \mathbf{u} + \mathbf{u}_r^1, \qquad (r = 1, 2, \ldots) \tag{11.95}$$

$$\rho_m \frac{Du_i}{Dt} - \sum_r N_r m_r a_i^r - [(\mathbf{j}^1 + \mathbf{j}^2) \times \mathbf{B}]_i + \frac{\partial}{\partial x_i}(NKT) + \frac{\partial p_{ik}^1}{\partial x_k} = 0 \tag{11.96}$$

$$\frac{D}{Dt}(\tfrac{3}{2}NKT) + \tfrac{5}{2}NKT \operatorname{div} \mathbf{u} + p_{ik}^1 \frac{\partial u_i}{\partial x_k} + \operatorname{div} \mathbf{q}^1 - \mathbf{j}^1 \cdot \mathbf{E}' = 0. \tag{11.97}$$

The continuity equations now exhibit diffusion. The momentum equation includes the divergence of a traceless tensor p^1_{ik}, and we interpret this as a generalized viscous term. The appearance of $\mathbf{j}^2$ is to be regarded in the same way as that of $\mathbf{j}^1$ in (11.88); for of the whole set of governing equations (11.96) is the only one in which $\mathbf{j}$ is required to $O(\varepsilon^2)$. The internal energy equation (11.97) has acquired three additional terms which have straightforward interpretations as the rate of working of the viscous forces, heat conduction, and ohmic dissipation respectively. The calculation of $\mathbf{u}^1_r$ also enables us to construct a formula for $\mathbf{j}^1$ as defined by (11.89); the result may be regarded as a generalized Ohm's Law, and it is this which provides the correction to the equation $\mathbf{E}' = 0$ which occurred at the lowest order. When combined with Maxwell's equations, an equation replacing the freezing-in equation is obtained by eliminating $\mathbf{E}$. Clearly the magnetic field lines will not be frozen into the fluid in the corrected theory.

As in the first stage, we invite the reader (Problem 8) to modify the argument on the supposition that $\mathbf{B}$ is $O(1)$ instead of $O(\varepsilon^{-1})$; the resulting hydrodynamic equations take the same form as given here except that the $\mathbf{j}^2$-term disappears from (11.96), and (11.91) lacks the terms involving $\mathbf{B}$ (though $\mathbf{u} \times \mathbf{B}$ is of course still present in the vectors $\mathbf{a}_r$). The formulas for p^1_{ik}, $\mathbf{u}^1_r$ and $\mathbf{q}^1$ are consequently much simplified, in a way which can be obtained by letting $\mathbf{B} \to 0$ in the results given below.

The assumption that (11.91) is indeed soluble for f^1_s thus leads to the existence of a set of equations which, when combined with Maxwell's equations, can be expected to form a closed set of governing equations for the plasma and electromagnetic field; for they give the first derivatives in time of N_r, $\mathbf{u}$, T, $\mathbf{E}$ and $\mathbf{B}$ together with equations for $\mathbf{u}^1_r$, p^1_{ij} and $\mathbf{q}^1$ (plus the straightforward definitions of ρ_m, ρ and $\mathbf{j}$). Continuing this optimistic assumption, it is also possible to foresee to some extent the form of the equations for $\mathbf{u}^1_r$, p^1_{ij} and $\mathbf{q}^1$ by inspection of (11.91) and by considerations of symmetry. We imagine each f^1_r split into an even and an odd part, $\frac{1}{2}[f^1_r(\mathbf{c}) \pm f^1_r(-\mathbf{c})]$. ("Even" and "odd" refer here to symmetry in $\mathbf{c}$-space.) The operation on the right of (11.91) is "even" in the sense that it preserves the even and odd parts of f^1_r; this is obvious for the magnetic term but it is true of the collisional term also. As f^0_r is even, it follows that the even part of f^1_r is related to the even terms within the bracket on the left, and the odd part related to the odd terms. But p^1_{ij} is constructed from even moments of the f^1_r, (11.72), while $\mathbf{u}^1_r$ and $\mathbf{q}^1$ are constructed from odd moments. Bearing in mind that our whole procedure is linear, we can expect that there will be a linear relation expressing p^1_{ij} in terms of $\partial u_k/\partial x_l$ only, and that $\mathbf{q}^1$, the $\mathbf{u}^1_r$'s (and consequently $\mathbf{j}^1$) will depend linearly on all the $\mathbf{d}_r$'s together with ∇T.

Further, we may, in (11.91), rewrite the first term by means of

$$(c_i c_k - \tfrac{1}{3}c^2\delta_{ik})\frac{\partial u_i}{\partial x_k} = c_i c_k s_{ik},$$

where

$$s_{ik} = \frac{1}{2}\left(\frac{\partial u_i}{\partial x_k} + \frac{\partial u_k}{\partial x_i}\right) - \tfrac{1}{3}\delta_{ik}\operatorname{div}\mathbf{u} \tag{11.98}$$

is the traceless part of the rate-of-strain tensor. So p_{ij}^1 must in fact be linear in s_{ij}. In the isotropic case one has (compare with (11.65)) $p_{ij}^1 = -2\mu s_{ij}$ where μ is the coefficient of viscosity. Here the relation must be more general, owing to the magnetic field; but it is restricted by the facts that the field is the only preferred direction in space, that reversing the field must leave the relation invariant except for sign changes due to the change of parity, and that p_{ij}^1 and s_{ij} are both traceless. One concludes (see Problem 9) that, with $\mathbf{B}$ along the z-axis, p_{ij}^1 must take the form

$$\begin{aligned}
p_{xx}^1 &= -\mu_0(s_{xx} + s_{yy}) - \mu_3(s_{xx} - s_{yy}) - 2\mu_4 s_{xy}\\
p_{xy}^1 &= -2\mu_3 s_{xy} + \mu_4(s_{xx} - s_{yy}) = p_{yx}^1\\
p_{yy}^1 &= -\mu_0(s_{xx} + s_{yy}) + \mu_3(s_{xx} - s_{yy}) + 2\mu_4 s_{xy}\\
p_{xz}^1 &= -2\mu_1 s_{xz} - 2\mu_2 s_{yz} = p_{zx}^1\\
p_{yz}^1 &= -2\mu_1 s_{yz} + 2\mu_2 s_{xz} = p_{zy}^1\\
p_{zz}^1 &= -2\mu_0 s_{zz},
\end{aligned} \tag{11.99}$$

where* the coefficients μ_0, μ_1, μ_2, μ_3, μ_4 have values given by the detailed calculations and such that if the sign of $\mathbf{B}$ is reversed μ_0, μ_1 and μ_3 are unchanged but μ_2 and μ_4 change sign. Actually μ_0 is independent of $\mathbf{B}$ and is just the viscosity in the absence of a field, to which μ_1 and μ_3 also tend as $\mathbf{B} \to 0$. The form of (11.99) and behavior under reversal of $\mathbf{B}$ are also consistent with the Onsager relations (see Problem 10).

So far as the heat flux is concerned, it is often convenient to write (omitting the superscript 1 from now on, as $\mathbf{q}^0 \equiv 0$)

$$\begin{aligned}
\mathbf{q} &= \sum_r \int \tfrac{1}{2}m_r c^2 \mathbf{c} f_r^1\, d^3c\\
&= \tfrac{5}{2}KT \sum_r N_r \mathbf{u}_r^1 + \sum_r \int (\tfrac{1}{2}m_r c^2 - \tfrac{5}{2}KT)\mathbf{c} f_r^1\, d^3c,
\end{aligned} \tag{11.100}$$

the additional term inserted within the integral being canceled by the new term in $\mathbf{u}_r^1$. Physically, this new term represents the transport of kinetic energy, relative to the mass-flow velocity $\mathbf{u}$, due to the diffusion of the species relative to $\mathbf{u}$, while the remaining (integral) term gives the heat flux for each species relative to its own mean velocity (see Problem 11).

Whether or not one takes this step, $\mathbf{q}$ and the $\mathbf{u}_r^1$'s are linear in ∇T and the $\mathbf{d}_r$'s. Superficially, one might have expected that the $\mathbf{u}_r^1$'s would depend only on the $\mathbf{d}_r$'s, without involving ∇T (except insofar as the $\mathbf{d}_r$'s involve ∇T).

---

* The use of the symbol μ_0 for this purpose, up to the end of §11.5.5, need cause no confusion with the vacuum permeability, since the latter quantity does not appear.

This is, however, not so; an additional term in ∇T does occur, an effect known as "thermal diffusion," discovered by S. Chapman in 1917 in connection with neutral gases. Similarly, the integral part of (11.100) depends not only on ∇T but upon all the $\mathbf{d}_r$'s in general.

Each contribution to these linear relations between vectors may be exemplified by an equation such as $\mathbf{a} = \boldsymbol{\lambda}\cdot\mathbf{b}$, where $\boldsymbol{\lambda}$ is a tensor which reduces to a simple scalar when $\mathbf{B} = 0$. As with (11.99), the form of $\boldsymbol{\lambda}$ is restricted by the symmetry, and with $\mathbf{B}$ along the z-axis again, one finds (Problem 9)

$$\boldsymbol{\lambda} = \begin{pmatrix} \lambda_1 & -\lambda_2 & 0 \\ \lambda_2 & \lambda_1 & 0 \\ 0 & 0 & \lambda_0 \end{pmatrix}, \tag{11.101}$$

or in vector notation

$$\mathbf{a} = \lambda_0 \mathbf{b}_{\parallel} + \lambda_1 \mathbf{b}_{\perp} + \lambda_2 \hat{\mathbf{B}} \times \mathbf{b}, \tag{11.102}$$

where $\mathbf{b}_{\parallel} = \hat{\mathbf{B}}(\mathbf{b}\cdot\hat{\mathbf{B}})$, $\mathbf{b}_{\perp} = \hat{\mathbf{B}} \times (\mathbf{b} \times \hat{\mathbf{B}})$ are the parts of $\mathbf{b}$ parallel and perpendicular to $\mathbf{B}$ respectively.* Here λ_0 is independent of $\mathbf{B}$, λ_1 is even in $\mathbf{B}$ and reduces to λ_0 as $\mathbf{B} \to 0$, while λ_2 changes sign with $\mathbf{B}$. The equations connecting $\mathbf{q}$, ∇T, the $\mathbf{u}_r^1$'s and $\mathbf{d}_r$'s will all take this form.

As a further simplification of the complete set of equations developed here, we revert to the point already made in §11.3.2 and elsewhere, namely that one can set $\rho = 0$ and ignore the displacement current provided the time scale is well in excess of the plasma period, on the grounds that space charge and displacement effects occur at or above the plasma frequency (except in special very thin regions such as sheaths). The role of Poisson's equation in this approximation must be considered carefully. When the full equation

$$\mu_0^{-1} \operatorname{curl} \mathbf{B} = \mathbf{j} + \varepsilon_0 \dot{\mathbf{E}}$$

is used, the situation, as is well known, is that by taking the divergence of this equation and using the continuity equation $\operatorname{div} \mathbf{j} = -\dot{\rho}$, one obtains the time derivative of Poisson's equation. So Poisson's equation itself is not a completely new equation but rather an initial condition. However, the approximation now under discussion severs this connection between the two. The governing equations form a closed set (subject to appropriate boundary conditions) without Poisson's equation, and the latter should not be used, i.e. one should not set $\operatorname{div} \mathbf{E} = 0$. At the end of a calculation (in which, so far, ρ has not appeared) one may evaluate $\operatorname{div} \mathbf{E}$ in order to verify that ρ really is small and could justifiably be omitted from the equations already used; similarly one could check that $\dot{\mathbf{E}}$ too was small. In this approximation, $\mathbf{j}^0 = \rho\mathbf{u} \approx 0$, so $\mathbf{j}^1$ is the leading contribution to $\mathbf{j}$. This looks strange, but of

* The suffixes 0, 1, 2 seem to be the notation most commonly used in this connection, but some writers use I, II, III, and some adopt the opposite sign convention or λ_2. Caution is therefore necessary!

course it remains true that *for each species* $\mathbf{u}$ is the lowest approximation to $\mathbf{u}_r$; it is only after summing over species that $\mathbf{u}$ ceases to contribute to $\mathbf{j}$.

A further reduction in complexity occurs if we restrict attention to the case where the positive ions occur in only one species; we shall take them to be singly charged and use the approximation $\rho = 0$ just mentioned. It is for such a case that most numerical calculations have been performed. Letting the labels i and e (ions and electrons) replace the suffix r, we have $N_i = N_e = n$, say. (Unlike earlier chapters we will not use N for this quantity, since here N denotes the total number density; so $N = 2n$.) Clearly $\mathbf{u}_i^1$ and $\mathbf{u}_e^1$ may be easily expressed solely in terms of $\mathbf{j}^1$, i.e. $\mathbf{j}$, and the two equations of continuity, (11.95), are equivalent to the equation of total mass conservation together with $\operatorname{div}\mathbf{j} = 0$. As the $\mathbf{u}^1$'s are not required elsewhere, it is sufficient to deal with $\mathbf{j}$ itself. What we need, therefore, are just the equations for p_{ij}^1, $\mathbf{q}$ and $\mathbf{j}$. (Where there is a mixture of species of positive ions, further equations are necessary to describe their relative diffusion.) If the only non-electromagnetic force is gravity, so $\mathbf{g}_i = \mathbf{g}_e = \mathbf{g}$, it drops out of the definition (11.92) of $\mathbf{d}_r$, and only appears in the obvious way in the momentum equation (11.96). We then find

$$\mathbf{d}_i = -\mathbf{d}_e = -(e/2KT)\mathbf{E}'', \tag{11.103}$$

where

$$\mathbf{E}'' = \mathbf{E}' + \frac{m_i - m_e}{m_i + m_e}\frac{1}{Ne}\nabla(NKT),$$

or, since $m_e \ll m_i$, and $\mathbf{E}' = \mathbf{E} + \mathbf{u} \times \mathbf{B}$,

$$\mathbf{E}'' = \mathbf{E} + \mathbf{u} \times \mathbf{B} + \frac{1}{Ne}\nabla(NKT). \tag{11.104}$$

$\mathbf{E}''$ is thus the electric field, modified not only by the motion of the plasma but also by the pressure gradient. The required transport equations will be (11.99) for p_{ij}^1, together with two vector equations, as typified by (11.102), which express $\mathbf{j}$ and $\mathbf{q}$ linearly in terms of ∇T and $\mathbf{E}''$. Thus

$$\mathbf{j} = \boldsymbol{\sigma}\cdot\mathbf{E}'' + \boldsymbol{\alpha}\cdot\nabla T = \sigma_0\mathbf{E}_{\parallel}'' + \sigma_1\mathbf{E}_{\perp}'' + \sigma_2\hat{\mathbf{B}} \times \mathbf{E}'' + \alpha_0\nabla_{\parallel}T + \alpha_1\nabla_{\perp}T + \alpha_2\hat{\mathbf{B}} \times \nabla T \tag{11.105}$$

$$\mathbf{q} = -\boldsymbol{\beta}\cdot\mathbf{E}'' - \boldsymbol{\kappa}\cdot\nabla T = -\beta_0\mathbf{E}_{\parallel}'' - \beta_1\mathbf{E}_{\perp}'' - \beta_2\hat{\mathbf{B}} \times \mathbf{E}'' - \kappa_0\nabla_{\parallel}T - \kappa_1\nabla_{\perp}T - \kappa_2\hat{\mathbf{B}} \times \nabla T. \tag{11.106}$$

In (11.105) the σ's are electrical conductivities, known as the direct, transverse and Hall conductivities respectively; the presence of the σ_2 term is known as the Hall effect. The α's express thermal diffusion. Equation (11.106) is just one of several ways of expressing $\mathbf{q}$; some writers prefer to express it in terms of $\mathbf{j}$ and ∇T. A further variant is to write separately the "convective" part, i.e. the first term of (11.100), here equal to $-5KT\mathbf{j}/2e$, and express the remainder of $\mathbf{q}$ in one of these ways. The term "thermal conductivity," meaning the coefficient of ∇T, is thus ambiguous and must be treated with caution; if (11.106) is adopted it means the κ's.

Another application of Onsager's relations (see Problem 10) yields the identity

$$\boldsymbol{\beta} = T\boldsymbol{\alpha} + \frac{5KT}{2e}\boldsymbol{\sigma}. \tag{11.107}$$

So it is sufficient to calculate the μ's, σ's, α's and κ's, a total of fourteen scalar coefficients, to specify completely the transport properties for a two-component plasma in a magnetic field.

11.5.5 Transport coefficients for a plasma

To obtain actual values for the transport coefficients one must carry out in detail the procedure described formally in the preceding section. Starting from explicit formulas for $(\partial f_r/\partial t)_{cs}$, one constructs the functionals C^r required in (11.91) and defined there. One must then solve (11.91) for f_r^1, or at least find enough about the solution to evaluate the integrals which give $\mathbf{u}_r^1$, p_{ij}^1 and $\mathbf{q}$. By giving convenient special values to ∇T, $\mathbf{d}_r$ and $\partial u_i/\partial x_k$ one can pick out the various coefficients in turn.

The problem of finding the expression for $(\partial f_r/\partial t)_{cs}$ for a plasma is itself a long and difficult matter, to which Chapter 12 is devoted. Until about 1960 the form of $(\partial f_r/\partial t)_{cs}$ used for the present purpose was the one given by Boltzmann, namely

$$\left(\frac{\partial f_r}{\partial t}\right)_{cs} = -\int [f_r(\mathbf{v})f_s(\mathbf{v}_2) - f_r(\mathbf{v}')f_s(\mathbf{v}_2')]\,|\,\mathbf{v} - \mathbf{v}_2\,|\,b\,db\,d\varepsilon\,d^3v_2. \tag{11.108}$$

This gives the rate of decrease of f_r due to collisions in which an r-particle with velocity $\mathbf{v}$ meets an s-particle with velocity $\mathbf{v}_2$, so that they emerge with velocities $\mathbf{v}'$ and $\mathbf{v}_2'$, together with the rate of increase in f_r due to the corresponding reverse collisions. The rate of occurrence of the collisions, deduced from the hypothesis of molecular chaos, is given by the final factors of (11.108); b is the closest distance to which the pair of particles would have approached if they were non-interacting, and ε is an angle indicating the orientation of the relative orbit. This form of collision operator was the one originally used by Chapman and by Enskog (following earlier attempts by Hilbert) in the theory of neutral monatomic gases, for which the method of this chapter was first developed; in that context it appears to be unobjectionable. Equation (11.108) is discussed further in §12.4.2. For an inverse square law of force it is unfortunately incorrect to construct $(\partial f/\partial t)_c$ on the basis of discrete collisions in which particles take part only two at once. For a thermal particle, all other particles within a Debye length exert some influence which should contribute to $(\partial f/\partial t)_c$. That (11.108) is inappropriate in that case is soon apparent in the divergence, at large values of b, of the integral with respect to b. To overcome this, the range of integration was "cut off" at an upper limit $b = b_0$. In the earlier work (for example Chapman and Cowling) it was usual to set $b_0 \approx N^{-1/3}$, the interparticle spacing, on the grounds that binary collisions could not be contemplated for a larger impact parameter. Later it came to be realized (for example by Cohen, Spitzer and Routly, 1950) that more accurate results are obtained by setting b_0 equal to the Debye length; extensive calculations on this basis were carried out by Marshall (1957). The correction is appreciable even though the results only vary with the logarithm of b_0. This procedure, which at first sight might be viewed with suspicion, gives results which are mainly in good agreement with those obtained from the Fokker–Planck form mentioned below; a physical discussion of this is given in §12.7.4.

As already mentioned in §9.5.3, it is now believed that for a plasma (11.108) should be replaced by a collision operator of the Fokker–Planck type

$$\left(\frac{\partial f_r}{\partial t}\right)_{cs} = \frac{\partial}{\partial v_i}\left[-A_i^{rs} f_r + \frac{1}{2}\frac{\partial}{\partial v_j}(B_{ij}^{rs} f_r)\right]. \qquad (11.109)$$

The coefficients A_i^{rs}, B_{ij}^{rs} are themselves complicated functionals of the distribution functions, which we obtain from first principles in Chapter 12. Provided the Debye length is less than both the macroscopic scale of the plasma and the ion Larmor radius, the coefficients are as given by (12.84), (12.85) and (12.82); and by means of a further excellent approximation they are reduced to (12.100) and (12.101), which we set out here:

$$A_i^{rs}(\mathbf{v}) = \frac{1}{4\pi}\left(\frac{e_r e_s}{\varepsilon_0 m_r}\right)^2 \frac{m_r + m_s}{m_s} \log \Lambda \frac{\partial}{\partial v_i} \int_{\mathbf{v}'} \frac{f_s(\mathbf{v}')}{|\mathbf{v} - \mathbf{v}'|}\, d^3v' \qquad (11.110)$$

$$B_{ij}^{rs}(\mathbf{v}) = \frac{1}{4\pi}\left(\frac{e_r e_s}{\varepsilon_0 m_r}\right)^2 \log \Lambda \frac{\partial^2}{\partial v_i \partial v_j} \int_{\mathbf{v}'} |\mathbf{v} - \mathbf{v}'|\, f_s(\mathbf{v}')\, d^3v', \qquad (11.111)$$

where

$$\Lambda = \frac{12\pi(\varepsilon_0 KT)^{3/2}}{N_e^{1/2} e^3} = 1.24 \times 10^7 T^{3/2} n^{-1/2} \qquad (11.112)$$

(T in °K, n in m^{-3}; the logarithm is the natural one). Starting with these equations, Robinson and Bernstein (1962) gave an extensive treatment of the Chapman–Enskog procedure for a two-component plasma, with tables of all transport coefficients.

Whichever form of $(\partial f_r/\partial t)_{cs}$ is used, it is a routine, though tedious, matter to construct C^r and to convert (11.91) into an equation for $\phi_r(\mathbf{c})$, where the substitution

$$f_r^1 = f_r^0\, \phi_r(\mathbf{c}) \qquad (11.113)$$

is customarily made. The method developed in Chapman and Cowling for solving this equation (for various laws of interaction) is to expand ϕ_r in a series of Sonine polynomials, and several workers (including Chapman and Cowling themselves) have applied this technique to the plasma case. An alternative method, used by Marshall (1957) and by Robinson and Bernstein (1962) is to show that the original equations imply a variational principle for the transport coefficients, and approximate the coefficients numerically by careful choice of trial functions. In a simple case, the variational principle is related to the maximization of the rate of entropy production. (An interesting feature of the model Fokker–Planck equation introduced in §9.5.3 is that if the Chapman–Enskog procedure is applied to it the solution of (11.91) can be achieved analytically without recourse to these approximate methods; see Dougherty, Watson and Hellberg, 1967.)

We do not propose here to enlarge on the details of these calculations, as our purpose has been to enable the reader to understand the principles of the Chapman–Enskog method and why the results take the form they do. We now proceed to quote the values obtained for the transport coefficients by Robinson and Bernstein (1962). Only the case where there is just one species of singly charged ions will be treated. First it is convenient to define a "collision time" to characterize the order of magnitude of $(\partial f_r/\partial t)_c$ in terms of that of f_r. Thinking in the first instance of collisions amongst like particles, equations (11.109)–(11.111) would suggest, for thermal particles, the quantity $4\pi\varepsilon_0^2 m_r^{1/2}(KT)^{3/2}/N_r e^4 \log \Lambda$ (r meaning i and e, in turn). However, the choice of numerical factor is arbitrary (within reason), and indeed various writers differ in the exact choice made. The outcome of the numerical work shows that the value just quoted is somewhat too small for convenience, but that an additional factor of 10 would be suitable. So we define

$$\tau_e = \frac{40\pi\,\varepsilon_0^2 m_e^{1/2}(KT)^{3/2}}{ne^4 \log \Lambda}, \qquad \tau_i = \left(\frac{m_i}{m_e}\right)^{1/2} \tau_e. \tag{11.114}$$

a) Viscosity. See (11.99). The main contribution to p_{ij}^1 is due to the ions, that of the electrons being smaller by the factor $(m_e/m_i)^{1/2}$, and ignored here. Then

$$\mu_0 = 0.513\, nKT\tau_i. \tag{11.115}$$

Also, though not obvious from the analysis we have given,

$$\mu_3(B) = \mu_1(2B), \qquad \mu_4(B) = \mu_2(2B). \tag{11.116}$$

The effect of the field on the ions is measured by Ω_i, so we plot in Fig. 11.1 the ratios μ_1/μ_0, μ_2/μ_0, μ_3/μ_0 and μ_4/μ_0 as functions of the dimensionless quantity $\Omega_i\tau_i$. We notice that as the field increases μ_1 and μ_3 decrease from the initial value μ_0, eventually tending to zero like B^{-2}, while μ_2 and μ_4 increase from zero to a maximum of the order $\frac{1}{2}\mu_0$, occurring where $\Omega_i\tau_i$ is

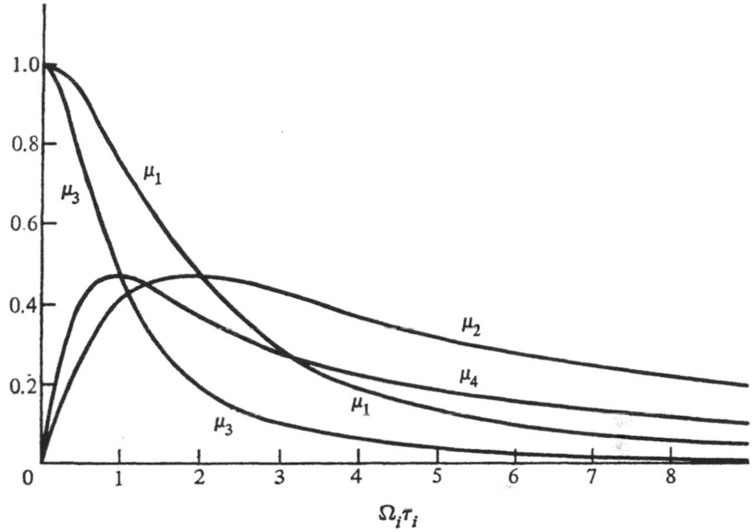

Fig. 11.1. Coefficients of viscosity for a two-component plasma.

about 2 for μ_2 and 1 for μ_4; then they decrease to zero like B^{-1}. In the limit $\Omega_i\tau_i \gg 1$, μ_1 and $\mu_3 \to 0$, while

$$\mu_2 \to nKT/\Omega_i, \qquad \mu_4 \to nKT/2\Omega_i. \tag{11.117}$$

b) Electrical conductivity. See (11.105). Here (by definition of the mean velocity **u**) the electrons make the major contribution, and we ignore the ions. The results are

$$\sigma_0 = 0.743\frac{ne^2\tau_e}{m_e} \tag{11.118}$$

$$\alpha_0 = 0.523\frac{K}{e}\frac{ne^2\tau_e}{m_e}. \tag{11.119}$$

We plot in Fig. 11.2 σ_1/σ_0 and σ_2/σ_0 as functions of $\Omega_e\tau_e$; their behavior is similar to that of μ_3 and μ_4, but it should be remembered that the field required to achieve $\Omega_e\tau_e \approx 1$ is smaller, by $(m_e/m_i)^{1/2}$, than that for $\Omega_i\tau_i \approx 1$. For $\Omega_e\tau_e \gg 1$, $\sigma_1 \propto B^{-2}$, while

$$\sigma_2 \to ne^2/m_e\Omega_e. \tag{11.120}$$

We also plot α_1/α_0 and α_2/α_0 on Fig. 11.2. These behave somewhat differently from the other coefficients, for α_1 changes sign at $\Omega_e\tau_e \approx 2.2$, having small negative values for larger values of the field, eventually tending to zero like B^{-2}, while $\alpha_2 \propto B^{-3}$ for large B. Thus in a strong field the phenomenon of thermal diffusion disappears.

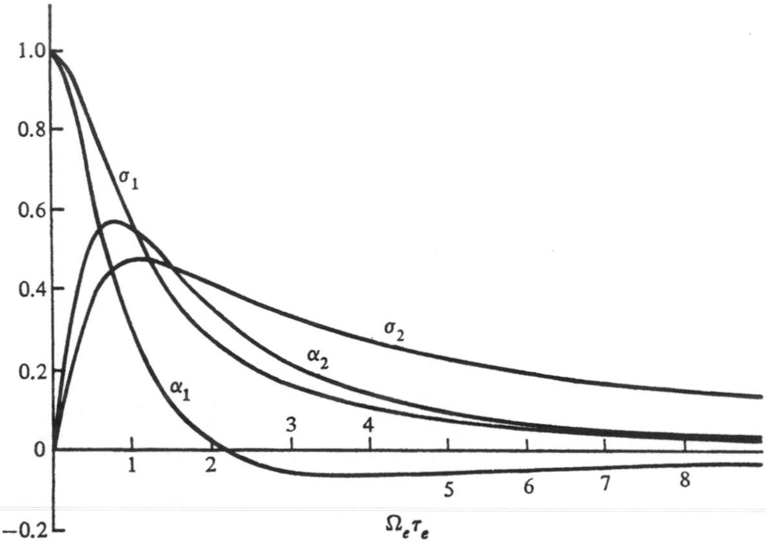

Fig. 11.2. Coefficients of electrical conductivity for a two-component plasma.

c) Thermal conductivity. See (11.106). The ions may again be neglected in computing the β's, but in the case of the κ's we need to give separate contributions κ^i and κ^e for the ions and electrons, the effective κ being the sum of the two. Then

$$\beta_0 = 2.38\frac{nKTe\tau_e}{m_e} \tag{11.121}$$

$$\kappa_0^e = 2.88\frac{nK^2T\tau_e}{m_e} \tag{11.122}$$

$$\kappa_0^i = 2.10\frac{nK^2T\tau_i}{m_i}\,. \tag{11.123}$$

Graphs of β_1/β_0, β_2/β_0, κ_1^e/κ_0^e, κ_2^e/κ_0^e as functions of $\Omega_e\tau_e$ are shown in Fig. 11.3. They present a now familiar appearance, and the quantities concerned are found to satisfy the Onsager relation (11.107). We plot κ_1^i/κ_0^i and $-\kappa_2^i/\kappa_0^i$ on Fig. 11.4, after noting that κ_2^i is negative. When $\Omega_e\tau_e \gg 1$, κ_1^e and β_1 are proportional to B^{-2}, and so is κ_1 when $\Omega_i\tau_i \gg 1$. The other coefficients tend, for large B, to limits

$$\beta_2 \to \frac{5}{2}\frac{nKTe}{m_e\Omega_e}, \qquad \kappa_2^e \to \frac{5}{2}\frac{nK^2T}{m_e\Omega_e}, \qquad \kappa_2^i \to -\frac{5}{2}\frac{nK^2T}{m_i\Omega_i}\,. \tag{11.124}$$

For small values of $\Omega_e\tau_e$, the contribution made by the ions to the heat flow is in fact negligible compared with that of the electrons; but in a strong field κ_2^i can be comparable with κ_2^e. As the conductivity is by then rather small

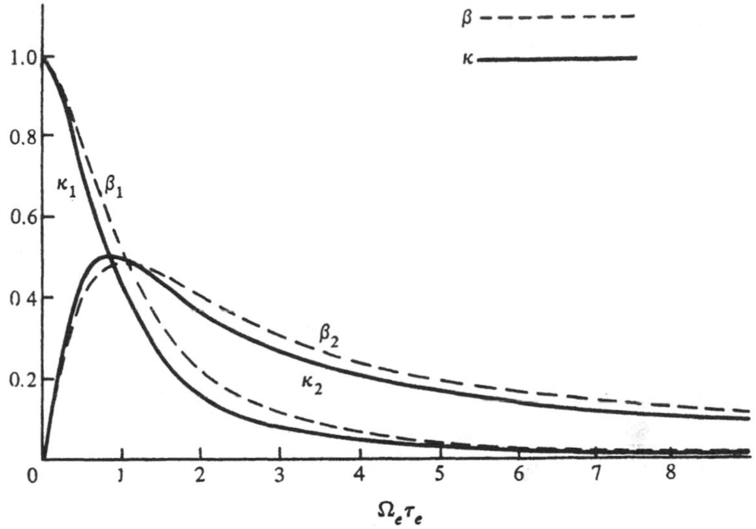

Fig. 11.3. Coefficients of thermal conductivity (electron component) for a two-component plasma.

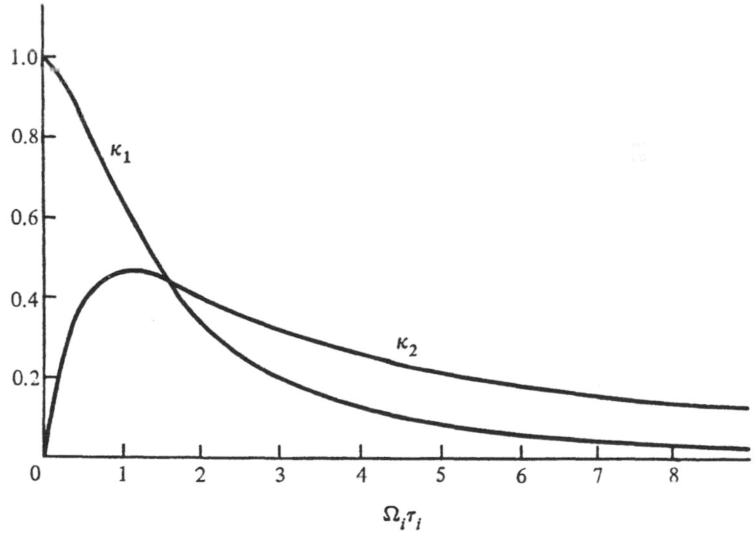

Fig. 11.4. Coefficients of thermal conductivity (ion component) for a two-component plasma.

anyway, this is not usually of much importance; indeed as the two contributions are of opposite sign, the effect of the ions is to make κ_2 even smaller. In the limit in which $\Omega_i\tau_i \gg 1$, the two contributions cancel, as is obvious from (11.124) since $m_e\Omega_e = m_i\Omega_i$.

d) General remarks. An interesting property of all the transport coefficients when $\mathbf{B} = 0$ is that when (11.114) is used to substitute for τ_i and τ_e explicitly, n cancels except for the mild dependence of $\log \Lambda$ upon n. In the case of electrical conductivity, for example, the physical picture is that an increase in n provides more carriers of charge but increases the number of collisions, and these two effects almost cancel.

The dependence on field strength of the normalized coefficients σ_1/σ_0 etc., (leaving aside α_1 and α_2), may be tolerably well represented by the functions

$$y_1 = \frac{1}{1 + x^2}, \qquad y_2 = \frac{x}{1 + x^2}, \tag{11.125}$$

where x is $\Omega_e\tau_e$ or $\Omega_i\tau_i$ as the case may demand, and y_1 applies for σ_1, β_1, κ and μ_3, y_2 for σ_2, β_2, κ_2 and μ_4 (μ_1 and μ_2 being given by equation 11.116) This approximation is interpreted roughly in Problem 12.

As emphasized by Robinson and Bernstein, the results quoted here should not be regarded as accurate to better than about 10%, because the derivation of the Fokker–Planck coefficients (11.110) and (11.111) involves uncertainty of this order. There is also a limitation that $\log \Lambda > 10$ since the same derivation is otherwise even more open to suspicion. However, the condition $\Lambda \gg 1$ is a natural one for plasmas, for it is simply the statement that the mutual potential energy for particles separated by $n^{-1/3}$ is much less than their thermal energy, and is equivalent to the condition $\alpha \ll 1$ discussed in §1.2. In fact, in almost all applications, $\log \Lambda$ lies between 10 and 20. Notice also that the rough "collision frequency," ν, introduced in §1.2, differs from τ^{-1}, as defined here, only by numerical factors together with $\log \Lambda$.

The reader needing more precise data than could be included in Figs. 11.1–11.4 is referred to our original source, Robinson and Bernstein (1962). It should be noted that their variable X is identical with $\Omega_e\tau_e/10$ in our units and notation, while their $X_I = 15\,Y_I/4$ corresponds to our $\Omega_i\tau_i/10$. Their ν_1 and ν_2 are our μ_3 and μ_4, and their σ_2, α_2, β_2 and κ_2 have opposite sign from ours. The notation is otherwise the same.

11.5.6 Finite Larmor-radius effects in a collisionless plasma

So far throughout §11.5 it has been assumed that the collision term is dominant in our ordering of terms, i.e. $(\partial f/\partial t)_c = O(\varepsilon^{-1})$. The Lorentz force $(\mathbf{c} \times \mathbf{B})\cdot\partial f/\partial \mathbf{v}$ may or may not be dominant also. The expansion procedure was thus the continuation of the process started with the adiabatic approximation of §11.2. It will naturally be asked whether there is an analagous continuation of the adiabatic approximation for a collisionless plasma in a strong magnetic field, given in §11.3. As explained in our introduction, the relevant circumstances would be $R_L \ll L_\perp \ll L_c$ where R_L is the Larmor radius, $L_\perp$ the scale for appreciable variation across the field lines (as fixed by the nature of the problem) and L_c a "mean free path," meaning $(KT/m)^{1/2}\tau$ where τ is the collision time. In addition, $L_\parallel$, the spatial scale along the field lines, must be large, though we have not properly quantified this. There is a corresponding ordering of the time scales, obtained by taking the thermal speed as characteristic.

Only a little consideration is required in order to realize that serious obstacles will stand in the way of such a development. The root of the difficulty is that the adiabatic approximation (§11.3) is mathematically incomplete in the collisionless case. For although we managed to obtain equations of a purely hydrodynamic type (that is, the distributions in velocity space were entirely eliminated), the adiabatic approximation did not in fact fix f^0 in terms of the hydrodynamic variables; it merely required f^0 to take the form (11.22), (axial symmetry in **c**-space) and to be such that the relevant moments N, $\mathbf{u}$, $p_\perp$ and $p_\parallel$ be correct. This should be contrasted with the collision-dominated case, where f^0 was determined completely as the appropriate Maxwellian distribution. As we saw (equation 11.91), one actually needs f^0 in order to deal with the first correction. Again, several further assumptions were introduced during §11.3. For example, we suppressed the possibility that $\mathbf{u}_\parallel$ might be different for the two species, and, most conspicuously, we assumed that the contributions from $q_\perp$ and $q_\parallel$ vanished in deriving (11.40) and (11.41), leading to the CGL equations for $p_\perp$ and $p_\parallel$. The need for these assumptions can be traced physically to the fact that the strong magnetic field results in localization perpendicular to the field only, so that the results are applicable only when there is near uniformity along the field lines. It should be emphasized that the desired results cannot be obtained directly from those of §§11.5.4–11.5.5 by the simple expedient of letting the collision term disappear while leaving the magnetic term as dominant.

Despite these difficulties, there have been attempts to improve on the theory of §11.3, with some limited success, especially in the purely two-dimensional case in which the field lines are straight and all motions and gradients are perpendicular to the field. In order to make progress, it has been found necessary to aim at a rather less ambitious program than the one just suggested. We may regard $\varepsilon = R_L/L_\perp$ as the effective expansion parameter in this work. The analogy with the collision-dominated case would lead us to expect the resulting fluid equations to be valid if a typical rate of change (or frequency) ω satisfies $\omega \lesssim \varepsilon\Omega$, so that $\omega L_\perp$ and ΩR_L are comparable, both being in fact the thermal speed. If, instead, the stronger restriction $\omega \lesssim \varepsilon^2\Omega$ is imposed, some of the difficulties just alluded to can be overcome. The characteristic speed $\omega L_\perp$ is then not the thermal speed (and still less the Alfvén or magnetosonic speeds, which for a low-β plasma are hypersonic), but ε times the thermal speed, which can be identified with the characteristic drift speeds occurring in (11.52), (i.e. those not involving $\mathbf{u}_E$). This means that we forgo the possibility of handling "fast" phenomena, such as the magnetosonic wave, in this approximation, though there are other problems for which the results are useful. It is also necessary to assume that $L_{\parallel} \gtrsim \varepsilon^{-1}L_{\perp}$.

As we saw in §11.4, the lowest approximation is closely related to the particle-orbit picture in which the gyrating particle is replaced by a magnetic moment at the guiding center, and the present scheme, being an attempt to improve on that, has been termed the "finite Larmor-radius" approximation. It should be remarked that even with the limitations just mentioned, it is not in general possible to eliminate completely the details of the velocity distribution.

The detailed development of this work is beyond the scope of the present book, and the reader may be referred particularly to the paper by Kennel and Greene (1966) for a full treatment.

11.6 APPLICATIONS OF MAGNETOHYDRODYNAMICS

11.6.1 Utilization of the MHD equations

A substantial proportion of the theoretical work on the dynamics of ionized gases, as applied to practical calculations, is carried out by means of the hydromagnetic formulation rather than Boltzmann's equation. The simplification achieved in the governing equations enables one to consider correspondingly more complicated physical situations, especially with regard to the geometry in configuration space. The solutions of Boltzmann's equation presented here (Chapters 7–10) have been almost entirely restricted to plane waves, though there is of course a considerable literature on other types of solution. The introduction of more interesting geometry is essential if one is to make any progress with problems such as plasma confinement, magnetism in stars, and many others, and so one will be prepared to pay the price of omitting for the time being the phenomena which are only revealed by Boltzmann's equation.

A book surveying the whole field of plasma dynamics would thus devote a substantial amount of space to the various solutions of the equations we have just derived. In the present text, however, we have aimed to deal mainly with particles and fields and the interactions between them as expressed by Boltzmann's equation and exhibited by means of waves. The derivation of MHD forms part of this program, but a detailed treatment of the many flows which have been worked out will not be attempted. The remarks made in the following few pages are intended only as a very brief introduction to the subject of the application of the MHD equations, though we give slightly more attention to the question of Alfvén waves in an anisotropic plasma as this part is related to our work in earlier chapters. We cite in the references several texts for further reading in MHD and in hydromagnetic stability.

11.6.2 Conditions for the neglect of transport effects

Notwithstanding the above comment about the simplification gained by use of the MHD approximation, the full system of MHD equations is still rather elaborate, especially when conditions are such that the transport coefficients are anisotropic. It is thus worth while to seek conditions which permit further reduction of the MHD equations.

A common simplification is to work in the adiabatic approximation. Thermal conductivity is assumed negligible, and so is the viscous contribution to the pressure tensor, while the main part of the pressure obeys the ordinary adiabatic law (or the CGL equations if appropriate). Ohm's Law is reduced to the simple equation $\mathbf{E} + \mathbf{u} \times \mathbf{B} = 0$, leading, as we have seen, to the freezing-in of field lines. This last step is indeed part of the adiabatic scheme in the collision-free (strong field) development, but its use in the collision-dominated case needs further justification. The results of §11.5.5 show that the coefficients of viscosity, heat conduction and electrical conductivity are all inversely proportional to an appropriate collision frequency (leaving aside possible anisotropy), yet it appears that we are regarding the first two as zero but the electrical conductivity as infinite. This can nevertheless be consistent. The point is that these coefficients are all finite quantities for a given plasma, and the question whether they are very small or large depends on how they will combine with the typical length scale L and typical fluid velocity U occurring in the problem. In the case of the viscosity, μ, the importance of the viscous term is measured by R^{-1}, where R is the usual Reynolds number $LU\rho_m/\mu$, so we require $R \gg 1$, i.e. $\mu \ll LU\rho_m$. This is fostered by large values of LU and of the collision frequency, as expected. For thermal conductivity, the discussion is similar. For electrical conductivity, we consider the simplified equation

$$\mathbf{E} + \mathbf{u} \times \mathbf{B} = \mathbf{j}/\sigma. \tag{11.126}$$

The question is whether the right-hand side introduces serious errors into the work of §11.3.3. Using Maxwell's equations in the form

$$\frac{\partial \mathbf{B}}{\partial t} = -\operatorname{curl} \mathbf{E}, \qquad \mathbf{j} = \mu_0^{-1} \operatorname{curl} \mathbf{B},$$

we eliminate $\mathbf{E}$ and $\mathbf{j}$ to obtain

$$\frac{\partial \mathbf{B}}{\partial t} = \operatorname{curl}(\mathbf{u} \times \mathbf{B}) + \frac{1}{\mu_0 \sigma} \nabla^2 \mathbf{B}, \tag{11.127}$$

taking the conductivity σ to be uniform for simplicity. The coefficient $(\mu_0\sigma)^{-1}$ is often called the "magnetic diffusivity." The freezing-in theorem holds if the last term may be neglected; its importance relative to the other term on the right is measured by R_M^{-1} where R_M is the "magnetic Reynolds number" $UL\mu_0\sigma$. So we need R_M to be large, and this again can be achieved by a sufficiently big value of LU. So once the collision frequencies have been fixed, all these conditions can be met by choosing LU appropriately large, and one should not be distracted by the fact that R and R_M depend in opposite ways on the collision frequencies. Needless to say, the astronomical context provides especially good opportunities for large values of L. The analogy between R and R_M is very close since the vorticity in a viscous fluid obeys an equation identical in form to (11.127), with the kinematic viscosity taking the place of the magnetic diffusivity.

11.6.3 Alfvén waves

We have already met Alfvén waves (waves with frequency $\omega \ll \Omega_i$) in the cold approximation (§5.4.2), with isotropic gas pressure included (§5.5.5) (with propagation at a general angle in each case), and also as the low frequency limit of the treatment by Boltzmann's equation (§§9.3.3 and 9.3.4), in which we only pursued the calculation for waves traveling along the field. This latter case revealed the firehose instability for sufficiently anisotropic distributions. The treatment for general angle of propagation could of course be carried out starting with equation (9.69), but, more simply, we can use collision-free hydromagnetics with the double adiabatic equations. The results will be applicable for wavelengths well in excess of the ion Larmor radius but less than the mean free path, provided the assumption of no heat flow along the field lines is justifiable. (For wavelengths greater than the mean free path ordinary hydromagnetics applies, as in §5.5.5.)

The governing equations are those of continuity (11.43), momentum (11.30), and freezing-in of field lines (11.34), together with the double adiabatic equations, which for this work are more conveniently expressed in their earlier form (11.41) and (11.42). We take a uniform static unperturbed situation and apply small perturbations proportional to $\exp[i(\omega t - k_\perp x - k_\parallel z)]$ in the familiar way, $\mathbf{B}$ being along the z-axis. In the linearized equations, the perturbation in mass density appears only in the equation of continuity, which may therefore simply be discarded. For the remainder, it becomes apparent that two possibilities arise. In the first, B_y and u_y are the only non-zero perturbation quantities, and we have waves which involve displacement of the field lines perpendicular to the unperturbed field and to $\mathbf{k}$. The dispersion relation in this case is

$$\frac{\omega^2}{k_\parallel^2} = \frac{1}{\rho_m}\left(\frac{B^2}{\mu_0} + p_\perp - p_\parallel\right), \tag{11.128}$$

irrespective of $k_\perp$, where B, $p_\perp$, $p_\parallel$ and ρ_m are the equilibrium values. This is one generalization of (9.73) and exhibits the firehose instability if $p_\parallel > p_\perp + B^2/\mu_0$, just as before. The second possibility is that $\mathbf{u}$ and the perturbation of $\mathbf{B}$ lie entirely in the xz-plane. In this case $p_\perp$, $p_\parallel$ and ρ_m are perturbed. Solving for $u_x : u_z$ leads to the determinantal equation

$$\begin{vmatrix} \omega^2\rho_m - k_\perp^2(2p_\perp + B^2/\mu_0) - k_\parallel^2(p_\perp - p_\parallel + B^2/\mu_0) & k_\perp k_\parallel p_\perp \\ k_\perp k_\parallel p_\perp & \omega^2\rho_m - 3k_\parallel^2 p_\parallel \end{vmatrix} = 0. \tag{11.129}$$

For $k_\perp = 0$ we find a solution identical with (11.128), together with a "sound wave" satisfying $\omega^2 = 3k_\parallel^2 p_\parallel/\rho_m$. The latter (exhibiting "$\gamma = 3$" as in 11.50) is, however, spurious as in reality Landau damping would be severe. For $k_\parallel = 0$, we find a "magnetosonic" wave, with

$$\frac{\omega^2}{k_\perp^2} = \frac{2(p_\perp + B^2/2\mu_0)}{\rho_m}, \tag{11.130}$$

which is just what we would expect from (11.48) and (11.49), ($\gamma = 2$); there is a second solution $\omega = 0$. At intermediate angles we have a quadratic for ω. An interesting result arises if we let $k_\parallel$ be small but non-zero; one solution is a slight modification of (11.130), but the other is approximately

$$\omega^2 = \frac{3k_\parallel^2 p_\parallel^2}{\rho_m} \frac{\dfrac{B^2}{2\mu_0} + p_\perp - \dfrac{p_\perp^2}{6p_\parallel}}{\dfrac{B^2}{2\mu_0} + p_\perp}. \tag{11.131}$$

A new possibility for instability occurs if the numerator is negative, i.e. if

$$\frac{1}{6}\frac{p_\perp^2}{p_\parallel} > p_\perp + \frac{B^2}{2\mu_0}. \tag{11.132}$$

This can happen if $p_\perp$ exceeds $p_\parallel$ to a sufficient extent, the anisotropy being thus in the opposite sense to that required for the firehose instability. This is known as the "mirror" instability. It was first noticed theoretically by several writers in the late 1950's, and has been observed experimentally. Unfortunately this simple derivation does not in fact lead to the correct result, owing to the neglect of heat conduction. It can be shown that for the anisotropic Maxwellian distribution a better approximation is obtained by omitting the factor $\frac{1}{6}$ on the left of (11.132). This follows, rather laboriously, from the Boltzmann treatment, and as mentioned above it is necessary to start from equation (9.69). In the corrected version, a lesser degree of anisotropy will achieve the instability than that demanded by (11.132).

The physical nature of these two instabilities is rather interesting, and can be revealed explicitly by working in the particle-orbit picture. The firehose instability can be explained by imagining small kinks in the lines of force. The particles still spiral along the deformed lines, and if $p_{\parallel}$ is large their speeds along the lines are rather high. This sets up a strong centrifugal force owing to the kinks, and using the freezing-in theorem the lines of force will be pulled still further from their equilibrium position; their electromagnetic tension and the lateral pressure will be unable to resist this if $p_{\parallel} > p_{\perp} + B^2/\mu_0$. The name "firehose" derives from a similar phenomenon in hoses carrying a fast flow of water. In the mirror instability a pattern of alternately diverging and converging lines of force is set up as in Fig. 11.5.

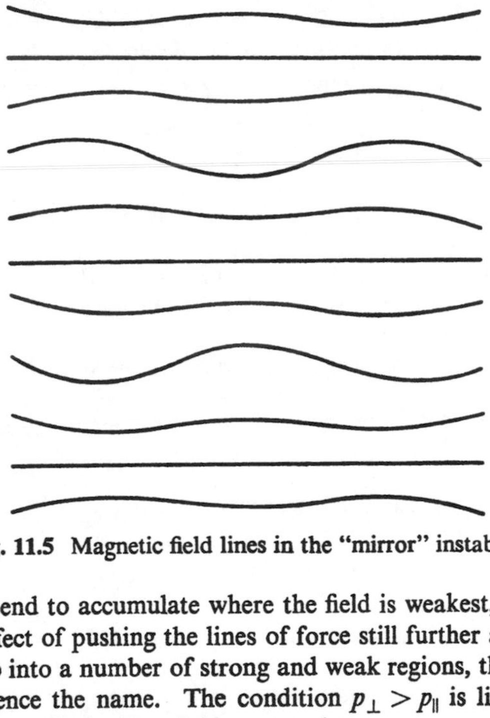

Fig. 11.5 Magnetic field lines in the "mirror" instability.

The particles tend to accumulate where the field is weakest, and if $p_{\perp} > p_{\parallel}$ this has the effect of pushing the lines of force still further apart. The field thus breaks up into a number of strong and weak regions, the former acting as mirrors—hence the name. The condition $p_{\perp} > p_{\parallel}$ is liable to occur if large externally applied mirror fields are used to attempt to confine a plasma. One could say that the mirror is liable to "shatter" into many small ones!

11.6.4 Laboratory applications

Much of the impetus for plasma research in recent years has originated in the idea that if hydrogen (or deuterium) could be converted into helium, enormous amounts of power could be produced. A temperature of about 10^8 °K is required for this, and to carry it out in a controlled manner it is necessary to confine the gas, which will be in plasma form, with suitable apparatus. Electromagnetic fields afford the only hope of doing this. The question is therefore whether a suitable equilibrium configuration can be devised. It must either be stable, or, if unstable, be such that the growth time is long enough for nuclear reactions to take place to a useful extent. In the early stages of design, both the equilibrium and its stability may be considered by hydromagnetic theory as this affords a more tractable approach than the Boltzmann equation. If the arrangement proposed is unstable hydromagnetically there is little hope that it will be stable in a more refined treatment. Most simple equilibria are actually unstable; for example, a self-pinched discharge in its most elementary state is subject to instabilities picturesquely described as "kinks," "sausages" and "flutes." Such equilibria can often be made stable by ingenious arrangements of external magnetic fields and conductors. Powerful variational methods (or "energy principles") have been developed for investigating hydromagnetic stability theoretically. Then must follow the more subtle enquiry into the possible microinstabilities, for which velocity-space methods must be used as explained in Chapter 10.

For a general account of the application of plasma physics to the nuclear fusion project, the reader may be referred to the text of Rose and Clark (1961).

Another application of great technological importance is that of the direct conversion of the energy of combustion into electrical power. Here the object is to avoid the use of rotating machinery by using a hydromagnetic device. In essence the idea is to force an ionized gas to flow at right angles to an applied magnetic field. The equation $\mathbf{E} + \mathbf{u} \times \mathbf{B} \approx 0$ shows that an electric field is induced. The aim is thus to arrange for a steady and stable flow whereby this electric field can be used as a source of power.

11.6.5 Natural plasmas

Most of the matter in the universe exists in the plasma state, planets forming a very peculiar (and cosmically unimportant) exception. Many of the electrodynamic properties of these natural plasmas can be described in hydromagnetic terms.

While the main features of stars and their evolution are not closely concerned with electromagnetism, there are several important matters in which it plays a large part. The existence of general magnetic fields in stars, and the connection with rotation, have received extensive study. The tendency of magnetic fields to inhibit convection is probably significant with reference to the transport to the surface of energy generated within, and sunspots are thought to be a manifestation of locally strong magnetic fields acting in this way. Solar flares, too, have been associated with electromagnetic effects. The condensation of stars from cosmic gas clouds is closely controlled by magnetic fields. The existence of radio stars and galaxies is of course an electromagnetic effect, and while the actual mechanism of emission is a matter for the kind of work described in Chapters 3 and 10, the provision of the background magnetic field to make this possible is a hydromagnetic problem. A similar remark applies to the Fermi acceleration theory of the origin of cosmic rays.

Closer to earth, many properties of the ionosphere and magnetosphere, and of the plasma in nearby space, are hydromagnetic in character. The flow of solar plasma from the sun (called the "solar wind"), and its interaction with the geomagnetic field, provides a remarkable instance of hydromagnetic behavior in spite of the virtually infinite mean free path. There is even a collision-free hydromagnetic shock wave, whose properties can be observed in some detail by rocket-borne equipment. The papers of S. Chapman and V. C. A. Ferraro in the early 1930's not only shed much light on this subject but also constituted significant pioneering work in MHD itself. The gross behavior of the magnetosphere is well described by hydromagnetic theory, again in the collision-free (field-dominated) version. For example, the reduction in the field strength observed at the earth's surface during the main phase of a magnetic storm has been attributed to the introduction of additional plasma in the outer reaches of the earth's magnetic field (say up to 10 earth's radii from the center). The increase in particle pressure tends to expand the magnetosphere, dragging the lines away from the earth. Equivalently, the extra particles follow orbits with the usual mirroring and precessing around the earth; in doing so they constitute a "ring current," whose sense is such as to decrease the field inside, and they also have a diamagnetic effect. That these two approaches lead to the same results is of course just the point at issue in §11.4. This subject is well reviewed in the text of Hines *et al.* (1965).

PROBLEMS

1. Derive equation (11.19).
2. Derive equation (11.30).
3. Derive equations (11.44) and (11.45).
4. Show from equation (11.49) that for steady two dimensional flow in double adiabatic MHD there is an analog of Bernoulli's theorem in which the constant is

$$\tfrac{1}{2}\mathbf{u}^2 + \frac{2}{\rho_m}(p_\perp + B^2/2\mu_0).$$

5. Show that in steady flow obeying the double adiabatic equations and having $\mathbf{u}$ everywhere parallel to $\mathbf{B}$, there is an analog of Bernoulli's theorem whereby

$$\tfrac{1}{2}\mathbf{u}^2 + \frac{1}{\rho_m}(p_\perp + \tfrac{3}{2}p_\parallel)$$

is constant along each streamline (or field line).

6. Supply the details omitted in deriving (11.63).

7. By contracting (11.82) with u_i and using (11.83) and (11.43), obtain the equation

$$\frac{\partial \mathscr{W}}{\partial t} + \operatorname{div}(\mathscr{W}\mathbf{u}) = -\operatorname{div}\mathbf{q} - \frac{\partial}{\partial x_k}(u_i p_{ik}) + \mathbf{j}\cdot\mathbf{E} + \mathbf{u}\cdot\mathbf{F}$$

where $\mathscr{W} = W + \frac{1}{2}\rho_m\mathbf{u}^2$, and $\mathbf{F} = \sum_r N_r m_r \mathbf{g}_r$. Give a physical interpretation of the result obtained by integrating this equation over a closed volume V bounded by a surface S and applying the divergence theorem where appropriate (transforming $\mathbf{j}\cdot\mathbf{E}$ by Poynting's theorem, too, if you wish).

8. Consider the amendments required throughout §§11.5.3–11.5.4 if the Lorentz force term is classified as $O(1)$ instead of $O(\varepsilon^{-1})$.

9. Two vectors $\mathbf{a}$ and $\mathbf{b}$ are to be linearly related in the form $\mathbf{a} = \boldsymbol{\lambda}\cdot\mathbf{b}$. The relation is to be invariant under rotation of the axes about the magnetic field (taken along the third axis), and a change of sign of the magnetic field is to be equivalent to the reversal of the z-axis. Show that $\boldsymbol{\lambda}$ must take the form (11.101), with λ_0 and λ_1 even functions of B and λ_2 an odd function of B.

Similarly, two symmetric tensors p_{ij} and s_{kl} are to be related by $p_{ij} = \mu_{ijkl}s_{kl}$ with the same requirements of symmetry in space.* If both p_{ij} and s_{kl} are traceless, show that the most general such relation can be put in the form (11.99) where μ_0, μ_1 and μ_3 are even, μ_2 and μ_4 are odd in B.

(The reader familiar with the (1, 0, −1) coordinates introduced in Chapter 9 should find it very easy to deal with this problem by noting that a rotation θ about the magnetic field is represented by a transformation $x^\nu \to e^{i\nu\theta}x^\nu$. In the vector case, the relation $a^\nu = \lambda^\nu_\sigma b^\sigma$ has the required symmetry if λ^ν_σ is diagonal, say $\lambda^{(\nu)}\delta^\nu_\sigma$, where $\lambda^{(\pm\nu)}$ are complex conjugates, and $\lambda^{(\pm\nu)}$ are also interchanged if the field is reversed. In the tensor case $p^{\nu\sigma} = \mu^{\nu\sigma}{}_{\alpha\beta}s^{\alpha\beta}$, and $\mu^{\nu\sigma}{}_{\alpha\beta}$ is non-zero only if $\nu + \sigma = \alpha + \beta$. Consideration of the trace conditions (noting that the trace is $\Sigma\, p^{\nu,-\nu}$) shows that only one distinct coefficient is needed when $\nu + \sigma = 0$, so in effect $\mu^{\nu\sigma}{}_{\alpha\beta} = f(\nu + \sigma)\delta^\nu_\alpha\delta^\sigma_\beta$, and the five quantities $f(-2), \ldots, f(+2)$ are the required coefficients. Translating these results into ordinary coordinates gives the formulas quoted. Further, the way in which the field enters the equation of motion of a particle suggests that $\lambda^{(\nu)}$ depends on the field in the combination νB, and $f(\nu + \sigma)$ in the combination $(\nu + \sigma)B$. This leads to the fact that the coefficients with suffix 0 are independent of B altogether, and to the identities (11.116).)

10. Onsager's reciprocal relations may be stated as follows. Let e be the rate of entropy production due to internal actions within a medium. Then for small

---

* A similar problem occurs in elasticity, in what is called the "transversely isotropic" case, where the stress tensor and strain tensor are to be related under the condition that the material has one axis of symmetry, with the further restriction $\mu_{ijkl} = \mu_{klij}$ imposed by considerations of energy. But of the five constants which then arise, two disappear if the two tensors to be related are traceless, leaving our μ_0, μ_1 and μ_3, while our two additional constants μ_2 and μ_4 are unknown in elasticity, being annulled by the energy condition; compare also with Problem 10.

departures from thermal equilibrium we can write

$$e = \sum_{\alpha} f_\alpha g_\alpha,$$

where g_α ($\alpha = 1, 2, \ldots$) are components of suitable forces or gradients, and f_α are resultant fluxes (the labels α are usually sets of Cartesian suffixes or pairs of suffixes). There is a general linear relation

$$f_\alpha = K_{\alpha\beta} g_\beta.$$

Then the coefficients $K_{\alpha\beta}$ must satisfy

$$K_{\alpha\beta}(\mathbf{B}) = K_{\beta\alpha}(-\mathbf{B}).$$

(See a text on irreversible thermodynamics; for instance de Groot, 1951.) For a two-component plasma, we have

$$e = -\frac{1}{T^2}(\nabla T)\cdot\mathbf{q}' + \frac{1}{T}(\mathbf{E}'\cdot\mathbf{j} - e_{ij}p^1_{ij}),$$

where $\mathbf{q}'$ is the heat flux from which the convective part has been omitted (see (11.100)), p^1_{ij} is the viscous part of the pressure tensor and e_{ij} is the symmetric part of the rate of strain tensor. Show that (11.99), (11.105), (11.106) and (11.107) are all consistent with Onsager's relations.

11. Show that the subdivision of $\mathbf{q}$ made in (11.100) has the physical significance stated there.

12. The following gives a rough interpretation of equations (11.104) and (11.105) (omitting thermal diffusion, however). Suppose the ion mean velocity may be identified with $\mathbf{u}$, the electrons having mean velocity $\mathbf{u}_e$, the effect of collisions being represented (as in §5.6) by a simple frictional force $-m_e\nu(\mathbf{u}_e - \mathbf{u})$, and the inertia of the electrons negligible. The momentum equation for the electrons, regarded as a fluid, is

$$m_e\nu(\mathbf{u}_e - \mathbf{u}) = -e(\mathbf{E} + \mathbf{u}_e \times \mathbf{B}) - \frac{1}{n}\nabla(nKT).$$

Show that this leads to

$$\mathbf{j} = \sigma_0\left(\mathbf{E}'' - \frac{1}{ne}\mathbf{j} \times \mathbf{B}\right),$$

where $\mathbf{E}''$ is as in (11.104) and $\sigma_0 = ne^2/m_e\nu$. The second term on the right expresses the anisotropy introduced by the field. Solve for $\mathbf{j}$ to obtain $\mathbf{j} = \boldsymbol{\sigma}\cdot\mathbf{E}''$ where $\boldsymbol{\sigma}$ is expressed by $(\sigma_0, \sigma_1, \sigma_2)$ in the usual way. Identifying ν with $1/\tau_e$ gives a result similar to (11.118), and gives σ_1/σ_0 and σ_2/σ_0 of exactly the form (11.125), where $x = \Omega_e/\nu$.

13. Show that the "magnetic Prandtl number," R_M/R, for a plasma (see §11.6.2) is approximately

$$11\frac{KT}{(m_e m_i)^{1/2}c^2}\left(\frac{\Lambda}{\log \Lambda}\right)^2$$

14. Carry out a detailed investigation of Alfvén waves as sketched in §11.6.3.

REFERENCES

The classic book on the kinetic theory of gases, containing an exhaustive treatment of the Chapman–Enskog method and its applications, together with numerous references to the original papers, is:

S. Chapman and T. G. Cowling, *The Mathematical Theory of Non-Uniform Gases*, 2nd edn, Cambridge University Press (1952).

Papers cited in the text of this chapter are:

G. F. Chew, M. L. Goldberger and F. E. Low, *Proc. Roy. Soc.*, **236**, 112 (1956).

R. S. Cohen, L. Spitzer and P. McR. Routly, *Phys. Rev.*, **80**, 230 (1950).

J. P. Dougherty, S. R. Watson and M. A. Hellberg, *J. Plasma Phys.*, **1**, 327 (1967).

C. F. Kennel and J. M. Greene, *Ann. Phys.* (*N.Y.*), **38**, 63 (1966).

W. Marshall, *Harwell Reports* 2247, 2352, 2419 (1957).

E. N. Parker, *Phys. Rev.*, **107**, 924 (1957).

B. B. Robinson and I. B. Bernstein, *Ann. Phys.* (*N.Y.*), **18**, 110 (1962).

For the Onsager relations:

S. R. de Groot, *Thermodynamics of Irreversible Processes*, North-Holland (1951).

L. Onsager, *Phys. Rev.*, **37**, 405 (1931); **38**, 2265 (1931).

Texts giving further details of applications of MHD:

H. Alfvén and C. G. Fälthammar, *Cosmical Electrodynamics*, 2nd edn, Clarendon Press (1963).

T. G. Cowling, *Magnetohydrodynamics*, Interscience (1957).

V. C. A. Ferraro and C. Plumpton, *An Introduction to Magneto-Fluid Mechanics*, Clarendon Press (1961).

C. O. Hines, I. Paghis, T. R. Hartz and J. A. Fejer, *Physics of the Earth's Upper Atmosphere*, Prentice-Hall (1965).

P. H. Roberts, *An Introduction to Magnetohydrodynamics*, Longmans (1967).

D. J. Rose and M. Clark, *Plasmas and Controlled Fusion*, M.I.T. Press and Wiley (1961).

For hydromagnetic stability:

S. Chandrasekhar, *Hydrodynamic and Hydromagnetic Stability*, Oxford University Press (1961).

CHAPTER 12

KINETIC EQUATIONS FOR PLASMAS

12.1 INTRODUCTION

Our treatment of plasmas in Chapters 7–10 was based mostly on the collision-free Boltzmann equation

$$\frac{\partial f}{\partial t} + \mathbf{v}\cdot\frac{\partial f}{\partial \mathbf{x}} + \mathbf{a}\cdot\frac{\partial f}{\partial \mathbf{v}} = 0 \tag{12.1}$$

where $\mathbf{a}$ is the acceleration due to the smoothed-out electric and magnetic fields together with any external fields. A self-consistent theory is obtained by combining (12.1) with Maxwell's equations. We introduced this procedure in a rather heuristic manner in §7.2, commenting that some approximation is involved since the smoothing process leads to an inaccurate treatment of the interactions between particles which are fairly close together. We supposed that an improved equation

$$\frac{\partial f}{\partial t} + \mathbf{v}\cdot\frac{\partial f}{\partial \mathbf{x}} + \mathbf{a}\cdot\frac{\partial f}{\partial \mathbf{v}} = \left(\frac{\partial f}{\partial t}\right)_c \tag{12.2}$$

would exist, the right-hand side denoting the corrections due to the close interactions, referred to as "collisions," though that word is rather misleading in the case of particles for which an inverse square law applies. However, we made little use of (12.2), and where we did (see §9.5) the formula for $(\partial f/\partial t)_c$ was a model one rather than a properly derived expression.

In a fundamental development of the kinetic theory of gases one must start from first principles, and find in what circumstances (if any) an equation of the form (12.2) is valid. The expression for the right-hand side should be derived, and the conditions under which (12.1) is a good approximation should become clear. Other matters, for example the statistical fluctuations of quantities hitherto regarded as definite, may also be treated by the techniques developed in the course of this work.

Substantial progress has been made in carrying out this program, but it is not yet complete. Serious difficulties seem to prevent a mathematically rigorous treatment. The analysis also becomes excessively cumbersome when one reaches matters of detail. The purpose of this chapter is to outline the general theory and to illustrate it with the detailed calculation of $(\partial f/\partial t)_c$ in a simple special case. It is hoped that this will introduce the reader to the extensive literature on this subject, notably the monographs by R. Balescu (1963) and Yu. L. Klimontovich (1966).

From a strictly logical point of view the theory developed here (or rather a more general treatment of it) ought to have been given before we made extensive use of (12.1) in Chapters 8–11; however, there seems little harm in postponing the detailed justification of the Vlasov equations since the rather heavy mathematical work involved can be handled more succinctly by means of a useful analogy with the mathematics used in Chapter 8 to solve (12.1) for plasma oscillations. Similarly, the detailed results given in §11.5 are accessible only when the proper form for $(\partial f/\partial t)_c$ for a plasma is known, so that we were obliged to make a forward reference to the required formulas. The logical structure of the whole will, however, become clear in the end, and we have left the present work until last merely for pedagogical reasons, it being somewhat more abstruse than anything encountered so far.

12.2 DEFINITIONS

12.2.1 Liouville's theorem and ensemble averages

The fundamental concept in statistical mechanics is the *ensemble*, a collection of numerous replicas of the system under consideration. The space constructed by ascribing one axis to each coordinate and each momentum, for all the particles, is Γ space, as introduced in §7.1. A single system is represented by one moving point of Γ-space, while an ensemble is represented by a probability density, ρ, in the space. In general, ρ will be evolving in time, in a way given by Liouville's theorem, equation (7.2).

Let us start with a gas of N point particles,* all of one species, in a vessel of volume V. The extension to a mixture of species is an unnecessary complication for the present.

In the case of charged particles, the electromagnetic field is treated in the electrostatic approximation, any magnetic field being purely external; otherwise further variables, describing electromagnetic oscillators, have to be introduced. The force of interaction between a pair of particles is given by the Coulomb law and depends only on their position at the present time.

As before, it is convenient to work solely in Cartesian coordinates, and to use the velocity rather than momentum. The jth particle is specified by $(\mathbf{x}_j, \mathbf{v}_j)$, which we may abbreviate as X_j. We have a Liouville density $\rho(X_j, t)$, where

$$\rho(X_1, \ldots, X_N, t)dX_1 \cdots dX_N = \text{the probability that the system is in the state } [(X_1, X_1 + dX_1), \ldots] \text{ at time } t. \tag{12.3}$$

* Note that throughout this chapter N denotes the *total* number of particles; the number density, previously called N, will now be written as n.

This is normalized by

$$\int \cdots \int \rho \, dX_1 \cdots dX_N = 1 \tag{12.4}$$

and satisfies the Liouville equation,*

$$\frac{\partial \rho}{\partial t} + \sum_{j=1}^{N} \mathbf{v}_j \cdot \frac{\partial \rho}{\partial \mathbf{x}_j} + \sum_{j=1}^{N} \mathbf{a}_j \cdot \frac{\partial \rho}{\partial \mathbf{v}_j} = 0. \tag{12.5}$$

Here $\mathbf{a}_j$ is the exact acceleration for the jth particle, due to all the other particles, without any smoothing. It is therefore a function of all the X's, not just X_j. Externally applied fields may also be included.

As (12.5) is symmetric with respect to interchanges of particles in the case of only one species, it is clear that if ρ is initially symmetric it remains so. Since we shall not wish to identify individual particles at any stage, we shall consider only solutions which have this complete symmetry. It should be noted that this is merely a convenient choice on our part; it has no connection with the restriction to symmetric or antisymmetric states imposed in quantum statistical mechanics.

An ensemble composed of a single copy of the system is represented by taking ρ to be a δ-function in Γ-space, whose location moves along the appropriate orbit as obtained by solving the exact equations of motion. Finding more general solutions of (12.5) is equivalent to constructing many such orbits, so is excessively difficult. A well-known formal solution is

$$\rho = C \exp(-W/KT), \tag{12.6}$$

where $W(X_i)$ is the total energy for the state X_i and C is a normalizing constant. This is the canonical ensemble of Gibbs, representing thermal equilibrium at temperature T, and satisfies (12.5) since W is a constant for each orbit. We know of no method for solving (12.5) explicitly in cases other than thermal equilibrium.

Let Q be a quantity which can be calculated if the state of the system is known. Thus $Q = Q(X_1, \ldots, X_N, t)$, allowing for possible explicit depend-

---

* Liouville's equation in the form (7.2) was derived by means of Hamiltonian mechanics; the momentum conjugate to $\mathbf{x}_j$ is therefore $m\mathbf{v}_j + e\mathbf{A}$ where $\mathbf{A}$ is the magnetic vector potential. However, Liouville's equation in the form (12.5), where we use $\mathbf{v}_j$ as the independent variable rather than $\mathbf{v}_j + e\mathbf{A}/m$, is also correct. This may be shown directly by starting with the equation of conservation of particles, which is simply (12.5) with $\mathbf{v}_j$ and $\mathbf{a}_j$ written following the differential operators instead of preceding them. Commuting these to recover (12.5) is trivial for the terms in the first summation and is justified for the second summation whenever the velocity-dependent part of $\mathbf{a}_j$ takes the form $\mathbf{v}_j \times \mathbf{B}$, a fact we used before when dealing with the Vlasov equation in §7.2.

ence on the time. By the *ensemble average* (or expectation value) of Q we mean

$$\langle Q(X_1, \ldots X_N, t)\rangle \equiv \int \cdots \int \rho(X_1, \ldots X_N, t) Q(X_1, \ldots X_N, t)\, dX_1 \cdots dX_N. \tag{12.7}$$

(More complicated cases where Q involves the state of the system at more than one time will be mentioned later, §12.9.3.) Examples of quantities Q might be the electric potential at a point $\mathbf{x}$ in space,

$$-\frac{e}{4\pi\varepsilon_0}\sum_i |\mathbf{x} - \mathbf{x}_i|^{-1}$$

(taking our particles to be electrons), or the number of particles within a prescribed closed surface, and so on.

Our object is to manipulate certain ensemble averages, which give the desired information about the gas, without calculating ρ, which would give the most complete information possible. For a system in thermodynamic equilibrium any such ensemble average may be calculated, in principle, by combining (12.6) and (12.7), and equilibrium statistical mechanics (so far as classical mechanics is concerned) is reduced to the problem of evaluating the required integrals. In the non-equilibrium case we try to construct "equations of motion" for the ensemble averages; this involves reducing (12.5) by integrating it with respect to most of its independent variables. The process is thus superficially similar to that of Chapter 11, where we reduced the Boltzmann equation to hydrodynamic equations. There is a further similarity. In both these topics one obtains a chain of equations which do not, as they stand, form a closed set, but we close them by means of approximations. The details of the work are, however, quite different in the two cases.

12.2.2 The hierarchy of distribution functions

For a gas in a completely defined state $X_1(t)$, $X_2(t), \ldots,$ we define the *exact one-particle distribution function* as

$$F(\mathbf{x}, \mathbf{v}, t) = \sum_i \delta[X - X_i(t)], \tag{12.8}$$

where X is $(\mathbf{x}, \mathbf{v})$, a typical point of μ-space (the phase space of one particle). This function simply gives the exact density of the gas in μ-space, and so contains all the information except for the identity of the particles. Then we define the *average one-particle distribution function* as

$$f_1(\mathbf{x}, \mathbf{v}, t) = \langle F(\mathbf{x}, \mathbf{v}, t)\rangle. \tag{12.9}$$

It is this which we shall identify as the ordinary Boltzmann function. By

(12.7) and (12.8) and the symmetry of ρ,

$$f_1(\mathbf{x}, \mathbf{v}, t) = \int \cdots \int \rho(X_1, \ldots, X_N, t) \sum_{i=1}^{N} \delta[X - X_i(t)] dX_1 \cdots dX_N$$

$$= N \int \cdots \int \rho(X, X_2, X_3 \ldots, X_N, t) dX_2\, dX_3 \cdots dX_N, \qquad (12.9a)$$

as all terms in the summation are equal. Thus f_1 is essentially ρ integrated with respect to the variables of all the particles except one. We may interpret $f_1(\mathbf{x}, \mathbf{v}, t)\, d^3x\, d^3v$ in either of two ways: (a) the probability that $d^3x\, d^3v$ is occupied by one particle at time t, provided $d^3x\, d^3v$ is so small that the probability of occupation by more than one particle is negligible; or (b) the mean number of particles in $d^3x\, d^3v$ at time t. In the latter case there will of course be fluctuations from one member of the ensemble to another, but if $d^3x\, d^3v$ is large enough for the number of particles to be large, the fluctuations are unimportant and we may think of f_1 deterministically, as we have done throughout Chapters 7–11.

Next we take two points X, X' of μ-space, and define the exact *two-particle distribution function* as

$$F(X, t)F(X', t),$$

where F is as in (12.8); clearly it is zero unless both X and X' are occupied by particles. Using (12.7) again and reducing integrals involving δ-functions, we have

$$\langle F(X, t)F(X', t)\rangle = f_2(X, X', t) + \delta(X - X')f_1(X, t), \qquad (12.10)$$

where

$$f_2(X, X', t) = N(N-1) \int \cdots \int \rho(X, X', X_3, X_4, \ldots, X_N, t) dX_3 \cdots dX_N. \qquad (12.11)$$

Equation (12.11) defines an average two-particle distribution, f_2. We interpret (12.10), when multiplied by $dX\, dX'$, as the probability that there is simultaneously a particle in dX and one in dX', assuming the probability of more than one such particle to be negligible. If dX and dX' do not overlap, this probability is given by $f_2\, dX\, dX'$, but if they do overlap there is an extra contribution corresponding to the observation of the same particle, so leading to the δ-function term. We may also interpret $f_2\, dX\, dX'$ as the mean product of the numbers of particles in dX and dX'.

Similarly we may consider three points X, X' and X'' in phase space, and show that

$$\begin{aligned}\langle F(X, t)F(X', t)F(X'', t)\rangle = {} & f_3(X, X', X'', t) + f_2(X, X'', t)\,\delta(X - X') \\ & + f_2(X'', X', t)\,\delta(X'' - X) + f_2(X', X, t)\,\delta(X' - X'') \\ & + f_1(X, t)\,\delta(X - X')\,\delta(X - X''), \qquad (12.12)\end{aligned}$$

where

$$f_3(X, X', X'', t) = N(N-1)(N-2)\int\cdots\int \rho(X, X', X'', X_4, \ldots, X_N)\, dX_4 \cdots dX_N \quad (12.13)$$

with similar interpretation.

Given s points $X, X', X'', \ldots, X^{(s)}$, it will clearly be convenient to define

$$f_s(X, X', \ldots, X^{(s)}, t) = \frac{N!}{(N-s)!}\int\cdots\int \rho(X, X', \ldots, X^{(s)}, X_{s+1}, \ldots, X_N, t)\, dX_{s+1} \cdots dX_N, \quad (12.14)$$

whose interpretation concerns the probability of finding particles simultaneously at $X, X', \ldots, X^{(s)}$, provided those points are all different. We shall only be concerned with $s \ll N$, so the factorials may be replaced by N^s. (Some writers normalize f_s with the factor V^s in place of our N^s.)

We now have a hierarchy of distribution functions $f_1, f_2, \ldots, f_{N-1}, \rho$, each containing more information than its predecessor. f_s is essentially ρ integrated with respect to the variables of $N - s$ particles.

12.3 THE HIERARCHY OF KINETIC EQUATIONS

The differential equation governing any f_s is obtained from the Liouville equation, (12.5), by integrating over the variables associated with $N-s$ particles. We write

$$\mathbf{a}_j = \sum_{i=0}^{N} \mathbf{a}_j^{(i)}, \quad (12.15)$$

where $\mathbf{a}_j^{(i)}$ is the acceleration experienced by particle j due to the presence of particle i, (zero if $i = j$), and $\mathbf{a}_j^{(0)}$ is the acceleration due to the external field of force. We multiply (12.5) by $N!/(N-s)!$ and integrate with respect to $X_{s+1}, X_{s+2}, \ldots, X_N$. The first term gives $\partial f_s/\partial t$. In the other two terms the contributions $s + 1 \leqslant j \leqslant N$ in the summations disappear by virtue of boundary conditions on the surface of V or at $\mathbf{v} \to \infty$ as the case may be. Each of the surviving contributions $(1 \leqslant j \leqslant s)$ in the acceleration term may be rewritten as a new summation over i by virtue of (12.15), and the terms $s + 1 \leqslant i \leqslant N$ are identical (for fixed j) by the symmetry of ρ. Writing all these terms as $i = s + 1$, one obtains

$$\frac{\partial f_s}{\partial t} + \sum_{j=1}^{s} \mathbf{v}_j \cdot \frac{\partial f_s}{\partial \mathbf{x}_j} + \sum_{j=1}^{s}\sum_{i=0}^{s} \mathbf{a}_j^{(i)} \cdot \frac{\partial f_s}{\partial \mathbf{v}_j} + \sum_{j=1}^{s}\int \mathbf{a}_j^{(s+1)} \cdot \frac{\partial f_{s+1}}{\partial \mathbf{v}_j}\, dX_{s+1} = 0. \quad (12.16)$$

Here the first three terms comprise the Liouville operator for s particles only, including the external forces and the forces they exert on each other.

The last term introduces the forces due to the remaining $N - s$ particles, though it involves only f_{s+1}.

This set of equations, for $s = 1, 2, \ldots, N$, is often called the "BBGKY chain," after N. N. Bogoliubov, M. Born, H. S. Green, J. G. Kirkwood and J. Yvon, who in various investigations developed this work. If we terminate the chain at any stage earlier than $s = N$, we are left with s equations and $s + 1$ unknown functions. Progress can only be made if, as a result of some approximation, f_{s+1} can be expressed in terms of $f_1, \ldots, f_s$. This is a situation very characteristic of non-equilibrium statistical mechanics. The Liouville equation determines $\rho(X_1, \ldots, X_N, t)$ at all times, so giving complete information about the ensemble (any lack of information about the plasma itself arising from our choice of ρ at $t = 0$). As soon as we elect to work with reduced information, say some f_s, we find this fails to determine its own future, except in trivial cases where the rejected information corresponds to degrees of freedom which are completely decoupled from those which are retained. A similar situation occurred in Chapter 11; given an equation governing f we tried to obtain hydrodynamic equations as the reduced description but could only do so by making special assumptions. (The analogy is not quite complete, as any equation in the BBGKY chain implies all the lower ones, which may be obtained by further integration, whereas no such statement applies to the chain of hydrodynamic equations.)

Normally the objective is to obtain from (12.16) an equation for f_1 only. If, then, f_1 is given at some instant, e.g. $t = 0$, any attempt to calculate f_1 in the future must assume (if only tacitly) some rule whereby f_1 fixes ρ initially. We imagine the Liouville equation solved to give the behavior of the underlying ensemble, and then f_1 recovered for future times. The rule will be of a statistical nature. For example, it might state that the unknown information is in some sense random, or that the events it describes are uncorrelated, and so on. Such a procedure is obviously unsatisfactory unless it can be shown that the rule is self-perpetuating, either exactly or at least to an acceptable approximation. If this is so, the ensemble is represented throughout by f_1, and we can expect that a governing equation for f_1 exists; but only a certain class of solutions of the Liouville equation, namely those for which ρ arises from f_1 by applying the rule, will have been considered. All other solutions will have been automatically excluded, and the hypothesis that they do not correspond with reality rests entirely on the agreement of the calculation with experiment. Restrictions are also imposed on f_1 if the approximation involved is to be valid; usually f_1 must be sufficiently slowly varying in space and time. A similar discussion applies to all situations where equations are supposed to determine reduced information. For example, in passing from the Boltzmann function to the hydrodynamic description, $f_1(\mathbf{x}, \mathbf{v}, t)$ is represented by $N(\mathbf{x}, t)$, $\mathbf{u}(\mathbf{x}, t)$ and $p(\mathbf{x}, t)$, and in the Chapman–Enskog expansion the lost information (i.e. the details of the distribution in velocity space) is supposed to be recovered to the first approximation by making f_1 the Maxwellian function consistent with the hydrodynamic variables (see §11.5).

For non-interacting particles this program can be carried out easily and exactly. In other cases we need an approximation which amounts to assuming that the interaction is small in some sense. However, the small parameter involved takes different forms in the cases of neutral gas and plasma, leading to quite dissimilar kinetic equations.

For future reference we record that for $s = 1$, (12.16) reduces to

$$\frac{\partial f_1}{\partial t} + \mathbf{v}_1 \cdot \frac{\partial f_1}{\partial \mathbf{x}_1} + \mathbf{a}_1^{(0)} \cdot \frac{\partial f_1}{\partial \mathbf{v}_1} + \int \mathbf{a}_1^{(2)} \cdot \frac{\partial f_2}{\partial \mathbf{v}_1} \, d^3x_2 \, d^3v_2 = 0. \tag{12.17}$$

This is the kinetic equation in its raw form; the procedure just outlined will give new forms for the integral term, replacing it by expressions involving only f_1.

12.4 UNCORRELATED PARTICLES

12.4.1 Non-interacting particles

When the particles of a gas are subject only to external forces, $\mathbf{a}_j$ reduces to $\mathbf{a}_j^{(0)}$ and depends only on $(\mathbf{x}_j, \mathbf{v}_j)$. The independence of the particles is expressed mathematically by the vanishing of the integral term in (12.16). The one-particle function, f_1, now satisfies

$$\frac{\partial f_1}{\partial t} + \mathbf{v}_1 \cdot \frac{\partial f_1}{\partial \mathbf{x}_1} + \mathbf{a}_1^{(0)} \cdot \frac{\partial f_1}{\partial \mathbf{v}_1} = 0, \tag{12.18}$$

which is the collision-free Boltzmann equation in the external force field. Given a solution of this equation (correctly normalized), we can solve the Liouville equation itself by writing

$$\rho(X_1, \ldots, X_N, t) = N^{-N} \prod_{j=1}^{N} f_1(X_j, t). \tag{12.19}$$

Thus ρ is simply the product of N identical one-particle distributions, and f_s would be a similar product of s such functions. For instance

$$f_2(X_1, X_2, t) = \frac{N-1}{N} f_1(X_1, t) f_1(X_2, t), \tag{12.20}$$

so that (12.10) reads

$$\langle F(X_1, t) F(X_2, t) \rangle = f_1(X_1, t) \left[\frac{N-1}{N} f_1(X_2, t) + \delta(X_1 - X_2) \right]. \tag{12.21}$$

Here the first factor gives the probability of finding a particle at X_1, while the second gives the conditional probability of finding a second one at X_2, given one already at X_1.

When ρ factorizes in this way we say the particles are *uncorrelated*, since the probability of finding the second particle is given by the same function as that for the first, apart from the trivial factor arising from the reduction in the total number of particles, and the usual δ-function term. The Liouville equation for independent particles does have other solutions not of this type, but if one assumes the particles to be uncorrelated at any one time, they remain so. Thus (12.19) is the "rule" referred to in the previous section, and it is indeed self-perpetuating. We shall regard it as the natural rule to adopt.

12.4.2 Binary collisions

For a gas composed of neutral molecules, the approximation to be used is based on the short-range nature of the forces together with the assumption that the gas is sufficiently dilute. There is then an *effective range*, r_0, such that molecules separated by more than r_0 are virtually non-interacting. For example, for perfectly elastic spheres, r_0 would simply be their diameter. A more realistic model is that of Lennard–Jones, in which $\mathbf{a}_i^{(j)}$ is derived from the potential

$$\phi(r) \propto [(r/r_0)^{-12} - (r/r_0)^{-6}], \tag{12.22}$$

where $r = |\mathbf{r}_i - \mathbf{r}_j|$ and r_0 is the distance at which ϕ vanishes. This potential leads to a very strong repulsive force up to a distance of about r_0 and a weak attractive force beyond, both of which are appropriate features for neutral molecules. By "dilute" we mean that $nr_0^3 \ll 1$ throughout the gas, where n is the number of particles per unit volume and is typically N/V if the gas is not excessively non-uniform.

Physically, the important characteristic of such a gas is this. A particle which is at the typical distance $n^{-1/3}$ from its nearest neighbor is negligibly influenced by it, as $n^{-1/3} \gg r_0$. The combined effect of the more distant (though more numerous) molecules is even smaller as the law of force decreases much faster than the inverse square. Such a particle is practically free, i.e. it is subject only to any external field. Only when two particles approach to a distance of about r_0 is the interaction appreciable, and we call such an event a "collision." The history of any one particle consists of "free paths" punctuated by collisions. The free paths occupy almost all the time, the collisions being virtually instantaneous. This suggests a development which may be sketched as follows.

In the limit $nr_0^3 \to 0$ we presume that the results of §12.4.1 hold, so the particles are uncorrelated. Further, the departures from these results may be taken to be in some sense small if nr_0^3 is small. The physics indicates the nature of the approximation. For instance (12.20) may continue to hold to extremely good approximation if the phase points X_1 and X_2 are well separated in physical space, i.e. $|\mathbf{x}_1 - \mathbf{x}_2| \gg r_0$, while if $|\mathbf{x}_1 - \mathbf{x}_2| \leqslant r_0$ we expect f_2 to be very small since the particles cannot overlap. The approximation of regarding the particles as uncorrelated is not therefore uniformly good; rather it is extremely good for almost all the space (X_1, X_2), but seriously wrong for special small regions of the space. Similar remarks apply to f_s for $s > 2$ and to ρ itself. However, in writing down (12.17), the important contributions to the integral term will arise from just those special regions, as $\mathbf{a}_1^{(2)}$ is elsewhere very small.

We turn to the second equation of the "chain," (12.16), which is explicitly

$$\frac{\partial f_2}{\partial t} + \mathbf{v}_1 \cdot \frac{\partial f_2}{\partial \mathbf{x}_1} + \mathbf{v}_2 \cdot \frac{\partial f_2}{\partial \mathbf{x}_2} + \mathbf{a}_1^{(0)} \cdot \frac{\partial f_2}{\partial \mathbf{v}_1} + \mathbf{a}_2^{(0)} \cdot \frac{\partial f_2}{\partial \mathbf{v}_2} + \mathbf{a}_1^{(2)} \cdot \frac{\partial f_2}{\partial \mathbf{v}_1} + \mathbf{a}_2^{(1)} \cdot \frac{\partial f_2}{\partial \mathbf{v}_2}$$
$$+ \int \left[\mathbf{a}_1^{(3)} \cdot \frac{\partial f_3}{\partial \mathbf{v}_1} + \mathbf{a}_2^{(3)} \cdot \frac{\partial f_3}{\partial \mathbf{v}_2} \right] d^3x_3\, d^3v_3 = 0. \quad (12.23)$$

As we have seen, only the first five terms of this are accounted for if we take f_2 to be given in terms of f_1 by (12.20) and take f_1 to satisfy the collision-free Boltzmann equation; however, to very good approximation the integral terms of (12.23) are also accounted for if we take f_2 and f_3 to be products of f_1's, with f_1 itself satisfying whatever is to be the corrected form of the Boltzmann equation, (12.17). The important question is thus that of the effect of the terms involving $\mathbf{a}_1^{(2)}$ and $\mathbf{a}_2^{(1)}$ in (12.23). Clearly these are important only when $|\mathbf{x}_1 - \mathbf{x}_2|$ is of order r_0, or in other words particles at X_1 and X_2 are colliding. The procedure is therefore to take, at each instant, f_2 to be fixed by (12.20) except where $|\mathbf{x}_1 - \mathbf{x}_2|$ is small, and to use this value of f_2 as a boundary condition enabling us to solve (12.23) within the regions of (X_1, X_2)-space corresponding to collisions. In solving (12.23) for the collision regions one neglects the $\mathbf{a}^{(0)}$ terms and the integral term, the former on the grounds that any external force is usually negligible compared with the intermolecular forces acting during a collision, the latter because triple collisions may be assumed to be very rare. Then (12.23) reduces to the Liouville equation for two molecules in otherwise empty space; its general solution (cf. Jeans's Theorem) corresponds to an enumeration of all the possible binary collisions which can occur (there are also solutions corresponding to all the possible bound states, which we disregard). Construction of the orbits for such binary collisions is comparatively easy.

Clearly f_2 varies rather rapidly in these collision regions (the time scale being the time of a collision and the length scale r_0), so that when we insert f_2 into (12.17) it will induce corresponding fine structure in f_1. This, however, is meaningless, for f_1 is intended to be smooth on a length scale of at least $n^{-1/3}$ and on the related time scale for thermal particles. We therefore average out the small fluctuations in the integral term of (12.17). This has the effect that the details of the orbits of the colliding particles are no longer needed, but only the relation between the initial velocities $\mathbf{v}_1$, $\mathbf{v}_2$, the final velocities $\mathbf{v}_1'$, $\mathbf{v}_2'$ and the impact parameters b (the closest distance to which the particles would have approached if non-interacting) and ε (an angle indicating the orientation of the relative orbit). After some manipulation, (12.17) assumes the standard form

$$\frac{\partial f_1}{\partial t} + \mathbf{v}_1 \cdot \frac{\partial f_1}{\partial \mathbf{x}_1} + \mathbf{a}_1^{(0)} \cdot \frac{\partial f_1}{\partial \mathbf{v}_1}$$

$$= -\int [f_1(\mathbf{v}_1) f_1(\mathbf{v}_2) - f_1(\mathbf{v}_1') f_1(\mathbf{v}_2')] \, | \mathbf{v}_1 - \mathbf{v}_2 | b \, db \, d\varepsilon \, d^3 v_2. \quad (12.24)$$

The right-hand side of this equation represents the rate of loss of particles of velocity $\mathbf{v}_1$ due to collisions with particles of velocity $\mathbf{v}_2$, together with the rate of gain due to inverse collisions. The collisions have come to be regarded as instantaneous by our approximation. They occur at rates given by appropriate values of f_1 itself; this is a consequence of the assumption that the particles are uncorrelated (i.e. (12.20) applies), and is simply the *hypothesis of molecular chaos* introduced in the early days of kinetic theory.

Formally, the calculation outlined here consists of an iteration in terms of the small parameter nr_0^3, with the particles uncorrelated as the lowest approximation. The details may be found in N. N. Bogoliubov's original book. The application of (12.24) to transport phenomena is treated by Chapman and Cowling (1952); see also §11.5.

For plasmas, an expansion of this type is appropriate for collisions between charged and neutral particles, though for reasons of space we do not pursue this any further. The expansion is not, by itself, valid for interactions between charged particles, for as we have seen such interactions have important effects when the separation exceeds $n^{-1/3}$. Collective effects, such as plasma oscillations, can occur on any length scale, and even a single particle gives rise to a disturbance extending to about a Debye length, as in the shielding effect described by the Debye–Hückel theory (§7.6). Equation (12.24) is therefore completely erroneous for charged particles, and this is manifest by the fact that on attempting to use it one finds the integration on the right-hand side diverging at the upper limit $b \to \infty$. Actually quite good results can be obtained by "cutting off" the divergent integral at an upper limit equal to the Debye length. This is not entirely fortuitous but it is only explicable in terms of the more correct development given below.

12.4.3 The self-consistent field for a plasma

As we have seen, the introduction of forces of interaction in a gas invalidates the solution of the type (12.19), representing uncorrelated particles. But remembering that the mutual potential energy is typically much smaller than the kinetic energy (see §1.2), we may suspect that for plasmas there will be circumstances where the particles may be regarded as uncorrelated to good approximation, and that a scheme for successive corrections may be set up. Such a development proceeds along a different line from the one appropriate for neutral gases, sketched in the preceding digression.

The first stage may be demonstrated very easily. Consider electrons with charge $-e$. We have

$$\mathbf{a}_j^{(0)} = -\frac{e}{m}[\mathbf{E}^{(0)} + \mathbf{v}_j \times \mathbf{B}^{(0)}], \qquad \mathbf{a}_j^{(i)} = \frac{e^2}{4\pi\varepsilon_0 m}\frac{\mathbf{x}_j - \mathbf{x}_i}{|\mathbf{x}_j - \mathbf{x}_i|^3} \tag{12.25}$$

where $\mathbf{E}^{(0)}$ and $\mathbf{B}^{(0)}$ are externally applied fields. We have supposed that N and V are both very large, with $n = N/V$ representing a typical value of the number density (actually the number density may be non-uniform throughout V but we assume it does not vary too violently). Now consider the limit in which $n \to \infty$, with $e \to 0$ and $m \to 0$ in such a way that the quantities ne, nm and e/m are held constant. This limit can be pictured as the result of subdividing the electrons until they are "pulverized" to form a continuous fluid, their total mass and charge being constant. In this limit we have $f_s = O(n^s)$, $\mathbf{a}_j^{(0)}$ is constant, and $\mathbf{a}_j^{(i)} = O(e^2/m) = O(n^{-1})$ for $i \neq 0$. In our chain of equations (12.16) the third term (the one involving the double summation) is one order less than the others, and so can be discarded, save for the term $i = 0$. Thus

$$\frac{\partial f_s}{\partial t} + \sum_{j=1}^{s} \mathbf{v}_j \cdot \frac{\partial f_s}{\partial \mathbf{x}_j} + \sum_{j=1}^{s} \mathbf{a}_j^{(0)} \cdot \frac{\partial f_s}{\partial \mathbf{v}_j} + \sum_{j=1}^{s} \int \mathbf{a}_j^{(s+1)} \cdot \frac{\partial f_{s+1}}{\partial \mathbf{v}_j}\, dX_{j+1} = 0. \tag{12.26}$$

(Actually each term of this equation tends to infinity as $n \to \infty$, but this could easily be avoided by renormalizing the functions f_s to absorb the factor n^s, so it is of no consequence.)

We now search for solutions of the type (12.19), representing uncorrelated electrons. In the limit $N \to \infty$ (12.14) leads to

$$f_s = \prod_{j=1}^{s} f_1(X_j, t), \tag{12.27}$$

and we find that all the equations (12.26) are exactly satisfied provided

$$\frac{\partial f_1(X_1, t)}{\partial t} + \mathbf{v}_1 \cdot \frac{\partial f_1(X_1, t)}{\partial \mathbf{x}_1} + \left[\mathbf{a}_1^{(0)} + \int \mathbf{a}_1^{(2)} f_1(X_2, t)\, dX_2\right] \cdot \frac{\partial f_1(X_1, t)}{\partial \mathbf{v}_1} = 0, \tag{12.28}$$

i.e., dropping the suffixes,

$$\frac{\partial f}{\partial t} + \mathbf{v} \cdot \frac{\partial f}{\partial \mathbf{x}} + \mathbf{a} \cdot \frac{\partial f}{\partial \mathbf{v}} = 0, \tag{12.28a}$$

where the acceleration **a** includes not only the externally applied fields, but also a self-consistent electrostatic field $\mathbf{E}^{(\text{self})}$, where

$$\mathbf{E}^{(\text{self})}(\mathbf{x}, t) = \frac{-e}{4\pi\varepsilon_0} \int \frac{\mathbf{x} - \mathbf{x}'}{|\mathbf{x} - \mathbf{x}'|^3} f(\mathbf{x}', \mathbf{v}', t)\, d^3x'\, d^3v'.$$

In other words it satisfies

$$\text{div}\, \mathbf{E}^{(\text{self})} = -\frac{e}{\varepsilon_0} \int f(\mathbf{x}, \mathbf{v}, t)\, d^3v, \qquad \text{curl}\, \mathbf{E}^{(\text{self})} = 0. \quad (12.29)$$

We have thus recovered the full set of Vlasov equations, in the electrostatic approximation. Equation (12.28) is "collision-free" in the sense that effects due to the interaction of individual particles have disappeared; however, unlike (12.18), it does now include the interactions of electrons regarded as a continuous fluid. As we have seen in Chapters 7–11 this set of equations (amended to remove the electrostatic approximation where appropriate) is capable of describing many of the collective phenomena which are particularly characteristic of plasmas.

The reader will find it a simple matter to extend this discussion to include a mixture of several species (e.g. positive ions). The result is an equation like (12.28a) for each species, **a** being derived from the common self-consistent field. The generalization necessary to make use of the full set of Maxwell's equations is more difficult but has also been carried out. The method is indicated in §12.6.3.

12.5 THE EFFECT OF CORRELATIONS IN A PLASMA

12.5.1 Closing the chain of equations

The Vlasov equation appeared in the previous section as the limiting result for perfectly smoothed out, or pulverized, electrons. We naturally expect this approximation to be the lowest stage of an expansion in terms of some small parameter. In the present section we ask how the expansion is to be continued.

Firstly, we must find what is the appropriate dimensionless small parameter in terms of which the expansion is to be carried out. We noted that in the Vlasov approximation the mean density n must be very large, with e and m small so that ne^2/m is finite, and further that $f_s n^{-s}$ remains finite. Clearly a suitable characteristic time is ω_p^{-1} where ω_p is the mean plasma frequency $(ne^2/\varepsilon_0 m)^{1/2}$. A characteristic speed is $(KT/m)^{1/2}$ where T is the mean temperature, for assuming that the actual temperature does not vary in an extreme fashion this speed is a measure of the extent of the region in velocity space throughout which f is appreciable. The corresponding typical length is, of course, the mean Debye length $h = (\varepsilon_0 KT/ne^2)^{1/2}$. We introduce dimensionless variables, denoted by bars, by writing

$$\mathbf{v} = h\omega_p \bar{\mathbf{v}}, \quad \mathbf{x} = h\bar{\mathbf{x}}, \quad t = \bar{t}/\omega_p, \quad \mathbf{a}^{(0)} = h\omega_p^2 \bar{\mathbf{a}}^{(0)},$$
$$n^s f_s = (h\omega_p)^{-3s} \bar{f}_s, \quad dX = h^6 \omega_p^3 \, d\bar{X}. \tag{12.30}$$

For the acceleration due to mutual interaction (12.25), it will be convenient to introduce

$$\bar{\mathbf{a}}_j^{(i)} = \frac{1}{4\pi} \frac{\bar{\mathbf{x}}_j - \bar{\mathbf{x}}_i}{|\bar{\mathbf{x}}_j - \bar{\mathbf{x}}_i|^3}. \tag{12.31}$$

When the chain of equations for $f_1, f_2, \ldots$, (12.16), is written in terms of these variables, there results

$$\frac{\partial \bar{f}_s}{\partial \bar{t}} + \sum_{j=1}^{s} \bar{\mathbf{v}}_j \cdot \frac{\partial \bar{f}_s}{\partial \bar{\mathbf{x}}_j} + \sum_{j=1}^{s} \bar{\mathbf{a}}_j{}^{(0)} \cdot \frac{\partial \bar{f}_s}{\partial \bar{\mathbf{v}}_j} + \frac{1}{nh^3} \sum_{i=1}^{s} \sum_{j=1}^{s} \bar{\mathbf{a}}_j{}^{(i)} \cdot \frac{\partial \bar{f}_s}{\partial \bar{\mathbf{v}}_j}$$
$$+ \sum_{j=1}^{s} \int \bar{\mathbf{a}}_j^{(s+1)} \cdot \frac{\partial \bar{f}_{s+1}}{\partial \bar{\mathbf{v}}_j} \, d\bar{X}_{s+1} = 0. \tag{12.32}$$

This is formally just the same as (12.16) itself except that the term involving the double summation has the additional coefficient $(nh^3)^{-1}$. It was this term which we neglected in the Vlasov approximation, and we can now see that this was possible because nh^3, the number of particles per cubic Debye length, tended to infinity in the limit there considered. So our next step is to regard $(nh^3)^{-1}$ as small but not zero, and work to the first order therein. We note in passing that

$$(nh^3)^{-1} = (n^{1/3} e^2/\varepsilon_0 KT)^{3/2}, \tag{12.33}$$

which, apart from a numerical factor of order unity, is $\alpha^{3/2}$ where α is the small parameter (introduced in §1.2) which compares typical potential and kinetic energy. We make no particular assumption as to the order of magnitude of the term involving $\mathbf{a}^{(0)}$.

We ask next whether the chain of equations can be solved correct to the first order in $(nh^3)^{-1}$ by assuming that any departures from the Vlasov solution are also of first order, i.e. the particles are nearly uncorrelated. So we add a correction to (12.27), but when substituting into (12.32) we omit the correction in the case of the double summation term. Subsequently we should check that the correction so obtained is indeed small. Once this is understood there is no particular need or advantage in using the normalized variables just referred to, so we revert to the ordinary physical quantities.

The appropriate form for the correction to (12.27) is

$$f_s = \prod_{j=1}^{s} f_1(X_j, t) + \sum_{i,j} \Big[g(X_i, X_j, t) \prod_k f_1(X_k, t) \Big], \tag{12.34}$$

where the sum extends over all pairs $i \neq j$ selected from 1, 2, . . ., s, while the product over k extends through 1, 2, . . ., s, omitting i and j. The *correlation* function $g(X_1, X_2, t)$ is to be symmetric in the phase variables X_1 and X_2 and represents the small correction. For $s = 2$,

$$f_2 = f_1(X_1, t)\, f_1(X_2, t) + g(X_1, X_2, t) \tag{12.35}$$

while for $s = 3$

$$f_3 = f_1(X_1, t)\, f_1(X_2, t)\, f_1(X_3, t) + f_1(X_1, t)\, g(X_2, X_3, t) + \text{two similar terms} \tag{12.36}$$

and so on. On substituting (12.34) into the chain of equations (12.16) and working to the first order in the small terms, one finds that all the equations are indeed satisfied provided the first two are, so justifying the choice (12.34). The first two equations, namely (12.17) and (12.23), then become the governing equations for f_1 and g. Omitting the laborious details of this, we may quote the results. The first equation* is

$$\frac{\partial f(X_1, t)}{\partial t} + \mathbf{v}_1 \cdot \frac{\partial f(X_1, t)}{\partial \mathbf{x}_1} + \left\{\mathbf{a}_1^{(0)} + \int \mathbf{a}_1^{(2)} f(X_2, t)\, dX_2\right\} \cdot \frac{\partial f(X_1, t)}{\partial \mathbf{v}_1} = -\int \mathbf{a}_1^{(2)} \cdot \frac{\partial g(X_1, X_2, t)}{\partial \mathbf{v}_1}\, dX_2. \tag{12.37}$$

Here the left-hand side is the usual Vlasov expression, and the right-hand side is the required correction $(\partial f/\partial t)_c$. The second equation may be reduced to an equation for $\partial g/\partial t$, making use of (12.37) to eliminate $\partial f/\partial t$. This equation is

$$\frac{\partial g}{\partial t} + V_1 g + V_2 g = S, \tag{12.38}$$

where $S(X_1, X_2, t)$ is defined by

$$S = -\mathbf{a}_1^{(2)} \cdot \frac{\partial f(X_1, t)}{\partial \mathbf{v}_1} f(X_2, t) - \mathbf{a}_2^{(1)} \cdot \frac{\partial f(X_2, t)}{\partial \mathbf{v}_2} f(X_1, t), \tag{12.39}$$

and V_1 is a linear integro-differential operator in X_1-space defined in relation to f as follows. For any $h(X_1, t)$

$$V_1 h = \mathbf{v}_1 \cdot \frac{\partial h(X_1, t)}{\partial \mathbf{x}_1} + \left\{\mathbf{a}_1^{(0)} + \int \mathbf{a}_1^{(3)} f(X_3, t)\, dX_3\right\} \cdot \frac{\partial h(X_1, t)}{\partial \mathbf{v}_1} + \left\{\int \mathbf{a}_1^{(3)} h(X_3, t)\, dX_3\right\} \cdot \frac{\partial f(X_1, t)}{\partial \mathbf{v}_1}. \tag{12.40}$$

The operator V_2 is similarly defined in X_2-space, and it should be noted that as X_1- and X_2-spaces involve completely separate sets of independent vari-

* From now on we may omit the suffix from f_1.

ables, these two operators commute when applied to a function such as $g(X_1, X_2, t)$ which involves both sets of phase variables. The significance of these operators can be understood by considering a case where the given function $f(X, t)$ is a known solution of the Vlasov equation (12.28). If one writes down the condition that $f + h$ is also a solution, where $h(X, t)$ is small, there results (on linearizing in the usual way) the equation

$$\frac{\partial h}{\partial t} + Vh = 0. \tag{12.41}$$

The operators V are thus simply those resulting from the linearized Vlasov equations in the corresponding phase spaces, and we can expect that the problem of solving (12.38) for g will be closely related to that of solving (12.41).

The two equations (12.37) and (12.38) completely determine the functions f and g subject to suitable initial conditions. We have therefore "closed" the BBGKY chain by means of our approximation. In a similar way we could work to the pth order in our parameter $(nh^3)^{-1}$, closing the chain after $p + 1$ equations, to yield a closed set of equations, though as yet nobody seems to have ventured beyond the stage $p = 1$ in this program.

12.5.2 Bogoliubov's hypothesis

Although we have closed the chain of equations formally, any attempt to solve (12.37) and (12.38) is liable to be extremely laborious; moreover, we have not completely realized our aim, which was to produce a differential equation, such as (12.2), involving f alone. Strictly, we could not expect a kinetic equation of that form, with $(\partial f/\partial t)_c$ calculable from the current value of f alone;* $(\partial f/\partial t)_c$ will depend on f at all earlier times, and on initial conditions for g, since these quantities enter into the solution of (12.38) for g, and so influence the right-hand side of (12.37).

This difficulty can be circumvented if it can be supposed that g, as determined by (12.38), tends to relax towards a steady value in a time much smaller than the time-scale of the variations of f. If this is so, we may solve (12.38) for the asymptotic value of g reached after a long time, assuming f to be time-independent. When this g is used in (12.37) we shall obtain a formula for $(\partial f/\partial t)_c$ which is a functional of f at the current time only. In this case the initial conditions for g also become irrelevant, since the asymptotic solution for g is independent of any initial conditions.

That this procedure is indeed valid is known as Bogoliubov's hypothesis, having been first clearly stated in the book referred to earlier. The same assumption is necessary in dealing with neutral gases; it was somewhat concealed in our outline of §12.4.2, but enters through the approximation that collisions occur practically instantaneously. As the duration of a collision is

* i.e. a "Markovian" equation.

shorter than the mean time between collisions, by a factor closely related to the small parameter involved in the expansion, nr_0^3, the hypothesis seems plausible. For Coulomb interactions Bogoliubov's hypothesis is rather more difficult to assess, and a detailed investigation of its validity does not seem to have been made. Roughly speaking, however, the time scale appropriate for g is the plasma period, for as already noted (12.38) is very similar to the equation for plasma oscillations. It therefore seems necessary that the time-scale for f should be substantially longer than the plasma period. So while the results which follow from Bogoliubov's hypothesis are valuable for handling slow phenomena like transport processes, they are unfortunately not reliable when applied, for instance, to the plasma oscillations themselves. We should stress that even at high frequency the closing of the chain of equations, leading to (12.37) and (12.38), is valid, at least for stable plasmas, and it is quite correct to regard the Vlasov equation as the appropriate first correction. It is only the applicability of Bogoliubov's hypothesis which is questionable.

This is a matter of deep significance because it is through the use of the asymptotic value, rather than the correct value, of g that irreversibility is introduced. It will be recalled that the Liouville equation is reversible, i.e. it is invariant under the transformation* $t \to -t, \mathbf{v} \to -\mathbf{v}$. The same is true of the Vlasov equation, and its improvement in the form (12.37) and (12.38). It is only this final step which treats time asymmetrically, so yielding a kinetic equation that is strictly irreversible.

12.5.3 Formal solution for the correlation function

Next we consider how to solve our equation (12.38)

$$\frac{\partial g}{\partial t} + V_1 g + V_2 g = S$$

for the correlation function g. As can be imagined this is in general quite intractable; however, in sufficiently simple cases it is possible to derive a compact form for the integral of g over $\mathbf{v}_2$-space, and this is all that is required for (12.37). This was first achieved in 1960 by R. Balescu and by A. Lenard, in separate investigations. We give here yet another method, due originally to T. H. Dupree (1961) and modified by later writers.

The first step is formally the same in all cases. We use the Laplace transform

$$\tilde{g}(p) = \int_0^\infty e^{-pt} g(t)\, dt \tag{12.42}$$

with the usual inversion formula

$$g(t) = \frac{1}{2\pi i} \int_C e^{pt} \tilde{g}(p)\, dp \tag{12.43}$$

* The sign of any external magnetic field must also be reversed.

where C is a contour parallel to the imaginary axis in the p-plane and to the right of all the singularities of $\tilde{g}(p)$. As V_1, V_2 and S are all time independent (by virtue of Bogoliubov's hypothesis) we have

$$\tilde{g}(p) = (p + V_1 + V_2)^{-1}[p^{-1}S + g(0)]. \tag{12.44}$$

(For the present we may handle the V operators as if they were numbers since they commute with each other and with the operations currently being carried out.) Now if the hypothesis is justified, $g(t)$ settles down to a steady value $g(\infty)$ as $t \to \infty$, so that any poles of $\tilde{g}(p)$ must have $\mathrm{Re}(p) < 0$ except possibly at $p = 0$. Deforming the contour to the left, (12.43) yields

$$g(\infty) = \lim_{p \to +0} p\tilde{g}(p), \tag{12.45}$$

the instruction $p \to +0$ arising from the need to continue $\tilde{g}(p)$ analytically from the region $\mathrm{Re}(p) > 0$ in cases where there is any ambiguity. Combining the last two equations

$$g(\infty) = \lim_{p \to +0} (p + V_1 + V_2)^{-1}S. \tag{12.46}$$

Remembering that e^{-at} has Laplace transform $(p + a)^{-1}$, we may use

$$\begin{aligned}(p + V_1 + V_2)^{-1} &= \int_0^\infty e^{-(p+V_1+V_2)t}\,dt \\ &= \frac{1}{(2\pi i)^2}\int_0^\infty e^{-pt}\int_{C_1}\frac{e^{p_1 t}\,dp_1}{p_1 + V_1}\int_{C_2}\frac{e^{p_2 t}\,dp_2}{p_2 + V_2}\,dt \\ &= \frac{1}{(2\pi i)^2}\int_{C_1}\int_{C_2}\frac{1}{p - p_1 - p_2}\frac{dp_1}{p_1 + V_1}\frac{dp_2}{p_2 + V_2}\end{aligned}$$

where $\mathrm{Re}(p) > \mathrm{Re}(p_1 + p_2)$ is required for the performance of the t-integral. So (12.46) becomes

$$g(\infty) = \frac{1}{(2\pi i)^2}\lim_{p \to +0}\int_{C_1}\int_{C_2}\frac{1}{(p - p_1 - p_2)(p_1 + V_1)(p_2 + V_2)}S\,dp_1\,dp_2, \tag{12.47}$$

where C_1 and C_2 are suitable Laplace inversion contours. From now on g will be understood to mean the asymptotic value as $t \to \infty$.

By this device we have reduced the problem of inverting the operator $p + V_1 + V_2$ in (X_1, X_2)-space to that of inverting operators of the type $p + V$ in a single phase space. This in turn is just the initial value problem for the linearized Vlasov equation (12.41), and can be solved, in principle, by the method of integrating along unperturbed trajectories as explained in §9.1.1.

12.5.4 Kinetic equation for a uniform plasma without magnetic field

Let us restrict attention to the case of a uniform plasma without magnetic field, and consider only a single species with the usual neutralizing background. It will be found necessary to assume that $f(X, t)$ is such that if treated by the Vlasov equation it is stable, that is, it satisfies the Penrose criterion (§10.2) for all directions of the wave vector $\mathbf{k}$. For the purposes of calculating g we may regard $f(X, t)$ as a function of $\mathbf{v}$ only. Writing $\boldsymbol{\xi} = \mathbf{x}_1 - \mathbf{x}_2$, (12.25) reads

$$\mathbf{a}_1^{(2)} = -\mathbf{a}_2^{(1)} = -\partial\phi/\partial\boldsymbol{\xi}, \tag{12.48}$$

where

$$\phi(\boldsymbol{\xi}) = e^2/4\pi\varepsilon_0 m\xi. \tag{12.49}$$

As the correlation function g will depend on $\mathbf{x}_1$ and $\mathbf{x}_2$ only through the combination $\boldsymbol{\xi}$, we introduce Fourier transforms with respect to this variable, for instance

$$g(\boldsymbol{\xi}, \mathbf{v}_1, \mathbf{v}_2) = \int \exp(-i\mathbf{k}\cdot\boldsymbol{\xi})\, g(\mathbf{k}, \mathbf{v}_1, \mathbf{v}_2)\, d^3k. \tag{12.50}$$

We shall need the identity (compare with 2.102)

$$\phi(\mathbf{k}) = e^2/(2\pi)^3\, \varepsilon_0 m k^2, \tag{12.51}$$

and the Fourier transform of the source function introduced in (12.39)

$$S(\mathbf{k}, \mathbf{v}_1, \mathbf{v}_2) = \phi(\mathbf{k}) i\mathbf{k}\cdot\left[\frac{\partial f(\mathbf{v}_2)}{\partial \mathbf{v}_2} f(\mathbf{v}_1) - \frac{\partial f(\mathbf{v}_1)}{\partial \mathbf{v}_1} f(\mathbf{v}_2)\right]. \tag{12.52}$$

The right-hand side of the kinetic equation (12.37) becomes

$$\left(\frac{\partial f(\mathbf{v}_1, t)}{\partial t}\right)_c = -\frac{\partial}{\partial \mathbf{v}_1}\cdot\mathbf{J}, \tag{12.53}$$

with

$$\mathbf{J} = \int \mathbf{a}_1^{(2)} g(X_1, X_2, t)\, dX_2 \tag{12.54}$$

and as already discussed $g(t)$ is to be replaced by $g(\infty)$ as fixed by the current value of $f(t)$. The spatial integral of (12.54) (transformed to one over $\boldsymbol{\xi}$) leads to the usual convolution when expressed in terms of the Fourier transforms, so that (compare with 2.73)

$$\mathbf{J} = (2\pi)^3 \iint -i\mathbf{k}\phi(\mathbf{k}) g(\mathbf{k}, \mathbf{v}_1, \mathbf{v}_2)\, d^3k\, d^3v_2. \tag{12.55}$$

Since we know that $\mathbf{J}$ is actually real, we observe that only the imaginary part of g need be included.

Next we must calculate the spatial Fourier components of g as given by (12.47). Referring to equations (8.31) and (8.32), restoring the three-dimensional formulation and adjusting the notation, we find for the inverse operator in $\mathbf{v}_1$-space, applied to a typical function $h(\mathbf{v}_1)$,

$$(p_1 + V_1)^{-1}h(\mathbf{v}_1) = \frac{1}{p_1 - i\mathbf{k}\cdot\mathbf{v}_1}\left\{h(\mathbf{v}_1) - \frac{(2\pi)^3\phi(\mathbf{k})}{\bar{\varepsilon}(\mathbf{k}, p)} i\mathbf{k}\cdot\frac{\partial f(\mathbf{v}_1)}{\partial \mathbf{v}_1}\int \frac{h(\mathbf{v}_1')\,d^3v_1'}{p_1 - i\mathbf{k}\cdot\mathbf{v}_1'}\right\}. \tag{12.56}$$

Here $\bar{\varepsilon}(\mathbf{k}, p)$ is the dielectric constant, in Laplace transform language, for longitudinal fields (see 8.17)

$$\bar{\varepsilon}(\mathbf{k}, p) = 1 + (2\pi)^3\phi(\mathbf{k})i\mathbf{k}\cdot\int \frac{\partial f/\partial \mathbf{v}}{p - i\mathbf{k}\cdot\mathbf{v}}\,d^3v \tag{12.57}$$

and is related to the more familiar $\varepsilon(\mathbf{k}, \omega)$ of Fourier transform language by the usual change of variable $p = i\omega$. In all these formulas the integrals over velocity space are of course carried out with the Landau prescription. The inverse of $p_2 + V_2$ is given similarly but with the sign of $\mathbf{k}$ reversed; however, we may take advantage of the fact that we shall not need $g(\mathbf{k}, \mathbf{v}_1, \mathbf{v}_2)$ explicitly, but only the quantity $\int g\,d^3v_2$. Comparing again with (8.30)–(8.32) we have

$$\int (p_2 + V_2)^{-1}h(\mathbf{v}_2)\,d^3v_2 = \frac{1}{\bar{\varepsilon}(-\mathbf{k}, p_2)}\int \frac{h(\mathbf{v}_2)\,d^3v_2}{p_2 + i\mathbf{k}\cdot\mathbf{v}_2} \tag{12.58}$$

(corresponding simply to the calculation of the number density when solving the Vlasov equation). We may now write down the Fourier transform of (12.47) and integrate over $\mathbf{v}_2$-space:

$$\begin{aligned}\int g(\mathbf{k}, \mathbf{v}_1, \mathbf{v}_2)\,d^3\mathbf{v}_2 &= \frac{1}{(2\pi i)^2}\lim_{p\to+0}\int_{\mathbf{v}_2}\int_{C_1}\int_{C_2}\frac{1}{p - p_1 - p_2}\,\frac{1}{\bar{\varepsilon}(-\mathbf{k}, p_2)}\,\frac{1}{p_2 + i\mathbf{k}\cdot\mathbf{v}_2}\\ &\quad\times (p_1 + V_1)^{-1}S(\mathbf{k}, \mathbf{v}_1, \mathbf{v}_2)\,d^3v_2\,dp_1\,dp_2.\end{aligned} \tag{12.59}$$

The kinetic equation is obtained by substituting this result into (12.55), noting (12.52) and (12.56). This gives an 8-fold integral (over p_1, p_2, $\mathbf{k}$ and $\mathbf{v}_2$, not counting the integrations involved in $(p_1 + V_1)^{-1}$ and $\bar{\varepsilon}$). Fortunately, the integrations over p_1 and p_2, and the limit $p \to +0$, may all be carried out without knowledge of f, and $\mathbf{J}$ thereby expressed in a much more compact form, namely equation (12.68). (The reader not interested in mere manipulation is recommended to continue from that point!)

The integration over p_2 is taken first. The contour C_2 runs parallel to the imaginary axis and is such that all the singularities arising from the operator

$(p_2 + V_2)^{-1}$ lie to the left. The relevant singularities are $p_2 = -i\mathbf{k}\cdot\mathbf{v}_2$ together with the zeros of $\bar{\varepsilon}(-\mathbf{k}, p_2)$. The former is on, and the latter (in the stable case) are to the left of, the imaginary axis. So C_2 may be taken to the right of the imaginary axis but very close to it, as in Fig. 12.1, and may be closed by a large semicircle in the right-hand half-plane. The only singularity captured on evaluating (12.59) is at $p_2 = p - p_1$; this is necessarily within the contour as $\text{Re}(p) > \text{Re}(p_1 + p_2)$. (In the unstable case it is not possible to let C_2 approach the imaginary axis in this way without capturing further poles and this creates difficulties when considering the limit $p \to +0$.) The contribution from the large semicircle clearly vanishes, leaving

$$\int g(\mathbf{k}, \mathbf{v}_1, \mathbf{v}_2)\, d^3v_2$$
$$= \frac{1}{2\pi i} \lim_{p\to+0} \int_{\mathbf{v}_2} \int_{C_1} \frac{1}{\bar{\varepsilon}(-\mathbf{k}, p - p_1)} \frac{1}{p - p_1 + i\mathbf{k}\cdot\mathbf{v}_2} \times (p_1 + V_1)^{-1} S(\mathbf{k}, \mathbf{v}_1, \mathbf{v}_2)\, d^3v_2\, dp_1. \tag{12.60}$$

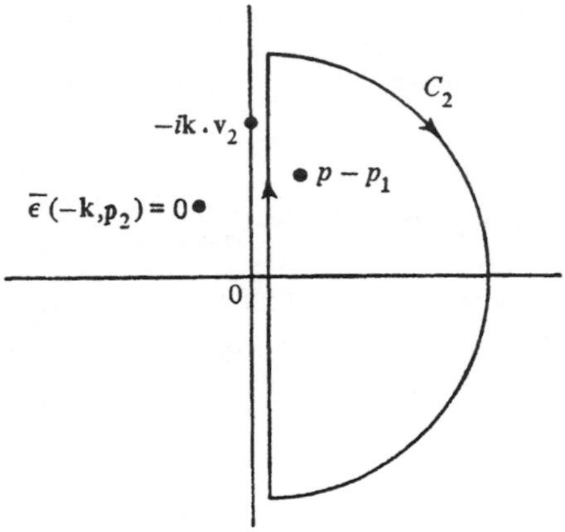

Fig. 12.1. Contour in the p_2-plane.

Before carrying out the integration over p_1 it is necessary to consider in explicit form the effect of the operations

$$(p_1 + V_1)^{-1} \quad \text{and} \quad \int (p - p_1 + i\mathbf{k}\cdot\mathbf{v}_2)^{-1}(\ldots)\, d^3v_2$$

applied to the functions f and $i\mathbf{k}\cdot\partial f/\partial\mathbf{v}$, as required when (12.52) is inserted as the expression for S. For the second of these operations the integrals over $\mathbf{v}_2$ arise in the first place with $\text{Re}(p - p_1) > 0$, so coincide with the Landau definition, and a similar remark applies to the integration arising in (12.56);

consequently integrals occur which may be identified with that of (12.57) and so expressed in terms of ε. These manipulations lead in a few steps to

$$\int g(\mathbf{k}, \mathbf{v}_1, \mathbf{v}_2)\, d^3v_2 = \frac{1}{2\pi i} \lim_{p\to+0} \int_{C_1} \frac{1}{p_1 - i\mathbf{k}\cdot\mathbf{v}_1} \Bigg\{ \Bigg[\underset{(a)}{\frac{1}{\bar\varepsilon(-\mathbf{k}, p - p_1)}} - \underset{(b)}{1} \Bigg]$$

$$\times \Bigg[\underset{(c)}{\frac{1}{(2\pi)^3} f(\mathbf{v}_1)} - \underset{(d)}{\frac{\phi(\mathbf{k}) i\mathbf{k}\cdot \dfrac{\partial f(\mathbf{v}_1)}{\partial \mathbf{v}_1}}{\bar\varepsilon(\mathbf{k}, p_1)} \int \frac{f(\mathbf{v}_2)\, d^3v_2}{p_1 - i\mathbf{k}\cdot\mathbf{v}_2}} \Bigg]$$

$$- \underset{(e)}{\frac{\phi(\mathbf{k}) i\mathbf{k}\cdot \dfrac{\partial f(\mathbf{v}_1)}{\partial \mathbf{v}_1}}{\bar\varepsilon(\mathbf{k}, p_1)\bar\varepsilon(-\mathbf{k}, p - p_1)} \int \frac{f(\mathbf{v}_2)\, d^3v_2}{p - p_1 + i\mathbf{k}\cdot\mathbf{v}_2}} \Bigg\} dp_1. \tag{12.61}$$

Here we have labeled the terms as (a), (b), . . . for easy reference; in (d) a change in the name of the variable of integration has occurred. The contour C_1 is shown in Fig. 12.2. Like C_2, it is a Laplace inversion contour with the singularities arising from $(p_1 + V_1)^{-1}$ to the left; these are $p_1 = i\mathbf{k}\cdot\mathbf{v}_1$ and the zeros of $\bar\varepsilon(\mathbf{k}, p_1)$, the latter being in the left half-plane for a stable plasma. The contour must also go to the right of $p_1 = i\mathbf{k}\cdot\mathbf{v}_2$ so that the term (d) is subject to the Landau prescription. The other possible singularities are $p_1 = i\mathbf{k}\cdot\mathbf{v}_2 + p$ and the zeros of $\bar\varepsilon(-\mathbf{k}, p - p_1)$; the former lies to the right of C_1, and so do the latter by the stability and the fact that $\mathrm{Re}(p) > 0$. For the limit $p \to +0$ we can let C_1 approach the imaginary axis, indented as shown.

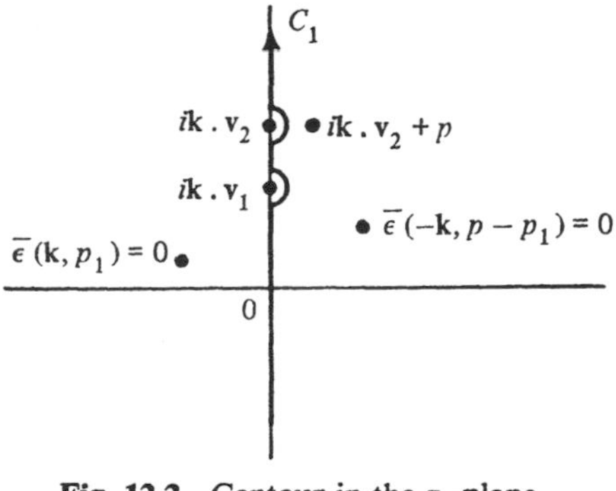

Fig. 12.2. Contour in the p_1-plane.

However, not every term of (12.61) has each of the singularities shown in the figure, so that in carrying out the integration over p_1 we may deal differently with the several terms. The guiding principle is to avoid evaluating a residue at a singularity due to the vanishing of $\bar{\varepsilon}$. For the term (c) we may complete C_1 by a large semicircle in the left half-plane provided we keep together the parts due to (a) and (b). The only pole enclosed is at $p_1 = i\mathbf{k}\cdot\mathbf{v}_1$. The resulting contribution to (12.61) is

$$\lim_{p\to+0}(2\pi)^{-3}\left[\frac{1}{\bar{\varepsilon}(-\mathbf{k}, p - i\mathbf{k}\cdot\mathbf{v}_1)} - 1\right]f(\mathbf{v}_1) \tag{12.62}$$

and we simply set $p = 0$ as $\bar{\varepsilon}$ is analytic and does not vanish in that limit. For the contribution due to the product of (b) and (d) we may complete C_1 by a semicircle in the right half-plane; no poles are enclosed and the result is zero. There remain (e) and the product of (a) and (d). Again, neither the indentations of C_1 nor the need to let $p \to +0$ are of any consequence for the factors involving $\bar{\varepsilon}$, as $\bar{\varepsilon}$ is analytic and does not vanish for p_1 on the imaginary axis. Noting that $\bar{\varepsilon}(\mathbf{k}, p_1)$ and $\bar{\varepsilon}(-\mathbf{k}, -p_1)$ are complex conjugates, we may write these remaining terms as

$$-\frac{1}{2\pi i}\lim_{p\to+0}\phi(\mathbf{k})i\mathbf{k}\cdot\frac{\partial f(\mathbf{v}_1)}{\partial \mathbf{v}_1}\int_{C_1}\int_{\mathbf{v}_2} f(\mathbf{v}_2)\frac{1}{(p_1 - i\mathbf{k}\cdot\mathbf{v}_1)\,|\,\bar{\varepsilon}(\mathbf{k}, p_1)\,|^2} \times\left(\frac{1}{p_1 - i\mathbf{k}\cdot\mathbf{v}_2} + \frac{1}{p - p_1 + i\mathbf{k}\cdot\mathbf{v}_2}\right)d^3v_2\,dp_1. \tag{12.63}$$

The final step is to take advantage of the fact that in evaluating $\mathbf{J}$ as given by (12.55) we need only the imaginary parts of these last two expressions. This considerably simplifies (12.63). Changing to Fourier transform language by writing $p_1 = i\omega$, C_1 becomes the indented real axis. For a factor such as $(\omega - a)^{-1}$ we write (compare with 8.15a)

$$\frac{1}{\omega - a \pm i\eta} = P\frac{1}{\omega - a} \mp i\pi\delta(\omega - a). \tag{12.64}$$

Here P denotes the principal part, and the indentation is represented by the small positive real quantity η. At the two poles converging on $\omega = \mathbf{k}\cdot\mathbf{v}_2$ the principal parts cancel and the δ-functions reinforce. The principal part at the other indentation disappears when the imaginary part is taken. The integration over ω is trivial owing to the δ-functions. The imaginary part of (12.63) reduces to

$$-i\pi\phi(\mathbf{k})\mathbf{k}\cdot\frac{\partial f(\mathbf{v}_1)}{\partial \mathbf{v}_1}\int\frac{f(\mathbf{v}_2)}{|\,\varepsilon(\mathbf{k}, \mathbf{k}\cdot\mathbf{v}_1)\,|^2}\delta[\mathbf{k}\cdot(\mathbf{v}_1 - \mathbf{v}_2)]\,d^3v_2. \tag{12.65}$$

The imaginary part of (12.62) is already in its most concise form, but it is usual to rewrite it as

$$\frac{f(\mathbf{v}_1)\,\mathrm{Im}[\varepsilon(\mathbf{k}, \mathbf{k}\cdot\mathbf{v}_1)]}{(2\pi)^3\,|\,\varepsilon(\mathbf{k}, \mathbf{k}\cdot\mathbf{v})\,|^2} \tag{12.66}$$

and to note that the imaginary part of ε for a real frequency is just the contribution from the indentation of a contour in (12.57). Using (12.64) again, (12.66) becomes

$$i\pi f(\mathbf{v}_1)\phi(\mathbf{k})\mathbf{k}\cdot\int \frac{\partial f(\mathbf{v}_2)/\partial\mathbf{v}_2}{|\,\varepsilon(\mathbf{k},\mathbf{k}\cdot\mathbf{v}_1)\,|^2}\,\delta[\mathbf{k}\cdot(\mathbf{v}_1-\mathbf{v}_2)]\,d^3v_2. \tag{12.67}$$

The sum of (12.65) and (12.67) give the imaginary part of $\int g(\mathbf{k}, \mathbf{v}_1, \mathbf{v}_2)\,d^3v_2$, and substituting into (12.55) we have

$$\mathbf{J} = 8\pi^4 \int_{\mathbf{k}}\int_{\mathbf{v}_2} \frac{[\phi(\mathbf{k})]^2\mathbf{k}\mathbf{k}\cdot\left[\dfrac{\partial f(\mathbf{v}_2)}{\partial\mathbf{v}_2}f(\mathbf{v}_1) - \dfrac{\partial f(\mathbf{v}_1)}{\partial\mathbf{v}_1}f(\mathbf{v}_2)\right]}{|\,\varepsilon(\mathbf{k},\mathbf{k}\cdot\mathbf{v}_1)\,|^2}\,\delta[\mathbf{k}\cdot(\mathbf{v}_1-\mathbf{v}_2)]\,d^3k\,d^3v_2. \tag{12.68}$$

This, when combined with (12.53), is the result obtained by R. Balescu (1960) and by A. Lenard (1960). It is often referred to as the Lenard–Balescu kinetic equation. The relevant expression, $-\partial\mathbf{J}/\partial\mathbf{v}_1$, is clearly of a form similar to what we called the Fokker–Planck equation (9.122), as $\mathbf{J}$ is linear in $f(\mathbf{v}_1)$ and its first derivative, though the coefficients are complicated functions of $\mathbf{v}_1$ and functionals of f itself. Further investigation of the properties of this kinetic equation is taken up in §12.7; meanwhile we describe some generalizations of it.

12.6 SOME GENERALIZATIONS

12.6.1 Mixture of species

It is a relatively simple matter to extend all our calculations in order to handle a gas composed of a mixture of species (as a plasma must be). The variables X_1, X_2, . . ., appearing in the Liouville density ρ, (12.3), are divided into groups, one group for each species, and the symmetry with respect to interchange of particles applies only within each group. We construct reduced distribution functions, like (12.14), and a BBGKY chain of equations governing them, just as before. For example, we introduce one-particle distribution functions $f_1^{(s)}(\mathbf{x}, \mathbf{v}, t)$ analogous to (12.9a); there is one such function for each species, hence the label s. Again, there is a two-particle distribution for every pair of species (whether like or unlike); these may be denoted

$$f_2^{(rs)}(X_1, X_2, t) \tag{12.69}$$

and give the joint probability of finding a particle of species r at the phase point X_1 and one of species s at X_2. In general such a distribution function is symmetric between X_1 and X_2 only if r and s are the same. Similarly there are distribution functions for three or more particles, but a rather cumbersome notation is necessary to write them down explicitly.

The Liouville equation (12.5), takes the same form as before, and one must again reduce it by integration over nearly all its variables, to obtain a chain of equations for the distribution functions of one, two, . . . particles. To close this chain, the parameter (12.33) is assumed small (it generally happens, in practice, that this parameter is of a similar order for each species). To the lowest order one obtains the Vlasov equations, the species interacting through the common self-consistent field in the usual way.

In the next order, an approximation analogous to (12.34) is required. All we need to state here is the analog of (12.35)

$$f_2^{(rs)} = f_1^{(r)}(X_1, t) f_1^{(s)}(X_2, t) + g^{(rs)}(X_1, X_2, t), \tag{12.70}$$

so introducing a correlation function $g^{(rs)}$ for every pair of species (r, s); $g^{(rs)}$ is to be small in the same sense as previously, and has the same symmetry properties as $f_2^{(rs)}$. There then results a kinetic equation, analogous to (12.37), for each one-particle distribution function. The left-hand side, being just the Vlasov expression, differs from that of (12.37) only in that the integral for the self-consistent field is now summed over species. The right-hand side is also summed over species; the equation for $f^{(r)}$ involves a sum over s in which the terms depend on $g^{(rs)}$ and the appropriate $\mathbf{a}_1^{(2)}$'s. There also results an equation for $g^{(rs)}$, analogous to (12.38), for every choice (r, s); it is formally the same provided V_1 and V_2 are written $V_1^{(r)}$ and $V_2^{(s)}$, denoting the linearized Vlasov operators for the respective species, and (12.39) is modified by writing f as $f^{(r)}$ or $f^{(s)}$ according as its argument is X_1 or X_2, and $\mathbf{a}_1^{(2)}$ is correctly interpreted. It should be noted that the Vlasov operators themselves involve all the species, owing to the self-consistent field.

The discussion of Bogoliubov's hypothesis takes a similar course as before, and the solution of the equation for g may be effected by exactly the same technique as in §12.5.3. The algebraic amendments to the work of §12.5.4 are merely tedious. The final result is the replacement of equations (12.53) and (12.68) by the following:

$$\left(\frac{\partial f^{(r)}(\mathbf{v}_1, t)}{\partial t}\right)_c = -\sum_s \frac{\partial}{\partial \mathbf{v}_1} \cdot \mathbf{J}^{(rs)}, \tag{12.71}$$

where

$$\mathbf{J}^{(rs)} = (2\pi)^3 \pi \int_{\mathbf{k}} \int_{\mathbf{v}_2} \frac{1}{m_r} \left[\frac{e_r e_s}{(2\pi)^3 \varepsilon_0 k^2}\right]^2 \mathbf{k}\mathbf{k} \cdot \left[\frac{1}{m_s} \frac{\partial f^{(s)}(\mathbf{v}_2)}{\partial \mathbf{v}_2} f^{(r)}(\mathbf{v}_1) - \frac{1}{m_r} \frac{\partial f^{(r)}(\mathbf{v}_1)}{\partial \mathbf{v}_1} f^{(s)}(\mathbf{v}_2)\right] \frac{\delta[\mathbf{k} \cdot (\mathbf{v}_1 - \mathbf{v}_2)]}{|\,\varepsilon(\mathbf{k}, \mathbf{k} \cdot \mathbf{v}_1)\,|^2} \, d^3k \, d^3v_2. \tag{12.72}$$

Here $\varepsilon(\mathbf{k}, \omega)$ is the dielectric constant for the plasma as a whole, so it is given by (12.57), but with a sum over species. The rth collision term thus appears as a sum of contributions due to all the species s, including r itself.

12.6.2 External magnetic field

Suppose next that there is a uniform externally applied magnetic field **B**. As in Chapter 9 we may suppose this directed along the third coordinate axis. For a single species, our treatment up to §12.5.4 was sufficiently general to include the effect of **B**. Modification first becomes necessary at equation (12.56), since the operator $(p + V)^{-1}$ now signifies the solution of the linearized Vlasov equation in the presence of a magnetic field. The technique for this has already been developed in Chapter 9, though we did not elaborate there the explicit form for $(p + V)^{-1}$ itself since we were concerned only with the charge and current densities. The required expression may, however, be readily constructed in the electrostatic approximation, and a development parallel to that of §12.5.4 follows. Retaining also the possibility of a mixture of species, the kinetic equation is

$$\left(\frac{\partial f}{\partial t}\right)_c = -(2\pi)^3\pi \sum_s \int_{\mathbf{k}} \int_{\mathbf{v}_2} \frac{1}{m_r}\left[\frac{e_r e_s}{(2\pi)^3 \varepsilon_0 k^2}\right]^2 \sum_{n=-\infty}^{\infty} \sum_{m=-\infty}^{\infty} D_n^{(r)}(\mathbf{v}_1)\, J_n^2\left(\frac{k_\perp v_{1\perp}}{\Omega_r}\right)$$
$$\times J_m^2\left(\frac{k_\perp v_{2\perp}}{\Omega_s}\right) \frac{(1/m_s) f^{(r)}(\mathbf{v}_1) D_m^{(s)}(\mathbf{v}_2) f^{(s)}(\mathbf{v}_2) - (1/m_r) f^{(s)}(\mathbf{v}_2)\, D_n^{(r)}(\mathbf{v}_1) f^{(r)}(\mathbf{v}_1)}{|\,\varepsilon(\mathbf{k}, k_z v_{1z} + n\Omega_r)\,|^2}$$
$$\times\, \delta[k_z v_{1z} + n\Omega_r - k_z v_{2z} - m\Omega_s]\, d^3k\, d^3v_2. \tag{12.73}$$

Here, as before, r and s are labels for the species, Ω_r and Ω_s being the corresponding gyro-frequencies; we use cylindrical polar coordinates for **k** and **v**. The operator $D_n^{(r)}(\mathbf{v})$ is defined for $n = \ldots, -1, 0, 1, \ldots$ by

$$D_n^{(r)}(\mathbf{v}) \equiv k_z \frac{\partial}{\partial v_z} + \frac{n\Omega_r}{v_\perp}\frac{\partial}{\partial v_\perp}, \tag{12.74}$$

and in effect replaces $\mathbf{k}\cdot\partial/\partial\mathbf{v}$ of the field-free theory. The expansions in Bessel functions enter in a way parallel to §9.4.1. The new dielectric constant $\varepsilon(\mathbf{k}, \omega)$ is given by (9.103) and is

$$\varepsilon(\mathbf{k}, \omega) = 1 + \frac{1}{k^2\varepsilon_0}\sum_r \frac{e_r^2}{m_r}\int_{\mathbf{v}} \sum_{n=-\infty}^{\infty} \frac{J_n^2\left(\dfrac{k_\perp v_\perp}{\Omega_r}\right) D_n^{(r)} f^{(r)}(\mathbf{v})}{\omega - k_z v_z - n\Omega_r}\, d^3v. \tag{12.75}$$

It should be noted that in order to preserve the formal comparison with the field-free case we have expressed the velocity-space integrals of (12.73) and (12.75) as triple integrals, though in fact as the integrands depend only upon $v_\perp$ and v_z in cylindrical coordinates the azimuthal integration may be trivially carried out, introducing the factor $2\pi v_\perp$. A similar remark applies to the integral over **k**.

The kinetic equation (12.73) was obtained by N. Rostoker (1960), but does not appear to have been used in practice, owing no doubt to its complexity. For values of $k_\perp$ and $v_\perp$ satisfying $k_\perp v_\perp \gg \Omega$ (i.e. transverse "wavelength" much less than the Larmor radius) we would expect the magnetic field to have little effect, so the integrand of (12.73) should revert to the field-free form; a mathematical demonstration of this is somewhat laborious. We shall see later (§12.7.4) that the main contributions to $(\partial f/\partial t)_c$ tend to arise from values of **k** greater than the inverse Debye length—in other words the interaction between particles separated by more than the Debye length is already well represented by the Vlasov equation. Consequently if (as is often the case) the Debye length is less than the Larmor radii for all thermal particles, the field-free version of $(\partial f/\partial t)_c$ is adequate. This is not to say that the magnetic field is in all respects negligible, for it may be very important in the Vlasov terms, that is in the left-hand side of (12.2); such a case has already been described in §9.5.3, and the same situation is assumed in §11.5.

12.6.3 Radiation

A further extension of these calculations is required if the "electrostatic approximation" is to be removed. The work is so lengthy that only the merest sketch will be attempted here. The additional complication arises essentially from the fact that the positions and momenta of the particles no longer form a complete set of variables; they need to be augmented in order to specify the transverse part of the electromagnetic field. One therefore divides the electromagnetic field into (a) the external part, if any, (b) the coulomb field of the particles, as if propagated instantaneously, and (c) the transverse part. The division into (b) and (c) corresponds to separation of the field into parts given by scalar and vector potentials, in the coulomb gauge. The vector potential may, at each instant, be expanded in terms of the eigenfunctions appropriate to a vacuum field in the volume V (which for convenience may be taken to be a rectangular box). The eigenfunctions may be regarded as describing oscillators, rather as in quantum field theory, and the amplitudes of the oscillators are the required additional coordinates. The Hamiltonian may then be written down. The Liouville density is now a function of the coordinates and momenta of the N particles, together with infinitely many coordinates and momenta corresponding to the oscillators. In quantum language, one is merely introducing the photons as an additional species. The hierarchy of equations is obtained by integrating out all but one particle, all but one oscillator, all but two particles, all but one particle and one oscillator, and so on. One then attempts an expansion in the discreteness parameter (12.33). In the lowest order the result can be shown to be equivalent to the usual Vlasov equation combined now with the full set of Maxwell's equations. To the next order one obtains a kinetic equation for each one-particle distribution function, together with equations governing the correlation between pairs of particles, and between single particles and single oscillators (no correlation arises between pairs of oscillators as there is no direct coupling between them). The correlations are assumed to adopt rapidly their asymptotic values, as in our earlier work. This procedure leads eventually to kinetic equations for both the plasma and the radiation. If one is actually interested only in the particles themselves, the corrections to our earlier work are negligible for non-relativistic plasmas. Physically this is due to the fact that strong interaction between transverse electromagnetic fields and plasmas tends to occur only at length scales greater or equal to c/ω_p (the "skin depth"), whereas correlations tend to be important only on scales up to the Debye length. It also emerges that, if there is no external magnetic field, the discreteness of the particles has little new effect on the oscillators. In other words such a plasma would not seem to radiate or absorb. (This is actually incorrect because radiation by the Bremsstrahlung process would occur, but the work outlined here omits that by excluding large-angle collisions.) An even more ambitious calculation, including an external magnetic field and treating the electron dynamics relativistically, is necessary if synchrotron radiation is to be featured in the theory.

Further details of this work may be found by consulting the papers of A. Simon (1961), N. Rostoker and A. Simon (1962), and of T. H. Dupree (1963, 1964).

12.6.4 Other extensions of the theory

There are several further respects in which the basic calculation of §12.5 can be generalized. As we have just mentioned, the large-angle collisions are treated inaccurately in the theory as it stands; this difficulty will be discussed in §12.7.1. Another possible source of trouble was noticed in §12.5.4, where we assumed that the one-particle distribution function, f_1, was stable in the Vlasov approximation. In cases where f_1 is unstable, that is when $\varepsilon(\mathbf{k}, \omega)$ has zeros in the lower half of the ω-plane for some real $\mathbf{k}$, the procedure we used for finding the asymptotic value of the correlation function $g(\infty)$ (see (12.46)) fails; indeed, there is no reason to believe that g will remain small and reach any asymptotic value, so that our whole approximation scheme, and Bogoliubov's hypothesis, must be abandoned. Physically, this is not really a disastrous shortcoming of the theory, since it is clearly difficult to set up a plasma in an initial state which admits of rapidly growing instabilities. A possibility which ought to be considered, however, is that f_1 corresponds to a marginal state, so that there are waves on the verge of instability, or perhaps even possessing very small growth rates. This could happen, for instance, when an electric field is applied to a plasma and a relative drift of ions and electrons builds up until the threshold for two-stream instability is reached. In these marginal cases the theory can be rescued, but takes on a different and more complicated form. The correlations (which may also be regarded as statistical fluctuations from the uniform state), though not growing indefinitely, acquire a much larger amplitude than in the stable case, and non-linear coupling between them becomes important. A lengthy account of this has been given by E. Frieman and P. H. Rutherford (1964). More recently, it has been suggested that one should give up the idea of searching for an equation governing f_1 alone, but work with f_1 and a fluctuating field spectrum which appears to be closely related to f_2 integrated over both velocity spaces. See Rogister and Oberman (1968, 1969).

Attempts have also been made to treat the case where f_1 is not spatially uniform. Our work up to the end of §12.5.3 applies equally well in those circumstances, but in continuing from there one needs to solve the linearized Vlasov equations for the non-uniform plasma in order to construct the operator $(p + V)^{-1}$. This in turn calls for the construction of, and integration along, the unperturbed orbits in the corresponding non-uniform field, a laborious task. To avoid this, it is commonly assumed that if the inhomogeneities of f_1 occur on a scale much larger than h, we may use a kinetic equation which incorporates the complete Vlasov terms on the left (as in (12.2)), and adopt local values for $(\partial f/\partial t)_c$ calculated as if f_1 were uniform. This is anticipated on the grounds that the correlation function g is extremely small for separation greater than h, so that $(\partial f/\partial t)_c$ at any point can depend only upon f_1 within the Debye sphere surrounding that point. This may, however, overlook subtle effects associated with "universal" instabilities (see §10.6).

12.7 DISCUSSION OF THE KINETIC EQUATION

12.7.1 Divergence of integrals

Let us first return to the simple case of a uniform electronic plasma with fixed positive ions and no magnetic field, so that the kinetic equation is given by (12.53) and (12.68). We must make a rather serious admission: the integral over $\mathbf{k}$ appearing in the latter equation is in fact divergent at infinity, so that, as it stands, our equation is meaningless! Specifically, we recall that $\phi \propto k^{-2}$ and note that, according to (12.57), $\varepsilon \to 1$ as $\mathbf{k} \to \infty$; allowing for a factor k^{-1} arising from the δ-function if the integral over $\mathbf{v}_2$ is taken first, then using polar coordinates in $\mathbf{k}$-space, the integrand is proportional to k^{-1}, and we have a logarithmic divergence. However, it is easy to check that the integral is convergent at $\mathbf{k} = 0$, contributions from k less than about h^{-1} being negligible. Thus the interactions on the large scale (small $\mathbf{k}$) give no trouble, but our treatment of the small scale (large $\mathbf{k}$) must be erroneous. Physically, this behavior is quite the opposite to that of the Boltzmann equation (12.24), which is satisfactory for close collisions (small b), but as noted earlier introduces a spurious divergence as $b \to \infty$.

The explanation of our present difficulty is that the approximation of regarding g as a small correction, as in (12.34)–(12.36), is not a uniformly valid one. Rather, it is a good approximation for most of the space (X_1, X_2) but not in certain regions. These are the regions for which $\xi = |\mathbf{x}_1 - \mathbf{x}_2|$ is very small, so they occupy a very small measure in the space as a whole, and of course correspond to large **k** in the Fourier transform. By "small" here we mean that the particles at $\mathbf{x}_1$, $\mathbf{x}_2$ interact strongly, so that despite the smallness of the parameter (12.33) the mutual potential energy may not be regarded as a small perturbation in solving the Liouville equation. The appropriate value of ξ may be deduced from equation (12.32) by asking when the term with coefficient $(nh^3)^{-1}$ can be of order unity instead of being smaller than the other terms. The answer depends on the speed of the particle concerned, but for a typical thermal particle $\partial \bar{f}_s/\partial \bar{\mathbf{v}}_j$ may be taken as unity, so we need $4\pi\bar{\xi}^2 nh^3 \lessdot 1$, where $\bar{\xi}$ is ξ on the normalized scale. In other words $\xi \lessdot \xi_0$ where

$$\xi_0^2 h = (4\pi n)^{-1}. \tag{12.76}$$

This distance is considerably smaller than $n^{-1/3}$, as $nh^3 \gg 1$, though actually it is not so small as the Landau distance, l, introduced in §1.2 and given by $lh^2 = (4\pi n)^{-1}$. The integrand of (12.68) is therefore discredited for values of k exceeding about ξ_0^{-1}, and the divergence is fictitious.

The non-uniformity in our approximation scheme is somewhat similar to that occurring in the neglect of viscosity for flow at high Reynolds number, where, however small the viscosity, there exists a thin boundary layer in which it cannot be neglected.

This discussion highlights the fundamental difference between gases whose molecules interact through short range forces, and plasmas where we have the inverse square law. For the former (cf. §12.4.2) the short range ensures that g is effectively zero except in the very small regions just mentioned, and that the collisions are practically instantaneous, so that g, although not small, can be handled quite easily in those regions. For the plasma, the smallness of the parameter (12.33) ensures that g is small throughout most of the region (the Debye sphere) in which it is needed, so that it can be found from a linearized calculation. However, for very close interactions the plasma does share some of the features of the neutral gas. Indeed for a pair of points $\mathbf{x}_1$ and $\mathbf{x}_2$ in a plasma three regions may be discerned. They are (i) $0 < \xi < \xi_0$, where we have binary collisions (ii) $\xi_0 < \xi < h$, in which g, although small, plays an important role, as do electrostatic effects generally (as for example in sheaths and the Debye–Hückel theory), and (iii) $h < \xi$, in which the interaction between the particles is adequately represented by the Vlasov approximation as g is negligible; at low frequency electric fields at this scale are very hard to set up owing to the tendency to neutrality (i.e. screening), but collective effects, involving space charge, can occur at the plasma frequency. (The addition of a strong magnetic field complicates this picture, as discussed in §12.6.2.)

This explanation of the divergence also suggests how it should be removed. We may expect that the kinetic equation should contain separate terms for the regions (i) and (ii), owing to their very different physical nature. For (i) we would expect a term of the Boltzmann type (12.24), with an upper limit of order ξ_0 for the integration over b, so avoiding the divergence of the Rutherford cross-section, while for (ii) we would expect a term given by (12.53), but with the integration with respect to $\mathbf{k}$ ranging only over $|\mathbf{k}| < k_0$ where $k_0 \sim \xi_0^{-1}$, again avoiding divergence. Such a heuristic procedure is contrary to the spirit of this chapter, where we aim to deduce all results from first principles with careful investigation of any approximations used; however, we will be content with only an outline of a more formal derivation. Returning for a moment to (12.23), we recall that our aim is to calculate f_2, and it is open to us to write

$$f_2 = f_2^{(c)} + f_2^{(d)} \quad \text{and} \quad \phi = \phi^{(c)} + \phi^{(d)}, \tag{12.77}$$

where "c" and "d" refer to close and distant, and ϕ is the potential for **a**. Here $f_2^{(c)}$ is to be small for $\xi > \xi_0$ and $f_2^{(d)}$ small for $\xi < \xi_0$, while $\phi^{(c)}$ and $\phi^{(d}$ are to be good representations of ϕ in the relevant regions. For example, but not necessarily, we may take

$$\begin{aligned} \phi^{(c)} &= \frac{e^2}{4\pi\varepsilon_0 m} \frac{\exp(-\xi/\lambda)}{\xi} \\ \phi^{(d)} &= \frac{e^2}{4\pi\varepsilon_0 m} \frac{1 - \exp(-\xi/\lambda)}{\xi}, \end{aligned} \tag{12.78}$$

which agrees with (12.49) for any choice of λ, and the above discussion would recommend $\lambda = \xi_0$. One then attempts to relate the close and distant parts separately. For $f_2^{(c)}$ we work in the way sketched in §12.4.2, so obtaining eventually a collision integral of the Boltzmann type, which will be nearly the same as the ordinary result for the inverse square law, with a cut off for impact parameter $b \sim \lambda$. For $f_2^{(d)}$ we work as in §12.5, writing f_2 in the form (12.35) and so on. This leads to the Lenard–Balescu equation (12.68) just as before, but with

$$\phi(\mathbf{k}) = \frac{e^2}{(2\pi)^3\varepsilon_0 m} \frac{1}{k^2(1 + \lambda^2 k^2)}, \tag{12.79}$$

with consequent change in $\varepsilon(\mathbf{k}, \omega)$ also. The divergence as $k \to \infty$ is thereby removed without changing the integrand appreciably for $k < \lambda^{-1}$. The final outcome is therefore approximately as we anticipated.

As the Boltzmann and Lenard–Balescu integrals both diverge logarithmically, one finds in practice that the results are fairly insensitive to the exact choice of cut off, or to the form it takes; (12.78) is thus merely illustrative. Lenard's procedure, adopted by us in §12.7.4, was to impose a simple cut off at about the Landau distance and to omit the Boltzmann term completely. This appears at first sight questionable, as it amounts to taking $\lambda = l$ in (12.78), yet interactions in the range $l < \xi < \xi_0$ would be better represented by the Boltzmann term. At the opposite extreme, one may take $\lambda = h$, so effectively removing the Lenard–Balescu term, which is therefore omitted. This too appears questionable since it extends the use of the Boltzmann term to impact parameters as large as the Debye length, so that the assumption of binary collisions is entirely unjustified. This was nevertheless the procedure adopted by several earlier writers (Landau, 1936; Chandrasekhar, 1943; Cohen, Spitzer and Routly, 1950; Spitzer and Härm, 1953), and appears to be satisfactory for some purposes, such as the calculation of transport coefficients by the Chapman–Enskog technique (see §11.5). Rosenbluth, MacDonald and Judd (1957), following this same procedure, have exploited the fact that nearly all the binary collisions involved are of very small angle, so permitting a further approximation. The result is identical with our equations (12.100) and (12.101), which we, however, will obtain as an approximation to the Lenard–Balescu equation. We interpret this later.

The argument of this section follows the general line given by D. E. Baldwin (1962). Other discussions of this problem are due to J. Hubbard (1961) and J. Weinstock (1964).

12.7.2 The Fokker–Planck equation and its physical interpretation

In this section we examine in more detail the form taken by the kinetic equation and interpret it physically. For completeness we take the case of a multispecies plasma, so we start at equations (12.71) and (12.72). The velocities $\mathbf{v}_1$ and $\mathbf{v}_2$ are rewritten as $\mathbf{v}$ and $\mathbf{v}'$, and suffixes such as i, j now have the usual tensor meaning, while r and s continue to label the species. So we have

$$\left[\frac{\partial f^{(r)}(\mathbf{v}, t)}{\partial t}\right]_c = -\sum_s \frac{\partial}{\partial v_i} J_i^{(rs)}(\mathbf{v}), \tag{12.80}$$

where

$$J_i^{(rs)} = \frac{1}{m_r}\left(\frac{e_r e_s}{\varepsilon_0}\right)^2 \int_{\mathbf{v}'} M_{ij}\left[\frac{1}{m_s}\frac{\partial f^{(s)}(\mathbf{v}')}{\partial v_j'} f^{(r)}(\mathbf{v}) - \frac{1}{m_r}\frac{\partial f^{(r)}(\mathbf{v})}{\partial v_j} f^{(s)}(\mathbf{v}')\right] d^3v' \tag{12.81}$$

and

$$M_{ij}(\mathbf{v}, \mathbf{v}') = \frac{1}{8\pi^2}\int_{\mathbf{k}} \frac{k_i k_j \delta[\mathbf{k}\cdot(\mathbf{v} - \mathbf{v}')]}{k^4 \,|\, \varepsilon(\mathbf{k}, \mathbf{k}\cdot\mathbf{v}') \,|^2}\, d^3k. \tag{12.82}$$

It will be observed that the matrix $\mathbf{M}$, besides being symmetric in the sense of the interchange of rows and columns, is also symmetric between $\mathbf{v}$ and $\mathbf{v}'$ and does not depend on the species.* The integral in (12.82) is cut off at large $\mathbf{k}$ as explained in the previous section.

Equation (12.81) may be cast in a different form as follows. In the second of the two terms within the square bracket, the order of the operations $\partial/\partial v_j$ and M_{ij} may be reversed, provided a compensating term is included. This new term involves $\partial M_{ij}/\partial v_j$, which by (12.82) may also be written $-\partial M_{ij}/\partial v_j'$. This in turn may be integrated by parts (with no contribution at infinity) to yield a similar term to the first part of (12.81), but with m_r in place of m_s. When this routine manipulation is completed, one finds

$$J_i^{(rs)} = A_i^{(rs)} f^{(r)}(\mathbf{v}) - \frac{1}{2}\frac{\partial}{\partial v_j}[B_{ij}^{(rs)} f^{(r)}(\mathbf{v})], \tag{12.83}$$

where

$$A_i^{(rs)}(\mathbf{v}) = \frac{1}{m_r}\left(\frac{1}{m_s} + \frac{1}{m_r}\right)\left(\frac{e_r e_s}{\varepsilon_0}\right)^2 \int_{\mathbf{v}'} M_{ij} \frac{\partial f^{(s)}(\mathbf{v}')}{\partial v_j'}\, d^3v' \tag{12.84}$$

$$B_{ij}^{(rs)}(\mathbf{v}) = \frac{2}{m_r^2}\left(\frac{e_r e_s}{\varepsilon_0}\right)^2 \int_{\mathbf{v}'} M_{ij} f^{(s)}(\mathbf{v}')\, d^3v'. \tag{12.85}$$

Combining (12.80) and (12.83), the kinetic equation for any species takes the Fokker–Planck form (compare with (9.122)),

$$\left(\frac{\partial f^{(r)}}{\partial t}\right)_c = -\frac{\partial}{\partial v_i}\left\{A_i^{(r)} f^{(r)} - \frac{1}{2}\frac{\partial}{\partial v_j}(B_{ij}^{(r)} f^{(r)})\right\}, \tag{12.86}$$

where the "friction coefficient" is the vector

$$\mathbf{A}^{(r)} = \sum_s \mathbf{A}^{(rs)} \tag{12.87}$$

and the "diffusion coefficient" is the matrix

$$\mathbf{B}^{(r)} = \sum_s \mathbf{B}^{(rs)}. \tag{12.88}$$

Equation (12.86) is of a form that was familiar long before the work of Lenard and Balescu, having arisen in connection with problems like Brownian motion and random walks; this supplies an illuminating physical interpretation of the equation. The discussion of the previous section suggests that if we fix attention on any one particle, the changes in its velocity to be represented by $(\partial f/\partial t)_c$ are due to the other particles at distances beyond the Landau length but within the Debye sphere. Such velocity changes are comparatively small, though occurring continually owing to the large number of

* Apart from some constant factors and a change of sign, **M** is the matrix designated as **Q** by Lenard and by several later writers; the new notation is more convenient when dealing with more than one species.

particles; hence the particle will pursue a random walk in velocity space. We consider a small time increment Δt and introduce $P(\mathbf{v}; \Delta\mathbf{v})$ as the transition probability that a particle with velocity $\mathbf{v}$ will experience a change $\Delta\mathbf{v}$ in that time. Here $\mathbf{v}$ and $\Delta\mathbf{v}$ are to be regarded as independent variables. Adopting the probability interpretation of the distribution function $f(\mathbf{v}, t)$ we may write

$$f(\mathbf{v}, t) = \int f(\mathbf{v} - \Delta\mathbf{v}, t - \Delta t)\, P(\mathbf{v} - \Delta\mathbf{v}; \Delta\mathbf{v})\, d^3(\Delta v). \tag{12.89}$$

This (known as the Chapman–Kolmogorov equation) follows merely from the meaning of conditional probabilities. Expanding in a Taylor series to the first order in Δt and to all orders in $\Delta\mathbf{v}$

$$\begin{aligned} f(\mathbf{v}, t) = \int \Bigg\{ \left(f - \Delta t \frac{\partial f}{\partial t}\right) P(\mathbf{v}; \Delta\mathbf{v}) - \Delta v_i \frac{\partial}{\partial v_i}[fP(\mathbf{v}; \Delta\mathbf{v})] \\ + \frac{1}{2!} \Delta v_i\, \Delta v_j \frac{\partial^2}{\partial v_i\, \partial v_j}[\ldots] \\ - \frac{1}{3!} \Delta v_i\, \Delta v_j\, \Delta v_k \frac{\partial^3}{\partial v_i\, \partial v_j\, \partial v_k}[\ldots] + \cdots \Bigg\}\, d^3(\Delta v). \end{aligned}$$

(Here f and its derivatives are evaluated at $\mathbf{v}$, t, and the square brackets are all identical.) But P is normalized by

$$\int P(\mathbf{v}; \Delta\mathbf{v})\, d^3(\Delta v) = 1$$

and we can interchange the orders of the operations of integration over $\Delta\mathbf{v}$ and differentiation with respect to $\mathbf{v}$ as these variables are independent. This leads to*

$$\frac{\partial f}{\partial t} = -\frac{\partial}{\partial v_i}[A_i f] + \frac{1}{2!}\frac{\partial^2}{\partial v_i\, \partial v_j}[B_{ij} f] - \frac{1}{3!}\frac{\partial^3}{\partial v_i\, \partial v_j\, \partial v_k}[C_{ijk} f] + \cdots \tag{12.90}$$

where

$$\begin{aligned} A_i &= \frac{1}{\Delta t}\int \Delta v_i\, P(\mathbf{v}; \Delta\mathbf{v})\, d^3(\Delta v) \\ B_{ij} &= \frac{1}{\Delta t}\int \Delta v_i\, \Delta v_j\, P(\mathbf{v}; \Delta\mathbf{v})\, d^3(\Delta v) \\ C_{ijk} &= \frac{1}{\Delta t}\int \Delta v_i\, \Delta v_j\, \Delta v_k\, P(\mathbf{v}; \Delta\mathbf{v})\, d^3(\Delta v) \end{aligned} \tag{12.91}$$

* For simplicity we have omitted the effect of spatial gradients and microscopic forces; it is easy to see that if they are included the left-hand side of (12.90) becomes the usual convective rate of change of f, as in (12.2).

and so on. These coefficients are often written $\langle \Delta v_i \rangle$, $\langle \Delta v_i \, \Delta v_j \rangle$, ... The argument so far is purely formal; the detailed physics enters when $P(\mathbf{v}; \Delta\mathbf{v})$ is considered explicitly. In proceeding only to the first order in Δt we assume that not all of $\mathbf{A}$, $\mathbf{B}$, . . . vanish identically in the limit $\Delta t \to 0$. For example, in the case where there is only an external force, and this gives an acceleration $\mathbf{a}(\mathbf{v})$, the change $\Delta\mathbf{v}$ is completely specified as $\mathbf{a}\Delta t$, and P is simply the appropriate δ-function; we find that $\mathbf{A} = \mathbf{a}$, and the higher coefficients all vanish, so that (12.90) is just an equation of continuity in velocity space. The next simplest case one could envisage would involve some dispersion in the values of $\Delta\mathbf{v}$ in such a way that $\mathbf{B} \neq 0$ but $\mathbf{C}$ and the higher coefficients vanish. In this case (12.90) is the Fokker–Planck equation and is a diffusion equation in velocity space. A group of particles with the same initial speed $\mathbf{v}$ are later spread out in velocity space to an extent governed by $\mathbf{B}$, hence the name "diffusion coefficient." ($\mathbf{A}$ is called the "friction coefficient" as it is often approximately antiparallel to $\mathbf{v}$.) Clearly the requirement for this case is that for small Δt the deflections $\Delta\mathbf{v}$ be small, so that $\mathbf{C}$ and the higher coefficients disappear in the limit. In fact $\Delta\mathbf{v}$ is typically of order $(\Delta t)^{1/2}$, but with mean of order Δt. This originates physically in the greater importance of the weak interactions with the numerous particles beyond the Landau distance, and would not be justified if close collisions were dominant, as in a neutral gas.

The point of view adopted in this chapter is that the real justification of the Fokker–Planck equation from first principles is to proceed via the BBGKY chain, using Bogoliubov's hypothesis, continuing through the derivations of Lenard or Balescu. The discussion just given merely provides an interpretation of the coefficients $\mathbf{A}$ and $\mathbf{B}$. An alternative plan is to consider the dynamical processes which determine $P(\mathbf{v}; \Delta\mathbf{v})$ for any test charge $\mathbf{v}$, and try to calculate the integrals (12.91) for small Δt. We have already alluded to the work of Landau (1936) and of Rosenbluth, MacDonald and Judd (1957), who (amongst others) treated the interactions as binary collisions and obtained $\mathbf{A}$ and $\mathbf{B}$ by expanding the Boltzmann collision operator for small deflections. Another approach, used by Thompson and Hubbard (1960) and Hubbard (1961), is to express $P(\mathbf{v}; \Delta\mathbf{v})$ in terms of the stochastic electric field $\mathbf{E}(\mathbf{x}, t)$ produced by particles within the Debye sphere. The coefficients $\mathbf{A}$ and $\mathbf{B}$ involve certain correlation functions of $\mathbf{E}$, which may be expressed as suitable ensemble averages. However, their calculation from first principles leads one back to the BBGKY chain, and as a matter of fact requires more apparatus than we have given so far, since it calls for ensemble averages of quantities involving two values of the time. Hubbard avoided lengthy calculations by a somewhat intuitive procedure later elaborated by Rostoker (1961) in his concept of "dressed" particles. The relation of this work to the development given here has recently been clarified by Dawson and Nakayama (1966), but further discussion would take us too far afield.

12.7.3 Elementary properties of the kinetic equation

Here we verify that the kinetic equation for the uniform field-free plasma possesses certain basic properties, without which it would be unacceptable. Several of these properties are implicit in the physics which formed our starting point, so their validity is no surprise; the point is that they have not been accidentally lost by approximations such as Bogoliubov's hypothesis. The treatment of this section is essentially that of Lenard (1960), with the slight extension to a multi-species plasma.

Firstly, suppose $f^{(r)}$ is at some instant everywhere non-negative for all species, as it must be to be physically acceptable. We expect the kinetic equation to predict that it will at no stage become negative. We show this by remarking that if it does so, there must be an instant at which the minimum value of $f^{(r)}$ first becomes negative. At this point $f^{(r)}$ and $\partial f^{(r)}/\partial \mathbf{v}$ vanish, $\partial^2 f^{(r)}/\partial v_i\,\partial v_j$ must be non-negative definite, and $\partial f^{(r)}/\partial t < 0$. By the first two of these conditions, (12.80) and (12.81) reduce to

$$\frac{\partial f^{(r)}(\mathbf{v})}{\partial t} = \left(\frac{\partial f^{(r)}(\mathbf{v})}{\partial t}\right)_c = \frac{1}{m_r^2}\sum_s \left(\frac{e_r e_s}{\varepsilon_0}\right)^2 \int_{\mathbf{v}'} M_{ij} f^{(s)}(\mathbf{v}')\,d^3v'\,\frac{\partial^2 f^{(r)}(\mathbf{v})}{\partial v_i\,\partial v_j}. \qquad (12.92)$$

But we can choose axes in which $\partial^2 f^{(r)}/\partial v_i\,\partial v_j$ is diagonal and with non-negative elements, while (12.82) shows that in any axes the diagonal components of M_{ij} must be positive. Thus the right-hand side of (12.92) cannot be negative, contradicting the requirement that $\partial f^{(r)}/\partial t < 0$.

Secondly, the kinetic equation must imply conservation of particles, momentum and energy (as already noted at equation 9.123). The first of these is immediate from (12.80), as each $(\partial f^{(r)}/\partial t)_c$ takes the form of the divergence of a vector which vanishes at infinity in velocity space. For momentum and energy, our concern is that the rate of transfer from species s to r due to the relevant term in $(\partial f^{(r)}/\partial t)_c$ shall be the negative of the rate in the opposite direction, or zero if $r = s$.

In the case of momentum, the rate of transfer of the i-component to species r is

$$\int m_r v_i \left(\frac{\partial f^{(r)}}{\partial t}\right)_c d^3v$$

and by (12.80) the contribution due to species s is

$$-\int m_r v_i \frac{\partial J_k^{(rs)}}{\partial v_k}\,d^3v = \int m_r J_i^{(rs)}\,d^3v$$

(integrating by parts). Using (12.81) this becomes a double integral over $\mathbf{v}$ and $\mathbf{v}'$. As $M_{ij}(\mathbf{v}, \mathbf{v}')$ is symmetric with respect to interchange of $\mathbf{v}$ and $\mathbf{v}'$, while the square bracket of (12.81) changes sign if both $(\mathbf{v}, \mathbf{v}')$ and (r, s) are interchanged, the momentum transfer has the required antisymmetry in (r, s).

Similarly, the rate of energy transfer to species r is

$$\int \tfrac{1}{2}m_r v^2 \left(\frac{\partial f^{(r)}}{\partial t}\right)_c d^3v$$

and the contribution due to species s is

$$-\int \tfrac{1}{2}m_r v^2 \frac{\partial J_k^{(rs)}}{\partial v_k} d^3v = \int m_r v_i J_i^{(rs)} d^3v$$

$$= \left(\frac{e_r e_s}{\varepsilon_0}\right)^2 \int_{\mathbf{v}}\int_{\mathbf{v}'} v_i M_{ij}\left[\frac{1}{m_s}\frac{\partial f^{(s)}(\mathbf{v}')}{\partial v_j'} f^{(r)}(\mathbf{v}) - \frac{1}{m_r}\frac{\partial f^{(r)}(\mathbf{v})}{\partial v_j} f^{(s)}(\mathbf{v}')\right] d^3v\, d^3v'.$$

We now write down the corresponding quantity with r and s interchanged and the dummy variables $\mathbf{v}$ and $\mathbf{v}'$ interchanged; this is just the same expression except that v_i becomes $-v_i'$. That the sum of these two integrals vanishes follows at once, for by (12.82) we have

$$(v_i - v_i')M_{ij}(\mathbf{v}, \mathbf{v}') = 0 \tag{12.93}$$

for all $\mathbf{v}$, $\mathbf{v}'$, owing to the identity $x\delta(x) = 0$.

A third general property of the kinetic equation concerns the special role of the Maxwellian distributions and the second law of thermodynamics. If $f(\mathbf{v})$ is a Maxwellian distribution with temperature T and drift velocity $\mathbf{u}$, one has

$$\frac{\partial f}{\partial v_j} = -\left(\frac{m}{KT}\right)(v_j - u_j)f. \tag{12.94}$$

Consequently if each species is in a Maxwellian distribution, and T and $\mathbf{u}$ are the same for each, we have

$$\frac{1}{m_s}\frac{\partial f^{(s)}(\mathbf{v}')}{\partial v_j'} f^{(r)}(\mathbf{v}) - \frac{1}{m_r}\frac{\partial f^{(r)}(\mathbf{v})}{\partial v_j} f^{(s)}(\mathbf{v}') = (KT)^{-1}(v_j - v_j')f^{(r)}(\mathbf{v})f^{(s)}(\mathbf{v}'),$$

and using (12.93) it follows that each $J_i^{(rs)}$, as given by (12.81), vanishes. Thus, as we would expect, $(\partial f/\partial t)_c = 0$ for each species.

Furthermore, no other distribution has this property, and, in the spatially homogeneous case, all distributions tend asymptotically to that Maxwellian state whose drift velocity and temperature are fixed by the conservation of momentum and energy. The proof given by Lenard is analogous to the "H-theorem" of classical kinetic theory. One considers the functional

$$H = \sum_r \int f_r(\mathbf{v}) \log f_r(\mathbf{v})\, d^3v$$

which is essentially the negative of the entropy for the gas (cf. §10.7). By a manipulation making use of the kinetic equation (as outlined in Problem 5) it may be shown that $dH/dt \leqslant 0$, with equality only in the case of Maxwellian distributions having drift velocity and temperature common to the species. This is, of course, just the law of increasing entropy. It is well known from elementary statistical mechanics that H is minimized by the appropriate Maxwellian distribution. The results just stated are therefore proved.

12.7.4 Approximate form for the kinetic equation

The kinetic equation as given by (12.80)–(12.82) is too complicated to be readily applied to practical calculations. Besides first and second derivatives with respect to $\mathbf{v}$, we have coefficients which are very awkward functionals of all the $f^{(s)}$, involving integrals over $\mathbf{v}'$ and $\mathbf{k}$, with the dielectric function ε also appearing in the formulas. Again following Lenard (1960), we may effect considerable simplification by carrying out the integrals over $\mathbf{k}$ to within a certain approximation, thereby reducing M_{ij} to a much more tractable form.

Referring to (12.57), we have

$$\varepsilon(\mathbf{k}, \mathbf{k}\cdot\mathbf{v}) = 1 + \frac{1}{h^2k^2}\psi(\mathbf{l}\cdot\mathbf{v}), \tag{12.95}$$

where $\mathbf{l}$ is the unit vector parallel to $\mathbf{k}$, and

$$\psi(\mathbf{l}, \mathbf{v}) = h^2 \sum_s \frac{c_s^2}{\varepsilon_0 m_s} \int_{\mathbf{v}''} \frac{\mathbf{l}\cdot\partial f^{(s)}(\mathbf{v}'')/\partial\mathbf{v}''}{\mathbf{l}\cdot(\mathbf{v}-\mathbf{v}'')}\, d^3v''. \tag{12.96}$$

Here h could mean anything, but ψ will be dimensionless, and nearly always of order unity, if we take h to be the Debye length, $(\varepsilon_0 KT/ne^2)^{1/2}$, for one species, say the electrons.

Integration of (12.82) is facilitated by the δ-function and by the fact that ψ does not depend on the magnitude of $\mathbf{k}$. Writing $\mathbf{w} = \mathbf{v} - \mathbf{v}'$ and $w = |\mathbf{w}|$, we may express $\mathbf{k}$ in polar coordinates (k, θ, ϕ) with the axis parallel to $\mathbf{w}$. Integration over θ yields

$$M_{ij} = \frac{1}{8\pi^2 w}\int_k \int_{\phi=0}^{2\pi} \frac{l_i l_j\, dk\, d\phi}{k\,|\,l + \psi/h^2k^2\,|^2},$$

where the unit vector $\mathbf{l}$ is now restricted to the plane $\theta = \frac{1}{2}\pi$ owing to the δ-function. The k-integration formally ranges over 0 to ∞ but we recall that a cut off at some $k = k_0$ is necessary. This leads to

$$M_{ij} = \frac{1}{16\pi^2 w}\int_0^{2\pi} l_i l_j \frac{\operatorname{Im}\{\psi \log(1 + h^2k_0^2/\psi)\}}{\operatorname{Im}(\psi)}\, d\phi. \tag{12.97}$$

The remaining integration may also be carried out explicitly if we are content with what is actually a very good approximation. For most values of $\mathbf{l}$ and $\mathbf{v}_1$, the function ψ as given by (12.96) is of order unity; but whatever the choice of cut off $h^2k_0^2$ will certainly be very large. Hence we may write $\log(1 + h^2k_0^2/\psi) \sim \log(h^2k_0^2)$. As this is real, ψ cancels altogether from the integrand of (12.97), so the complicated dependence of ψ on $\mathbf{l}$ becomes irrelevant. Noting that

$$\int_0^{2\pi} l_i l_j \, d\phi = \pi(\delta_{ij} - w_i w_j/w^2),$$

we reach finally

$$M_{ij} = \frac{\log \Lambda}{8\pi} \frac{w^2\delta_{ij} - w_i w_j}{w^3}, \tag{12.98}$$

where (following convention) we have written $\Lambda = hk_0$. According to the discussion of §12.7.1, the best choice for k_0 (if we include no term of the Boltzmann type) is of the order l^{-1} where l is the Landau distance $4\pi\varepsilon_0 KT/e^2$; the value usually quoted is actually $3l^{-1}$, leading to

$$\Lambda = \frac{12\pi(\varepsilon_0 KT)^{3/2}}{n^{1/2}e^3}, \tag{12.99}$$

where T, n and e are assumed to refer to the electrons. In nearly all practical cases, $\log \Lambda \gtrsim 10$, so the approximation made above is fairly well justified (except when $\mathbf{v}$ has a large (suprathermal) component perpendicular to $\mathbf{w}$, but as there are few such particles this is often unimportant). The reference to a particular species (the electrons) in choosing k_0 may seem rather arbitrary, but as $\log \Lambda$ is so insensitive to the choice, and as there is really no satisfactory basis for an exact value of k_0, we prefer to let (12.99) stand. Some writers give a formula for Λ which depends on the pair (r, s) of species whose interaction is under consideration, so that M_{ij}, for example in (12.81), depends on (r, s).

The reader will easily verify that those properties of $\mathbf{M}$ which were used in obtaining the results of §12.7.3 are also possessed by $\mathbf{M}$ in its approximate form (12.98); those results are not, therefore, vitiated by the approximation developed here. If, however, $\mathbf{M}$ is calculated with a different cut off for each pair of species, care is needed to ensure that the conservation of momentum and energy still hold!

The derivation just given also shows explicitly that correlations on a scale greater than h (i.e. Fourier components with $hk \ll 1$) make little contribution to $(\partial f/\partial t)_c$. This gives a rough justification for using the kinetic equation in the non-uniform case provided the inhomogeneities occur on a scale well in excess of h.

Next we may substitute the simplified M_{ij}, (12.98), into our general kinetic equation, for which purpose we use the form (12.83). The coefficients (12.84) and (12.85) become, after some further manipulation (Problem 6)

$$A_i^{(rs)}(\mathbf{v}) = \frac{1}{4\pi}\left(\frac{e_r e_s}{\varepsilon_0 m_r}\right)^2 \frac{m_r + m_s}{m_s} \log \Lambda \frac{\partial}{\partial v_i} \int_{\mathbf{v}'} \frac{f^{(s)}(\mathbf{v}')}{|\mathbf{v} - \mathbf{v}'|} d^3v' \tag{12.100}$$

$$B_{ij}^{(rs)}(\mathbf{v}) = \frac{1}{4\pi}\left(\frac{e_r e_s}{\varepsilon_0 m_r}\right)^2 \log \Lambda \frac{\partial^2}{\partial v_i \partial v_j} \int_{\mathbf{v}'} |\mathbf{v} - \mathbf{v}'| f^{(s)}(\mathbf{v}')\, d^3v'. \quad (12.101)$$

These are precisely the results of Rosenbluth, MacDonald and Judd (1957), who obtained them by the quite different route of expanding the Boltzmann collision operator under the approximation that all the deflections are of small angle, and cutting off the impact parameter at about the Debye length (see §12.7.1). At first sight the success of that method is surprising, as any treatment dealing with binary collisions would seem to be discredited. The physical reason for the agreement is that, for the majority of particles, there is little difference between a succession of numerous small-angle collisions (regarded as instantaneous and occurring at random) and the stochastic deflections due to the presence of many nearby particles continually exerting weak forces. These two pictures of the dynamics are of course represented respectively by the Boltzmann and the Fokker–Planck expressions. A complete connection with the Boltzmann approach is thus established, apart from some uncertainty in the value of Λ.

Obviously we have here made substantial progress in simplifying the Lenard–Balescu equation; even so the Fokker–Planck coefficients are still rather complicated functionals of the $f^{(s)}$ owing to the integrals over $\mathbf{v}'$ in (12.100) and (12.101). As Rosenbluth, MacDonald and Judd put it: "In the general case this fourth-order, three-dimensional, time-dependent and non-linear partial differential equation seems quite difficult to handle"! For this reason, little progress has been made in finding solutions of the equation, except in connection with transport coefficients as discussed in the next section. It was for this reason, too, that we considered the model Fokker–Planck equation of §9.5.3. Actually, if $\mathbf{v}$ is modest compared with most of the values of $\mathbf{v}'$ that are important in the above integrals, **A** and **B** do take the form proposed in the model equation (see Problem 7). Of course, the lack of progress in solving the kinetic equation is not, in itself, any worse than in the case of neutral gases; but there are, for plasmas, many more practical circumstances in which something better than the hydrodynamic treatment is desirable. Often, the collision-free theory may take its place, or even the adiabatic theory of particle orbits. But in other cases, one really is faced with the immense difficulty of working with the Fokker–Planck equation.

12.8 THE HYDRODYNAMIC APPROXIMATION

We now have expressions of various degrees of sophistication and accuracy for $(\partial f/\partial t)_c$ in a plasma. As we know already, one of the main applications of the kinetic equations

$$\frac{\partial f}{\partial t} + \mathbf{v}\cdot\frac{\partial f}{\partial \mathbf{x}} + \mathbf{a}\cdot\frac{\partial f}{\partial \mathbf{v}} = \left(\frac{\partial f}{\partial t}\right)_c \quad (12.102)$$

(one for each species) is the derivation of magnetohydrodynamics, complete with expressions for the transport coefficients. The technique for doing this—the Chapman–Enskog expansion—has already been described in §11.5, and some of the results obtained for plasmas have been sketched there.

The purpose of the present section is to discuss the philosophy underlying that work rather than enlarge on the mathematical details. The derivation of the hydrodynamic approximation, both for neutral gas and for plasma, is to be regarded as the second part of a two-stage approximation. The stages correspond to the successive weakening of the description of the gas. We pass from the Liouville function, ρ, to the one-particle distribution function f (first stage), and from f to the hydrodynamic variables $\mathbf{u}$, etc. (second stage). This reduction can be achieved only by approximation, the first stage by the closure of the BBGKY chain, as discussed in this chapter, and the second by the Chapman–Enskog expansion as developed in the preceding chapter. If we wish, we can stop half way and tackle problems in terms of the Boltzmann or Fokker–Planck equation, but if, as is now supposed, we accept both stages, the question arises whether the assumptions made therein are mutually consistent, and we need to know in what range of physical conditions are the results applicable.

This is not an entirely perfunctory matter. The reader may have noticed that throughout this chapter the interaction between particles has been regarded as small (the exact meaning of this depending on the law of force), leading to the notion that $(\partial f/\partial t)_c$ is in some sense small; yet the Chapman–Enskog procedure begins with the assumption that $(\partial f/\partial t)_c$ dominates (12.102). Can we really maintain both these assumptions?

The answer to this question involves careful consideration of the approximations we introduced in the respective stages, bearing in mind especially the following two aspects. The first is the assumed limitation on rates of change in space, velocity and time. The second is that, at both stages, the approximation is not simply a matter of ignoring certain terms in differential equations; it involves the complete removal of some of the independent variables and the construction of fresh equations working with "reduced" information, even though this is not strictly possible.

It may clarify the discussion to begin with the case of gases with short range interaction (i.e. neutral gases). The reduction from the Liouville equation to the Boltzmann equation is achieved (see §12.4.2) by expanding in terms of the diluteness parameter nr_0^3 and by assuming that the two-particle distribution function is a product of one-particle distribution functions except when $|\mathbf{x}_1 - \mathbf{x}_2| < r_0$. The smallness of the interaction is reflected in the fact that these exceptional regions form a very small proportion of $(\mathbf{x}_1, \mathbf{x}_2)$-space; yet it is they which give rise to $(\partial f/\partial t)_c$, and we obtain it by treating these regions in terms of binary collisions, assuming that variations of f occur only on a scale well in excess of r_0. We also smooth out the rapid variations which occur on a time scale $r_0/\bar{v}$, that is we pick out that part of the change in f which would be regarded as *secular* on that time scale. In this way we relinquish many of the possibilities contained in the original Liouville equation. However, we are entitled to consider any solutions of the new governing equation (the Boltzmann equation) provided they are consistent with the assumptions used in its derivation. We may therefore envisage solutions of (12.102) in which spatial gradients and external forces are so small that $(\partial f/\partial t)_c$ is the principal contribution to $\partial f/\partial t$. So long as $\partial f/\partial t$ is not too large (it should be much less than $\bar{v}f/r_0$) this is still valid, and consistent with the notion that the interaction is small. We may say that $(\partial f/\partial t)_c$ is indeed small, but is important in the long run since it is a secular effect. For a thermal particle, the order of magnitude of the terms in $(\partial f/\partial t)_c$ is $nr_0^2\bar{v}f$, so we introduce a *collision time* $(nr_0^2\bar{v})^{-1}$ and a *mean free path* $(nr_0^2)^{-1}$. If we take an initial f with spatial variations only on a scale greater than the mean free path, and with sufficiently small external force, the kinetic equation tells us (via the H-theorem) that f tends to a local Maxwellian distribution in, roughly, the collision time, and that thereafter it changes much more slowly. If we are prepared to smooth out variations on that time scale, taking account only of still slower secular changes, f will always be close to the appropriate local Maxwellian, and we obtain the so-called "normal" solutions, which can be represented by hydrodynamic variables alone; this is just the Chapman–Enskog procedure. As the collision time and mean free path are much greater (by a factor $(nr_0^3)^{-1}$) than $r_0/\bar{v}$ and r_0 respectively, the conditions for the second stage imply *a fortiori* the validity of the first. The two stages are thus quite distinct, and mutually compatible.

It is interesting to add that, once the hydrodynamic equations have been obtained in this way, there is no implication about the relative magnitude of the various terms (just as with the Boltzmann equation). All that is necessary is that the conditions used in the derivation should be satisfied. For instance, one can discuss situations with either high or low Reynolds number. Perfect gas behavior arises when conditions are such that viscosity and thermal conductivity do not have time to be effective by comparison with other causes of change; in this sense the time scale must not be too long, though the word "adiabatic," frequently used here, refers to the need for the time scale to be much longer than the collision time, which is implicit in any hydrodynamic situation.

For a plasma the discussion follows a similar course. The treatment of the first stage needs amendment in several respects. The relevant small parameter is now $(n^{1/3}e^2/\varepsilon_0 KT)^{3/2}$. The assumption that particles are uncorrelated must be corrected (by the function g) for separation up to the Debye length, h, and this is quite different from the neutral gas, as $h \gg n^{-1/3}$; however, the correction is small enough for g to be handled by linearization except within the Landau distance, which is $\ll n^{-1/3}$. The appropriate time scale for this stage is the plasma period (see §12.5.2). Provided f varies slowly over the Debye length and during a plasma period, we can obtain a kinetic equation with $(\partial f/\partial t)_c$ expressed as a functional of f. (The fact that this has only been done properly for the spatially uniform case, and the complications mentioned in §12.7.1, are technical difficulties rather than matters of principle.) The kinetic equation may be used in any circumstances that do not violate the conditions for its derivation, so in particular we can proceed to stage two and take $(\partial f/\partial t)_c$ to be dominant, and the distributions all close to Maxwellian. The time scale for relaxation towards the Maxwellian may be estimated from the coefficients (12.100) or (12.101), and the best value for it emerged from our calculations of Chapter 11, namely (11.114)

$$\tau_c = \frac{10}{n}\left(\frac{4\pi\varepsilon_0}{e^2}\right)^2 \frac{m^{1/2}(KT)^{3/2}}{4\pi \log \Lambda} \tag{12.103}$$

and the associated distance is

$$L_c = \frac{10}{n}\left(\frac{4\pi\varepsilon_0 KT}{e^2}\right)^2 \frac{1}{4\pi \log \Lambda}. \tag{12.104}$$

Here we have ignored the complications arising from the presence of several species; in order to justify the hydrodynamic approximation for the plasma as a whole, m, e and n should refer to the most numerous of the heavy species. The quantities τ_c and L_c are again much larger than those of stage one. They may be loosely called the "collision time" and "mean free path." They are actually the same as those obtained from a crude treatment of large angle collisions (see §1.2), except for a numerical factor of about $10/(4\pi \log \Lambda)$; the fact that this is substantially less than unity indicates of course that the "collisional" effect is much enhanced by the correlations extending to the Debye sphere. Provided spatial gradients are supposed to occur only on a scale larger than L_c, and assuming a limitation on the macroscopic acceleration $\mathbf{a}$, stage two is justified. The proviso about $\mathbf{a}$ needs more care for a plasma, as at least a part of it is due to the self-consistent field of the plasma itself. If our equations imply appreciable space charge we are led to plasma oscillations, which violate stage one, so cannot possibly be treated hydrodynamically.

The basic ideas underlying the two stages, and the justification for combining them, are thus quite similar for neutral gases and for plasmas, even though the mathematical details of both stages are very different.

As a final remark it should be noted that the "double-adiabatic" hydrodynamic approximation (see §11.3) is based on a quite different conception of the two stages from that discussed here, and applies only to cases where the collisions (or rather, correlations) are not so dominant that a near-Maxwellian state is maintained. This can arise only if the Larmor radius is much smaller than the mean free path. In neglecting $(\partial f/\partial t)_c$ altogether it is assumed in effect that the phenomena to be treated have spatial scale (e.g. size of vessel, wavelength, etc.) intermediate between those two lengths. There are also special assumptions in connection with heat flow along field lines. The collision-free Boltzmann equation is taken as the result of the first stage. The second stage, while formally similar to the Chapman–Enskog method, uses the magnetic term as the dominant one. This suffers from the difficulty that the lowest approximation does not completely fix f in terms of the hydrodynamic variables (in the way that the collision-dominated theory requires f to be the relevant Maxwellian distribution), so leading to the ambiguity concerning heat flow parallel to the field (see §11.5.6). The physical basis for the double-adiabatic hydrodynamics is thus very different from that of ordinary hydrodynamics, and one needs to be more cautious in using the theory of §11.3. But we considered it worth developing, both for its practical applications and for its illuminating connection with particle-orbit theory.

12.9 FURTHER APPLICATIONS OF CORRELATION FUNCTIONS

The main objective of this chapter, the construction of a kinetic equation corrected for collisional effects, has now been achieved to the extent that space allows. But the techniques developed in this work are useful in other connections, some of which even have practical applications, and our account of the subject would be incomplete without a brief mention of these. In this final section (really an addendum to the chapter), these additional applications will be illustrated by obtaining one or two results which are easily accessible from the theory already given, and still further extensions will be outlined.

12.9.1 Fluctuations and their spatial correlation

In §12.2.1 we introduced the idea of an observable quantity Q and its ensemble average $\langle Q\rangle$, defined by (12.7); for the present, Q is supposed to involve the state of the system at one time only. By the *fluctuation* of Q we mean the quantity

$$\Delta Q = Q - \langle Q\rangle. \tag{12.105}$$

Clearly $\langle \Delta Q\rangle = 0$. For macroscopic systems we are accustomed to thinking that ΔQ is usually extremely small for the type of observation that can be made in practice. It is nevertheless of interest to discuss the evaluation of other quantities involving ΔQ, and whether in simple cases these could actually be measured. Such quantities must, of course, be ensemble averages themselves; one might consider, for instance, the variance $\langle \Delta Q^2\rangle$. More generally, Q may itself be a function of new "free" variables rather than just a single measurement; for example, the one-particle distribution function $F(X, t)$, defined by (12.8) depends on the "free" phase point X as well as on the phase points of all the particles, $X_1, X_2, \ldots$. If so, one can consider a *correlation function* for the fluctuations, defined in that example by

$$C(X, X', t) = \langle \Delta F(X, t)\, \Delta F(X', t) \rangle \tag{12.106}$$

for all pairs of phase points (X, X'). This is only the very simplest of all the quantities that could be contemplated, but it is the one which we shall now investigate.

The positions of individual particles will not appear in the analysis below, so without confusion we may rename the phase points X and X' as X_1 and X_2. Then since by (12.105) and (12.9) $\Delta F = F - f_1$ and $\langle \Delta F \rangle = 0$, the quantity (12.106) becomes

$$C(X_1, X_2, t) = \langle F(X_1, t)\, F(X_2, t) \rangle - f_1(X_1, t)\, f_1(X_2, t). \tag{12.107}$$

Here the first term may be expressed in terms of f_2 by (12.10), and using the definition (12.35) of $g(X_1, X_2, t)$, we may write (12.107) as

$$C(X_1, X_2, t) = g(X_1, X_2, t) + \delta(X_1 - X_2)\, f_1(X_1, t). \tag{12.108}$$

Provided we accept the expansion procedure for closing the BBGKY chain, together with Bogoliubov's hypothesis, our earlier work supplies (in principle) an expression for g in terms of f_1, namely the asymptotic value $g(\infty)$, equation (12.47). It follows that C may be expressed in terms of f_1. And, to this same approximation, all higher distributions f_s $(s > 2)$ may also be expressed in terms of f_1 and g by (12.34); so ensemble averages like (12.106) but involving a product of s factors evaluated at s phase points could be handled in the same way.

To give a tractable example, let us restrict attention to the case where f_1 is uniform in space; then we showed in §12.5.4 how to obtain the spatial Fourier transform of g, though we did not compute it in detail since only the integral of g with respect to $\mathbf{v}_2$ was needed for the calculation then in progress. Here we will simplify matters still further by integrating over $\mathbf{v}_1$ also. These integrations over $\mathbf{v}_1$ and $\mathbf{v}_2$ commute with the operation of taking ensemble averages. Letting n denote density in physical space, and $\Delta n = n - \langle n \rangle$, equations (12.106) and (12.108) lead to

$$\langle \Delta n(\mathbf{x}_1, t)\, \Delta n(\mathbf{x}_2, t) \rangle = \int g(\mathbf{x}_1, \mathbf{v}_1, \mathbf{x}_2, \mathbf{v}_2, t)\, d^3v_1\, d^3v_2 + \delta^3(\boldsymbol{\xi}) n_0, \tag{12.109}$$

where $\boldsymbol{\xi} = \mathbf{x}_1 - \mathbf{x}_2$, $n_0 = \langle n(\mathbf{x}, t) \rangle$, and g now depends on $\mathbf{x}_1$ and $\mathbf{x}_2$ only in the combination $\boldsymbol{\xi}$. The Fourier transform of g with respect to $\boldsymbol{\xi}$ (see (12.50)), integrated over $\mathbf{v}_1$ and $\mathbf{v}_2$, may be written as an integral like (12.59) but with the further simplification that (12.58) may be used for the suffix 1 operation also. Thus

$$\iint g(\mathbf{k}, \mathbf{v}_1, \mathbf{v}_2)\, d^3v_1\, d^3v_2 = \frac{1}{(2\pi i)^2} \lim_{p\to+0} \int_{\mathbf{v}_1}\int_{\mathbf{v}_2}\int_{C_1}\int_{C_2} \frac{1}{p - p_1 - p_2}$$

$$\times \frac{1}{\bar{\varepsilon}(\mathbf{k}, p_1)\bar{\varepsilon}(-\mathbf{k}, p_2)} \frac{1}{p_1 - i\mathbf{k}\cdot\mathbf{v}_1} \frac{1}{p_2 + i\mathbf{k}\cdot\mathbf{v}_2} S\, d^3v_1\, d^3v_2\, dp_1\, dp_2. \quad (12.110)$$

Here C_1 and C_2 are the Laplace inversion contours used before and S is as in (12.52). By handling the p_1 and p_2 integrals in a way analogous to §12.5.4, (12.110) may be reduced to integrals over $\mathbf{v}_1$ and $\mathbf{v}_2$, which in turn can be evaluated if $f_1(\mathbf{v})$ is specified.

In the important case where the plasma is in thermal equilibrium, so f_1 is the Maxwellian distribution, all the integrations in (12.110) may be completed analytically by routine methods outlined below. Accepting for the moment the result

$$\iint g(\mathbf{k}, \mathbf{v}_1, \mathbf{v}_2)\, d^3v_1\, d^3v_2 = -\frac{n_0}{(2\pi)^3} \frac{1}{1 + h^2k^2}, \quad (12.111)$$

we may invert the spatial Fourier transform to obtain

$$\iint g(\boldsymbol{\xi}, \mathbf{v}_1, \mathbf{v}_2)\, d^3v_1\, d^3v_2 = -\frac{n_0}{4\pi h^2 \xi} e^{-\xi/h}, \quad (12.112)$$

so that (12.109) reads

$$\langle \Delta n(\mathbf{x} + \boldsymbol{\xi}, t)\, \Delta n(\mathbf{x}, t)\rangle = n_0\left[\delta^3(\boldsymbol{\xi}) - \frac{1}{4\pi h^2 \xi} e^{-\xi/h}\right]. \quad (12.113)$$

As one would expect, this implies that the fluctuations in the number of particles counted in a small volume would be proportional to the square root of the mean number (Problem 8). We also have

$$\langle n(\mathbf{x} + \boldsymbol{\xi}, t) n(\mathbf{x}, t)\rangle = n_0\left[\delta^3(\boldsymbol{\xi}) + n_0 - \frac{1}{4\pi h^2 \xi} e^{-\xi/h}\right]. \quad (12.114)$$

Comparing with equation (12.21) and the discussion given there, it becomes clear that the second factor on the right is the conditional probability of finding an electron at $\mathbf{x} + \boldsymbol{\xi}$, given that there is one at $\mathbf{x}$. After omitting the δ-function term (which has the usual interpretation), what remains is just what would be expected from the elementary Debye–Hückel theory. For rewriting (7.34) in the present notation, with $|e\phi| \ll KT$, and using (7.39),

$$n(\boldsymbol{\xi}) \approx n_0\left[1 - \frac{e^2}{4\pi\varepsilon_0 KT} \frac{e^{-\lambda\xi}}{\xi}\right], \quad (12.115)$$

which is the same expression provided λ^{-1} is h (see also Chapter 8, Problem 5). Actually in §7.6.4 λ^{-1} denoted $h/\sqrt{2}$, the difference arising from the fact that there we allowed for the dynamics of the ions, which were also in thermal equilibrium at the same temperature, whereas here the ions were supposed to form a fixed uniform background. When (singly charged) ions are included in the theory, as outlined in §12.6.1, but one calculates the electron–electron correlation function, (12.111) is modified as

$$\iint g(\mathbf{k}, \mathbf{v}_1, \mathbf{v}_2)\, d^3v_1\, d^3v_2 = -\frac{n_0}{(2\pi)^3}\frac{1}{2 + h^2k^2}, \tag{12.116}$$

and this leads to the required correction.

The evaluation of (12.110) for the Maxwellian distribution may be effected as follows. The two terms which comprise S, (equation 12.52), transform into each other if we interchange 1 and 2 and reverse the sign of $\mathbf{k}$. The same is true of the operations to be carried out in (12.110). So we may concentrate on the first term. We will perform the $\mathbf{v}_1$ and $\mathbf{v}_2$ integrations first; these may be handled separately since our integrand factorizes. So we require

$$\phi(\mathbf{k}) i\mathbf{k}\cdot \int \frac{\partial f(\mathbf{v}_2)/\partial \mathbf{v}_2}{p_2 + i\mathbf{k}\cdot\mathbf{v}_2}\, d^3v_2 \quad \text{and} \quad \int \frac{f(\mathbf{v}_1)}{p_1 - i\mathbf{k}\cdot\mathbf{v}_1}\, d^3v_1.$$

Here p_1 and p_2 start with positive real part and the results are to be continued analytically; this implies the Landau rule for the $\mathbf{v}$-integrations. Comparing with the definition of $\bar{\varepsilon}$, (12.57), the first of these integrals is just $[1 - \bar{\varepsilon}(-\mathbf{k}, p_2)]/(2\pi)^3$. For the second integral we exploit the choice of a Maxwellian distribution, for which $\mathbf{v}f(\mathbf{v}) = -(KT/m)\partial f/\partial \mathbf{v}$. By writing

$$\frac{1}{p_1 - i\mathbf{k}\cdot\mathbf{v}_1} = \frac{1}{p_1}\left(1 + \frac{i\mathbf{k}\cdot\mathbf{v}_1}{p_1 - i\mathbf{k}\cdot\mathbf{v}_1}\right)$$

and making use of (12.57) again, this second integral may eventually be expressed as $1 + h^2k^2[1 - \bar{\varepsilon}(\mathbf{k}, p_1)]$. We therefore reach

$$\frac{n_0}{(2\pi)^3}\frac{1}{(2\pi i)^2}\lim_{p\to 0}\int_{C_1}\int_{C_2}\frac{1}{p - p_1 - p_2}\,\frac{1 - \bar{\varepsilon}(-\mathbf{k}, p_2)}{\bar{\varepsilon}(-\mathbf{k}, p_2)} \times \frac{1 + h^2k^2[1 - \bar{\varepsilon}(\mathbf{k}, p_1)]}{p_1\bar{\varepsilon}(\mathbf{k}, p_1)}\, dp_1\, dp_2.$$

The p_2-integration may now be carried out using the contour of Fig. 12.1, the contribution on the large semicircle vanishing (noting $\bar{\varepsilon} \to 1$ as $p \to \infty$), the only singularity being at $p_2 = p - p_1$. This done, we may let $p \to +0$. This gives

$$\frac{n_0}{(2\pi)^3}\frac{1}{2\pi i}\int_{C_1}\left(\frac{1}{\bar{\varepsilon}(-\mathbf{k}, -p_1)} - 1\right)\left(\frac{1 + h^2k^2}{\bar{\varepsilon}(\mathbf{k}, p_1)} - h^2k^2\right)\frac{dp_1}{p_1}. \tag{12.117}$$

The contour C_1 is as in Fig. 12.2, but the singularities involving $\mathbf{v}_1$ or $\mathbf{v}_2$ do not now arise. On the other hand, there appears to be a new singularity at p_1. But for the Maxwellian distribution we have noted before (Chapter 8, Problem 5) that

$$\bar{\varepsilon}(\mathbf{k}, 0) = 1 + \frac{1}{h^2k^2} \tag{12.118}$$

which shows that the second factor of the integrand vanishes as $p_1 \to 0$, so removing the pole. It does not, therefore, matter how C_1 passes $p_1 = 0$, and the same is true if we split the integrand into two parts provided each part is treated in the same way. We choose to rewrite (12.117) as

$$\frac{n_0}{(2\pi)^3}\frac{1}{2\pi i}\int_{C_1}\left\{\left(\frac{1}{\bar{\varepsilon}(-\mathbf{k}, -p_1)} - 1\right) + (1 + h^2k^2)\left|\frac{1}{\bar{\varepsilon}(\mathbf{k}, p_1)} - 1\right|^2\right\}\frac{dp_1}{p_1} \tag{12.119}$$

(remembering that $\bar{\varepsilon}(\mathbf{k}, p_1)$ and $\bar{\varepsilon}(-\mathbf{k}, p_1)$ are complex conjugates), let C_1 pass through $p_1 = 0$, and take the principal part there. For the first term we close the contour by a semi-circle in the left half-plane as $\bar{\varepsilon} \to 1$ at infinity, and the only pole enclosed is that at the origin, counted at half value as we want the principal part. Using (12.118) again this gives

$$-\frac{n_0}{(2\pi)^3}\frac{1}{2}\frac{1}{1 + h^2k^2}.$$

As this is even in $\mathbf{k}$, the corresponding contribution from the other part of S is the same; these combine to give already the result quoted, (12.111). The second term of (12.119), together with the corresponding term with the sign of $\mathbf{k}$ changed, give a zero principal part by symmetry.

12.9.2 Scattering of electromagnetic radiation

It will be asked how the rather delicate quantity $\langle \Delta n(\mathbf{x} + \boldsymbol{\xi}, t)\, \Delta n(\mathbf{x}, t)\rangle$ is to be observed in practice. A way of doing so is to allow electromagnetic radiation to be incident on the plasma and measure the power scattered in a direction other than the incident. This is sometimes called *incoherent scattering*. The ordinary theory of electromagnetic wave propagation in a plasma (Chapter 5) does not predict any such scattering; a plane wave can proceed along the incident direction without deformation. This is because the electrons are regarded as a continuous fluid, and fluctuations, which are precisely due to the discrete nature of the electrons, are ignored. For a wave frequency well above the plasma frequency, the fluctuations $\Delta n(\mathbf{x}, t)$ may be treated by working out the consequent fluctuations in the refractive index (which is, of course, close to unity). The scattering which results from these fluctuations is normally very weak and may be treated in the Born approximation. Suppose we Fourier-analyse Δn in space, leading to $\Delta n(\mathbf{k}, t)$. Then if $\mathbf{k}_1$ is the propagation vector for the incident radiation, and $\mathbf{k}_2$ that for a scattered wave (so $|\mathbf{k}_1| = |\mathbf{k}_2|$) the cross-section (per unit solid angle per unit volume) for scattering along $\mathbf{k}_2$ can be shown to be

$$8\pi^3\sigma_e\langle|\Delta n(\mathbf{k}, t)|^2\rangle, \tag{12.120}$$

where σ_e is the Thomson scattering cross-section $(e^2/4\pi\varepsilon_0 m_e c^2)^2$, and $\mathbf{k} = \mathbf{k}_2 - \mathbf{k}_1$. Such an observation thus yields directly the quantity $\langle|\Delta n(\mathbf{k}, t)|^2\rangle$ for some fixed $\mathbf{k}$. But this is just the Fourier transform of the correlation function $\langle\Delta n(\mathbf{x} + \boldsymbol{\xi}, t)\,\Delta n(\mathbf{x}, t)\rangle$, given by (12.109). The δ-function contributes $n_0/(2\pi)^3$. For the Maxwellian case (12.111) (fixed ions) and (12.116) (allowing for ion dynamics) lead respectively to

$$n_0\sigma_e\frac{h^2k^2}{1 + h^2k^2} \quad \text{and} \quad n_0\sigma_e\frac{1 + h^2k^2}{2 + h^2k^2} \tag{12.121}$$

as the value of (12.120). For $hk \gg 1$ (very short wavelength), these tend to $n_0\sigma_e$. In this limit the electrostatic interaction between particles is negligible, and the scattering is the same as that of free particles, each with cross-section σ_e, distributed in random positions (so that owing to the random phases, it is the *power* scattered which is additive). In the opposite limit ($hk \ll 1$) the electrostatic coupling is strong; with fixed uniform ions the fluctuations are much inhibited, while with free ions the fluctuations of ions and electrons are closely correlated, and this leads to the reduction of the cross-section by a factor of $\frac{1}{2}$. More generally, if the ions have charge Ze and density n_0/Z, (12.121) is replaced by

$$n_0\sigma_e\frac{Z + h^2k^2}{Z + 1 + h^2k^2} \tag{12.122}$$

(the case of a fixed smooth background of ions corresponding to $Z = 0$).

Before describing the experimental circumstances to which this work applies, we will discuss a further generalization.

12.9.3 Temporal correlations of fluctuations: the frequency spectrum in incoherent scattering

The scattering of electromagnetic radiation by a single electron is obviously subject to a Doppler shift if the electron is moving. Likewise the scattering due to a gas of electrons exhibits a Doppler broadening about the incident frequency. To deal with this we Fourier-analyse $\Delta n(\mathbf{x}, t)$ with respect to both space and time, yielding $\Delta n(\mathbf{k}, \omega)$, say. Then an incident wave described by $(\mathbf{k}_1, \omega_1)$ will be scattered into $(\mathbf{k}_2, \omega_2)$, and the cross-section for this process is given, by analogy with (12.120), by

$$\sigma(\omega_1 + \omega) = (2\pi)^3\sigma_e\langle|\Delta n(\mathbf{k}, \omega)|^2\rangle, \tag{12.123}$$

where $\mathbf{k} = \mathbf{k}_2 - \mathbf{k}_1$ and $\omega = \omega_2 - \omega_1$. The ensemble average appearing here is the Fourier transform with respect to ξ and τ of the correlation function

$$\langle\Delta n(\mathbf{x} + \boldsymbol{\xi}, t + \tau)\,\Delta n(\mathbf{x}, t)\rangle. \tag{12.124}$$

By measuring the frequency spectrum of the radiation scattered along a fixed direction, one will observe directly the quantity (12.123) as a function of ω, at fixed $\mathbf{k}$ (actually $\mathbf{k}$ should be slightly corrected since $|\mathbf{k}_2| \approx \omega_2/c$ will vary slightly; in practice the Doppler shift is so slight that this is negligible). The integral of (12.123) over ω should of course give our previous result for the total cross-section.

To find an expression for (12.124) is, unfortunately, beyond the scope of our work so far. It involves an ensemble average of quantities which relate to two different times, whereas our Liouville function, ρ, deals with probabilities at only one time. The general method for treating this problem is to return to first principles and introduce a new Liouville function $\rho(X_1, X_2, \ldots, t; X_1', X_2', \ldots, t')$. This gives the probability of observing the system in the microstate $(X_1, X_2, \ldots)$ at time t *and* of observing it in the state $(X_1', X_2', \ldots)$ at time t'. In general these two states will not lie on the same orbit in Γ-space, so the probability is zero; we expect ρ to be the product of a very large number of δ-functions multiplying the single-time (conventional) Liouville density. The new ρ will also satisfy an equation like Liouville's equation, with respect to both the primed and unprimed variables. One must then repeat much of the work of this chapter, obtaining a BBGKY chain for the new ρ, closing it by approximation and applying a generalization of Bogoliubov's hypothesis. This will involve a two-time correlation function $g(\mathbf{x}_1, \mathbf{v}_1, t_1, \mathbf{x}_2, \mathbf{v}_2, t_2)$, which finally reduces to the required correlation by integration over $\mathbf{v}_1$ and $\mathbf{v}_2$. A general theory of this type has been given by Rostoker (1961), and the application to incoherent scattering treated by Rosenbluth and Rostoker (1962), but it is not proposed to give here a detailed exposition of this work.

In the special case where the plasma is uniform and in thermal equilibrium, the required quantity may be obtained without the lengthy theoretical development just indicated. This is made possible by the *fluctuation–dissipation* or *generalized Nyquist* theorem. The frequency spectrum of the voltage fluctuations (or noise) in an electrical circuit can be obtained from the real part of the impedance without a detailed consideration of the dynamics of the electrons within the circuit, and similarly the fluctuations occurring in a plasma can be related to the appropriate conductivity. The general theorem is given by Landau and Lifshitz (1958) (§123). When the theorem is applied to a single spatial Fourier component, $\mathbf{k}$, there results, for the electrons (but taking account of the ions as an internal mechanism),

$$\langle |\Delta n(\mathbf{k}, \omega)|^2 \rangle = \frac{KTk^2}{\pi\omega^2}(2\pi)^{-3}\,\mathrm{Re}\left\{\frac{\sigma^{(i)} + \varepsilon_0 i\omega}{\sigma^{(i)} + \sigma^{(e)} + \varepsilon_0 i\omega}\,\sigma^{(e)}\right\}. \qquad (12.125)$$

Here $\sigma^{(i)}$ and $\sigma^{(e)}$ are the longitudinal conductivities for ions and electrons, as calculated in §8.2. It is interesting to note that for a cold plasma, the real part required in (12.125) would vanish, as there is no dissipation; the reason why it does not do so when there is a thermal distribution is the existence of Landau damping. The denominator of (12.125) is of course just the expression whose vanishing gives the dispersion relation. For a Maxwellian plasma these conductivities are easily expressible in terms of the G-function, being σ_{11} of equation (8.67). Equations (12.123) and (12.125) thus provide the required results. The integration over ω required to demonstrate that the total cross-section agrees with the second formula of (12.121) can be carried out by completing with a semi-circle in the lower half-plane and evaluating the residue at $\omega = 0$. (This is merely the same procedure as for the Kramers–Kronig relations; see Landau and Lifshitz (1958) §122.)

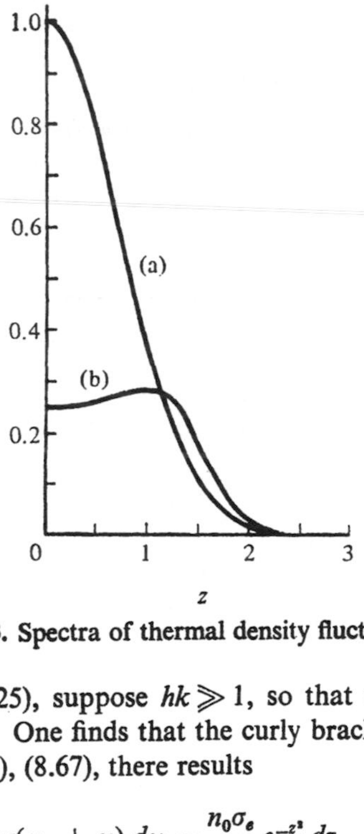

Fig. 12.3. Spectra of thermal density fluctuations.

To illustrate (12.125), suppose $hk \gg 1$, so that the electrostatic interaction is unimportant. One finds that the curly bracket in (12.125) reduces to $\sigma^{(e)}$, and using (8.63), (8.67), there results

$$\sigma(\omega_1 + \omega)\, d\omega = \frac{n_0 \sigma_e}{\pi^{1/2}} e^{-z_e^2} dz_e \tag{12.126}$$

Here z_e is the usual variable for the G-function, $\dfrac{\omega}{k}\left(\dfrac{m_e}{2KT}\right)^{1/2}$, as appropriate for the *electrons*. The influence of the ions has disappeared. The function e^{-z^2} is shown as curve (a) of Fig. 12.3. This result has the very simple interpretation that the electrons scatter from random positions, so that the powers of their individual contributions are added to give the total scattered signal;

the Doppler broadening is essentially a direct observation of the Maxwell distribution of their velocities.

Now suppose $hk \ll 1$, and that $m_e \ll m_i$ as usual. The above results then lead to

$$\sigma(\omega_1 + \omega)\, d\omega = \frac{n_0 \sigma_e}{\pi^{1/2}} \frac{e^{-z_i^2}\, dz_i}{\pi z_i^2 e^{-2z_i^2} + 4\left[1 - z_i e^{-z_i^2} \int_0^{z_i} e^{p^2}\, dp\right]^2}. \qquad (12.127)$$

In this limit the fluctuations are strongly coupled to those of the ions, hence the dependence on the *ion* variable z_i. This spectrum is shown as (*b*) on Fig. 12.3, z now meaning z_i. On this scale, curve (a) would be the spectrum for broadening associated with the ion thermal distribution. The remarkable aspect of the case $hk \ll 1$ is that the Doppler broadening is drastically reduced (by a factor of order $(m_e/m_i)^{1/2}$). It is of course still true that it is the electrons that are doing the scattering, and that they have the appropriate thermal distribution. But the fluctuations in electron density are so contrived that the highly Doppler-shifted scattered wavelets almost perfectly cancel one another, while the very slightly shifted ones reinforce to an abnormal extent. One other feature of this case, not shown in the figure, is that at the (very large) value of z_i corresponding to the plasma frequency there is a very sharp and narrow peak (or "spike") on the spectrum. This arises since but for the very slight Landau damping the denominator of (12.125) would vanish. Physically, this corresponds to scattering by plasma oscillations which are maintained at a thermal level within the plasma. The total cross-section due to this peak is however small. More details of this subject may be found in Dougherty and Farley (1960).

The calculations of Rosenbluth and Rostoker (1962) do not rely on the assumption of thermal equilibrium, and they are thus able to consider the case where the ions and electrons have a relative drift. As the drift is increased the fluctuations grow in magnitude, eventually diverging when the threshold for the two-stream instability is reached (cf. §12.6.4). The strong scattering which occurs in the marginal state is somewhat similar to what is termed *critical opalescence* in a vapor at the critical point (see Landau and Lifshitz, (1958), §116).

The phenomenon of incoherent scattering was first observed, by using a very powerful radar transmitter beamed at the ionosphere, by K. L. Bowles in 1958. The situation is such that $hk \ll 1$ (h of order a millimeter, k^{-1} a few meters), so (11.127) should apply, and this has indeed been verified. The experiment has been developed into a standard method for observing the ionosphere (see Gordon, 1967, for a review). It has also been found possible to observe the effect in laboratory plasmas, using laser light as the radiation to be scattered, so furnishing a valuable diagnostic technique.

These experiments afford a pleasing contact between real plasmas, which are so often badly behaved, and kinetic theory, which is often so abstruse.

PROBLEMS

1. Carry out the derivation of (12.16) as indicated.
2. Show that if it is desired to continue the expansion (12.34) to the next order in the small parameter, (12.35) should be left unaltered, and a new term of the form $h(X_1, X_2, X_3, t)$ be added to (12.36). What is the new form of (12.34)? Show how to carry the expansion still further. (The enumeration of the various terms appearing in the summation is conveniently carried out by diagram methods—see for example Balescu (1963).)
3. Supply the details in the derivation of (12.37) and (12.38).
4. Obtain the results given by equations (12.83)–(12.85).
5. To establish the H-theorem (see §12.7.3), start with the definition
$$H = \sum_r f_r(\mathbf{v}) \log f_r(\mathbf{v})\, d^3v$$
and use the kinetic equation in the form (12.80)–(12.82) to show that
$$\frac{dH}{dt} = -\frac{1}{2}\sum_{r,s}\left(\frac{e_r e_s}{\varepsilon_0}\right)^2 \int_{\mathbf{v}}\int_{\mathbf{v}'} \left(\frac{1}{m_s}\frac{\partial \log f_s(\mathbf{v}')}{\partial v_i'} - \frac{1}{m_r}\frac{\partial \log f_r(\mathbf{v})}{\partial v_i}\right) M_{ij} \left(\frac{1}{m_s}\frac{\partial \log f_s(\mathbf{v}')}{\partial v_j'} - \frac{1}{m_r}\frac{\partial \log f_r(\mathbf{v})}{\partial v_j}\right) d^3v\, d^3v'.$$
(Use, in turn, the conservation of particles, integration by parts with respect to $\mathbf{v}$, and the fact that r and s may be interchanged in your expression.) Next, show from (12.82) that the quadratic form $M_{ij}x_i x_j$ is $\geqslant 0$, with equality only if $\mathbf{x}$ is parallel to $\mathbf{v} - \mathbf{v}'$. Thus $dH/dt \leqslant 0$. Equality can only be achieved if
$$\frac{1}{m_s}\frac{\partial \log f_s(\mathbf{v}')}{\partial \mathbf{v}'} - \frac{1}{m_r}\frac{\partial \log f_r(\mathbf{v})}{\partial \mathbf{v}} = \mu_{rs}(\mathbf{v}, \mathbf{v}')(\mathbf{v} - \mathbf{v}'),$$
since the expression on the left has to be everywhere parallel to $\mathbf{v} - \mathbf{v}'$. Show that the scalar μ must in fact be independent of $\mathbf{v}$, $\mathbf{v}'$, r and s. Hence for each species
$$\frac{1}{m}\frac{\partial \log f}{\partial \mathbf{v}} = -\mu(\mathbf{v} - \mathbf{u})$$
say, where the constant $\mathbf{u}$ is the same for all species. This implies the Maxwellian distribution, with common drift and temperature. Show, by slight modification of the proof, that the result still holds if $\mathbf{M}$ is given by the approximate formula (12.98).
6. Obtain equations (12.100) and (12.101).
7. Investigate the circumstances in which the coefficients (12.100) and (12.101) can be approximated by the model form of §9.5.3. (This is rather lengthy; see Dougherty and Watson (1967) for a solution.)

8. For a uniform plasma in thermal equilibrium, write down the mean number of electrons in a sphere of radius a. Use (12.113) to show that the variance of this number is

$$\frac{4\pi a^3}{3} n_0^2 \left(1 + \frac{a}{h}\right) e^{-a/h}.$$

Comment on this result.

REFERENCES

Two texts which devote a substantial amount of space to the subject of this chapter are

R. Balescu, *The Statistical Mechanics of Charged Particles*, Interscience (1963).

D. C. Montgomery and D. A. Tidman, *Plasma Kinetic Theory*, McGraw–Hill (1964).

The original BBGKY sources are

N. N. Bogoliubov, *Problems of a Dynamical Theory in Statistical Physics* (State Technical Press, Moscow, 1946). English translation in *Studies in Statistical Mechanics*, vol. **1**. Ed. de Boer and Uhlenbeck, North Holland Publishing Co. (1962).

M. Born and H. S. Green, *A General Kinetic Theory of Liquids*, Cambridge University Press (1949).

J. G. Kirkwood, *J. Chem. Phys.*, **14**, 180 (1946) and **15**, 72 (1947).

J. Yvon, *La théorie des fluides et l'équation d'état*, Hermann et Cie, Paris (1935).

Other references mentioned in the text are

D. E. Baldwin, *Phys. Fluids*, **5**, 1523 (1962).

R. Balescu, *Phys. Fluids*, **3**, 52 (1960).

S. Chandrasekhar, *Revs. Modern Phys.*, **15**, 1 (1943).

S. Chapman and T. G. Cowling, *The Mathematical Theory of Non-Uniform Gases*, 2nd edn, Cambridge University Press (1952).

R. S. Cohen, L. Spitzer and P. McR. Routly, *Phys. Rev.*, **80**, 230 (1950).

J. M. Dawson and T. Nakayama, *Phys. Fluids*, **9**, 252 (1966).

J. P. Dougherty and D. T. Farley, *Proc. Roy. Soc.*, A **259**, 79 (1960).

J. P. Dougherty and S. R. Watson, *J. Plasma Phys.*, **1**, 317 (1967).

T. H. Dupree, *Phys. Fluids*, **4**, 696 (1961).

T. H. Dupree, *Phys. Fluids*, **6**, 1714 (1963).

T. H. Dupree, *Phys. Fluids*, **7**, 923 (1964).

E. Frieman and P. H. Rutherford, *Ann. Phys.* (*N.Y.*), **28**, 134 (1964).

W. E. Gordon, *Revs. Geophys.*, **5**, 191 (1967).

J. Hubbard, *Proc. Roy. Soc.*, A **260**, 114 (1961) and **261**, 371 (1961).

Yu. L. Klimontovich, *The Statistical Theory of Non-Equilibrium Processes in a Plasma*, Pergamon (1966).

L. D. Landau, *Physik Z. Sowjetunion*, **10**, 154 (1936).

L. D. Landau and E. M. Lifshitz, *Statistical Physics*, Pergamon (1958).

A. Lenard, *Ann. Phys.* (*N.Y.*), **10**, 390 (1960).

A. Rogister and C. Oberman, *J. Plasma Phys.*, **2,** 33 (1968) and **3,** 119 (1969).
M. N. Rosenbluth, W. MacDonald and D. Judd, *Phys. Rev.*, **107,** 1 (1957).
M. N. Rosenbluth and N. Rostoker, *Phys. Fluids*, **5,** 776 (1962).
N. Rostoker, *Phys. Fluids*, **3,** 922 (1960).
N. Rostoker, *Nuclear Fusion*, **1,** 101 (1961).
N. Rostoker and A. Simon, *Nuclear Fusion Supplement*, **1,** 761 (1962).
A. Simon in *Radiation and Waves in Plasmas* (Ed. M. Mitchner), Stanford University Press (1961).
L. Spitzer and R. Härm, *Phys. Rev.*, **89,** 997 (1953).
W. B. Thompson and J. Hubbard, *Revs. Modern Phys.*, **32,** 714 (1960).
J. Weinstock, *Phys. Rev.*, **133A,** 673 (1964).

INDEX